SCIENTIFIC FARM ANIMAL PRODUCTION
AN INTRODUCTION TO ANIMAL SCIENCE

SCIENTIFIC FARM ANIMAL PRODUCTION

AN INTRODUCTION TO ANIMAL SCIENCE

ELEVENTH EDITION

Thomas G. Field
University of Nebraska

Robert E. Taylor

PEARSON

Boston   Columbus   Hoboken   Indianapolis   New York   San Francisco
Amsterdam   Cape Town   Dubai   London   Madrid   Milan   Munich   Paris   Montreal   Toronto
Delhi   Mexico City   São Paulo   Sydney   Hong Kong   Seoul   Singapore   Taipei   Tokyo

Executive Editor: Daryl Fox

Editorial Director: Andrew Gilfillan

Editorial Assistant: Lara Dimmick

Team Lead of Program Management, Workforce Readiness:
Laura Weaver

Program Manager: Susan Watkins

SVP Field Marketing, NA: David Gesell

Executive Marketing Manager, Careers and Associations:
Ramona Elmer

Senior Marketing Manager, Careers and Associations:
Darcy Betts

Senior Marketing Coordinator: Alicia Wozniak

Marketing Coordinator: Les Roberts

Team Lead of Project Management, Workforce Readiness:
JoEllen Gohr

Project Manager: Kris Roach

Digital Studio Team Lead, Careers: Rachel Collett

Multimedia Production Coordinator: April Cleland

Senior Art Director: Diane Y. Ernsberger

Cover Art: Coleman Locke

Cover Designer: Studio Montage

Procurement Specialist: Deidra Skahill

Full-Service Project Management: Lumina Datamatics

Composition: Lumina Datamatics

Printer/Binder: RR Donnelley

Cover Printer: Phoenix Color

Text Font: 11/13 Adobe Garamond Pro

Credits and acknowledgments borrowed from other sources and reproduced, with permission, in this textbook appear on the appropriate page within text. Any photographs not credited on page are owned by the author,

Microsoft® and Windows® are registered trademarks of the Microsoft Corporation in the U.S.A. and other countries. Screen shots and icons reprinted with permission from the Microsoft Corporation. This book is not sponsored or endorsed by or affiliated with the Microsoft Corporation.

Library of Congress Cataloging-in-Publication Data

Field, Thomas G. (Thomas Gordon)
   Scientific farm animal production : an introduction to animal science / Thomas G. Field, University of Nebraska.—Eleventh edition.
     pages cm
   Includes bibliographical references and index.
   ISBN 978-0-13-376720-9 (alk. paper)—ISBN 0-13-376720-5 (alk. paper) 1. Livestock.
I. Title.
   SF61.T39 2014
   636—dc23
                                                       2014035179

10 9 8 7 6 5 4 3 2 1

SUSTAINABLE FORESTRY INITIATIVE | **Certified Sourcing**
www.sfiprogram.org
SFI-01042

PEARSON

ISBN 10:   0-13-376720-5
ISBN 13: 978-0-13-376720-9

This book is inspired by the men and women who make their living from the land and from applying not only the principles of science but the art of husbandry to their role as stewards of land, livestock, and communities. It is offered as a resource to the many students and teachers who daily invest their time, energy and talent into the process of improving animal agriculture in the hope that one day humanity might be free of hunger.

This work is dedicated to my wife Laura and children Justin, Sean, Trae, Kate, and Coleman who have contributed their talent and support in its creation.

Contents

9

Visual Evaluation of Market Animals 135

10

Reproduction 147

11

Artificial Insemination, Estrous Synchronization, and Embryo Transfer 171

12

Genetics 186

13

Genetic Change Through Selection 205

14

Mating Systems 223

15

Nutrients and Their Functions 236

16

Digestion and Absorption of Feed 252

22

Animal Behavior 333

23

Issues in Animal Agriculture 347

24

Beef Cattle Breeds and Breeding 380

33
Feeding and Managing Horses 546

34
Poultry Breeding, Feeding, and Management 563

Preface

Scientific Farm Animal Production is distinguished by an appropriate coverage of both breadth and depth of livestock and poultry production and their respective industries. The book gives an overview of the biological principles applicable to the animal sciences with chapters on reproduction, genetics, nutrition, lactation, consumer products, and other subjects. The book also covers the breeding, feeding, and management of beef cattle, dairy cattle, horses, sheep and goats, swine, and poultry. Although books have been written on each of these separate topics, the author has highlighted the significant biological principles, scientific relationships, and management practices in a condensed but informative manner.

TARGET AUDIENCE

This book is designed as a text for the introductory animal science course typically taught at universities and community colleges. It is also a valuable reference book for livestock producers, vocational agriculture instructors, and others desiring an overview of livestock production principles and management. The book is appropriate for the urban student with limited livestock experience, yet challenging for the student who has a livestock production background.

KEY FEATURES

Chapters 1 through 9 cover animal enterprises and products; Chapters 10 through 22 discuss the biological principles that are utilized to improve livestock and poultry production and the issues facing animal agriculture; while livestock and poultry management systems are presented in Chapters 23 through 34.

The glossary of terms used throughout the book has been expanded so that students can readily become familiar with animal science terminology. Many of the Key Terms in the text are included in the glossary. Additionally, key words are provided at the end of each chapter as an aid to student learning.

Photographs and figures are used throughout the book to communicate key points and major relationships. The visual aspects of the text should help students expand their global and macro view of the livestock industry as well as better understanding how theory is put into practice.

At the end of each chapter, a set of questions are provided that are designed to facilitate an in-depth understanding of the material. Students are encouraged to utilize the questions to assist them in making connections between concepts and to better integrate relationships to allow for not only listing the facts but creating a framework for the application of knowledge.

NEW TO THIS EDITION

This text continues to blend the various disciplines of science with contemporary management practices and industry trends to build a cohesive discussion of animal agriculture. The following improvements have been made to this edition:

- The input of nearly 20 reviewers was utilized to assure accuracy, clarity, and effective delivery of material.
- Demographic, industry data, and consumer trends have been updated.

- Photos and illustrations have been upgraded to enhance the reader experience.
- Management chapters have been revised to reflect the most current protocols and technologies used by the industry.
- More attention has been given to the issues and challenges confronting the livestock and poultry industry.
- Significant revision has been accomplished to provide a comprehensive but more clear communication of science based principles and relationships.
- Financial and enterprise-based cost and return data has been integrated to facilitate better understanding of the economic consequences of management decisions.
- The text effectively balances science and practice as it applies to the livestock and poultry industry.

INSTRUCTOR'S RESOURCES

An online Instructor's Manual, PowerPoint slides, and TestGen are available to Instructors at www.pearsonhighered.com. Instructors can search for a text by author, title, ISBN, or by selecting the appropriate discipline from the pull-down menu at the top of the catalog home page. To access supplementary materials online, instructors need to request an instructor access code. Go to www.pearsonhighered.com, click the Instructor Resource Center link, and then click Register Today for an instructor access code. Within 48 hours of registering, you will receive a confirmation e-mail including an instructor access code. Once you have received your code, go to the site and log on for full instructions on downloading the materials you wish to use.

ACKNOWLEDGMENTS

Appreciation is expressed to the reviewers of the eleventh edition, who offered suggestions to strengthen the book. They are Bonnie Ballard, Gwinnett Technical College; Angela Beal, Bradford School, Vet Tech Institute; Dennis Brink, University of Nebraska-Lincoln; Anne Duffy, Kirkwood Community College; Brian Hoefs, Globe University; Chip Lemieux, McNeese State University; Farabee McCarthy, The University of Findlay; Kasey Moyes, University of Maryland; Mary O'Horo-Loomis, State University of New York at Canton; Margi Sirois, Wright Career College; Bonnie Snyder, Central New Mexico Community College; Melissa Stacy, Rockford Career College; Brett VanLear, Blue Ridge Community College; Peg Villanueva, Vet Tech Institute @ International Business College—Indianapolis; Elizabeth Walker, Missouri State University; Julie Weathers, Southeast Missouri State University; Jennifer Wells, University of Cincinnati; Cynthia Wood, Virginia Tech; and Brenda Woodard, Northwestern State University of Louisiana.

About the Author

Dr. Thomas G. Field serves as the director of the Engler Agribusiness Entrepreneurship Program and holder of the Engler Chair in Agribusiness Entrepreneurship at the University of Nebraska—Lincoln.

He is also a noted agricultural author and a frequent speaker at agricultural events in the United States and abroad. He has consulted with a number of agricultural enterprises and organizations, and has served on numerous boards related to education, agriculture, and athletics. He is the co-owner of Field Land and Cattle Company, LLC in Colorado.

Dr. Field was raised on a Colorado cow–calf and seedstock enterprise. He managed a seedstock herd of cattle after completing his B.S. degree. A competitive horseman as a youth, he has had practical experience with seedstock cattle, commercial cow–calf production, stockers, and horses. He has a B.S., M.S., and Ph.D. in animal science from Colorado State University.

Dr. Field has received teaching awards from the USDA National Excellence in Teaching program, the National Association of Colleges and Teachers of Agriculture, the American Society of Animal Sciences, Colorado State University, and the University of Nebraska.

He is married to Laura and father to Justin, Sean, Trae, Kate, and Coleman.

1

Animal Contributions to Human Needs

In many ways the history of civilization is told in the application of human creativity to the task of feeding, clothing, and raising the standard of living for the world's various societies via animal agriculture. Over the ages the relationship between humans and domesticated animals have shaped history, impacted economies, altered the outcomes of war, sped the exploration and settlement of new territories, revolutionized agriculture and transportation, provided entertainment through sport, and etched itself into nearly every aspect of civilization. For example, the domestication of horses transformed the range of land movement for human beings. And with the stability that agriculture provided human communities came time for intellectual and cultural pursuits. Livestock have been incorporated into the telling of the human story through expression in the form of art, literature, and music.

Domestication of livestock depends on the animal reproducing within the management decisions of human beings and the creation of a complex mutually beneficial relationship founded on the ancient concept of the good shepard. This "contract" offers the animal protection from predators and a more consistent supply of nutrients, to name a few of the benefits, in exchange for food, fiber, draft power, and companionship as contributions to the well-being of humans. This relationship in which domesticated species and humans seem to have chosen each other is still the basis for sound management and husbandry of livestock and has allowed domesticated animals far greater survival rates than those in the wild. The timeline of livestock and poultry domestication provides context for the relationship between humans and animals of agricultural importance (Fig. 1.1).

Table 1.1 outlines the major domesticated livestock species, their approximate numbers, and their primary uses. Chickens are the most numerous (20.7 billion), followed by cattle (1.43 billion), sheep (1.09 billion), ducks (1.08 billion), and swine (967 million).

CONTRIBUTIONS TO FOOD NEEDS

When opportunity exists, most humans consume both plant and animal products (Fig. 1.2).

The contribution of animal products to the **per-capita calorie and protein supply** in food is shown in Table 1.2. Animal products constitute approximately 16% of the calories, 37% of the protein, and 45% of the fat in the total world food supply. Large differences exist between developed countries and developing countries in total daily supply of calories, protein, and fat.

learning objectives

- Describe the global distribution of livestock
- Quantify the role of animal products in the global food supply
- Evaluate differences in food production and agricultural productivity between developed and developing nations
- Compare food expenditures for at-home and away-from-home consumption in the United States
- Compare food consumption across diverse nations and cultures
- Describe changes in the U.S. agricultural productivity
- Describe the nonfood contributions of livestock

Figure 1.1
Timeline for the domestication of livestock and poultry.

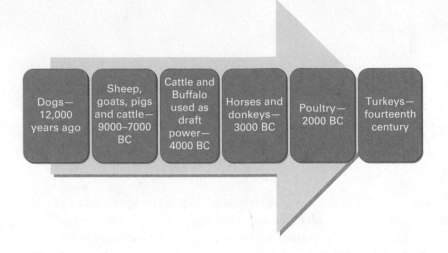

For example, consumers in developed nations derive 26% of their calories from animal products with just over one-half of their total protein and fat supply from animal products. Consumers in developing nations derive 13% of their calorie supply, 29% of their protein, and 41% of their fat from animal products. The United States ranks higher than the world average for percent of calories and protein from animal sources but about average for percent of fat from animal products.

Table 1.1
MAJOR DOMESTICATED ANIMAL SPECIES—THEIR NUMBERS AND USES IN THE WORLD

Animal Species	World Numbers (mil)	Leading Countries or Areas with Numbers[a] (mil)	Primary Uses
Ruminants			
Cattle	1,426	Brazil (213), India (211), United States (93), China (83), Ethiopia (53)	Meat, milk, hides
Sheep	1,093	China (139), India (74), Australia (73), Iran (49), Sudan (39)	Wool, meat, milk, hides
Goats	924	India (157), China (142), Pakistan (61), Nigeria (57) Bangladesh (53)	Milk, meat, hair, hides
Buffalo	195	India (113), Pakistan (32), China (24)	Draft, milk, meat, hides
Camels	27	Somalia (7), Sudan (5),Kenya (3), Niger (2)	Packing, transport, draft, meat, milk, hides
Nonruminants			
Chickens	20,708	China (5,230), United States (2,080), Indonesia (1,427), Brazil (1,268), Iran (900)	Meat, eggs, feathers
Swine	967	China (464), United States (66), Viet Nam (27), Germany (27), Spain (25)	Meat
Turkeys	468	United States (248), Chile (32), France (24), Italy (24), Russian Fed. (17)	Meat, eggs, feathers
Ducks	1,108	China (809), Vietnam (98), Indonesia (49), Malaysia (49), Bangladesh (44),	Meat, eggs, feathers
Horses	58	United States (10), China (7), Mexico (6), Brazil (5), Argentina (3)	Draft, riding, sport, occasionally meat
Donkeys and Mules	54	China (9), Ethiopia (7), Mexico (6), Pakistan (5), Egypt (3)	Draft, transport

Source: Adapted from USDA and FAO.

Figure 1.2
Meat, milk, and eggs are nutrient dense foods that meet the needs of both domestic and global consumers. The livestock industry and food supply chain must align with consumer demand to assure continuation of a successful business model.
Source: Kirill Kedrinski/Fotolia.

Table 1.2
ANIMAL PRODUCT CONTRIBUTION TO PER-CAPITA CALORIE, PROTEIN, AND FAT SUPPLY

Country	Total Kilo Calories	Animal Products Kilo Cal	%	Total Protein (g/day)	Animal Products g/Day	%	Total Fat	Animal Products g/Day	%
Australia	3,176	1,049	33	107	71	66	138	72	52
Bangladesh	2,103	67	3	45	6	13	22	4	18
Brazil	2,985	615	21	80	41	51	89	42	47
China	3,029	583	19	85	30	35	84	49	58
Egypt	3,346	256	8	93	18	19	60	18	30
Germany	3,451	1,035	30	95	57	60	152	82	54
India	2,428	194	8	57	10	17	48	13	27
Japan	2,762	569	21	92	51	55	83	35	42
Kenya	1,965	234	12	50	15	30	47	15	32
Mexico	3,165	583	18	88	38	43	89	4	46
Nigeria	2,850	87	3	65	8	12	68	6	7
United Kingdom	3,334	1,002	30	98	55	56	145	77	53
United States	3,772	1,043	27	114	73	64	151	71	47
Developed	3,260	857	26	99	55	55	119	62	52
Developing	2,679	348	13	69	22	29	63	26	41
World Average	2,805	459	16	76	28	37	75	34	45

Source: Adapted from USDA, FAO.

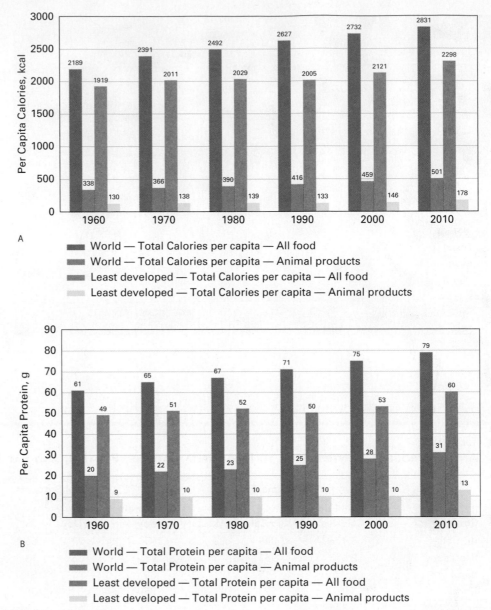

Figure 1.3

Caloric and protein intake from animal products.

Source: Adapted from USDA.

Changes in per-capita calorie supply and protein supply during the past 40 years are shown in Figure 1.3. Per-capita caloric supplies of both calories and protein have increased in most areas of the world. The contribution of animal products to the per-capita protein supply has increased in most of the world. The large differences among countries in the importance of animal products in their food supply can be partially explained by available resources and development of those resources. Most countries with only a small percentage of their population involved in agriculture have higher standards of living and a higher per-capita consumption of animal products. Comparing Table 1.3 with Table 1.2, note that the countries in Table 1.3 are listed by percentage of population involved in agriculture.

Agriculture **mechanization** (Fig. 1.4) has been largely responsible for increased food production and allowing people to turn their attention to professions other than production agriculture. This facilitates the provision of many goods and services, raises standards of living, and allows for the creation of more diverse economies.

Table 1.3
POPULATION INVOLVED IN AGRICULTURE IN SELECTED COUNTRIES

Country	Population (mil)	Population in Agriculture[a] (mil)	Percent of Economically Active Population in Agriculture[b]
United States	316	5	1
United Kingdom	63	1	1
Germany	82	2	2
Australia	23	1	4
Japan	126	2	1
Brazil	198	20	10
Mexico	116	20	17
Nigeria	167	39	23
China	1,385	825	59
India	1,258	597	47
Bangladesh	152	66	43
Kenya	43	30	70
Least developed nations	871	552	63
Low-income, food-deficit nations	2,874	1,360	47
World Total	7,052	2,621	37

Source: Adapted from USDA and FAO.
[a]*Agricultural population* is defined as all persons depending for their livelihood on agriculture. This comprises all persons actively engaged in agriculture and their nonworking dependents.
[b]Includes all economically active persons engaged principally in agriculture, forestry, hunting, or fishing.

Figure 1.4
The mechanization of agriculture has enabled a relatively small proportion of the human population to provide for a growing world market. Source: Tom Field

Note that 50% of the people in developing nations are engaged in agriculture while only 7% of the citizens in developed countries are active in the agricultural sector.

The tremendous increase in the productivity of U.S. agriculture (Table 1.4) has lowered the relative cost of food as vividly demonstrated in Table 1.5. Historical data show that agricultural productivity doubled in the 100-year span of 1820–1920. For example, at the turn of the century a team of horses, one handler, and a moldboard

Table 1.4
PRODUCTIVITY CHANGES IN SEVERAL FARM ANIMAL SPECIES IN THE UNITED STATES

Species and Measure of Productivity	1925	1950	1975	2000	2012
Beef cattle	955	976	1,039	1,210	1,280
Average liveweight at finishing (lbs)					
Sheep					
Average liveweight at finishing (lb)	86	94	102	133	141
Dairy cattle					
Milk marketed per breeding female (lb)	4,189	5,313	10,500	17,192	21,148
Swine					
Average liveweight at finishing (lb)	235	243	245	259	275
Broiler chickens[a, b]					
Liveweight at marketing (lb)	2.8	3.1	3.8	5.0	5.7
Turkeys[a, b]					
Liveweight at marketing (lb)	13.0	18.6	18.4	25.8	29.5
Laying hens[a]					
Eggs per hen per year (no.)	112	174	232	257	271

Source: Adapted from USDA Annual Agricultural Statistics.
[a]Feed required per lb. of weight gain or per dozen eggs was reduced by more than half over the same time period.
[b]Time to market was reduced by more than half over the same time period.

Table 1.5
EXPENDITURES FOR FOOD IN THE UNITED STATES (GROSS DOLLARS AND AS PERCENT OF PERSONAL DISPOSABLE INCOME)

Year	At Home ($ bil)	At Home (%)	Away from Home ($ bil)	Away from Home %	Total ($ bil)	Total (%)
1930	15.8	21	2.3	3	18.1	24
1940	13.5	18	2.4	3	15.9	21
1950	35.7	17	7.6	4	43.3	21
1960	51.5	14	12.6	3	64.0	17
1970	75.5	10	26.4	4	102.0	14
1980	180.8	9	85.2	4	266.0	13
1990	314.5	7	175.2	4	489.6	11
2000	431.6	6	292.9	4	724.4	10
2010	622.3	6	454.9	4	1,077.2	9.9
2012	677.5	6	512.4	4	1,189.9	10

Source: USDA.

plow could plow 2 acres per day. Today, one tractor pulling three plows, each with five moldboards, plows 110 acres per day, accomplishing the work that once required 110 horses and 55 workers.

Livestock productivity since 1925 has progressively increased to extraordinary levels. The mix of animal enterprises on U.S. farms has shifted from a typical situation involving a vast number of species being raised on an average farm in the 1920s to contemporary scenarios where animal agriculture is considerably more specialized. These improvements in productivity have occurred primarily because people had an incentive to progress under a free-enterprise system.

Table 1.6
CONTRIBUTIONS OF VARIOUS FOOD GROUPS TO THE WORLD FOOD SUPPLY

Food Group	Calories (%)	Protein (%)
Cereals	50	45
Roots, tubers, pulses	8	7
Nuts, oils, vegetable fats	11	4
Sugar and sugar products	8	2
Vegetables and fruits	7	5
All animal products	16	37
Meat	7	16
Eggs	1	3
Fish	1	7
Milk and dairy	5	10
Other	2	1

Source: Adapted from USDA and FAO.

In the United States, releasing people from producing their own food has given them the opportunity to improve their per-capita incomes. Increased per-capita income associated with an abundance of animal products has resulted in reduced relative costs of many animal products with time. U.S. consumers allocate a smaller share of their disposable income for food than do people in many other countries. For example, per-capita expenditures for food as a percent of household expenses in Canada, France, Mexico, South Africa, and China are 9.2, 13.7, 24.2, 20.6, and 34.9%, respectively.

Table 1.6 shows that cereal grains are the most important source of energy in world diets. The energy derived from cereal grains, however, is twice as important in developing countries (as a group; there are exceptions) as in developed countries. Table 1.6 also illustrates that meat and milk are the major animal products contributing to the world supply of calories and protein.

Most of the world meat supply comes from cattle, swine, sheep, goats, chickens, and turkeys. There are, however, twenty or more additional species that collectively contribute about 6.5 billion pounds of edible protein per year or approximately 10% of the estimated total protein from all meats. These include the alpaca, llama, yak, horse, deer, elk, antelope, kangaroo, rabbit, guinea pig, capybara, fowl other than chicken (duck, turkey, goose, guinea fowl, pigeon), and wild game exclusive of birds. For example, the Russian Federation cans more than 110 million pounds of reindeer meat per year, and in Germany the annual per-capita consumption of venison exceeds 3 pounds. Peru derives more than 5% of its meat from the guinea pig.

Meat is important as a food for two scientifically based reasons. The first is that the assortment of amino acids in animal protein more closely matches the needs of the human body than does the assortment of amino acids in plant protein. The second is that vitamin B_{12}, which is required in human nutrition, may be obtained in adequate quantities from consumption of meat or other animal products but not from consumption of plants.

Milk is one of the largest single sources of food from animals. In the United States, 99% of the milk supply comes from cattle, but on a worldwide basis, milk from other species is important. Domestic buffalo, sheep, goat, alpaca, camel, reindeer, and yak supply significant amounts of milk in some countries. Milk and products made from milk contribute protein, energy, vitamins, and minerals for humans.

Figure 1.5

Ruminant animals produce food for humans by utilizing grass, crop residues, and other forages from land that cannot produce crops to be consumed directly by humans. (A) Cattle grazing stubble in New South Wales, Australia. (B) Cattle grazing hillsides in Georgia. (C) Cattle grazing native range in Arizona. (D) Sheep grazing native range. Source: 1.5 a-c: Tom Field. 1.5 d: Dalajlama/Fotolia.

Besides the nutritional advantages, a major reason for human use of animals for food is that most countries have land areas unsuitable for growing cultivated crops. Approximately two-thirds of the world's agricultural land is permanent pasture, range, and meadow; of this, about 60% is unsuitable for producing cultivated crops that would be consumed directly by humans. This land, however, can produce feed in the form of grass and other vegetation that is digestible by grazing ruminant animals, the most important of which are cattle and sheep (Fig. 1.5). These animals can harvest and convert the vegetation, which is for the most part indigestible by humans, to high-quality protein food. In the United States, about 385 million acres of rangeland and forest, representing 44% of the total land area, are used for grazing. Although this acreage now supports only about 40% of the total cattle population, it could carry twice this amount if developed and managed intensively.

Ruminant **animal agriculture** therefore does not compete with human use for production of most land used as permanent pasture, range, and meadow. On the contrary, the use of animals as intermediaries provides a means by which land that is otherwise unproductive for humans can be made productive (Fig. 1.6).

People are concerned about energy, protein, population pressures (Fig. 1.7), and land resources as they relate to animal agriculture. Quantities of energy and

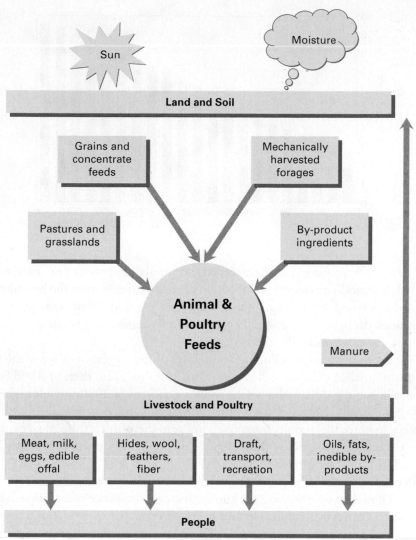

Figure 1.6

A graphic illustration of the land-plant-ruminant-animal-human relationship.

protein present in foods from animals are smaller than quantities consumed by animals in their feed because animals are inefficient in the ratio of nutrients used to nutrients produced. More acres of cropland are required per person for diets high in foods from animals than for diets including only plant products. As a consequence, animal agriculture has been criticized for wasting food and land resources that could otherwise be used to provide persons with adequate diets. Consideration must be given to economic systems and consumer preferences to understand why agriculture perpetuates what critics perceive as resource-inefficient practices. These practices relate primarily to providing food-producing animals with feed that could be eaten by humans and using land resources to produce crops specifically for animals instead of producing crops that could be consumed by humans.

Hunger continues to be a challenge in some regions of the world. The factors that contribute to the hunger problem are varied and complex. Hunger takes two forms—chronic persistent hunger and famine. **Chronic persistent hunger (CPH)** results from a combination of poverty, climatic change, political instability, water shortages, loss of soil fertility, poor infrastructure (transportation, storage facilities, banking services, etc.), and illiteracy. Note that food scarcity is not a significant contributing factor to CPH.

Figure 1.7
Past, present, and projected world population.
Source: U.S. Census Bureau.

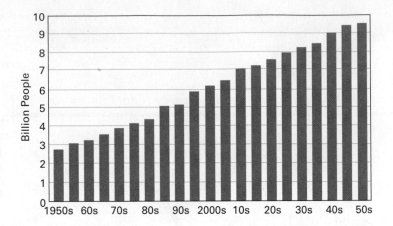

In fact, global food production has exceeded the population growth rate. **Famine**, unlike CPH, is typically a relatively short period of crisis resulting from the breakdown in food production and distribution infrastructure resulting from catastrophic events such as hurricanes, drought, or civil war. The international community is relatively adept at reacting to and minimizing the effects of famine.

The International Food Policy Research Institute suggests that while the number of malnourished children will decline from 1993 to 2020, there will still be 150 million babies and toddlers who will be insufficiently fed in 2020. An additional 500 million people will also suffer from hunger. Africa, Latin America, the Caribbean, and West Asia are the regions most likely to bear the brunt of the problem in the future. Of the 11 countries with daily per-capita consumption of less than 2,000 calories, 10 of these are located in Africa.

Per-capita food availability is estimated to increase by nearly 7% by 2020, with China and East Asia experiencing the greatest increase. Evidence of a decline in global population increases is becoming apparent. However, slowing population increases is a gradual process, and for the next several decades approximately 80 million people will be added to the global population annually. Over 90% of this increase will occur in developing nations in or near urban areas. The 70 most susceptible countries to the effects of hunger are also the world's poorest nations. Sub-Saharan African nations have a per-capita income of approximately $380 per year.

Conquering hunger in developing nations involves a multifaceted strategy that includes increasing literacy rates, particularly in women; reducing poverty; improving health care; enhancing agricultural production; and improving the total food system infrastructure. As demand for food increases, some regions of the world will become more import-dependent while others will become more export-focused.

Even developed nations are not immune to the effects of hunger for part of the population. For example, 85% of U.S. households are categorized as **food secure**, meaning all members of the household have access to enough food for a healthy lifestyle and sufficient financial resources to acquire food, as it is needed. Households deemed **food insecure** (15%) are further divided into categories of without hunger (9%) or with hunger (6%). Food insecure households without hunger are able to gain access to groceries via food assistance programs or community outreach programs. Fortunately, only 0.7% of children in the United States live in food insecure environments coupled with periods of hunger. Those households considered food insecure typically have incomes below the poverty line of $19,000, and are comprised of single adults or single parents with children.

There will be annual demand increases of 1–1.5% for cereal grains, while worldwide demand for meat is expected to approach 2% per annum. Demand for meat will increase at the highest rate in developing countries (approximately 3%) with developed nations only accounting for less than 1% of annual demand growth. In the final analysis, the ability to pay will dictate food distribution. The need for economic growth in developing regions is of paramount importance.

Agriculture producers generate what consumers want to eat as reflected in the prices consumers are able and willing to pay. Eighty-five percent of the world's population desires food of animal origin in its diets, perhaps because foods of animal origin are considered more palatable than foods from plants. In most countries, as per-capita income rises, consumers tend to increase their consumption of meat and animal products.

If many consumers in countries where animal products are consumed at a high rate were to decide to eat only food of plant origin, consumption and price of foods from plants would increase, and consumption and price of foods from animals would decrease. Agriculture would then adjust to produce greater quantities of food from plants and lesser quantities of food from animals. Ruminant animals can produce large amounts of meat without grain feeding. The amount of grain feeding in the future will be determined by cost of grain and the price consumers are willing to pay for meat.

Some people advocate shifting from the consumption of foods from animals to foods from plants. They see this primarily as a moral issue, believing it is unethical to let people elsewhere in the world starve when our own food needs could be met by eating foods from plants rather than feeding plants to animals. The balance of plant-derived foods could then be sent abroad. These people believe that grain can be shipped with comparative ease because a surplus of grain exists in many developed nations and because any surplus should be provided at no cost. Providing free food to other countries has met with limited success in the past. In some situations, it upsets their own agricultural production, and in many cases the food cannot be adequately distributed in the recipient country because transportation and marketing systems are poorly developed.

There are strong feelings that the United States has a moral obligation to share its abundance with other people in the world, particularly those in developing countries. It appears that sharing our time and technology can best do this. However, people need to have self-motivation to improve, access to knowledge and appropriate technology, and sufficient resources to develop agricultural productivity and infrastructure aligned to their own cultural values.

Advances in agricultural production and related topics must be shared to minimize the effects of hunger on civilization. These achievements have been built on knowledge gained through experience and research, the extension of knowledge to producers, and the development of an industry to provide transportation, processing, and marketing in addition to production. Dwindling dollars currently being spent to support agricultural research and extension of knowledge may not provide the technology needed for future food demands. The next generation of agricultural leaders should view the decline in resources allocated to agricultural research, extension, and education as an emerging crisis.

About 20% of the world human population and 32% of the ruminant animal population live in developed regions of the world, but ruminants of these same regions produce two-thirds of the world's meat and 80% of the world's milk. In developed regions, a higher percentage of animals are used as food producers, and these animals are more productive on a per-animal basis than animals in developing regions. This is the primary reason for the higher level of human nutrition in developed countries of the world.

Possibly many developing regions of the world could achieve levels of plant and animal food productivity similar to those of developed regions. Except perhaps in India, abundant world supplies of animal feed resources that do not compete with production of food for people are available to support expansion of animal populations and production. It has been estimated that through changes in resource allocation, an additional 8 billion acres of arable land (twice what is now being used) and 9.2 billion acres of permanent pasture and meadow (23% more than is now being used) could be put into production in the world. These estimates, plus the potential increase in productivity per acre and per animal in developed countries, demonstrate the magnitude of world food-production potential. This potential cannot be realized, however, without coordinated planning and increased incentive to individual producers.

Fortunately progress can be made in reducing hunger. For example, Asia reduced hunger (percent of population consuming less than 2,100 calories per day) by 30% in the 10-year period from 1994 to 2004. During the same time, Bangladesh reduced the number of hungry people by 70% by making significant strides in food production and distribution. This is particularly impressive given that Bangladesh was once considered the epicenter of famine and chronic persistent hunger. Interestingly, significant changes in governmental policy focused not only on increasing food production but also on enhancing exports as a means to infuse foreign exchange into the economy. Furthermore, government policy focused on private-sector investment in irrigation systems, seeds, and fertilizer to stimulate food production. These policies increased irrigated acreage by 50% from 1994 to 2004.

In the long run, each nation must assume the responsibility of producing its own food supply by efficient production, barter, or purchase and by keeping future food-production technology ahead of population increases and demand. Extensive untapped resources that can greatly enhance food production exist throughout the world, including an ample supply of animal products. The greatest resource is the human being, who can, through self-motivation, become more productive and self-reliant.

CONTRIBUTIONS TO CLOTHING AND OTHER NONFOOD PRODUCTS

Products other than food from ruminants include wool, hair, hides, and pelts. Synthetic materials have made significant inroads into markets for these products. For example, the world's production of wool peaked in 1990 but since has declined to 40-year lows. It is important to note that in more than 100 countries, ruminant fibers are used in domestic production and cottage industries for clothing, bedding, housing, and carpets.

Annual production of animal wastes from ruminants contains millions of tons of nitrogen, phosphorus, and potassium. The annual value of these wastes for fertilizer is estimated at more than $1 billion.

Inedible tallow and greases are animal **by-products** used primarily in soaps and animal feeds and as sources of fatty acids for lubricants and industrial use. Additional tallow and grease by-products are used in the manufacture of pharmaceuticals, candles, cosmetics, leather goods, woolen fabrics, and tin plating. The individual fatty acids can be used to produce synthetic rubber, food emulsifiers, plasticizers, floor waxes, candles, paints, varnishes, printing inks, and pharmaceuticals.

Gelatin is obtained from hides, skins, and bones and can be used in foods, films, and glues. Collagen, obtained primarily from hides, is used to make sausage casings.

CONTRIBUTIONS TO WORK AND POWER NEEDS

The early history of the developed world abounds with examples of the importance of animals as a source of work energy through draft work, packing, and human transport. The horse made significant contributions to winning wars and exploration of the unknown regions of the world.

In the United States during the 1920s, approximately 25 million horses and mules were used, primarily for **draft purposes**. The tractor has replaced all but a few of these draft animals. In parts of the developing world, however, animals provide as much as 99% of the power for agriculture even today.

In more than half the countries of the world, animals—mostly buffalo and cattle, but also horses, mules, camels, and llamas—are kept primarily for work and draft purposes (Fig. 1.8). Approximately 20% of the world's human population depends largely or entirely on animals for moving goods. According to the Food and Agriculture Organization of the United Nations, in developing countries animals provide 52% of the cultivation power, with an additional 26% derived from human labor. Developed countries, in contrast, use tractors for 82% of the cultivation, with animals and humans providing 11 and 7% of the power, respectively. There are more than three times as many tractors and harvesting machines and twice as many milking machines in use in developed nations as compared to developing countries. It is estimated that India alone would have to spend more than $1 billion annually for gasoline to replace the animal energy it uses in agriculture.

A

B

Figure 1.8

Animals provide significant contributions to the draft and transportation needs of countries lacking mechanization in their agricultural technology. In developed countries, the use of draft animals is more oriented to recreation than necessity. (A) Donkey pulling a cart as an example of draft power. (B) Carriage horses provide a leisurely experience that harkens to times before widespread mechanization. Source: 1.8 a: Africa/Fotolia, 1.8 b: Pink candy/Fotolia.

Figure 1.9
(A) Many people enjoy the sport of horse racing. (B) Hunter jumpers are a key attraction in many equine shows and competitions. (C) Horse showing is a popular sport with increasing participation on the amateur and professional levels. (D) Horseback experiences provide high-quality recreational experiences for many. (E) Polo is an action-packed sport enjoyed by many. (F) Horses still play an integral role on many ranches. Source: 1.9a: Donna/Fotolia; 1.9b: Kseniya Abramova/Fotolia; 1.9c: JJAVA/Fotolia; 1.9d: Yanlev/Fotolia; 1.9e: MrSegui/Fotolia; 1.9f: PROMA/Fotolia.

A

B

C

D

E

F

ANIMALS FOR COMPANIONSHIP, RECREATION, AND CREATIVITY

Estimates of the number of **companion animals** in the world are unavailable. There are an estimated 59 million family-owned dogs and 75 million family-owned cats in the United States, in addition to the animals identified in Table 1.1. Approximately one-half of all U.S. households have at least one pet dog or cat. The U.S. pet food industry annually processes more than 3 million tons of cat and dog food valued at more than $1 billion. Many species of animals would qualify as companions where people derive pleasure from them. The contribution of animals as companions, especially to the young and elderly, is meaningful, even though it is difficult to quantify the emotional value.

Animals used in rodeos, equestrian sports, livestock shows, and other venues provide income for thousands of people and recreational entertainment for millions (Fig. 1.9). Numerous people who have made money from nonagricultural businesses have invested in land and animals for recreational and emotional fulfillment.

Our livestock heritage involves the interactions of humans with animals over the centuries. Historically, animals have been highly respected, revered, and even worshiped by humans. Early humans expressed the sacred, mysterious qualities of some animals through art on cave walls between 15,000 and 30,000 B.C. Thus, through these early paintings and sculptures, animals found their way into an expression of the things of humans—the humanities. Several historians have noted that an extremely high form of art is the intelligent manipulation of animal life—the modeling and molding of different types through the application of breeding principles.

ADDITIONAL ANIMAL CONTRIBUTIONS

The use of livestock species in human health research is also of significance. Most human health research involving livestock is focused on smaller animals such as miniature pigs, swine, and sheep. Biomedical research using these species has focused on such topics as human aging, diabetes, arteriosclerosis, and development of replacement joints. This research is conducted under strict federal guidelines governing the care and use of animals in laboratory settings.

Transgenic technologies also offer significant potential in terms of utilizing livestock to produce specialized proteins for use in the creation of therapeutic drugs and other medical applications. This approach is often referred to as "pharming" or the use of agricultural animals to produce pharmaceuticals. Transgenic dairy cows and goats have been utilized to produce hepatitis B antigens, tissue plasminogen activator for treatment of heart disease, and the clotting agent antithrombin III for example.

CHAPTER SUMMARY

- Domesticated animals (>16 billion) contribute to the well-being of 5.6 billion humans throughout the world by providing food, clothing, shelter, power, recreation, and companionship.

- Animal products contribute significantly to the world's human protein needs and energy supply.

- As people increase their standard of living, the per-capita consumption of animal products also increases.

- Human health is improved via the continuing research utilizing domestic animals.

KEY WORDS

per-capita calorie supply
per-capita protein supply
mechanization
animal agriculture
chronic persistent hunger (CPH)
famine

food secure
food insecure
by-products
gelatin
draft purposes
companion animals

REVIEW QUESTIONS

1. Compare and contrast the global distribution of livestock and poultry.
2. Describe the role of animal products in the diets of consumers in developed and developing countries.
3. Describe the role of mechanization in changing agricultural practices, productivity and the experience of consumers.
4. Discuss the trends in the livestock and poultry productivity.
5. How have expenditures for food purchased at-home, away-from-home, and on a per capita basis changed over time?
6. Describe the role of animal products to the global caloric and protein supply.
7. How does livestock production allow humans to capture value from land resources.
8. Compare and contrast chronic persistent hunger and famine.
9. Describe the hunger situation in the United States.
10. Define the value of nonfood values of the livestock and poultry industries.

SELECTED REFERENCES

Baldwin, R. L. 1980. *Animals, Feed, Food and People: An Analysis of the Role of Animals in Food Production.* Boulder, CO: Westview Press.

Caras, R. A. 1996. *A Perfect Harmony; the Intertwining of Lives of Animals and Humans throughout History.* New York: Simon & Schuster.

Devendra, C. 1980. Potential of sheep and goats in less developed countries. *J. Anim. Sci.* 51:461.

Economic Research Service: USDA. 2006. *Food Security in the United States: Conditions and Trends.* Washington, DC.

Meade, B., S. Rosen, and S. Shapouri. 2006. *Food Security Assessment, 2005.* Washington, DC: Economic Research Service: USDA.

National Research Council. 1983. *Little-Known Asian Animals with a Promising Economic Future.* Washington, DC: National Academy Press.

Pearson, R. A. 1994. Draft animal power. *Encyclopedia of Agriculture Science.* San Diego, CA: Academic Press.

Pinstrup-Andersen, P., R. Pandya-Lorch, and M. W. Rosegrant. 1997. The world food situation: Recent developments, emerging issues, and long-term prospects. *Food Policy Report.* Washington, DC: The International Food Policy Research Institute.

Willham, R. L. 1985. *The Legacy of the Stockman.* Morrilton, AR: Winrock International.

2

An Overview of the Livestock and Poultry Industries

It is important to see the broad picture of the livestock and poultry industries before studying the specific biological and economic principles that explain animal function and production. These industries are typically described with numbers of animals, pounds produced, production systems, prices, products, people (producers and consumers), and profitability. Products and consumers are covered in the next several chapters.

An understanding of the animal industries begins with basic terminology, especially the various species and sex classifications (Table 2.1). Refer to the glossary for definitions of many commonly used terms in the livestock industries.

learning objectives

- Quantify the economic impact of the U.S. livestock industry
- Describe the role of international trade in livestock and livestock products
- Overview the international and domestic beef, dairy, horse, poultry, sheep, goat, and swine industries
- Overview nontraditional livestock enterprises

U.S. ANIMAL INDUSTRIES: AN OVERVIEW

Historically, farms in the United States were highly diversified in both crop and livestock enterprises. For example, during the depression most farms had cattle, chickens, hogs, and horses. Over time, however, as fewer people chose to work in agriculture production, the industry had to develop more specialization in order to create the productivity required to meet consumer demand. The livestock and poultry industries in the United States must generate large volumes of output to meet the high animal product preference of more than 316 million U.S. consumers and also to supply the growing export market. Table 2.2 shows the number of livestock and poultry producers and the inventory number for each species.

Cash Receipts

An evaluation of farm **cash receipts** from the sale of animals and animal products provides another perspective of U.S. animal industries. Table 2.3 shows the cash receipts for animal commodities ranked against all agricultural commodities. The top five states for each commodity are also shown in this table. Note the cash receipts for all livestock products comprise 44% of all agricultural commodities in the United States.

Figure 2.1 shows the farm cash receipts from livestock and poultry products for states that have annual cash receipts in excess of $2 billion.

World Trade

World trade influences profitability of U.S. animal industries. Table 2.4 shows the export and import markets for several animal commodities. While world trade for all U.S. products shows a deficit,

Table 2.1

SEX AND SEX-CONDITION TERMINOLOGY FOR LIVESTOCK AND POULTRY

Species	Female Young[a]	Female Mature[b]	Uncastrated Male Young[a]	Uncastrated Male Mature[b]	Castrated Male Young[a]	Castrated Male Mature[b]
Cattle	Heifer[c]	Cow	Bull[c]	Bull	Steer	Stag
Chicken	Chick/Pullet	Hen	Chick/Cockerel	Cock/Rooster	Capon	—
Goat	Doe	Doe	Buck	Buck	Wether	—
Horse[d]	Filly	Mare	Colt[e]	Stallion	Gelding	—
Sheep	Ewe[f]	Ewe	Ram[f]	Ram/Buck	Wether	Stag
Swine	Gilt	Sow	Boar[g]	Boar	Barrow	Stag
Turkey	Young Hen/Poult[h]	Hen	Young Tom/Poult[h]	Tom	—	—

[a]Young: generally prior to puberty or sexual maturity and before the development of secondary sex characteristics.
[b]Mature: generally after puberty and before the development of secondary sex characteristics.
[c]Referred to as a *bull calf* (under 1 year of age); *heifer* can be a heifer calf, yearling heifer, or first calf heifer (after the first calf is born and prior to the birth of the second calf).
[d]A close relative of the horse is the *donkey*, also known as an *ass* or a *burro*. The male ass is referred to as a *jack*; the female is known as a *jennet*. A *mule* is produced by crossing a *jackass* with a mare. A *hinny* is produced by crossing a stallion with a jennet. Mules and hinnys are reproductively sterile, although their visual sexual characteristics appear to be normal.
[e]Under 3 years of age.
[f]Referred to as a *ewe lamb* or a *ram lamb* (under 1 year of age).
[g]Referred to as a *boar pig* (under 6 months of age).
[h]Under 10 weeks of age. Chicks or poults are newly hatched or a few days old.

Table 2.2

NUMBERS OF PRODUCERS AND ANIMALS IN THE U.S. LIVESTOCK AND POULTRY INDUSTRIES

Species	Number of Producers (1,000)	Number of Animals (mil head)
Beef cows	734	29.9
Swine	69	66.4
Dairy cattle	60	9.2
Broilers	NA	8,607.0
Layers	NA	338.5
Breeding ewes	80	4.0
Turkeys	NA	248.5
Goats—meat/milk/fiber	144	2.8[a]

[a]85% are meat-type.
Source: Adapted from USDA.

agricultural and animal products show a positive trade balance. The leading destinations for U.S. agricultural exports are China ($19.9 bil), Canada ($18.6 bil), Mexico ($17.7 bil), Japan ($13.9 bil), and the EU-27 ($10.2 bil).

Commodity Prices

The profitability of U.S. animal industries is partly influenced by the prices paid to producers for animals and animal products. Prices can fluctuate monthly, weekly, or even daily. These price changes are influenced primarily by supply and demand.

Table 2.3
LEADING U.S. STATES FOR FARM CASH RECEIPTS

Commodity	Rank[a]	Value ($ bil)	Five Leading States ($ bil)				
			1	2	3	4	5
All commodities		395.1	CA	IA	NE	TX	MN
			$44.7	$31.9	$24.5	$22.7	$20.6
Cattle and calves	2	67.9	TX	NE	KS	IA	CO
			10.5	10.4	7.9	3.8	3.7
Dairy products	4	37.0	CA	WI	NY	ID	PA
			6.9	5.3	2.5	2.4	2.1
Broilers	5	24.7	GA	AR	NC	AL	MS
			3.8	2.9	2.8	2.8	2.2
Hogs	6	22.1	IA	MN	NC	IL	IN
			6.9	2.8	2.5	1.5	1.2
Chicken eggs	9	7.8	IA	GA	PA	OH	IN
			1.0	0.53	0.52	0.52	0.47
Turkeys	12	5.4	NC	MN	IN	AR	MO
			0.85	0.84	0.44	0.42	0.41

[a]Ranking is in comparison to all agricultural commodities. Corn is #1 and soybeans #3.
Source: Adapted from USDA, 2012 data.

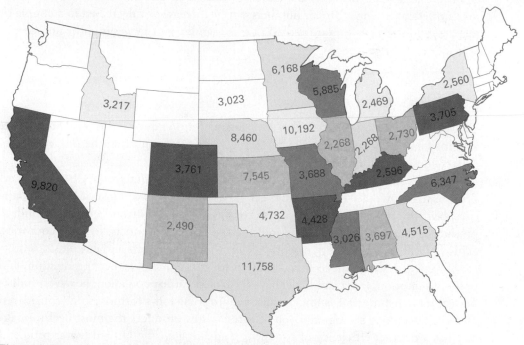

Figure 2.1
Cash receipts for U.S. livestock and products ($ mil). Source: Adapted from USDA.

Those individuals interested in animal profitability should know the average prices of animal products. In addition, an understanding of the prices for specific classes and grades of animals, of what causes prices to fluctuate, and of how high prices are obtained is also necessary. Figure 2.2 compares market prices of several

Table 2.4
U.S. EXPORTS AND IMPORTS OF MAJOR ANIMAL PRODUCTS

Commodity	Exports Value ($ mil)	Imports Value ($ mil)
Live animals	689	2,092
Red meat and products	11,412	5,305
Poultry meat and products	5,484	519
Dairy products	4,609	2,856
Hides and skin	2,603	197
Fats, oils, and greases	1,303	166
Total for agricultural products	137,374	94,487

Source: Adapted from USDA, 2012 data.

livestock and poultry commodities. This overview shows average prices and variability in prices producers receive for the animals and products they sell. Details of how some factors influence prices are discussed in later chapters.

Biological Differences in Meeting Market Demand

Changes in consumer demand, feed prices, weather, and other factors dictate the need to increase or decrease animal numbers and amount of product produced. Figure 2.3 shows the large differences between some farm animal species and how quickly or slowly inventories can change. Broiler numbers can be increased or decreased in a couple of months, while several years are needed to make significant changes in cattle numbers.

BEEF INDUSTRY

Global Perspective

Cattle were probably domesticated in Asia and Europe during the New Stone Age. Humped cattle (*Bos indicus*) were developed in tropical countries; the *Bos taurus* cattle were developed in more temperate zones.

Cattle, including the domestic water buffalo, contribute food, fiber, fuel, and draft animal power to the 6.8 billion people of the world. For most developed countries, beef (meat) is a primary product. For developing countries, beef is a secondary product as draft animal power and milk are the primary products. In some countries, cattle are still a mode of currency or a focus of religious beliefs and customs.

Table 2.5 shows the leading countries for cattle numbers, beef production, and beef consumption. India and Brazil have the largest cattle population; however, India's per-capita consumption is low because religious customs forbid cattle (considered sacred) from being slaughtered. The United States produces the most beef tonnage and has a distinct efficiency of production advantage. The United States produces approximately 20% of the world's beef with only 7% of the cattle. Brazil with 13% of the global cattle inventory produces 13% of the total beef while Australia accounts for 2% of the cattle inventory and 4% of the beef tonnage.

Countries with a high cattle population relative to their human population typically have a high per-capita consumption of beef and high export tonnage. For example, there are approximately 1.5 cattle per person in Australia and Argentina. Both countries rank high in per-capita consumption of beef. Leading export markets for U.S. beef include Japan, Canada, Hong Kong, Mexico, and Korea (Table 2.6).

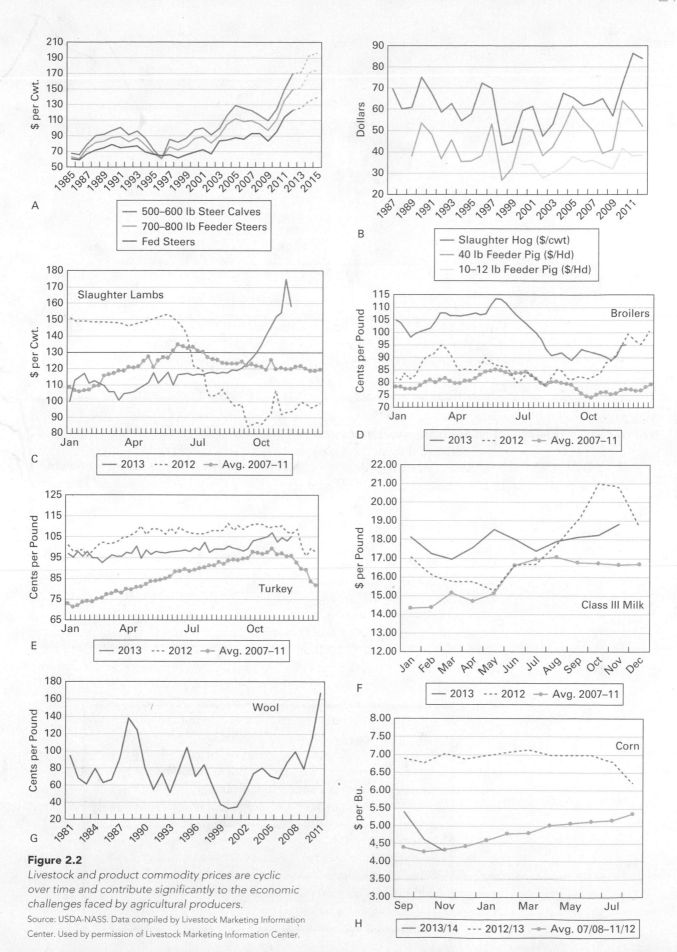

Figure 2.2

Livestock and product commodity prices are cyclic over time and contribute significantly to the economic challenges faced by agricultural producers.

Source: USDA-NASS. Data compiled by Livestock Marketing Information Center. Used by permission of Livestock Marketing Information Center.

Figure 2.3
Production cycle in several species of livestock and poultry.

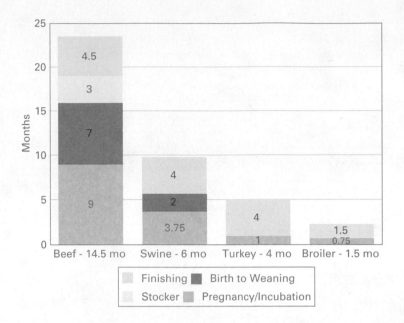

Table 2.5
WORLD CATTLE NUMBERS, PRODUCTION, AND CONSUMPTION

Country	No. Cattle (mil head)	Country	Production (mil tons) [a,b]	Carcass Weight (lb/hd)
1. Brazil	213	1. United States	10.9	752
2. India	211	2. Brazil	8.2	509
3. United States	93	3. China	5.6	315
4. China	83	4. Argentina	4.4	491
5. Ethiopia	53	5. Australia	3.8	580
World Total	1,426	World Total	57	468

[a]Does not include buffalo meat.
[b]Carcass weight.
Source: Adapted from USDA and FAO.

Table 2.6
U.S. BEEF TRADE

Exports		Imports	
Country	($ mil)	Country	($ mil)
1. Japan	$1,181	1. Australia	$1,029
2. Canada	973	2. New Zealand	817
3. Hong Kong	784	3. Canada	797
4. Mexico	698	4. Mexico	547
5. South Korea	558	5. Brazil	228
World Total	5,208	World Total	3,760

Source: Adapted from USDA: FAS, 2013 data.

Table 2.7

AN OVERVIEW OF THE BEEF INDUSTRY SEGMENTS IN THE UNITED STATES

Segment of the Beef Industry	Products Produced or Utilized	Approximate Number
Seedstock Producers ↓ ↑ ª↓ ↑ᵇ	Breeding stock—primarily bulls (14–30 months old) and some heifers and cows. Some steers and heifers for feeding. Slaughter cows and bulls.	50,000 breeders, 3 major AI studs
Commercial Cow-Calf Producers ↓ ↑ ↓ ↑ ↓ ↑	Calves (6–10 months old), weighing 300–700 lb. Slaughter cows and bulls with majority over 5 years of age, weighing 800–1,500 lb (cows) and 1,000–2,500 lb (bulls).	751,000 producersᶜ
Yearling or Stocker Operator ↓ ↑ ↓ ↑	Feeder steers and heifers (most of them 12–20 months old, weighing 500–900 lb).	
Feeders ↓ ↑ ↓ ↑ ↓ ↑	Market steers, heifers, cows, and bulls (mostly steers and heifers 16–30 months old, weighing 900–1,400 lb).	130 feeding companies with capacity >1,000 head
Packers ↓ ↑ ↓ ↑ ↓ ↑	Carcasses (approx. 600–800 lb). Boxed beef (carcasses into subprimal cuts).	700 packing plants
Retailers ↓ ↑ ↓ ↑ ↓ ↑	Retail cuts. By-products.	37,000 grocery stores with sales of $2 million or more per year
Consumers	Cooked products. By-products (leather, pharmaceuticals, variety meats, etc.).	317 million consumers (U.S.); 7.1 billion consumers (world)

ªAnimal and/or product flow.
ᵇDemands and expectations.
ᶜIn addition, there are approximately 60,000 dairy farms that produce about 20% of U.S. beef.
Source: Adapted from Field and Taylor, 2007.

United States

The U.S. **beef industry** is made up of a series of producing, processing, and consuming segments that relate to each other but that operate independently. Table 2.7 identifies the various segments and the products they produce.

Figure 2.4 shows historic trends for total cattle and amount of carcass beef produced in the United States. Interestingly, the U.S. beef industry has been able to retain high levels of beef production despite the recent cattle inventory losses. There are several reasons for this relatively high production of beef from fewer numbers of cattle: (1) the average carcass weight has increased from 579 lb in 1975 to 776 lb in 2007, (2) an increased number of cattle are fed per feedlot (2.4 times the feedlot capacity), (3) the market age of fed cattle has decreased, (4) more crossbreeding and faster-gaining European breeds (e.g., Simmental, Limousin, and Charolais) are being used in commercial breeding programs, and (5) more Canadian and Mexican calves are imported and fed in the United States when market conditions are favorable.

Cattle Production

Most commercial beef cattle production occurs in three phases: the **cow-calf**, **stocker-yearling**, and **feedlot** operations. The cow-calf operator raises the young calf

Figure 2.4
*Total U.S. cattle inventory
versus commercial beef
production.* Source: USDA-NASS.
Data compiled by Livestock
Marketing Information Center. Used
by permission of Livestock Marketing
Information Center.

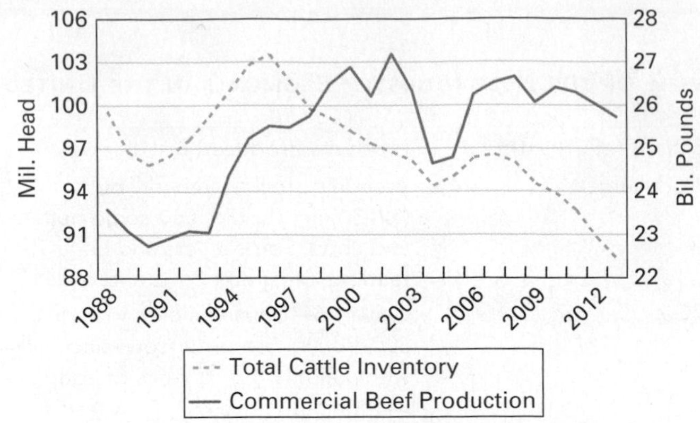

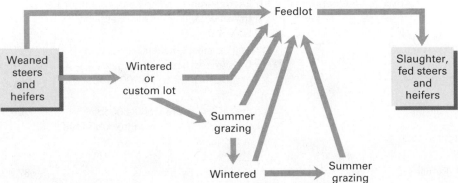

Figure 2.5
*Alternative production and
marketing strategies in the
U.S. beef industry.*

from birth to 6–10 months of age (400–650 lb). The stocker-yearling operator then grows the calf to 600–850 lb, primarily on roughage. Finally, the feedlot operator uses high-energy rations to finish the cattle to a desirable slaughter weight, approximately 900–1,300 lb. Most fed steers and slaughter heifers are between 15 and 24 months of age when marketed.

However, there are alternatives to the typical three-phase operation. In an integrated operation, for instance, the cattle may have a single owner from cow-calf to feedlot, or ownership may change several times before the cattle are ready for slaughter. Alternative production and marketing strategies are diagrammed in Figure 2.5.

Cow-Calf Production.   U.S. cow-calf production involves some 32.5 million head of beef cows that are distributed throughout the country. Most of the cows are concentrated in areas where forage is abundant. As Figure 2.6 shows, 14 states each have over 700,000 head of cows (75% of the U.S. total), most of them located in the Plains, Corn Belt, and southeastern states. Approximately 68% of the beef cow operations have less than 50 cows per operation. However, more than 70% of the beef cow inventory is in operations with more than 100 cows (Table 2.8). Cow numbers fluctuate over the years, depending on drought, beef prices, and land prices.

There are two kinds of cow-calf producers. Commercial cow-calf producers raise most of the potential slaughter steers and heifers. Seedstock breeders—specialized cow-calf producers—produce primarily breeding cattle and semen.

Stocker-Yearling Production.   Stocker-yearling producers feed cattle for growth prior to their going into a feedlot for finishing. Replacement heifers intended for the breeding herd are typically included in the stocker-yearling category. Our focus here, however, is on steers and heifers grown for later feedlot finishing.

Several alternate stocker-yearling production programs are identified in Figure 2.5. In some programs, a single producer owns the calves from birth through

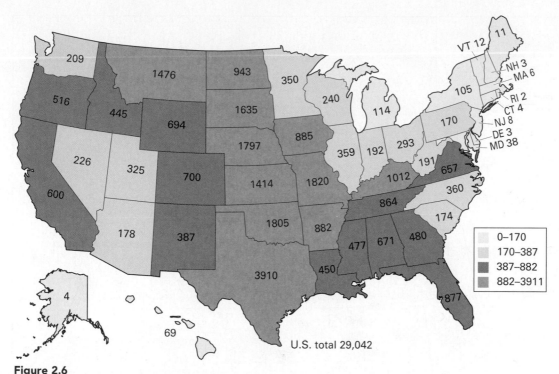

Figure 2.6

Beef cow inventory by state (1,000 head). Source: USDA-NASS. Data compiled by Livestock Marketing Information Center. Used by permission of Livestock Marketing Information Center.

Table 2.8
U.S. BEEF COW OPERATIONS AND INVENTORY

Herd Size (No. Cows)	Operations Percent of Total	Inventory Percent of Total
1–49	68	28
50–99	14	17
100–499	15	38
500+	3	17

Source: Adapted from USDA.

the feedlot-finishing phase, and the cattle are raised on the same farm or ranch. In other programs, one operator retains ownership, but the cattle are custom-fed during the growth and finishing phases. In still other programs, the cattle are bought and sold once or several times.

The primary basis of the stocker-yearling operation is to market available forage and high-roughage feeds, such as grass, crop residues (e.g., corn stalks, grain stubble, and beet tops), wheat pasture, and silage. Stocker-yearling operations also make use of summer-only grazing areas that are not suitable for the production of supplemental winter feed.

Stocker-yearling operations are desirable for early-maturing cattle. These cattle need slower gains to achieve heavier slaughter weights without being excessively finished. Larger-framed, later-maturing cattle usually are more efficient and profitable if they go directly to the feedlot after weaning.

Feedlot Cattle Production.   Feedlot cattle are fed in pens or fenced areas, where harvested feed is brought to them. Some cattle are finished for market on pasture, but

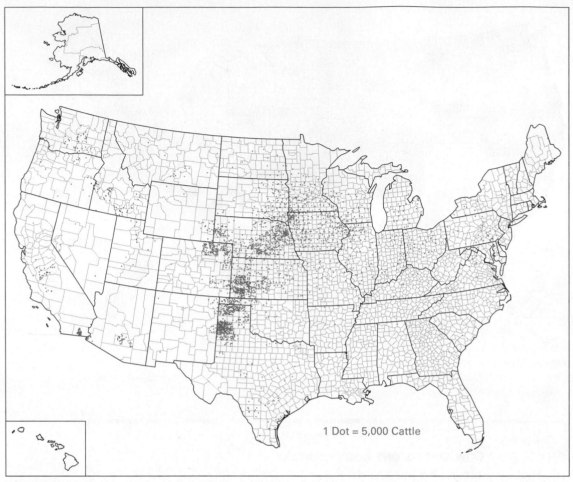

1 Dot = 5,000 Cattle

Figure 2.7

Cattle-feeding areas in the United States. Note that cattle feeding is concentrated in the southern and central region of the Great Plains. Source: USDA.

they represent only 10–15% of the slaughter steers and heifers. They are sometimes referred to as *nonfed cattle* because they are fed little, if any, grain or concentrate feeds. The cattle-feeding areas in the United States (Fig. 2.7) correspond to the primary feed-producing areas where cultivated grains and roughage are grown. These locations are determined primarily by soil type, growing season, and amount of rainfall or irrigation water. Figure 2.8 shows where the approximately 22 million feedlot cattle are fed in the various states. The total number of fed cattle marketed in the leading states is shown in Figure 2.9. Marketing by the leading states represent 95% of the 21.6 million head of fed cattle in all states. By contrasting Figures 2.8 and 2.9, the number of cattle marketed for each state is considerably higher than the number on feed. This is because most commercial feedlots feed more than 1.7 times their one-time capacity in a given year.

Cattle Feeding

The two basic types of cattle-feeding operations are (1) **commercial feeders** (Fig. 2.10) and (2) **farmer-feeders.** The two operations are distinguished by type of ownership and size of feedlot.

The farmer-feeder operation is usually owned and operated by an individual or a family and has a feedlot capacity of less than 1,000 heads. An individual or partnership sometimes owns the commercial feedlot, but more often a corporation owns it, especially as feedlot size increases. It has a feedlot capacity of over 1,000 head.

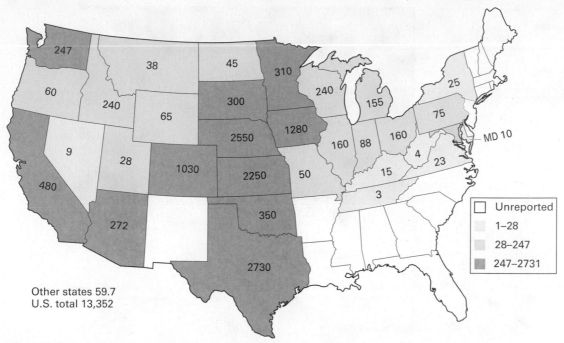

Figure 2.8
Cattle on feed by state (1,000 head). Source: USDA: NASS compiled by Livestock Marketing Information Center.

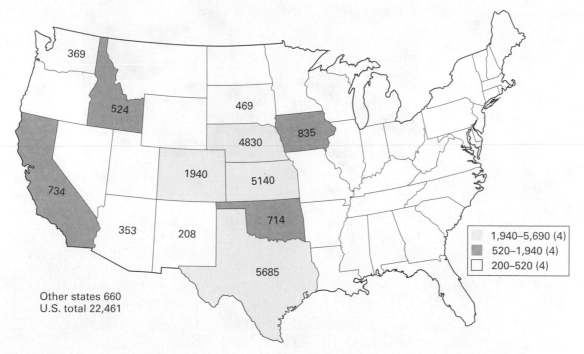

Figure 2.9
Fed cattle marketings 1,000+ head feedlots (1,000 head). Source: USDA-NASS. Data compiled by Livestock Marketing Information Center. Used by permission of Livestock Marketing Information Center.

Approximately 95% of fed cattle are fed in feedlots with over 1,000-head capacity. A number of U.S. commercial feedlots have capacities of 40,000 head or higher, and a few have capacities of over 100,000 head. Some commercial feedlots custom-feed cattle; that is, the commercial feedlot provides the feed and feeding service to the cattle owner.

Figure 2.10
The feed mill associated with a large commercial cattle-feeding enterprise. Rations can be formulated and delivered to meet the needs of thousands of head of cattle. Source: Tom Field.

Figure 2.11
Ethanol production has created greater volatility in grain markets and thus impacts the feedyard profit margins of both commercial and farmer-feeder enterprises. Source: Tom Field.

Each type of feeding operation has its advantages and disadvantages. What is an advantage to one type of feedlot is usually a disadvantage to the other type. The large commercial feedlot usually enjoys economic advantages associated with its size as well as professional expertise in nutrition, health, marketing, and financing. The farmer-feeder has the advantages of distributing labor over several enterprises by using high-roughage feeds effectively, creating a market of homegrown feeds through cattle, and more easily closing down the feeding operation during times of unprofitable returns. However, all feedyards are affected by feed price volatility (Fig. 2.11).

THE DAIRY CATTLE INDUSTRY

Global Perspective

Milk and milk products are produced and consumed by most countries of the world. Although buffalo, goat, and other types of milk are important in some areas, our focus here is on cow's milk and its products.

Table 2.9 shows world dairy cattle numbers, fluid milk production, and per-capita consumption of milk. Of the 260 million dairy cows in the world, India leads

Table 2.9
WORLD DAIRY CATTLE NUMBERS, MILK PRODUCTION, AND CONSUMPTION

Country	No. Dairy Cattle (mil head)	Country	Fluid Milk Production (mil ton)	Country	Production per Cow[a] (lb)
1. India	45	1. United States	81	1. Israel	25,117
2. Brazil	23	2. India	52	2. Saudi Arabia	23,424
3. China	12	3. China	33	3. South Korea	22,153
4. Ethiopia	11	4. Brazil	29	4. United States	21,336
5. Pakistan	10	5. Russian Fed.	28	5. Canada	19,179
World Total	260	World Total	557	World Average	5,210

[a]Whole and skim milk.
Source: Adapted from USDA and FAO.

Table 2.10
VALUE OF U.S. DAIRY PRODUCT IMPORTS AND EXPORTS

Exports Country	($ mil)	Imports Country	($ mil)
Mexico	1,429	New Zealand	585
China	706	Canada	438
Canada	569	Italy	352
Philippines	364	France	229
Indonesia	316	Mexico	177
Total	6,719	Total	3,018

Source: Adapted from USDA: FAS, 2013.

all other countries with its 45 million. The average annual world production per cow is 5,210 lb, with production in Israel the highest at nearly 25,000 lb. Countries with high milk-production levels per cow have excellent breeding, feeding, and health-management programs.

World butter production is in excess of 21 billion lb. The European Union, India, the United States, and Pakistan are the leaders in total butter production, while New Zealand, the European Union, and Switzerland have the highest per-capita consumption of butter.

The leaders in cheese production are the European Union and the United States—jointly accounting for 70% of the world's cheese production. Consumers in Iceland, the European Union, Switzerland, and the United States have the highest per-capita consumption of cheese.

Fluid milk is produced worldwide but is rarely traded. Manufactured dairy products dominate dairy trade. Manufactured products traded internationally account for about 5% of the global fluid milk production on a milk-equivalent basis. Trade restrictions keep trade volume and prices lower than if trade was more liberalized. Most developed countries, including the United States, extensively regulate their dairy industries by subsidizing production and often exportation.

Table 2.10 shows the value of U.S. import and export trade in dairy products. Cheese and nonfat dry milk are the most significant export items with cheese going to the Pacific Rim and NAFTA partners and dry milk to countries with minimal dairy

Figure 2.12
Dairies in the United States continue to increase in herd size. Many dairy cow herds are intensively managed in excellent facilities. Photo by Justin Field.

production. Cheese and milk components are imported from countries in Europe and Oceania.

United States

The U.S. **dairy industry** has changed dramatically since the days of the family milk cow. Today it is a highly specialized industry that includes the production, processing, and distribution of milk. A large investment is required in cows, machinery, barns, and milking parlors where cows are milked. Dairy operators who produce their own feed need additional investments for land on which to grow feed. They also require machinery to produce, harvest, and process the crops.

Although the size of a dairy operation can vary from less than 30 milking cows to more than 5,000 milking cows (Fig. 2.12), the average U.S. dairy has approximately 135 mature milking cows. Average dairy producers farm 200–300 acres of land, raise much of the forage, and market the milk through member-owned cooperatives. Producers sell about 4,100 lb of milk daily, or about 1.5 million lb annually, valued at about $200,000. Their average total capital investment may exceed $500,000. The average dairy producer has a partnership (with a family member or another person) to make the management of time and resources easier.

Herds of less than 100 cows comprise nearly 75% of all dairy herds but only produce 14% of U.S. milk production. Six percent of dairy herds have more than 500 cows and provide almost two-thirds of total domestic milk production.

New technologies have required extensive capital investment thus creating a competitive advantage for larger dairies capable of generating sufficient cash flow. Since 1977, farms with fewer than 100 milk cows have declined continuously as a share of all farms with milk cows. The share of farms having 500 or more milk cows is increasing in share of all farms with milk cows. The largest farms are typically found in the West and Southwest. The traditional milk-producing states of the Northeast and Great Lakes states have lost share of milk production.

One way to stress the importance of dairying in the United States is to examine the number of cows in the states (Fig. 2.13). The five leading states in thousands of dairy cows are, respectively, California (1,780), Wisconsin (1,270), New York (610), Idaho (580), and Pennsylvania (535). The five leading states in milk production per cow are, respectively, New Mexico (24,800 lb), Washington (23,700 lb), followed by Colorado, California, and Arizona (23,400). As Figure 2.14 shows, today's 9.2 million dairy cows are approximately one-third the number of cows 50 years ago, yet total milk production continues to increase due

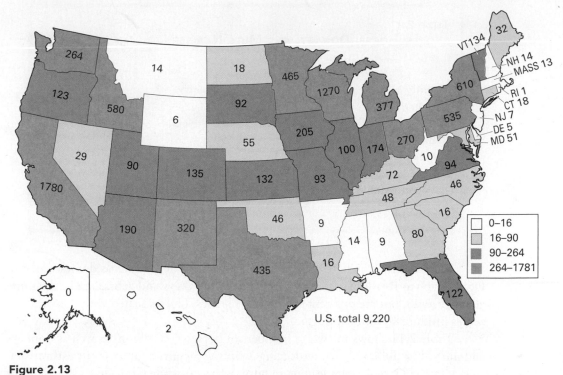

Figure 2.13
U.S. milk cow numbers by state (1,000 head). Source: USDA-NASS. Data compiled by Livestock Marketing Information Center. Used by permission of Livestock Marketing Information Center.

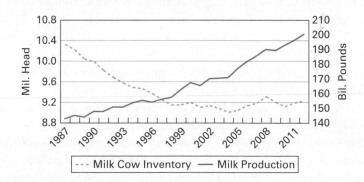

Figure 2.14
Trends in milk cow numbers and milk production.
Source: USDA: NASS compiled by Livestock Marketing Information Center.

to increased production per cow, which is 21,345 lb of milk per cow. This marked improvement is the result of effective breeding, feeding, health, and management programs.

THE HORSE INDUSTRY

Global Perspective

The horse was domesticated about 5,000 years ago. It was one of the last farm animals to be domesticated. Horses were first used as food, then for war and sports, and also for draft purposes. They were used for transporting people swiftly and for moving heavy loads. In addition, horses became important in farming, mining, and forestry.

Table 2.11
WORLD HORSE, DONKEY, AND MULE NUMBERS

Horses		Donkeys and Mules	
Country	(mil head)	Country	(mil head)
1. United States	10	1. China	9
2. China	7	2. Ethiopia	7
3. Brazil	6	3. Mexico	6
4. Mexico	5	4. Pakistan	5
5. Argentina	3	5. Egypt	3
World Total	58	World Total	54

Source: Adapted from FAO.

The donkey, which descended from the wild ass of Africa, was domesticated in Egypt prior to the domestication of the horse. Donkeys and zebras are in the same genus (*Equus*), but are different species from horses. Horses mated with donkeys and zebras produce sterile offspring.

Table 2.11 shows the world numbers of horses (58 million) as well as donkeys and mules (54 million). The five leading countries are given for each of the three species, with the United States leading in horse numbers while China leads in inventory of donkeys and mules.

Horses have been companions for people since their domestication. Once important in wars, mail delivery, farming, forest harvesting, and mining, the horse today is used in shows, in racing, in the handling of livestock, and for companionship, recreation, and exercise.

United States

There were no horses on the North American continent when Columbus arrived. However, there is fossil evidence that the early ancestor of the horse was here some 50–60 million years ago. The **Eohippus** (or dawn horse), a four-toed animal less than a foot high, is believed to be the oldest relative of the horse.

The early ancestors of the horse disappeared in pre-Columbian times, supposedly by crossing from Alaska into Siberia. It is from these animals that horses may have evolved in Asia and Europe. The draft horses and Shetland ponies developed in Europe, whereas the lighter, more agile horses developed in Asia and the Middle East. The Spaniards and colonists introduced modern horses to the Americas.

In the early 1900s, there were approximately 25 million horses and mules in the United States. Shortly after World War I, however, horse numbers began a rapid decline. The war stimulated the development and use of motor-powered equipment, such as automobiles, trucks, tractors, and bulldozers. Railroads were heavily used for transporting people and for moving freight long distances. By the early 1960s, horses and mules had declined to a mere 3 million in the United States.

With the shorter workweek and greater affluence of the working class, there has been more time and money available for recreation. An estimated 2 million American horse owners and an additional estimated 5 million people acting as service providers, employees, and volunteers are involved in the industry. Approximately 3.5% of consumer expenditures for recreation are spent in the **horse industry.** The U.S. horse business generates more than $25 billion in goods and services annually.

The American Horse Council estimates that there are 9.2 million horses in the United States, with about 43% utilized in recreational activities, 29% as show

Table 2.12
HORSE POPULATION IN SELECTED STATES AND PRIMARY USE

State	Horses 1,000(N)	Racing (%)	Showing (%)	Recreation (%)	Other[a] (%)
Texas	979	11	19	27	44
California	698	11	30	43	17
Florida	500	12	52	18	14
Oklahoma	326	20	34	10	36
Colorado	256	10	26	29	35
Ohio	326	21	34	35	9
Kentucky	320	45	21	25	10
New York	202	18	27	37	18
U.S. Total	1,200	10	28	43	18

[a]Includes agricultural, other sports, nonspecific breeding, etc.
Source: Adapted from American Horse Council.

animals, 10% involved in racing, and the remainder utilized in rodeo, polo, ranch work, or other activities. The leading states for horse numbers are identified in Table 2.12.

Horse owners tend to be well educated, upper-middle income (median income of $60,000), and middle-aged. However, approximately one-third of horse owners have an annual income of less than $50,000. The annual cost of horse maintenance ranges widely from $1,000 to $15,000 per horse depending on region and use. Monthly boarding costs (stall space, bedding, and feed) are typically $150 to $250 per head. Training and riding lessons would typically run from $300 to $600 per month and $20 to $50 per hour, respectively. Horse owners typically pay $48 to $75 per horse for farrier service (horseshoeing and hoof care). The average cost of participating in a horse show ranges from $100 to $540 per show.

Horse numbers in the United States increased rapidly from 1960 to 1976, with slower growth after that time period. Horse numbers declined in the mid-1980s when applicable tax laws (that removed some of the advantages of horse ownership) were passed by Congress. Horse numbers are slowly rebuilding at the present time. Interest in diverse horse activities has risen as evidenced by increases in membership of the American Driving Society, the U.S. Combined Training Association, and the U.S. Dressage Federation by 4%, 14%, and 23%, respectively, over the past half-decade.

THE POULTRY INDUSTRY

Global Perspective

The term *poultry* applies to chickens, turkeys, geese, ducks, pigeons, peafowls, and guineas. Chickens, ducks, and turkeys dominate the world poultry industry. In parts of Asia, ducks are commercially more important than broilers (young chickens), and in areas of Europe, there are more geese than other poultry because they are more economically important.

Chickens originated in Southeast Asia and were kept in China as early as 1400 B.C. Charles Darwin concluded in 1868 that domestic chickens originated from the Red Junglefowl, although three other Jungle fowl species were known to exist.

Table 2.13 shows the leading countries in world poultry production.

Poultry exports account for more than $5 billion (Table 2.14).

Table 2.13
POULTRY AND EGG PRODUCTION

Broiler	(mil ton)	Turkey	(mil ton)	Eggs	(bil eggs)
U.S.	15.5	U.S.	2.4	China	478
China	10.5	Brazil	0.44	United States	92
Brazil	10.4	Germany	0.42	India	63
Russian Fed.	2.6	France	0.36	Mexico	48
Mexico	2.5	Canada	0.11	Russian Fed.	41
World	81.6	World	4.9	World	1,126

Source: Adapted from USDA and FAO.

Table 2.14
POULTRY AND EGG IMPORT AND EXPORT VALUES FOR THE UNITED STATES

Country	Exports (mil $)	Country	Imports (mil $)
Mexico	1,393	Canada	327
Canada	766	China	164
China	503	Chile	76
Russia	323	Hungary	33
Hong Kong	307	Mexico	14
World	6,465		669

[a]All poultry meat combined.
Source: Adapted from USDA: FAS, 2013.

World governments influence world poultry trade by controlling production and pricing and by placing **tariffs** on incoming goods—all barriers to free international trade. Trade liberalization by countries with industrial market economies would likely increase the trade of and decrease the price of poultry meat. Countries with efficient producers (such as the United States, Brazil, and Thailand), combined with consumers from countries with considerable trade protection (such as Japan, Canada, and the European Union), would benefit most from liberalized trade.

Poultry is the fastest-growing source of meat for people. The industrialized countries produce and export more than 50% of the poultry meat in the world.

United States

The annual U.S. income from broilers, turkeys, and eggs exceeds $35 billion (broilers, $23.2 billion; eggs, $7.4 billion; and turkeys, $4.9 billion). Figure 2.15 shows the leading states in broiler, turkey, and egg production. The U.S. poultry industry is concentrated primarily in the southeastern area of the country.

From 1900 to 1940, the primary concerns of the U.S. poultry industry were egg production by chickens and meat production by turkeys and waterfowl. Meat production by chickens was largely a by-product of the egg-producing enterprises. The **broiler** industry, as it is known today, was not yet established. Egg production was well established near large population centers, but the quality of eggs was often low because of seasonal production, poor storage, and the absence of laws to control grading standards.

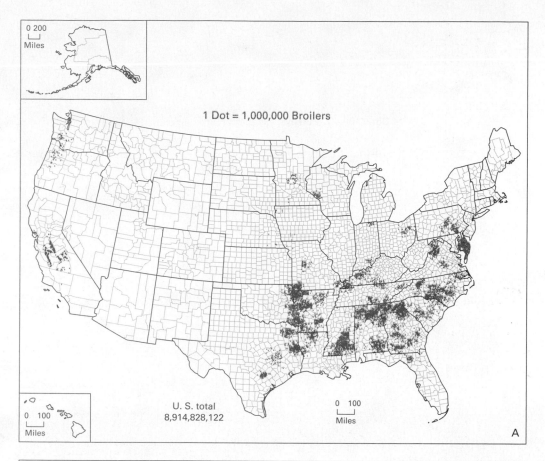

1 Dot = 1,000,000 Broilers

U. S. total
8,914,828,122

A

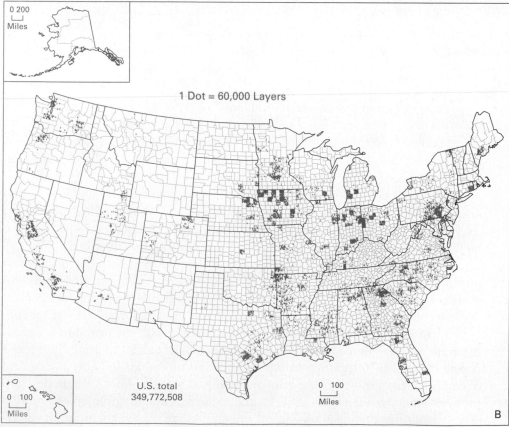

1 Dot = 60,000 Layers

U.S. total
349,772,508

B

Figure 2.15

Concentration of broiler (A), egg (B), and turkey (C) production in the United States (million).

Source: Adapted from USDA.

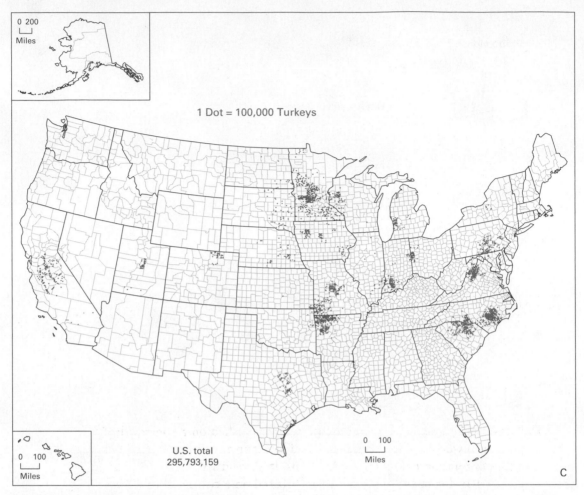

1 Dot = 100,000 Turkeys

U.S. total
295,793,159

Figure 2.15 (continued)

Before 1940, large numbers of small farm flocks existed in the United States, but the management practices applied today were practically unknown. The modern mechanized poultry industry of the United States emerged during the late 1950s as the number of poultry farms and hatcheries decreased and the number of birds per installation dramatically increased. Larger cage-type layer operations appeared, and egg-production units grew, with production geared to provide consumers with eggs of uniform size and high quality. Large broiler farms that provided consumers with fresh meat throughout the year were established, and large dressing plants capable of dressing 50,000 or more broilers daily were built. The U.S. poultry industry was thus revolutionized.

One of the most striking achievements of the poultry industry is its increased production of eggs and meat per hour of labor. Poultry operations with 1 million birds at one location are not uncommon. Automatic feeding, watering, egg collecting, egg packing, and manure removal have accomplished significant labor reduction.

Dramatic changes came between 1955 and 1975 with the introduction of integration. It began in the broiler industry and is applied to a lesser degree in egg and turkey operations. Integration brings all phases of an enterprise under the control of one head, frequently the corporate ownership of breeding flocks, hatcheries, feed mills, raising, dressing plants, services, and marketing and distribution of products (Fig. 2.16). After 1975, poultry entered a new period of consolidation, in which vertically integrated companies purchased, acquired, or merged with each other, creating a small number of superintegrated companies.

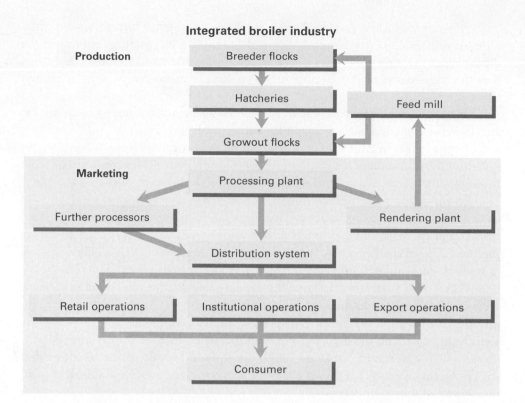

Figure 2.16
Structure of the integrated broiler industry.

The actual raising of broilers is sometimes accomplished on a contract basis between the person who owns the houses and equipment and furnishes the necessary labor, and the corporation that furnishes the birds, feed, field service, dressing, and marketing. Payment for raising birds is generally based on a certain price for each bird reared to market age. Bonuses are usually paid to those who do a commendable job of raising birds. An important advantage to the integration system is that all phases are synchronized to ensure the utmost efficiency.

Broiler production is concentrated in the southern and southeastern states. California and Texas are the only states with large production located outside the "broiler belt." The "broiler belt" includes Alabama, Arkansas, Georgia, Mississippi, North Carolina, and Virginia. This region produces two-thirds of U.S. broilers.

Broiler production has increased tremendously over the past three decades, from 3.7 million lb of ready-to-cook broilers in 1960 to 49.7 million lb in 2012. The U.S. broiler industry is concentrating its grow-out operations into fewer but larger farms.

Although the number of laying hens in the United States has declined over the years, egg production is higher because of improved performance of the individual hen. In 1880, for example, the average laying hen produced 100 eggs per year; in 1950, the average was 175; in 1986, it increased to 250; and in 2012, it was 271 eggs.

The United States Department of Agriculture (USDA) estimates that 85% of the market eggs produced in the United States originate from large commercial producers (those that maintain 1–11 million birds). The 45 largest egg-producing companies in the United States have more than 97 million layers, which is 35.5% of the nation's total.

Important changes have also occurred in U.S. turkey production. Fifty years ago, turkeys were raised in small numbers on many farms; today, they are raised in larger numbers on fewer farms. The average turkey farmer is producing well over 50,000 turkeys per year. The leading turkey-producing states are Minnesota, North Carolina, Arkansas, and Indiana.

Most of the turkeys produced today are the heavy or large, broad-breasted white type. At one time, the fryer-roaster-type turkey (5–9 lb) was popular; it bridged the gap between the turkey and the broiler chicken. Today, the broiler industry has taken over the fryer-roaster market because it can produce fryer-roaster chickens at a much lower price.

It takes 16 weeks for turkey hens and 19 weeks for turkey toms to reach market size. Because there is a heavy demand for turkey meat in November and December, turkey eggs are set in large numbers in April, May, and June. However, eggs are set year-round because there is a constant, though lower, demand for fresh turkey meat throughout the year. The normal incubation period for turkey eggs is 28 days.

In the recent past, most consumers considered turkey a seasonal product, with consumption occurring primarily at Thanksgiving. Increasingly, turkey is consumed throughout the year, and consumers have available to them a large variety of turkey products. The greater demand for turkey has meant that approximately 1 billion lb more of ready-to-cook turkey is produced today than was produced in the early 1980s.

The export market has become increasingly important to the poultry industry with 16% of broiler production, 12% of turkey, and 4% of eggs sold to customers outside of the United States. However, disruptions in foreign markets can have significant impact on the industry. For example, when Russian buyers stopped purchasing U.S. chicken in 2001 and 2009, not only did broiler prices go down but the competing meats also experienced price slippage as the supply of animal protein overwhelmed short-term demand.

THE SHEEP AND GOAT INDUSTRY

Global Perspective

Sheep and goats are closely related, with both originating in Europe and the cooler regions of Asia. Sheep are distinguished from goats by the absence of a beard, less odor (males only), and glands in all four feet. The horns are spiral in different directions; goat horns spiral to the left, while sheep horns spiral to the right (like a corkscrew).

Sheep and goats are important ruminants in temperate and tropical agriculture. They provide fibers, milk, hides, and meat, making them versatile and efficient, especially for developing countries. Sheep and goats are better adapted than cattle to arid tropics, probably because of their superior water and nitrogen economy. Cattle, sheep, and goats often are grazed together because they utilize different plants. Goats graze **browse** (shrubs) and some **forbs** (broad-leaved plants), cattle graze tall grasses and some forbs, and sheep graze short grasses and some forbs. Many of the forbs in grazing areas are broadleaf weeds.

More than 60% of all sheep are in temperate zones and fewer than 40% are in tropical zones. Goats, though, are mostly (80%) in the tropical or subtropical zones (0–40°N). Temperature and type of vegetation are the primary factors encouraging sheep production in temperate zones and goat production in tropical zones.

Sheep originated in the dry, alternately hot-and-cold climate of Southwest Asia. To succeed in tropical areas, then, sheep had to adapt the abilities to lose body heat, resist diseases, and survive in an adverse nutritional environment. For sheep to lose heat easily, they require a large body surface-to-mass ratio. Such a sheep is a small, long-legged animal. In addition, a hairy coat allows ventilation and protects the skin from the sun and abrasions. In temperate areas, sheep with large, compact bodies, a heavy fleece covering, and storage of subcutaneous fat have an advantage. Sheep can also constrict or relax blood vessels to the face, legs, and ears for control of heat loss.

Table 2.15
WORLD SHEEP NUMBERS, PRODUCTION, AND CONSUMPTION

Country	No. Sheep (mil head)	Country	Production (mil tons)[a]	Country	Carcass Weight (lb/head)
1. China	139	1. China	1.9	1. Syria	88
2. India	74	2. Australia	0.46	2. Montenegro	87
3. Australia	73	3. New Zealand	0.42	3. Barbados	79
4. Iran	49	4. United Kingdom	0.26	4. United States	68
5. Sudan	39	5. Algeria	0.23	5. Oman	660
World Total	1,094	World Total	7.4	World Average	35

[a]Carcass weight of lamb and mutton.
Sources: Adapted from USDA and FAO.

The productivity of sheep is much greater in temperate areas than it is in tropical environments. This difference is the result not only of a more favorable environment (temperature and feed supply) but also of a greater selection emphasis on growth rate, milk production, lambing percentage, and fleece weight. Sheep from temperate environments do not adapt well to tropical environments; therefore, it may be more effective to select sheep in the production environment, rather than introduce sheep from different environments.

World sheep numbers of nearly 1.1 billion head are the highest on record. China, India, Australia, and Iran are the leading sheep producing countries (Table 2.15). Mongolia, Iceland, and New Zealand have the highest per-capita consumption of mutton, lamb, and goat meat.

The United States exports a relatively small amount of lamb and mutton with most of its export in the form of mutton from cull ewes. However, Australia and New Zealand provide a significant amount of lamb to the U.S. market. In fact, the U.S. imports more lamb than it produces domestically. The 924 million head of goats in the world are concentrated primarily in India and China, with Pakistan, Afghanistan, and Bangladesh also having large goat populations. In many countries, goats are important for both milk and meat production. Table 2.16 shows countries with the greatest goat meat and goat milk production.

United States

The number of sheep in the United States reached a high of 56 million in 1942, thereafter declining to approximately 5 million today (Fig. 2.17). U.S. sheep are

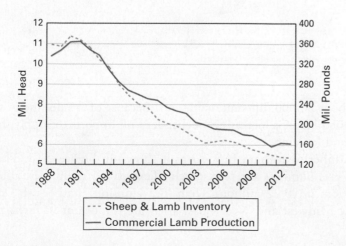

Figure 2.17
Total sheep and lamb population trends in the United States.
Source: USDA-NASS. Data compiled by Livestock Marketing Information Center. Used by permission of Livestock Marketing Information Center.

Table 2.16
WORLD GOAT NUMBERS, MEAT PRODUCTION, AND GOAT MILK PRODUCTION

Country	No. Goats (mil head)	Country	Goat Meat Production (mil ton)[a]	Country	Milk Production (mil ton)
1. India	157	1. China	1.7	1. India	4.3
2. China	142	2. India	0.54	2. Indonesia	4.3
3. Pakistan	61	3. Nigeria	0.26	3. Bangladesh	2.3
4. Nigeria	57	4. Pakistan	0.25	4. China	1.7
5. Bangladesh	53	5. Iran	0.13	5. Sudan	0.97
World Total	924	World Total	4.7	World Total	15.5

[a]Carcass weight.
Sources: Adapted from USDA and FAO.

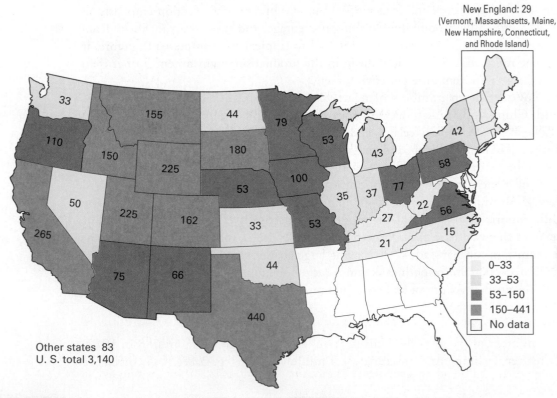

Figure 2.18
Breeding ewes, 1 year of age and older by state (1,000 head). Source: USDA-NASS. Data compiled by Livestock Marketing Information Center. Used by permission of Livestock Marketing Information Center.

located on about 80,000 different operations. Because of the decline in sheep numbers, the United States is increasingly dependent on imports to supply the relatively limited demand for lamb.

The number of ewes on U.S. farms is shown in Figure 2.18. The five leading states in ewe numbers are, respectively, Texas (0.50 million), California (0.27 million), Wyoming (0.22 million), Utah (0.21 million), and South Dakota (0.17 million). The Midwest and West Coast areas are important in **farm-flock** sheep

Table 2.17

AVERAGE EWE FLOCK SIZE IN SELECTED STATES IN THE UNITED STATES

Range Flocks		Farm Flocks	
State	Ewes (N)	State	Ewes (N)
Wyoming	508	Minnesota	60
Colorado	287	Virginia	40
Nevada	280	Ohio	37
Utah	190	Pennsylvania	23
South Dakota	167	New England	15

Source: USDA.

production, whereas other western states are primarily **range-flock** areas. The Midwestern states generally raise more lambs per ewe than the western states do, though Washington, Oregon, and Idaho resemble the Midwestern states in this regard.

Table 2.17 shows average ewe flock sizes in selected states. Most of the decline in U.S. sheep numbers during the past 50 years has occurred in the West. However, this region, with its extensive public and private rangelands, still produces 80% of the sheep raised in the country today. Most U.S. sheep growers have small flocks (50 or fewer sheep) and raise sheep as a secondary enterprise. About 40% of the West's sheep producers maintain flocks of more than 50 sheep; these flocks contain about 93% of the sheep in that region. Only about one-third of the operators in the West who have flocks of 50 or more sheep specialize in sheep; the other two-thirds have diversified livestock operations.

About 23% of all sheep born in the western United States and 13% of those born in the North-Central region are lost before they are marketed. In the West, predators, especially coyotes, and weather are the most frequent causes of death. In the North-Central region, the primary causes of death include weather and disease before docking, and disease and attacks by dogs after docking.

The range operator produces some slaughter lambs if good mountain range is available, but most range-produced lambs are feeders (e.g., lambs that must have additional feed before they are slaughtered). Feeder lambs are fed by producers, sold to feedlot operators, or fed on contract by feedlot operators. Some feedlots have a capacity for 20,000 or more lambs on feed at one time, and many feedlots that have a capacity of 5,000 are in operation. The leading lamb-feeding states are Colorado, California, Texas, Wyoming, South Dakota, and Kansas.

Purebred Breeder

Some producers raise purebred sheep and sell rams for breeding. **Purebred breeders** have an important responsibility to the sheep industry: They determine the genetic productivity of commercial sheep. The purebred breeder should rigidly select breeding animals and then sell only those rams that will improve productivity and profitability for the commercial producer.

Records are essential to indicate which ram is bred to each of the ewes and which ewe is the mother of each lamb. All ram lambs are usually kept together to identify those that are desirable for sale. The purebred breeder often retains the ram lambs until they are 1 year of age, at which time they are offered for sale.

Commercial Market Lamb Producers

Commercial lamb producers whose pastures are productive can produce market wether and ewe lambs on pasture, whereas those whose pastures are less desirable produce feeder lambs. Lambs that are neither sufficiently fat nor large for market at weaning time usually go to feedlot operators, where they are fed to market weight and condition. However, most feedlot lambs are obtained from producers of range sheep. The producer of market lambs, whose feed and pasture conditions are favorable, strives to raise lambs that finish at 90–120 days of age, weighing approximately 120–130 lb each.

Commercial Feedlot Operator

Lambs are sent to the feedlot if they need additional weight prior to slaughter. The feedlot operator hopes to profit by increasing the value of the lambs per unit of weight and by increasing their weight. Lambs on feed that gain about 0.5–0.8 lb per head per day are considered satisfactory gainers. Feeder buyers prefer feeder lambs weighing 70–80 lb over larger lambs because of the need to put 20–30 lb of additional weight on the lambs to finish them. A feeding period of 40–60 days is usually sufficient for finishing healthy lambs.

The inventory of goats in the United States is almost 2.8 million head (2.4 million meat goats, 360,000 dairy goats, and 149,000 Angoras). Nearly 500,000 meat goats are harvested annually under federal meat inspection. Texas is the leading producer of both meat and Angora goats while California is the leading state for dairy goat production.

THE SWINE INDUSTRY

Global Perspective

Swine are widely distributed throughout the world, even though 50% of the world's pigs are located in China (Table 2.18). China far exceeds all other countries in number of pigs and total pork production. However, the swine herd in the United States is more productive on a per-head basis. Other swine numbers and production leaders include Brazil, Germany, and the Russian Federation (Table 2.18).

Table 2.19 shows the import and export of pork for the United States. Japan is the primary customer for the U.S. industry while the largest supplier to the United States is Canada.

United States

Cash receipts from the **swine industry** total nearly $18 billion (Table 2.3). Swine production in the United States is concentrated heavily in the nation's midsection known as the **Corn Belt.** This area produces most of the nation's corn, the principal feed used for swine. Many swine producers farm several hundred acres of land and have other livestock enterprises. The Corn Belt states of Iowa, Missouri, Illinois, Indiana, Ohio, northwest Kentucky, southern Wisconsin, southern Minnesota, eastern South Dakota, and eastern Nebraska produce nearly 70% of the nation's swine, with Iowa alone accounting for approximately 25% (Fig. 2.19). Oklahoma and North Carolina have experienced dramatic growth in swine numbers since the early 1990s.

Table 2.18
WORLD SWINE NUMBERS, PRODUCTION, AND CONSUMPTION

Country	No. Swine (mil head)	Country	Production (mil ton)[a]	Country	Carcass Weight (lb/head)
1. China	465	1. China	46	1. Puerto Rico	359
2. United States	66	2. U.S.	9	2. Thailand	312
3. Brazil	39	3. Germany	5	3. Italy	270
4. Germany	27	4. Spain	3	4. South Africa	270
5. Russian Fed.	17	5. Brazil	3	5. Chile	221
World Total	967	World Total	100	World Average	176

[a]Carcass weight.
Sources: Adapted from USDA and FAO.

Table 2.19
U.S. PORK IMPORT AND EXPORT VALUES

Country	Exports (mil $)	Country	Imports (mil $)
Japan	1,810	Canada	1,033
Mexico	853	Denmark	147
Canada	432	Italy	97
China	326	Poland	60
South Korea	234	Netherlands	25
Total	4,429		1,463

Source: Adapted from USDA: FAS, 2013.

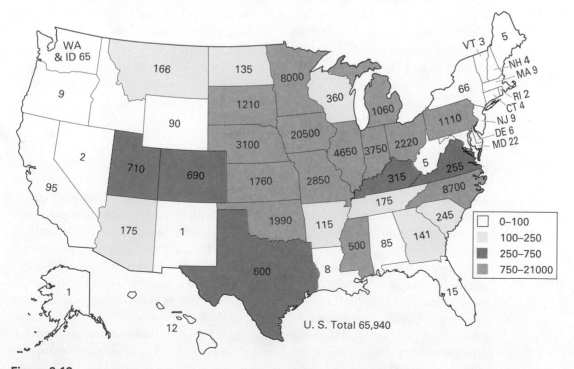

Figure 2.19
Total hogs and pigs inventory by state (1,000 head). Source: USDA-NASS. Data compiled by Livestock Marketing Information Center. Used by permission of Livestock Marketing Information Center.

Figure 2.20
Pork production versus breeding hog inventory.
Source: USDA-NASS. Data compiled by Livestock Marketing Information Center. Used by permission of Livestock Marketing Information Center.

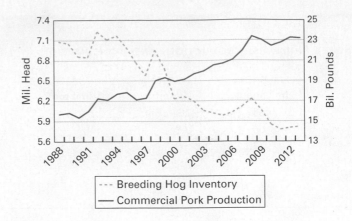

Table 2.20		
PERCENT OF PRODUCERS AND HOG INVENTORY IN VARIOUS SIZED ENTERPRISES		
Herd Size	% Operations	% Inventory
1–99	72	0.8
100–499	7	2.0
500–999	3	2.7
1,000–1,999	5	7.5
2,000–4,999	8	24.9
5,000+	5	62.1

Figure 2.20 shows that during the past 20 years the number of hog farms has significantly decreased. In fact, fewer than half of the farms that raised pigs in 1987 still did so by 2010. However, during this same period of time, pork production has continued to increase. This implies that the number of hogs raised per farm has increased. Production costs, based on per hundredweight of hogs produced, decrease as number of hog production per farm increase. Table 2.20 provides the breakdown of the percent of producers and hog inventory held by various sized enterprise classes.

OTHER ANIMAL INDUSTRIES

Aquaculture

Aquaculture is the farming or rearing of aquatic plants and animals. In a sense, it is underwater agriculture. Farming implies some form of intervention in the rearing process to enhance production such as stocking, feeding, and protection from predators. The most common aquatic animals raised in the United States include finfish (catfish, trout, and salmon), crustaceans (shrimp and crawfish), and mollusks (oysters and clams).

Global fish production has expanded at such a large rate that the per-capita supply has nearly doubled in the last half century to a current level of nearly 38 lb per person. The U.S. level of aquaculture production is sufficient to rank 13th in the world. Worldwide fish consumption accounts for one-sixth of all animal protein intake. Fish production also provides feed ingredients for livestock and aquaculture. Aquaculture produces 26% of the world's supply of food fish with approximately 60% originating from inland waters and 40% from marine environments.

Aquaculture sales in the United States exceed $1.3 billion with about two-thirds accounted for by food fish sales (catfish, salmon, trout, etc.). Production has increased significantly over the past several years with 365,566 acres involved in aquaculture. Catfish farming is the largest aquaculture activity in the United States with annual sales of $423 million involving approximately 334 million lb. Mississippi produces about one-half of the U.S. total. Other major catfish-producing states are Alabama and Arkansas.

Trout production is the second largest sector in the U.S. aquaculture industry, with Idaho accounting for approximately one-half of the $85 million annual sales.

The third largest sector in the U.S. aquaculture industry is crawfish production. Concentrated mostly in Louisiana, the industry produces around 36 million lb a year valued at $21 million.

There has been increased interest in producing other aquatic species, although most are in the experimental stages. Two species that have garnered publicity are hybrid striped bass and tilapia. Tilapia is a warmwater fish, native to Africa and the Middle East, but its production is expanding throughout the world. The Chinese have dominated the world's production of fish and they have claimed 26% of the world's production. In so doing, per-capita fish production has increased three times over the past decade. In terms of imported fish, shrimp and crayfish tonnage is significant. For example, over $1 billion in shrimp is imported from Thailand and Indonesia. Crayfish imports grew so fast that a substantial tariff was implemented to protect domestic producers.

One of the most attractive aspects of aquatic farming is its low feed conversion ratio. Because fish are suspended in water, they don't expend many calories to move about, and because they are cold-blooded, they are more efficient to produce than any of the major land-raised meats. Today, fish farmers are achieving average feed conversion ratios as low as 1.3.

U.S. aquaculture producers have a number of competitive advantages. First, the United States is one of the world's largest seafood markets, and the domestic live market is one that pays premium prices. Second, although most aquaculture firms are in rural areas, they still are close to good transportation facilities. Third, U.S. producers have access to a growing network of researchers and companies working on aquaculture. Fourth, the United States is a major producer of grains and has large supplies of livestock by-products for use in fish feeds, which can account for up to 50% of variable costs. Fifth, the United States has a wide variety of climates and long coastlines for marine aquaculture, in addition to large freshwater resources.

On the other hand, U.S. growers also have a number of disadvantages. First, many of the biggest farm-raised species are tropical and, with the exception of Hawaii, the United States does not have any tropical areas. Second, land costs are generally higher in the United States than in less developed countries, especially the coastal properties needed for marine aquaculture. Third, in most cases, labor costs are considerably higher in the United States than in competing countries. Fourth, many countries have fewer regulations controlling aquaculture production and processing practices.

Bison

Over 200,000 **bison** are found on 4,500 private farms and ranches in the United States. An additional 20,000 wild bison can be found on public lands in the United States and Canada. In 2012, 41,000 head of bison were processed under federal inspection. The National Bison Association has focused its promotional efforts to increase consumption of buffalo meat via merchandising to chefs, food editors, and special events.

Development of a bison enterprise requires that owners/managers build a suitable set of facilities with fences and handling equipment well adapted to the behavioral patterns of the buffalo. Just as is the case with other livestock enterprises, a carefully managed nutritional program is required for high levels of reproductive and growth performance. Selection tends to be focused on fertility testing, weaning weight, growth rate, carcass performance, and maternal traits.

Elk

The farming of elk and deer is a relatively new enterprise to the United States. In addition to the sale of breeding animals and meat, antlers or velvet from males are sold. A two-year-old bull elk yields approximately 5–6 lb of velvet while mature males may produce 35–40 lb per year. A majority of the velvet is sold into the Asian market for use in traditional, holistic medicines.

The breeding market is the dominant focus for most elk and deer breeders. U.S. elk farms produced 68,000 head in 2012 from 1,900 enterprises. The three leading states are Texas, Wisconsin, and Minnesota. As is often the case with new enterprises, speculation often creates unsustainable high markets. The use of domesticated elk and deer as food is typically found only in limited venues, such as specialty meat shops or in upscale restaurants that offer game meats as a menu item. Approximately 3 million lb of venison from farmed deer is consumed in the United States.

Ostrich and Emu Farming and Ranching

Ostrich and **emu**, native to Africa and Australia, respectively, belong to the ratite family of flightless birds. The ostrich is the largest living bird. A general description of emu and ostrich is provided in Table 2.21.

Emu farming in the United States is relatively new with Texas, California, and Kansas as the leading states for production. In 1989, Texas emu farmers started a national association. Fewer than 1,200 ostriches are processed under federal inspection in the mid-1990s. However, 61,634 were inspected in 1997 with 100,000 birds processed in 1998. In 2002, 1,643 farms raised approximately 20,500 birds. However, the industry has not been able to establish significant market demand and as such by 2007, only 714 farms reported production of just over 11,000 birds. An additional 3,600 farms had at least one emu on inventory and these enterprises produced approximately 6,500 birds. The industry fills a small, specialized niche with little evidence of significant growth on the horizon.

Llama and Alpaca Production

The introduction of South American camelids (Llama and Alpaca) have been a successful new venture in the United States. More than 100,000 Alpacas are on inventory in the United States where they are predominately selected for fiber production

---

Table 2.21
PRODUCTIVITY OF THE EMU AND OSTRICH

| Name | Adult Size (male) | | No. Eggs Per Year | Egg Weight (lb) | Products |
	Height (ft)	Weight (lb)			
Emu	5–6	150	30–40	1–1 1/2	Meat, leather, oil, feathers, carved eggs
Ostrich	up to 8	300	30–50	3–3 1/2	Meat, hides, feathers, egg shells

Source: Adapted from Hermes, 1996.

and show quality. Llamas are utilized as recreational and show animals, as pack animals, to control predators, and to a lesser extent for their fiber.

CHAPTER SUMMARY

- The animal industries are described by numbers, cash receipts, per-capita product consumption, and their contribution to world trade.

- U.S. farm cash receipts were $395.1 billion with livestock contributing $171 billion. Cash receipts for the major livestock species were cattle ($67.9 bil), dairy products ($37 bil), broilers ($24.7 bil), hogs ($22.1 bil), and turkeys ($5.4 bil).

- The major animal industries are becoming more integrated and concentrated. The poultry industry is highly integrated and swine production units continue to become larger. In the beef cattle industry, four major packers slaughter most of the cattle; however, most of the cattle are produced in herd sizes of less than 100 cows.

- Sheep numbers have shown a significant decline in the past 50 years. Dairy cow numbers have decreased; however, milk production has remained about the same. Milk production per cow continues to increase.

- Aquaculture and other types of animal farming (e.g., ostrich and emu) are increasing in the United States.

KEY WORDS

cash receipts	broiler
world trade	browse
beef industry	forbs
cow-calf	sheep industry
stocker-yearling	farm flock
feedlot	range flock
commercial feeders	purebred breeder
farmer-feeders	swine industry
dairy industry	Corn Belt
Eohippus	aquaculture
horse industry	bison
poultry industry	ostrich
tariff	emu

REVIEW QUESTIONS

1. Recite and spell the correct terminology to describe young and mature males, females, and castrates by farm animal species.
2. Rank the major U.S. livestock industries for number of producers, animal inventory, and volume of cash receipts.
3. Describe the regional concentration of U.S. livestock and poultry production.
4. Discuss the importance of international trade in livestock, poultry, and products to the United States.
5. Describe the importance of productivity differences on national animal inventories on a global basis.
6. What are the primary factors that allowed the increase in per animal productivity in the U.S. beef industry?
7. Describe the basic stages of production for each of the primary livestock and poultry industries in the United States.
8. Compare and contrast the beef and dairy industries.
9. Compare and contrast the swine and poultry industries.
10. Describe the U.S. horse industry.
11. Briefly describe the "new" livestock industries in the United States.

SELECTED REFERENCES

American Poultry History. 1996. Mount Morris, IL: Watt Publishing Co.

Carolan, M. 2002. *Summary of U.S. Census of Agriculture Data for Iowa Farms*. Dept. of Sociology. Ames, IA: Iowa State University.

Directions: A 21st Century Vision. National Cattlemen, July, 2006. Englewood, CO: National Cattlemen's Association.

Field, T. G. and R. E. Taylor. 2007. *Beef Production and Management Decision*. 5th Edition. Upper Saddle River, NJ: Pearson Prentice Hall.

Hermes, J. C. 1996. Raising Ratites. Pacific Northwest Coop. Ext. Service. Publ. 494-e.

Horse Industry Directory. 2008–09. Washington, DC: American Horse Council, Inc.

Rankings of States and Commodities by Cash Receipts. 2013. Washington, DC: USDA.

Stillman, R., T. Crawford, and L. Aldrich. 1990. *The U.S. Sheep Industry*. Washington, DC: USDA, ERS.

USDA. 2014. *Economic Indicators of the Farm Sector: Costs of Production—Livestock and Dairy*. Washington, DC.

USDA. 2014. *Foreign Agricultural Trade of the United States*. Washington, DC.

USDA. 2014. *Livestock and Poultry Situation and Outlook Report*. Washington, DC.

3

Red Meat Products

Red meat products consumed on a global scale come primarily from cattle, swine, sheep, and goats. Poultry meat, sometimes called *white meat*, is discussed later in Chapter 4. Pork while advertised as "the other white meat" is a red meat.

Red meats are named according to their source: **beef** is typically from cattle over a year of age; **veal** is from calves 3 months of age or younger (veal carcasses are distinguished from beef by the grayish-pink color of their lean); **pork** is from swine; **mutton** is from mature sheep; **lamb** is from young sheep; **chevon** is from goats, but it is more commonly called **goat meat**.

PRODUCTION

World **red meat** production is approximately 250 million tons, with China, the United States, and Brazil leading all other countries. The leading 15 countries for red meat production account for nearly 85% of all production.

Figure 3.1 shows red meat production trends in the United States. Beef and pork comprise most of the annual production of nearly 50 billion pounds. The total production has been relatively stable over the past several years. Some red meat has to be imported into the United States to fill U.S. production deficits. For example, due to smaller cattle inventories, a shortage of beef for grinding into hamburger has given rise to the need to import lean beef.

Table 3.1 outlines top 10 states for cattle, hogs, and sheep harvest in the United States. Sixty percent of U.S. cattle harvest is accounted for by the leading three states—Nebraska, Texas, and Kansas. Iowa alone harvests 27% of the nation's hogs while Colorado is responsible for nearly 45% of the total lamb harvest.

Meat animals are processed in packing plants located near areas of live animal production. This reduces the transportation costs because carcasses, wholesale cuts, and retail cuts can be transported at less cost than live animals. The five leading meat packers/processors in the United States are identified in Table 3.2.

PROCESSING

Livestock are harvested at packing plants for eventual processing into various products and by-products for distribution into the marketplace. During the initial processing stage, the animals are made unconscious by using carbon dioxide gas or by stunning (electrical or mechanical). The jugular vein and/or carotid artery is then cut to drain the blood from the

learning objectives
- Describe trends in red meat production
- Define the major wholesale cuts of beef, veal, pork, and lamb
- Identify the physical structures of meat and muscle
- Compare the nutrient value of beef, pork, and lamb
- Quantify domestic consumption patterns of red meat
- Explain the marketing of red meat

Figure 3.1

Annual commercial red meat production by type of meat. Source: USDA-NASS. Data compiled by Livestock Marketing Information Center. Used by permission of Livestock Marketing Information Center.

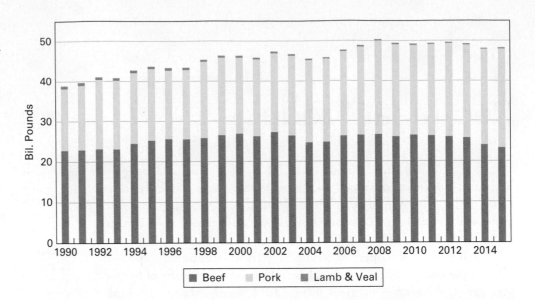

Table 3.1

LEADING STATES FOR COMMERCIAL HARVEST OF CATTLE, HOGS, AND SHEEP

Cattle Harvest		Hog Harvest		Sheep Harvest	
State	1,000 hd	State	1,000 hd	State	1,000 hd
NE	6,865	IA	29,744	CO	953
TX	6,833	NC	11,204	CA	295
KS	6,418	MN	10,365	MI	167
CO	2,500	IL	9,693	IL	135
IA	NA[a]	IN	8,459	NJ	116
CA	1,763	MO	8,076	TX	71
WI	1,746	NE	7,712	IA	51
WA	1,074	OK	5,576	PA	40
PA	989	SD	4,725	NY	40
MN	895	CA	2,509	IN	36
U.S.	34,087	U.S.	110,860	U.S.	2,164

[a]Disclosure limited to protect single company data.
Source: USDA: NASS, 2013.

animal. After bleeding, the hides are removed from cattle and sheep. Hogs are scaled to remove the hair, but the skin is usually left on the carcass. A few packers (especially small plants) skin hogs, as it is more energy efficient than leaving the skin on.

After the hide or hair is removed, the viscera, sometimes referred to as the drop or offal, is separated from the carcass. The primary components of the drop are the head, hide, hair, shanks (lower parts of legs and feet), and internal organs.

Dressing percentage (sometimes referred to as yield) is the relation of hot or cold carcass weight to live weight. It is calculated as follows:

$$\text{Dressing percentage} = \frac{\text{Hot or cold carcass weight}}{\text{Live weight}} \times 100$$

Table 3.3 shows the average live slaughter weights, carcass weights, and dressing percentages for cattle, sheep, and hogs. Rule of thumb average dressing percentages for cattle, hogs and lambs are 64%, 73%, and 51%, respectively. However, dressing percentage can vary several percentage points. The primary factors affecting dressing

Table 3.2
LEADING RED MEAT PACKING AND PROCESSING COMPANIES

Rank	Company (location)	Sales($ bil)	No. of Plants[a]	No. of Employees (1,000)	Products[b]
1	Tyson Foods, Inc. (Springdale, AR)	$33.3	100	96	B-P-PL
2	JBS, USA (Greeley, CO)	29.1	35	61	B-P
3	Cargill Meat Solutions (Wichita, KS)	20.0	13	34	B-P-T
4	SYSCO Corporation (Houston, TX)	12.2	NA	46	B-P-PL-T
5	Smithfield Foods, Inc. (Smithfield, VA)	11.6	40	36	P

[a]Includes harvest, processing, rendering, etc.
[b]B = beef; P = pork; PL = poultry; T = turkey
Source: Adapted from multiple sources.

Table 3.3
U.S. LIVESTOCK SLAUGHTER AND MEAT PRODUCTION, 2008

Species/Class	Product	No. Head (mil)	Average Live Weight (lb)	Total Carcass Weight (bil lb)	Average Carcass Weight (lb)	Average Dressing Percentage
Cattle	Beef	31.9	1,314[a]	25.7	841[a]	64
Calves[b]	Veal	0.85	263	0.13	154	58
Hogs	Pork	107.9	275	23.2	203	73
Sheep/lamb	Mutton/lamb	2.1	138	0.15	71	51

[a]Fed steers.
[b]Part of these are classified as calf carcasses, which are intermediate in age and size between veal and beef carcasses.
Source: Adapted from USDA.

percentage are fill (contents of digestive tract), fatness, muscling, weight of hide, and, in sheep, weight of wool. Fatter livestock or those with above average muscularity will typically have higher dressing percentages while those with excessive fill or heavy hides (fleeces) will have below average dressing percentages.

Beef and pork carcasses are split down the center of the backbone (sheep carcasses are not split), giving two sides approximately equal in weight. After the carcasses are washed, they are put into a cooler to chill. The cooler temperature of approximately 28–32°F removes the heat from the carcass and keeps the meat from spoiling. Carcasses can be stored in coolers at 32–35°F for several weeks; however, this is uncommon in large packing plants as profitability is dependent on processing live animals one day and shipping carcasses or products out by the next day or two. Smaller packers may keep carcasses in the coolers for as long as 2 weeks to allow the carcasses to age. This aging process increases meat tenderness through an enzyme activity that decreases the tensile strength of muscle fibers. Research has shown that injection of $CaCl_2$ (calcium chloride) into beef carcasses improves tenderness.

In the past, most red meat was shipped in carcass form, but today most packing plants process the carcasses into **wholesale** (Table 3.4) or primal **cuts**. The latter process involves shipping meat in boxes; thus the terms **boxed beef, boxed pork, and boxed lamb**. Some plants fabricate carcasses into case ready cuts as a means to capture marketing and transportation efficiencies. Retail cuts for beef, pork, and lamb fabricated from the various wholesale cuts are listed in Table 3.5. The goal of the red meat supply chain is to provide consumers a highly enjoyable eating experience (Fig. 3.2).

Table 3.4
WHOLESALE CUTS OF BEEF, VEAL, PORK, AND LAMB

Beef	Veal	Pork	Lamb
Round	Leg/round	Leg/ham	Leg
Sirloin	Sirloin	Loin	Loin
Short loin	Loin	Blade shoulder	Rib
Rib	Rib	Jowl	Shoulder
Chuck	Shoulder	Arm shoulder	Neck
Foreshank	Foreshank	Spareribs	Foreshank
Brisket	Breast	Side	Breast
Short plate			
Flank			

Table 3.5
RETAIL CUTS OF MEAT FROM THE PRIMALS OF BEEF, PORK, AND LAMB CARCASSES

Beef	Pork	Lamb
Chuck	*Shoulder (Blade and Arm)*	*Shoulder*
Chuck Eye Roast, Blade Steak, Pot Roast, Mock Tender, Blade Roast, Short Ribs, Flat Iron Steak, Cross Rib Roast, Denver Steak, Sierra Steak, Country Style Ribs	Blade Roast, Blade Steak, Shoulder Roll, Boneless Picnic, Smoked Picnic, and Smoked Hocks	Square Cut Shoulder Roast, Shoulder Roast, Blade Chop, Arm Chop
Rib	*Side*	*Rib*
Ribeye Steak, Prime Rib, Ribeye Roast, Cowboy Steak, Rib Steak, Shortribs, Backribs, Chef Cut Ribeye	Spareribs, Bacon	Rib Roast, Rib Chop, French Style Rib Chop, Crown Roast
Loin/Sirloin	*Loin*	*Loin*
Strip Steak, Filet, T-Bone Steak, Porterhouse Steak, Filet Mignon, Hanging Tender, Top Sirloin Steak, Tri-Tip Steak, Ball Tip Steak, Bottom Sirloin Steak, Strip Roast	Blade Chop, Rib Chop, Loin Chop, Butterfly Chop, Sirloin Chop, Country Style Ribs, Sirloin Cutlet, Tenderloin, Back Ribs, Center Rib Roast, Blade Roast, Boneless Sirloin, Canadian Style Bacon, Top Loin Roast, Crown Roast	Loin Roast, Loin Chop, Double Loin Chop
Round	*Ham*	*Leg*
Rump Roast, London Broil, Top Round Roast, Petite Tender Steak, Sirloin Roast, Butterfly Top Round Steak, Eye of Round Steak, Bottom Round Roast	Leg Cutlet, Top Leg Roast, Smoked Ham, Center Slice Ham Steak, Canned Ham, Smoked Ham (shank), Smoked Ham (rump), Country Style Ham	Whole Leg, Short Cut Leg, Shank Roast, Center Leg Roast, American-Style Roast, French-Style Roast, Boneless Leg Roast, Sirloin Chop, Sirloin Roast, Hind Shank

KOSHER AND MUSLIM MEATS

Traditional Jewish and Muslim laws have specific requirements for the slaughter of religiously acceptable animals. The primary difference from the general practices in the United States is that the animals are not stunned prior to slaughter.

Figure 3.2
*A thick, tender, juicy steak
is the most highly preferred
beef product.*
Source: Sattriani/Fotolia.

All **kosher meat** comes from animals with split hooves, that chew their cud, and that have been slaughtered in a manner prescribed by the Halakha (Orthodox Jewish law). Meat from undesirable animals or from animals not properly slaughtered or with imperfections is called nonkosher (trefah). In addition to the specification of which animals are considered kosher, the Talmud prohibits blood and the mixing of meat and milk. Foods that are kosher bear the endorsement symbol ∪.

Kosher slaughter is performed by a person specifically trained, known as a *Shochet*. The trachea and esophagus of the animal must be cut in swift continuous strokes with a perfectly smooth, sharp knife—the *chalef*—causing instant death and minimum pain to the animal. Various internal organs are inspected for defects and adhesions.

The Torah forbids the eating of blood, including blood of arteries, veins, and muscle tissue. For this reason, certain arteries and veins, as well as the prohibited sciatic nerve must be removed from cattle; neck veins of poultry must be severed to allow the free flow of blood. Blood is extracted from meat by broiling and salting. There are specific steps required in proper broiling and salting to endorse the meat as kosher. Purchases of kosher foods in the United States exceed $100 million annually.

The Muslim code is found in the Quran. Any Muslim may slaughter an animal while invoking the name of Allah. In cases where Muslims cannot kill their own animals, they may use meat killed by a "person of the book" (e.g., a Christian or a Jew). Usually this Halal slaughter is done by regular plant workers while Muslim religious leaders are present and reciting the appropriate prayers.

COMPOSITION

Meat consumption can be defined in either physical or chemical terms. Physical composition is observed visually and with objective measurements; chemical composition is determined by chemical analysis.

Physical Composition

The major physical components of meat are lean (muscle), fat, bone (Fig. 3.3), and connective tissue. The proportions of fat, lean, and bone change from birth

Figure 3.3
The fundamental structure of meat and muscle in the beef carcass.

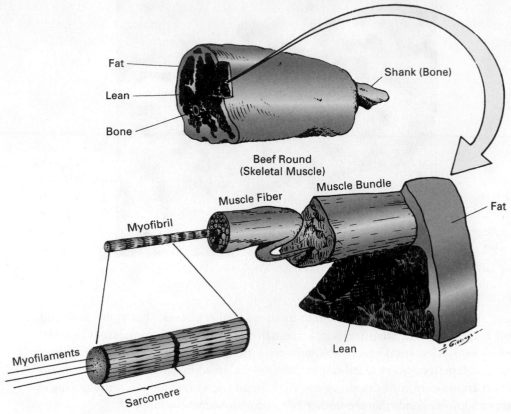

to slaughter time. Chapter 18 discusses these compositional changes in more detail. Connective tissue, which to a large extent determines meat tenderness, exists in several different forms and locations. For example, tendons are composed of connective tissue (collagen) that attaches muscle to bone. Other collagenous connective tissues hold muscle bundles together and provide covering to each muscle fiber. Figure 3.3 shows the physical structure of muscle: myofibrils are component parts of muscle fibers, muscle fibers combined together make up a muscle bundle, and groups of muscle bundles are components of whole muscles or muscle systems. Within the myofibrils are two types of **myofilaments**—namely, thick (myosin) and thin (actin) filaments.

Chemical Composition

Since muscle or lean meat is the primary carcass component consumed, only its chemical composition is discussed here. Chemical composition is important because it largely determines the nutritive value of meat.

Muscle consists of approximately 65–75% water, 15–20% protein, 2–12% fat, and 1% minerals (ash). As the animal increases in weight, water and protein percentages decrease and fat percentage increases.

Fat-soluble vitamins (A, D, E, and K) are contained in the fat component of meat. Most B vitamins (water-soluble) are abundant in muscle.

The major protein in muscle is **actomyosin**, a globulin consisting of the two proteins **actin** and **myosin**. Most of the other nitrogenous extracts in meats are relatively unimportant nutritionally. However, these other extracts provide aroma and flavor in meat, which stimulate the flow of gastric juices.

Simple carbohydrates in muscle are less than 1%. Glucose and glycogen are concentrated in the liver. They are not too important nutritionally, but they do have an important effect on meat quality, particularly muscle color and water-holding capacity.

NUTRITIONAL CONSIDERATIONS

Nutritive Value

Heightened health awareness in the United States has made today's consumers more astute food buyers than ever before (Fig. 3.4). Consumers look for foods that meet their requirements for a nutritious, balanced diet as well as for convenience, versatility, economy, and ease of preparation. Processed meat products have long been among Americans' favorite foods. Consumers today can choose from more than 200 styles or forms of meat, including over 50 cold cuts, dozens of sausages, and a wide variety of hams and bacon. Nutrient-rich red meat—beef, pork, lamb, and veal—are primary ingredients in these products. Meat products in all delicious and unique varieties contain the same nutrients as the fresh meats from which they are made (Table 3.6).

Meat is nutrient-dense and is among the most valuable sources of important vitamins and minerals in a balanced diet. **Nutrient density**, a measurement of food value, compares the amount of essential nutrients to the number of calories in food.

Red meats are abundant sources of iron, a mineral necessary to build and maintain blood hemoglobin, which carries oxygen to body cells. Heme iron, one form of iron found in red meat, is used by the body most effectively, and enhances the absorption of nonheme iron from meat and other food sources. Red meat is also a rich source of zinc, an essential trace element that contributes to tissue development, growth, and wound healing.

Red meats contain a rich supply of the B vitamins. Pork is a particularly rich source of thiamine, which is essential to convert carbohydrates into energy. Vitamin B_{12} occurs only in animal products, such as meat, fish, poultry, and milk, and is essential for

Figure 3.4

Examples of Nutrition Facts labels for ground beef and ham.

Table 3.6
NUTRIENT CONTENTS OF RED MEATS (3 OZ OF COOKED, LEAN PORTIONS) AND THEIR CONTRIBUTION TO RDAs (RECOMMENDED DAILY ALLOWANCES)[a]

Nutrient	Beef Amount	Beef % RDA	Pork Amount	Pork % RDA	Lamb Amount	Lamb % RDA
Calories	192	10	198	10	176	9
Protein	25 g	56	23 g	51	24 g	54
Fat[b]	9.4 g	14	11.1 g	17	8.1 g	12
Cholesterol[b]	73 mg	24	79 mg	26	78 mg	26
Sodium	57 mg	2	59 mg	2	71 mg	2
Iron	2.6 mg	14	1.1 mg	6	1.7 mg	10
Zinc	5.9 mg	39	3.0 mg	20	4.5 mg	30
Thiamin	0.08 mg	6	0.59 mg	39	0.09 mg	6
Niacin	3.4 mg	17	4.3 mg	22	5.3 mg	27
B_{12}	2.4 mg	79	0.7 mg	24	2.2 mg	37

[a]Based on a 2,000-calorie diet and the RDA for women aged 23–51.
[b]The American Heart Association recommends that fat contribute not more than 30% of total daily calories and that cholesterol be limited to 300 mg/day.
Source: Adapted from USDA.

Table 3.7
BEEF CUTS THAT MEET USDA GUIDELINES FOR LEAN[a]

Beef Cut	Calories	Saturated Fat (g)	Total Fat (g)
Top round	157	1.6	4.6
Top sirloin steak	156	1.9	4.9
95% Lean ground beef	139	2.4	5.1
Round steak	154	1.9	5.3
Chuck pot roast	147	1.8	5.7
Western griller steak	155	2.2	6.0
Strip steak	161	2.3	6.0

[a]A representative sample of 29 cuts is provided based on 3-ounce serving.

the protection of nerve cells and the formation of blood cells in bone marrow. Niacin, riboflavin, and vitamin B_6 are also found in significant amounts in red meats.

Finally, red meats are an excellent source of protein, a fundamental component of all living cells. Fresh and processed meats contain high-quality or complete protein with all essential amino acids in quantity. The variety of retail cuts that can be generated from a beef or pork carcass offer greater menu options and flexibility to consumers (Table 3.7).

New product development can have a significant impact both on consumer demand and profitability. For example, the Beef Value Cuts project, funded by beef checkoff dollars has resulted in the development of new products such as the flat iron steak, the western griller steak, western tip, ranch steak, and the petite tender. These products are the result of better muscle separation from wholesale cuts that are traditionally undervalued. The introduction of the flat iron steak contributed to a 60% increase in the value of beef chucks in the five-year period following market introduction.

Issues related to red meat consumption and human diet and health are presented in Chapter 23.

CONSUMPTION

A number of factors impact the level of meat consumption in the United States. However, taste (palatability), price, and perception of quality are the primary drivers of consumer demand.

Tables 3.8 and 3.9 quantify **per-capita disappearance** of red meat, poultry, and fish. Data is presented in three forms—carcass weight, retail weight, and boneless. Actual consumption is difficult to measure due to losses in storage, cooking, and plate waste. Carcass weight basis disappearance is based on a chilled carcass basis, retail weight comparisons are based on fabrication yields to retail ready cuts, which includes some bone-in styles, and boneless data comes closest to actual consumption. However, even the boneless data do not reflect the actual amount consumed, as there is fat trim, cooking loss, and plate waste. Surveys have shown that most retail cuts of beef, pork, and lamb have 0.1 in. or less of external fat covering the lean muscle tissue.

Because actual consumption data is difficult to obtain, the use of production data coupled with population size has served as a proxy for per-capita intake of a given product. In essence, much of the information provided as per-capita consumption is actually a better barometer of per-capita production because it is

Table 3.8
Per-Capita Disappearance of Beef, Pork, and Lamb[a]

	Beef (lb)			Pork (lb)			Lamb (lb)		
Year	Carcass	Retail	Boneless	Carcass	Retail	Boneless	Carcass	Retail	Boneless
1970	114	84	80	72	55	48	3.1	2.8	2.1
1980	103	76	72	73	57	52	1.5	1.4	1.1
1990	96	67	64	64	49	46	1.6	1.4	1.0
2000	96	67	64	65	51	48	1.3	1.1	0.8
2005	93	65	62	64	50	47	1.2	1.1	0.8
2010	85	65	56	61	47	44	1.0	0.9	0.7
2012	81	57	54	58	45	42	0.9	0.8	0.6

[a]Multipliers for converting carcass weight to retail and boneless equivalents are: beef (.70/.67); pork (.78/.73); lamb (.89/.66).
Source: Adapted from USDA.

Table 3.9
Per-Capita Disappearance of Red Meat, Broilers, and Fish

	Red Meat (lb)		Poultry (lb)		Fish (lb)	Total Meat (lb)	
Year	Retail	Boneless	Retail	Boneless	Retail	Retail	Boneless
1970	145	132	49	34	12	204	177
1980	137	126	59	41	12	205	180
1990	120	112	80	57	15	210	183
2000	120	114	77	68	15	213	197
2005	118	112	84	73	16	219	201
2010	108	102	83	71	16	207	189
2012	103	98	83	71	15	201	184

Source: Adapted from USDA.

Table 3.10
DISPOSABLE PER-CAPITA INCOME AND MEAT EXPENDITURES

Year	Annual Per-Household Food Expenditures ($)	Annual Per-Household Meat, Poultry, Fish and Eggs Expenditures ($)	Expenditure for Food as a Percent of Disposable Income	
			At-Home (%)	Food Service (%)
1985	3,477	579	7.5	4.1
1990	4,296	668	7.0	4.2
1995	4,505	752	6.4	4.2
2000	5,158	795	5.8	4.0
2005	5,931	825	5.4	4.1
2010	6,129	784	5.6	4.1
2012	6,599	852	5.7	4.3

Source: Adapted from USDA, Bureau of Labor.

Figure 3.5
Annual retail meat and poultry prices (nominal basis).
Source: USDA: ERS compiled by Livestock Marketing Information Center.

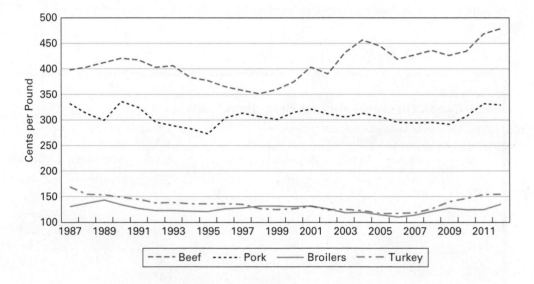

the net value of changes in cold storage, imports, and exports. Thus, data on the consumption and disappearance of meat and other food items can be misleading. What people say they eat and what they purchase does not determine the amount actually consumed. In one study, Dr. Bill Rathje (University of Arizona) evaluated garbage in the Tucson area in an attempt to determine what people eat by what they throw away. Over 10 years of his research data show that (1) people throw away approximately 15% of all the solid food they purchase; and (2) when questioned, consumers in middle- and low-income brackets consistently overestimate their meat consumption, while those in higher-income brackets underestimate their meat consumption.

Per-capita disposable income does affect the total amount of money spent on meat and the amount of product consumed (Table 3.10). However, consumers are spending a smaller percentage of their disposable income on meat and food, indicating that these items are relatively inexpensive compared to other purchases.

Lower poultry prices relative to red meat prices have influenced some consumers to shift their preference from red meat to poultry (Fig. 3.5). However, as Table 3.11 demonstrates, beef has held the dominant position in share of per capita meat spending for nearly four decades.

Table 3.11
MARKET SHARE OF MEAT SPENDING (%)

	Beef	Pork	Poultry
1980	54	27	19
1985	50	26	23
1990	45	27	27
1995	40	29	30
2000	43	28	29
2005	47	25	28
2010	48	27	25
2013	47	27	26

Source: Adapted from USDA.

DEMAND

Demand is a more important measure than is per-capita consumption (disappearance) because it deals not only with quantity but also the price at which consumers will buy a specific quantity all other factors being equal. Demand is a response curve of the quantities of product consumers will buy at various prices. Demand is based on a complex set of determinants such as prices of competing products, levels of per-capita disposable income, cultural preferences, consumer perceptions, and a host of other demographic factors.

Demand should not be confused with per-capita disappearance, share of expenditures, or share of consumption as none of these measures incorporates a schedule of prices into the equation. Demand can be more readily interpreted by evaluation of quantity and price relationships (Figs. 3.6 and 3.7). In both cases, there are periods of demand increase when consumers are willing to buy more products at higher prices (demand gain) and periods of time when consumption could only be increased by lowering prices (demand loss).

MARKETING

Meat production originates at the farm level and moves through feeder/finishers, marketing points, packers/processors, and food retailers or food service enterprises before reaching consumers. On hog farms, most pigs are raised to harvest weight. Young cattle and sheep are typically managed in off-farm feedlots to add weight under higher-energy diets prior to being harvested.

Packers and processors purchase the animals through several different marketing channels (Table 3.12). Direct purchases can occur at the packing plant, the feedlot, or a buying station near where the animals are produced.

Approximately 2,000 auctions, sometimes called **sale barns**, are located throughout the United States. The auction company has pens, a sale ring, and an auctioneer. Auctioneers sell the livestock to the highest bidder among the buyers attending the auction. Producers pay a marketing fee to the auction company for the sale service.

Most cattle, sheep, and swine have traditionally been purchased on a live weight basis, with the buyer estimating the value of the carcass and other products. Increasingly, animals are purchased on the basis of their carcass weight and the desirability

Figure 3.6
Price–quantity relationship for beef and pork—annual retail weight (deflated price basis). Source: USDA: ERS compiled by Livestock Marketing Information Center.

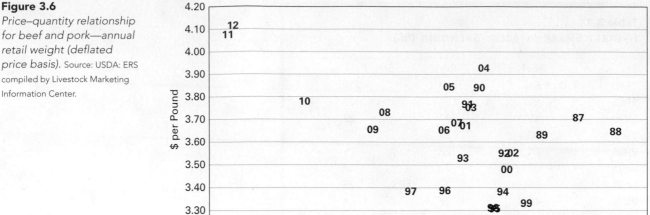

Figure 3.7
Price–quantity relationship for chicken and turkey—annual retail weight (deflated price basis).
Source: USDA: ERS compiled by Livestock Marketing Information Center.

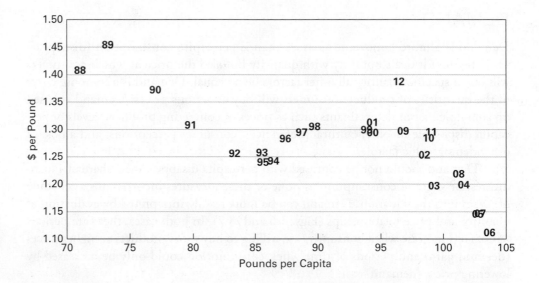

Table 3.12
PRICING PROTOCOLS USED BY PACKERS TO ACQUIRE MARKET CATTLE AND HOGS (%)

Strategy	Fed Cattle (%)	Market Hogs (%)
Packer owned	6	28
Forward contract/formula	60	68
Negotiated	34	4

Source: Adapted from USDA, *Packers and Stockyards' Statistical Report.*

of the carcass produced. For cattle, hogs, and sheep there has been an increased number of purchases on a carcass basis during the past several years (Table 3.13). In this method of marketing, sometimes called **grade and yield**, the seller is paid on a carcass-merit basis. The advantage of this trend is that price is established one step closer to the consumer. However, critics suggest that increased price determination at the packer undercuts the viability of the live market.

Table 3.13
LIVESTOCK PURCHASED BY PACKERS ON CARCASS GRADE AND WEIGHT BASIS

Year	Cattle Percent of Total Purchases (%)	Hogs Percent of Total Purchases (%)	Sheep Percent of Total Purchases (%)
1975	23	9	10
1980	27	11	28
1985	30	16	37
1990	38	12	30
1995	46	43	46
2000	52	74	56
2005	56	79	52
2010	59	76	41

Source: Adapted from USDA, *Packers and Stockyards' Statistical Report.*

The USDA enforces the Packers and Stockyards Act, originally passed in 1921. The act provides for uniform and fair marketing practices. Federal supervision is given at market outlets, including packing plants, for fee and yardage charges, testing of scales, commission charges, and other marketing transactions. Carcass grade and weight sales are covered, as well as live animal sales. Marketing protection is provided to both the buyer and the seller.

The sale of food in both food service and retail formats has undergone dramatic shifts with food service sales capturing an increasing share of the consumer dollar. Growth of the food service industry has put additional demands on livestock industries to improve the consistency and value-added attributes of their product offer.

Simultaneously, national chains have captured a relative high market share of food sales (Table 3.14). These marketing entities will be increasingly focused on developing source-verified supply chain approaches. As such, the livestock industry will have to develop quality assurance and total quality management formats to assure access to markets.

The retail grocery business has undergone a dramatic shift from the 1950s when the independently owned corner grocery store dominated the market. Increasingly the supercenters, dollar stores, and club store formats have taken sales away from more traditional grocery and mass merchandising stores. Sales of meat, fish, and poultry account for approximately 14% of gross sales, dairy products are responsible for 9% while the deli brings in an additional 5% of gross revenue. The retail grocery segment comprises the following categories:

Conventional supermarkets—full offering of grocery, meat, and produce as well as an in-store deli and bakery, generating at least $2 million in sales, inventory of 15,000 items

Superstore—larger version of conventional supermarket, inventory of 25,000 items, and more than 40,000 ft^2 of floor space

Combination food/drug store—superstore format with more than 85% of sales generated from food products

Warehouse store—low-margin grocery with reduced variety, lower service levels, and lower prices

Super warehouse—high volume, more upscale combination of a superstore and warehouse format

Table 3.14
RETAIL FOOD CHAINS IN NORTH AMERICA

Company	Sales ($ bil)	Total Outlets (N)
Wal-Mart Stores, Inc.	274	4,177
Kroger Co.	97	3,226
Costco Wholesale Group	87	511
Target Corp.	73	1,788
Safeway	44	1,644

Source: Adapted from multiple sources

Limited-assortment food store—low priced, limited service format with less than 2,000 product inventory
Specialty/gourmet retailer—single theme focus such as organic/natural, ethnic, or locally grown
Supercenter—large food-drug combination store and mass merchandiser under a single roof, 40% of shelf space devoted to grocery, more than 170,000 ft² of floor space
Wholesale club—bulk sales format in combination membership/wholesale store with 120,000 ft² or more of floor space, approximately one-third of revenues from food sales
Mass merchandiser—predominately household items, electronic goods, and apparel retailers offering limited packaged food items
Dollar store—limited assortment format offering general merchandise and food at industry lowest prices

Regardless of store format, consumers want grocery outlets to provide stores that are clean and conveniently formatted with high-quality meat and produce offerings. Consumers are increasingly time constrained and the food chain must acknowledge the following trends:

• Just over half of all food prepared away from home is carried out for at-home consumption and nearly 50% of consumers consider meals prepared away from home as essential to their lifestyle.
• One-half of white-collar workers eat one meal daily at their desk.
• One-quarter of calories are consumed in the form of snack foods.
• Ten percent of all food is consumed in a vehicle.
• The meat case confuses consumers in terms of what to buy and how to best prepare their selections.
• Three-quarters of consumers have not made an evening meal plan by the end of the workday.

Growth in the restaurant and food service has fueled increased demand for red meat and poultry. With $600 billion in annual sales, the food service sector employs 13.1 million people, hosts more than 130 million patrons per day, and provides more than 70 billion eating experiences per year.

Understanding the consumer market is not an easy task, as human attitudes and perceptions are diverse and in a constant state of flux. **Consumer** preferences are subject to change and can be affected by factors such as income, size of household, gender, age, geographic region, and ethnicity.

Table 3.15
CONSUMER PREFERENCES FOR FOOD ARE BASED ON MULTIDIMENSIONAL EXPECTATIONS

Source of comfort and well-being

Simplicity in preparation—45 min. of preparation time/meal on weekends, 30 min. on weekdays

Important element of family and social gatherings

Ethnic cuisine in high demand

Twenty-four-hour per day availability, take-out from traditional sit-down restaurants growing in demand

Diet is upgraded as incomes rise

Quality is multifaceted (style, taste, image, freshness, packaging, etc.)

Value-added foods in highest demand, desire high-quality recipes

Consumers losing culinary skills thus demand is high for meal solution kits or food service

Quick-fix diets

Expects safe food supply

In the early 1980s, the livestock industries made a concentrated effort to produce leaner products. The beef industry, for example, responded to the market demand for leaner products with genetic changes, processing innovations, and a shift to closely trimmed subprimals. However, consumer demand actually slipped despite these efforts. As products became leaner, the taste of meat changed. Only when the industry optimized leanness with flavor did consumption begin to rise, especially in the case of the beef industry. Consumers also wanted better convenience, packaging, menu versatility, and value-added attributes. As food producers work to evolve into consumer-driven industries, they must act on emerging consumer trends (Table 3.15).

CHAPTER SUMMARY

- Most of the 50 billion pounds of red meat production in the United States is beef (25.9 bil lb) and pork (22.9 bil lb).

- The major contributions of red meats to human nutrition are the abundant sources of protein, iron, and vitamin B_{12}.

- Per-capita red meat consumption in the United States is 102 lbs of retail weight with beef (57 lb) and pork (45 lb) being the primary contributors.

KEY WORDS

beef
veal
pork
mutton
lamb
chevon
goat meat
red meat
dressing percentage
wholesale
cuts
boxed beef, lamb, and pork

kosher meat
myofilaments
actomyosin
actin
myosin
nutrient density
per-capita consumption (disappearance)
sale barns
grade and yield
consumer

REVIEW QUESTIONS

1. Describe the global and domestic distribution of red meat production.
2. Describe the importance of dressing percentage, its calculation, and the factors that influence it.
3. What are the wholesale cuts of beef, pork, and lamb?
4. Compare the volume, average live and carcass weights, and dressing percent by species.
5. What are the impacts on meat merchandising due to the variety of retail cuts that originate from beef, lamb, and pork?
6. Describe the structure of muscle and its composition.
7. How does beef, pork, and lamb meet nutritional requirements (RDA, nutrient density, etc.)?
8. Describe consumption and household expenditure trends of red meat.
9. Compare the pricing protocols of beef and pork processors.
10. Describe the market channels for cattle, hogs and sheep as well as beef, pork, and lamb.
11. Describe the various categories of retail food outlets.
12. What are the primary factors affecting demand for red meat?

SELECTED REFERENCES

Boggs, D. L. and R. A. Merkel. 1993. *Live Animal, Carcass Evaluation and Selection Manual*. Dubuque, IA: Kendall-Hunt.

Cassens, R. G. 1994. Meat processing. *Encyclopedia of Agricultural Science*. San Diego, CA: Academic Press, Inc.

Eating in America Today. 1994. Chicago, IL: National Live Stock and Meat Board.

Hedrick, H. B., E. D. Aberle, J. C. Forrest, M. D. Judge, and R. A. Merkel. 1993. *Principles of Meat Science*. Dubuque, IA: Kendall/Hunt.

Kashruth. Handbook for Home and School. 1972. New York: Union of Orthodox Jewish Congregations of America.

Meat and Poultry Facts. 2009. Washington, DC: American Meat Institute.

Regenstein, J. M., and T. Grandin. 1992. Religious slaughter and animal welfare—An introduction for animal scientists. *Reciprocal Meat Conference Proceedings* 45:155.

Romans, J. R., K. W. Jones, W. J. Costello, and C. W. Carlson. 2001. *The Meat We Eat*. 14th Edition. Danville, IL: Interstate Printers and Publishers.

Poultry and Egg Products

Poultry meat and eggs are nutritious and relatively inexpensive animal products used by humans throughout the world. Feathers, down, livers, and other **offal** are additional useful products and by-products obtained from poultry.

Application of genetics, nutrition, and disease control, along with sound business practices, has advanced the commercial poultry industry to the point where eggs and poultry meat can be produced very efficiently. The industry has developed two specialized and different types of chickens—one for meat production and one for egg production. Broiler lines are superior in efficient, economical meat production but have a lower egg-producing ability than the egg production lines. Egg-type birds have been selected to produce large numbers of eggs profitably. They have been bred to mature at lightweights and they are therefore slower growing and less efficient meat producers.

POULTRY MEAT AND EGG PRODUCTION

Broiler chickens provide most of the world's production and consumption of poultry meat. Turkeys and roaster chickens, mature laying hens (fowl), ducks, geese, pigeons, and guinea hens are consumed in smaller quantities, although some of these are important food sources in some areas of the world. World poultry and egg production is presented in Chapter 2. The more common kinds of poultry used for meat production in the United States are shown in Table 4.1. The top five broiler producers are listed in Table 4.2.

Meat-type chickens are classified by size, weight, and age at the time of processing. As the demands of various market channels have differentiated, the industry has responded by producing birds specific to the needs of each market outlet. The common classes of chickens used for meat are as follows:

Broiler—a chicken raised specifically for meat as opposed to a *layer* that is utilized for egg production.

Poussin—a one-pound or less bird that is less than 24 days of age.

Cornish game hens—birds less than 30 days of age and weighing 2 lbs.

Fast-food oriented broiler—birds weighing between 2 lbs 6 ounces and 2 lbs 14 ounces, less than 42 days of age, delivered cut-up without necks, giblets, tail fat, and leaf fat.

3's and up—birds weighing 3 to 4.75 lbs and between 40 and 45 days of age, delivered in whole, parts, and cut-up forms. This is the standard bird sold at retail.

learning objectives
- Describe the composition of poultry meat and eggs
- Describe the processing of poultry and eggs
- Explain the nutritive value of poultry meat and eggs
- Quantify poultry and egg consumption trends
- Overview the marketing of poultry and eggs

Table 4.1
AGES AND WEIGHTS OF VARIOUS KINDS OF POULTRY USED FOR MEAT PRODUCTION

Classification	Typical Age (mo)	Average Slaughter Weight (lb)	Average Dressed Weight (lb)	Dressing (%)
Young chickens				
Broiler (fryer)	1.5–2	4–5	3–4	78
Rock Cornish	1.0	2	1.5	75
Roaster	3–5	6–8	5	75
Fowl (mature for stewing)	19	5	3.8	75
Turkeys	4–6	10–25	8–20	83
Ducks	<2	6–7	4–5	70
Geese	6	11	8–9	75

Sources: USDA and various others.

Table 4.2
TOP FIVE INTEGRATED BROILER PRODUCING COMPANIES

Rank	Company	Average Weekly Production (mil lbs—ready to cook)	Facilities[a]	Employees	Sales (bil $)
1	Tyson Foods, Inc.	168	S-33, P-9, H-35, F-30	115,000	33.3[b]
2	Pilgrim's Inc.	149	S-25, H-40, F-11	38,000	8.1
3	Perdue Farms, Inc.	56	S-12, P-3, H-17, F-11	19,301	3.6
4	Koch Foods, Inc.	48	S-8, P-4, H-7, F-5	12,500	3.0
5	Sanderson Farms, Inc	58	S-9, H-8, F-7		2.7

[a]S = slaughter plants, P = processing plants, H = hatcheries, F = feed mills.
[b]Red meat sales included.
Source: Adapted from WATTAgNet, *National Provisioner* and National Broiler Council.

Roaster—birds weighing 5 to 8 lbs and between 55 and 60 days of age.

Broilers for deboning—males weighing 7 to 9 lbs, 47 to 56 days of age, processed to make nuggets, patties, and other value-added products.

Capon—once the standard, these are surgically castrated males approximately 14–15 weeks of age weighing 7 to 9 lbs. This product is very desirable in plumpness and tenderness but is considered a specialty item.

Heavy hens—these 15-month-old birds are spent breeders (culled from the laying flock) and typically weigh 5–5.5 lbs. These birds are used to produce cooked, diced, or pulled meat for inclusion in precooked and canned items such as soups and stews.

COMPOSITION

Meat

Table 4.1 shows that broilers yield approximately 78% of their live weight as carcass weight, whereas the dressing percent for turkeys is 83%. The proportion of edible raw meat from carcasses of chickens and turkeys is approximately 67% and 78%, respectively.

The gross chemical composition of chickens and turkeys is shown in Table 4.3. **Dark meat** is higher in calories, lower in protein, and higher in cholesterol than **light (white) meat** for both chicken and turkey. If the skin remains on poultry, it usually adds cholesterol and total fat.

Table 4.3
COMPOSITION OF CHICKEN AND TURKEY PER 100 g, EDIBLE PORTION

	Water (g)	Food Energy (calories)	Protein (g)	Total Fat (g)	Cholesterol (mg)	Ash (g)
Young chicken						
Light meat (no skin)	74.9	114	23.2	1.6	58	0.98
Dark meat (no skin)	76.0	125	20.1	4.3	80	0.94
Turkey (hen)						
Light meat (no skin)	73.6	116	23.6	1.7	58	1.0
Dark meat (no skin)	74.0	130	20.1	4.9	62	0.95

Source: Adapted from USDA.

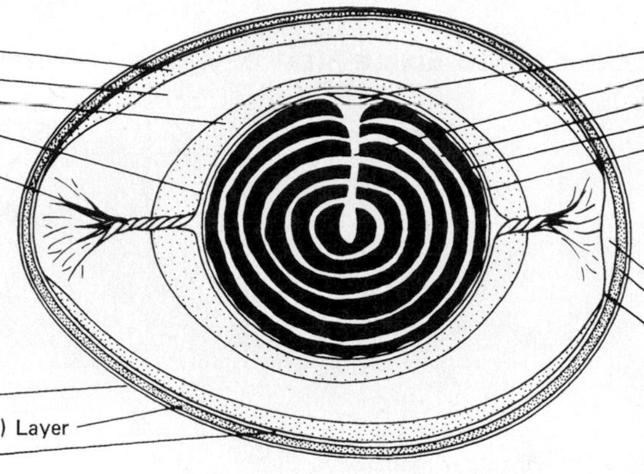

Figure 4.1
Longitudinal section of a hen's egg. Source: USDA.

Eggs

An egg has a spherical shape, with one end being rather blunt and larger than the other smaller, more pointed end. Figure 4.1 shows a longitudinal section of a hen's egg. The mineralized shell surrounds the contents of the egg. Immediately inside the shell are two membranes; one is attached to the shell itself and the other tightly encloses the content of the egg. The air cell usually forms between the membranes in the blunt end of an egg shortly after it is laid, as the contents of the egg cool and contract.

Seven thousand to 17,000 tiny pores are distributed over the shell surface, with a greater number at the larger end. As the egg ages, these tiny holes permit moisture and carbon dioxide to move out and air to move in to form the air cell. The shell is covered with a protective covering called the **cuticle** or **bloom**. By blocking the pores, the cuticle helps to preserve freshness and prevent microbial contamination of the contents.

The egg white, or **albumen**, exists in four distinct layers. One of these layers surrounds the yolk and keeps the yolk in the center of the egg. The spiral movement of the developing egg in the magnum causes the mucin fibers of the albumen to draw together into strands that form the chalaziferous layer and chalazae. The twisting and drawing together of these mucin strands tend to squeeze out the thin albumen to form an inner thin albumen layer.

The yellowish-colored yolk is in the center of the egg, its contents surrounded by a thin, transparent membrane called the **vitelline membrane**.

Egg weight varies from 1.5 to 2.5 oz, with the average egg weighing approximately 2 oz. The component parts of an egg are shell and membranes (11%),

Table 4.4
CHEMICAL COMPOSITION OF THE EGG, INCLUDING THE SHELL

	Percent of Total	Water (%)	Protein (%)	Fat (%)	Ash (%)
Whole egg	100	65.5	11.8	11.0	11.7
White	58	88.0	11.0	0.2	0.8
Yolk	31	48.0	17.5	32.5	2.0
	Percent of Total	Calcium Carbonate (%)	Magnesium Carbonate (%)	Calcium Phosphate (%)	Organic Matter (%)
Shell and shell membranes	12	94.0	1.0	1.0	4.0

Source: Adapted from USDA.

Table 4.5
THE RETAIL PARTS AND COOKED EDIBLE MEAT OF A BROILER-FRYER CHICKEN

Part	% of Carcass	Edible[a]
Breast	28	63
Thighs	18	60
Drumsticks	17	53
Wings	14	30
Back and neck	18	27
Giblets	5	100
Total	100	53

[a]Net average yield of meat is 67% without neck and giblets.
Source: Adapted from USDA; and National Broiler Council.

albumen (58%), and yolk (31%). The gross composition of an egg is shown in Table 4.4. The mineral content of the shell is approximately 94% calcium carbonate. The yolk has a relatively high percentage of fat and protein, although there is more total protein in the albumen.

The weight classes and grades of eggs are discussed in Chapter 8. Egg formation is presented in Chapter 10.

POULTRY PRODUCTS

Meat

More than 60% of all broilers leave the processing plant as cut-up chickens or selected parts rather than whole birds (compared to less than 30% in 1970). The percent of carcass and edible portion of each retail cut is outlined in Table 4.5.

Approximately 60% of the broilers, fowl (mature chickens), and turkeys are further processed beyond the ready-to-cook carcass or parts stage. The meat is separated from the bone and formed into products (e.g., turkey roll fingers and nuggets), cut and diced (e.g., for preparation of salads and casseroles), canned, or ground. Some of the cured and smoked ground products are frankfurters, bologna, pastrami, turkey ham, and salami. An even higher percentage of broiler production is expected to be further processed in future years. The poultry industry has led the meat business in terms of understanding and responding to the demands of consumers for further processed, value-added products. In 1970, 75% of broiler meat was merchandised through grocery stores, today approximately 55% moves through retail grocery channels with the remainder sold via food service. About half of all food service chicken is sold via fast food outlets.

Eggs

Eggs are unique, prepackaged food products that are ready to cook in their natural state. Eggs find their way into the human diet in numerous ways. They are most popular as a breakfast entrée; however, they are commonly used in sandwiches, salads, beverages, desserts, and several main dishes.

Shell eggs can be further processed into pasteurized liquid eggs and dried eggs and other egg products. Pasteurized liquid egg can be blast-frozen and sold to bakeries and other food-processing plants as an ingredient in other fabricated food products, or transported to users as a liquid product in refrigerated tank trucks. The dried egg products are sold to food manufacturers to be used in the production of products such as cake mixes, candy, and pasta.

The initial step in making egg products is breaking the shells and separating the yolks, whites, and shells. This is done in most egg-breaking plants by equipment completely automated to remove eggs from egg-filler flats, wash and sanitize the shells, break the eggs for individual inspection of the contents by an operator, and separate the yolks from the whites. Processing equipment can also be used to produce various white and yolk products or mixtures of them. Present automatic systems are capable of handling 18,000–50,000 eggs per hour (Fig. 4.2).

A

B

Figure 4.2

Automated egg-handling equipment. (A) Quality control on production line. (B) Eggs are sanitized, weighed and sorted into uniform marketing groups.

Source: 4.2a: Chalabala/Fotolia; 4.2b: Nikkytok/Fotolia.

Some breeds of chickens lay white-shelled eggs and other breeds produce brown-shelled eggs. The brown shell color comes from a reddish-brown pigment (ooporphyrin) derived from hemoglobin in the blood. Although there are shell color preferences by consumers in certain parts of the United States, there are no significant nutritional differences related to eggshell color.

Feathers and Down

In some areas of the world, ducks and geese are raised primarily for feathers and down. **Down** is a layer of small, soft **feathers** found beneath the outer feathers of ducks and geese. Feathers and down provide stuffing for pillows, quilts, and upholstery. They are also used in the manufacture of hats, clothing (outdoor wear), sleeping bags, fishing flies, and brushes.

The duck and goose industry in the United States is too small to meet the manufacturing demand for down. Thus, approximately 90% of raw feathers and down are imported, primarily from China, France, Switzerland, Poland, and several other European and Asian countries.

Breeder geese are often used for down production, as the older birds produce the best down. Four ducks or three geese will produce 1 lb of feather-and-down mixture, of which 15–25% is down. Down and feathers are separated by machine, then washed and dried.

Other Products and By-Products

Goose livers are considered a gourmet product in European countries. Force-feeding corn to geese three times a day for 4–8 weeks produces these enlarged fatty livers, weighing as much as 2 lb each. The livers are used to make a flavored paste called *pâté de foie gras* (for hors d'oeuvres or sandwiches).

Poultry by-products include hydrolyzed feather meal and poultry meat and bone scraps. These by-products are occasionally included in livestock and poultry rations as protein sources (refer to Chapter 7 for more discussion on these by-products).

NUTRITION CONSIDERATIONS

Nutritive Value of Poultry Meat

The white meat from fowl is approximately 33% protein, and the dark meat is about 28% protein. This protein is easily digested and is of high quality, containing all the essential amino acids. The fat content of poultry meat is lower than that found in many other meats. Poultry meat is also an excellent source of vitamin A, thiamin, riboflavin, and niacin.

Nutritive Value of Eggs

The egg (Fig. 4.3) contains many essential nutrients, it is recognized as one of the important foods in the major food groups recommended in *Dietary Guidelines for Americans*. These guidelines were developed by the U.S. Department of Agriculture to use in planning diets contributing to good health.

Eggs have a high nutrient density in that they provide excellent protein and a wide range of vitamins and minerals and have a low calorie count. The abundance of readily digestible protein, containing large amounts of essential amino acids, makes eggs a least-cost source of high-quality protein.

Figure 4.3
"The incredible edible egg." Source: Vankad/Fotolia.

Table 4.6

NUTRIENT VALUES FOR VARIOUS SIZED EGGS AND PERCENT OF RECOMMENDED DAILY ALLOWANCE MET BY CONSUMING ONE LARGE EGG

Nutrient	Large Egg 50 g	Extra Large Egg 56 g	Jumbo Egg 63 g	Percent of RDA from 1 Lg. Egg
Energy (kcal)	72	80	90	3
Protein (g)	6.3	7.0	7.9	12
Fat (g)	4.8	5.3	6.0	8
Cholesterol (mg)	186	208	234	300
Vitamin A (IU)	270	302	340	6
Vitamin D (IU)	41	46	52	10
Folate (µg)	24	26	30	6
Riboflavin (mg)	.23	.26	.29	15
Phosphorus (mg)	99	111	125	10
Zinc (mg)	.64	.72	.81	4

Source: USDA.

Table 4.6 lists the amounts of selected nutrients found in eggs. When the values of these nutrients are compared with the dietary allowances recommended by the Food and Nutrition Board of the National Research Council (1980), it is found that one large egg supplies a relatively high percentage of essential nutrients (Table 4.6). Eggs are an especially rich source of high-quality protein, vitamins (A, D, E, folic acid, riboflavin, B12, and pantothenic acid), and minerals (phosphorus, iodine, iron, and zinc). Some consumers believe that dark yellow yolks are of higher nutrient value than paler yolks, but there is no evidence to support this belief.

Table 4.7
FORM IN WHICH BROILERS ARE MARKETED IN RETAIL GROCERIES

Year	Whole (%)	Cut-Up (%)	Further Processed (%)
1970	70	26	4
1980	50	40	10
1990	18	56	26
2000	10	44	46
2005	7	41	52
2010	12	43	45
2013	11	41	48

Source: Adapted from USDA and National Chicken Council.

CONSUMPTION

Meat

Worldwide annual poultry and egg consumption averages 24 lbs and 18 hen eggs per capita. The annual per-capita consumption of chicken and turkey in the United States is approximately 100 lb (Fig. 4.4). Poultry consumption is approximately one-half of the total per-capita consumption of red meat and poultry. In recent years, poultry consumption has increased at the expense of red meat consumption on a weight basis. This is due in part to the price differential existing between the various meats.

An important reason for the marked increase in chicken consumption has been the speed of preparing and serving chicken. The Kentucky Fried Chicken Corporation (KFC) led the way in preparing and serving chicken in a short time, and now chicken has become a major item in fast-food service. The KFC restaurants are found in 100 countries. In recent years, other fast-food restaurants have featured chicken or added more chicken items to their menus. Nearly 90% of consumers surveyed stated that they didn't plan to change their chicken consumption, suggesting the market is maturing. Taste, health considerations, and convenience were the primary demand drivers. Price was relatively low on the list of reasons for consuming chicken.

Heavy hens, the type used to produce hatching eggs for broiler production, are sometimes available in retail stores as "stewing hens" or "baking hens." These account for about 10% of the poultry meat consumed by Americans each year. Consumption of turkey meat per person has increased from 1.7 lb in 1935 and 6.1 lb in 1960 to 8.0 lb in 1970 and 18 lb in 2001. Most of this increase in turkey consumption was stimulated by development of processed products such as turkey rolls, roasts, potpies, and frozen dinners. The selling of prepackaged turkey parts in small packages has also increased consumption.

Eggs

Per-capita egg production in the United States is approximately 251 eggs per year resulting from a total production of 224 million cases of eggs produced by 56 egg companies with more than 1 million layers per firm. Fifty-five percent are sold via retail grocers, 32% via breakers who further process the eggs, 9% to institutional trade, and 4% as exports. Most of these are consumed as shell eggs (70%) rather than processed eggs (Fig. 4.5). Per-capita consumption is nearly the number of eggs (254) produced by the average hen per year. Egg consumption reached its low in 1991 at 234 eggs per capita. The growth in consumption over the past 15 years has been the result of an

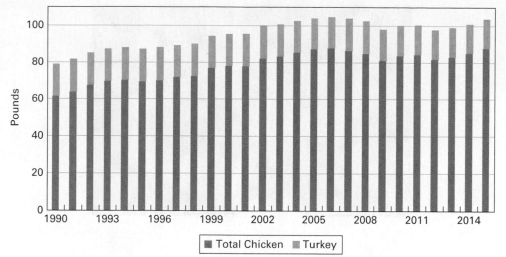

Figure 4.4
U.S. poultry consumption per capita, retail weight, annual. Source: Bureau of Economic Analysis & USDA-ERS. Data compiled by Livestock Marketing Information Center. Used by permission of Livestock Marketing Information Center.

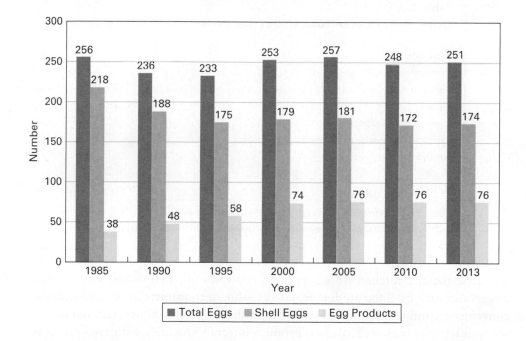

Figure 4.5
Per-capita production of eggs. Source: Adapted from USDA.

aggressive marketing and education campaign promoting the versatility, affordability, healthfulness, and value of eggs and egg products. Eggs have a 93% market penetration in U.S. households.

MARKETING

In the United States, large amounts of poultry meat are processed in commercial facilities (Fig. 4.6) and then transported under refrigeration to local and world markets. Chickens are increasingly marketed in further processed form as consumers demand more convenience, variety, and meal solution approaches (Table 4.7). The primary drivers of demand for chicken consumption include taste, convenience of preparation, and perceived health benefits (Table 4.8). The poultry industry has aggressively worked to meet the demands of consumers with a variety of value-added strategies that have resulted in slow and steady growth in consumption. Consumer studies have shown that among heavy, medium, and light buyers of chicken, the leading products were salads using chicken as an ingredient, chicken strips, ready-to-eat

Figure 4.6
Broiler processing line in a modern plant. Source: USDA.

Table 4.8
CONSUMER PERCEPTIONS ABOUT CHICKEN

Response	%
Primary reason for eating chicken	
Taste	38
Health	28
Ease of preparation	10
Like specific parts (white meat, legs, breast, etc.)	7
Method of preparation	7
Price	5
Other	5

Source: Adapted from Broiler Industry and National Broiler Council.

rotisserie-roasted chicken from supermarkets, spicy chicken wings, and chicken nuggets. In each case, the product represents value to consumers in versatility, flavor, convenience, and price. Another study found that boneless, skinless chicken breasts were clearly the preferred chicken product in retail groceries with two-thirds of chicken consumers usually buying chicken in this form.

Broilers are the primary poultry meat exported, valued at more than $2.1 billion. Approximately 62 million cases (dozen per case) of eggs valued at $38 million as well as processed egg products worth an additional $123 million are exported annually from the United States. The share of broiler meat exported is shown in Figure 4.7.

Chicken and turkey meat is marketed in a variety of forms. Most of the broilers are sold to consumers through retail stores. Fresh chicken can reach the retail market counter the day following slaughter. The "sell-by" date on each package is usually 7 days after processing and is the last date recommended for sale of chicken. However, with proper refrigeration (28–32°F), shelf life can be extended up to 3 days longer. Some processors guarantee 14 days of shelf life for their broilers.

Federal or state government employees inspect almost all commercially available chicken and turkey meat for wholesomeness. The USDA quality grades about two-thirds of the chicken harvested in the United States.

In an effort to differentiate product lines to meet the unique desires of niche markets, poultry companies have utilized labeling strategies that align production methods with the desire of some consumers for products touted as organic, natural,

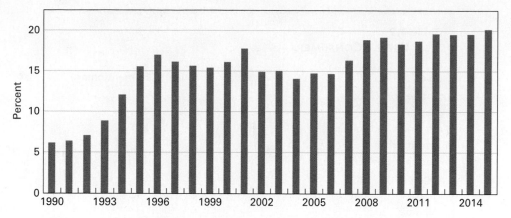

Figure 4.7
U.S. Broiler exports as a percentage of broiler production. Source: USDA-AMS. Data compiled by Livestock Marketing Information Center. Used by permission of Livestock Marketing Information Center.

and free range. The use of labeling claims has in reality created confusion among some consumers. For clarification, the basic definitions for these "attribute"-based labeling categories are as follows:

Free range—birds have access to an enclosed pen outside of the poultry house. Because feed and water are provided in the house structure birds are often reluctant to utilize the pen for long periods of time. Less than 1% of chickens are raised under free-range conditions.

Natural—this is perhaps the most misunderstood term in food production. USDA defines "natural" meats as minimally processed and having no added ingredients. Almost all fresh poultry meets these standards.

Organic—this production system prohibits the use of antibiotics, pesticides, and many farm chemicals and specifies utilization of feedstuffs made from organic ingredients.

Produced without hormones—this label is misleading in that no artificial or added hormones are used by the poultry industry in accordance with FDA regulations.

Antibiotic-free—this label indicates that the production system did not incorporate any animal health care products classified as antibiotics in either a preventative or treatment form. Health care products not classified as antibiotics may still be incorporated.

Fresh—this term means that the product has not been chilled to temperatures below 26 °F.

In-store promotions are frequently utilized to enhance sales of food products in retail outlets. Reducing the price of the lead item in a menu (chicken breast for example) and buy one–get one free offers are the most effective strategies.

Before and during the 1950s, the marketing of turkey was highly seasonal as 90% of the turkeys were consumed during the last quarter of the year for the Thanksgiving and Christmas holidays. Today less than 40% of turkey sales are made during that period. Current trends indicate that turkey is being consumed more on a year-round basis and has become a meat staple. However, turkey consumption has leveled off in the past decade.

The egg industry has invested significant resources into expanding the consumption of eggs beyond the traditional use of eggs in breakfast items (Table 4.9). Consumers rank freshness, product integrity (no broken/cracked eggs), cleanliness, and price as leading factors affecting purchase of table eggs at retail. Consumers typically group purchases around a meal concept and, thus, by increasing demand for an item such as eggs, the sale of other products (many of animal origin) can

Table 4.9
EGG USE BY CONSUMERS

Use	Consumers (% mentioning)
Entrees	31
Cooking	30
Breakfast and cooking	28
Breakfast	23
Recipes for meals	11
Ingredient	8
Salads	4
All ways	3

Source: Adapted from American Egg Board, 2002.

be increased. Breakfast meats, cheese, milk, and butter are frequently purchased in conjunction with egg sales.

Retail sales strategies in the future will include cooperative merchandising efforts by diverse food providers to enhance the overall sales of several items. For example, pork processors and egg wholesalers will find mutually beneficial approaches in selling their products. Furthermore, it will be increasingly important for the livestock industry to correctly interpret consumer signals, implement effective sales strategies, and develop profitable partnerships with both food retailers and food service companies as a means to assure market access.

CHAPTER SUMMARY

- Poultry meat and egg products are nutritious and relatively inexpensive animal products used by humans throughout the world.

- Poultry meat and eggs are excellent sources of protein, vitamin A, and several B vitamins for human nutrition.

- Annual per-capita consumption of broilers is 84 lb and has been increasing rapidly over the past several years. Per-capita turkey consumption is 18 lb and stabilizing. Egg consumption has been increasing and currently is 257 eggs per year.

- Increases in demand for chicken and turkey have been created largely by price competitiveness compared to other meats, ease of preparation, and convenience.

KEY WORDS

offal
broiler
poussin
cornish game hens
fast-food oriented broiler
3's and up
roaster
broilers for deboning
capon

heavy hens
dark meat
light (white) meat
cuticle (bloom)
albumen
vitelline membrane
down
feathers

REVIEW QUESTIONS

1. Describe the classes of meat-type chickens.
2. Compare the age, average live and carcass weights, and dressing percentage of meat-type poultry.
3. Compare and contrast the leading poultry processing companies with the leading red meat processors from Chapter 3.
4. Compare the composition of light and dark poultry meat as well as eggs.
5. Describe the physical structure of the egg.
6. Describe the merchandising of poultry meat and eggs.
7. Define the nutritional value of poultry meat and eggs.
8. Describe the consumption trends of poultry and eggs.
9. Define the various "attribute" labels of poultry meat.

SELECTED REFERENCES

Eggcyclopedia. 2010. American Egg Board, Park Ridge, IL 60068.

Martinez, S. W. 1999. *Vertical Coordination in the Pork and Broiler Industries: Implications for Pork and Chicken Products*. ERS: USDA.

Quality of U.S. Agricultural Products (poultry, p. 195). 1996. Ames, IA: Council for Agricultural Science and Technology.

Scanes, C. G., G. Brandt, and G. E. Ensminger. 2004. *Poultry Production*. Upper Saddle River, NJ: Pearson Prentice Hall.

Scanes, A. R. 1994. Poultry processing and products. *Encyclopedia of Agricultural Science*. San Diego, CA: Academic Press.

Stadelman, W. J. 1994. Egg production, processing, and products. *Encyclopedia of Agricultural Science*. San Diego, CA: Academic Press.

5

Milk and Milk Products

The world's population obtains most of its milk and milk products from cows, water buffalo, goats, and sheep (Fig. 5.1). Horses, donkeys, reindeer, yaks, camels, and sows contribute a smaller amount to the total milk supply. Milk, with its well-balanced assortment of nutrients, is sometimes called "nature's most nearly perfect food." While milk is an excellent food product in many ways, it is not perfect, nor is any other food. Milk and numerous milk products (e.g., cheese, butter, ice cream, and cottage cheese) are major components of the human diet in many countries.

Changes in dietary preferences and milk marketing are some of the challenges facing the dairy industry. Health considerations and new products are changing the consumption patterns of milk and milk products. Milk surpluses in the United States and the world result in an abundance of consumer products, but the surpluses represent economic challenges to milk producers.

This chapter focuses primarily on milk as human food, but the importance of milk in nourishing suckling farm mammals should not be overlooked. When the term *milk* is used, reference is to milk from dairy cows unless otherwise specified.

MILK PRODUCTION

Table 5.1 shows the most important sources of milk for human consumption. Total world milk production has increased during the past 25 years, but not at the same rate as world human population. Dairy cows produce over 80% of the world fluid milk supply (Fig. 5.2). Goat's milk can be processed into numerous value-added cheeses (Fig. 5.3).

Leading sources of global milk supply are also shown in Table 5.1.

The national dairy herd in the United States has decreased from over 20 million cows in 1956 to approximately 9.2 million in 2013. Yet total milk production has increased significantly. Reduction in cow numbers has been offset by an increased production per cow—from 5,800 lb in 1956 to more than 21,345 lb today. The 21,345 lb per cow converts to 2,476 gal (1 gal = 8.62 lb), which would supply the annual per-capita fluid milk consumption for more than 100 people in the United States. However, since part of the milk is made into other dairy products (Fig. 5.4), the milk from one cow provides milk and manufactured products (cheese, frozen dairy products, and others) for more than 60 people.

Fluid milk sales are important to the U.S. economy. Consumer expenditures for dairy products in the United States are in excess of $90 billion annually.

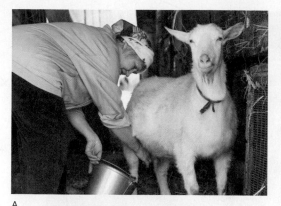

Figure 5.1
*Milk plays an important role in human nutrition throughout the world.
(A) Goats are an important source of milk in many regions of the world.
(B) Lack of infrastructure makes developing a modern dairy supply chain
difficult in some regions of the world. However, local demand provides
market opportunities. (C) Milking cows in a modern U.S. dairy.* Source: 5.1a:
Jenoche/Fotolia; 5.1b: Klodien/Fotolia; 5.1c Justin Field.

Table 5.1
MAJOR SOURCES OF MILK PRODUCTION

Species	Total Production (bil lb)	Leading Regions (Milk from Cows) (bil lb)
Cows	1,379	EU—333
Buffalo	215	U.S.—200
Goats	39	India—119
Sheep	22	China—82
Camel	6	Russia—69
World Total	1,661	

Source: Adapted from multiple sources.

MILK COMPOSITION

Milk is a colloidal suspension of solids in liquid. Fluid whole milk is approximately
88% water, 8.6% **solids-not-fat (SNF)**, and 3–4% milk fat. The SNF is the total
solids minus the milk fat. It contains protein, lactose, and minerals.

Figure 5.2
Dairy cows have the ability to convert feedstuffs into milk. Milk is an important source of nutrients to humans throughout the world. Photo by Justin Field.

Figure 5.3
Dairy goat milk can be processed into a variety of high-quality cheeses.
Source: Mirabelle Pictures/Fotolia.

Even though milk is a liquid, its 12% total solid is similar to the solids content of many solid foods. Differences in milk composition for several species of animals are given in Chapter 19.

The first milk a female produces after the young is born is called **colostrum**. True colostrum is milk obtained during the first milking. For the next several

Figure 5.4

Milk, cheese, and yogurt are among the most highly preferred milk products. Source: 5.4a: Miguel Garcia Saaved/Fotolia; 5.4b: Aitormmfoto/Fotolia; 5.4c: Viktorija.

milkings, the milk is called *transitional milk,* and it is not legally salable until the 11th milking. Colostrum differs greatly in composition from milk. Colostrum is higher in protein, minerals, and milk fat, but it contains less lactose than milk. The most remarkable difference between colostrum and milk produced later is the extremely high immunoglobulin content of colostrum. These protein compounds accumulate in the mammary gland and they are the antibodies that are transferred through milk to the suckling young. People seldom use colostrum from animals because of its unpleasant appearance and odor, although a few may use it for pudding. Additionally, the high globulin content causes a precipitation of proteins when colostrum is pasteurized.

Milk contains a high nutrient density—more than 100 milk components have been identified. **Nutrient density** refers to the concentration of major nutrients in relation to the caloric value of the food. In other words, milk contains important amounts of several nutrients while being relatively low in calories. The major components of milk solids can be grouped into nutrient categories including proteins, fats, carbohydrates, minerals, and vitamins.

Milk Fat

In whole milk, the approximate 3–4% **milk fat** is a mixture of lipids existing as microscopic globules suspended in the milk. The fat contributes about 48% of the total calories in whole milk. Fat-soluble vitamins (A, D, E, and K) are normal components in milk fat. Milk fat contains most of the flavor components of milk, so when milk fat is decreased, there may be a concurrent reduction in flavor.

Approximately 500 different fatty acids and fatty-acid derivatives from 2 to 26 carbon atoms in length have been identified in milk. Milk fat from ruminant animals contains both short-chain and long-chain fatty acids.

Carbohydrates

Lactose, the predominant carbohydrate in milk, is synthesized in the mammary gland. Approximately 4.8% of cow's milk is lactose. It accounts for approximately 54% of the SNF content in milk. Lactose is only about one-sixth as sweet as sucrose, and it is less soluble in water than other sugars. It contributes about 30% of the total calories in milk. Milk is the only natural source of lactose.

Proteins

Milk contains approximately 3.3% protein. Protein accounts for about 38% of total SNF and about 22% of the calories of whole milk. The proteins of milk are of high quality. They contain varying amounts of all amino acids required by humans. A surplus of the amino acid lysine offsets the low lysine content of vegetable proteins and particularly cereals.

Casein, a protein found only in milk, is approximately 82% of the total milk protein. Whey proteins, primarily lactalbumin and lactoglobulin, constitute the remaining 18%. Immunoglobulins, the antibody proteins of colostral milk, were discussed earlier in the chapter.

Vitamins

All vitamins essential in human nutrition are found in milk. Fat-soluble vitamins are in the milk fat portion of milk, and water-soluble vitamins are in the nonfat portion. Milk is usually fortified with vitamin D during processing.

Milk fat from Jersey and Guernsey cows has a rich, yellow color due to carotene (a precursor of vitamin A). Milk fat from Holstein cows has a pale yellow color, and goat milk fat is white. Carotene can be split into two molecules of vitamin A. The milk fat of Holstein cows is pale yellow because most of the carotene has been split, and the milk fat of goats is white because all of the carotene has been changed to colorless vitamin A.

Water-soluble vitamins (C and B) are relatively constant in milk and are not greatly influenced by the vitamin content of the cow's ration. Rumen microorganisms produce the B vitamins, and vitamin C is formed by healthy epithelial tissue in most animals (excluding humans, other primates, and guinea pigs).

Minerals

Milk is a rich source of calcium for the human diet and a reasonably good source of phosphorus and zinc. Calcium and vitamin D are needed in combination to contribute to bone growth in young humans and to prevent **osteoporosis** in adults, particularly women. Milk is not a good source of iron, and the iodine content of milk varies with the iodine content of the animal's feed.

MILK PRODUCTS IN THE UNITED STATES

Tables 5.2 and 5.3 show how the milk supply is used to produce various milk products. More than 80% of the milk produced is marketed as fluid milk, cream, cheese, and butter.

The amount of milk needed to produce each product depends primarily on the milk fat content of the milk. Table 5.4 gives the approximate amount of milk used to produce each product.

Table 5.2
PRIMARY MILK AND MILK PRODUCTS RESULTING FROM TOTAL U.S. MILK SUPPLY

Product	Total Production (bil lb)
Fluid milk and cream	57.1
Cheese	10.6
Yogurt	4.3
Butter	1.8
Ice cream	0.8
Nonfat dry milk	1.5

Source: Adapted from USDA.

Table 5.3
DAIRY PRODUCTS AND BY-PRODUCTS

Primary Products		
Milk	**Spreadable**	**Cream**
Whole	Butter	Table
2%	Low-fat butter	Heavy
1%	Whipped butter	Half & Half
Skim	Ghee (clarified)	Light
Chocolate (flavored)		Whipping
Eggnog		Coffee flavoring
Ultra high temperature (UHT)		
Ultra pasteurized (UP)		
Evaporated		
Condensed (sweet and unsweet)		
Dried (whole and low fat)		
Cheese	**Cultured**	**Frozen Desserts**
Soft (fresh)	Yogurt (flavored, low fat, fat free)	Ice cream (flavored, low fat, nonfat)
Soft (ripened)	Buttermilk	Sherbet
Semi-soft	Sour cream	Ice milk
Blue-veined		Frozen yogurt (flavored, low fat, nonfat)
Hard		Gelato
Very hard		Frozen custard
Pasta filata		
By-Products Whey	**Protein-Enriched (non-whey)**	
Lactose	Casein	
Whey paste	Caseinates	
Condensed whey		
Milk protein		
Whey protein concentrates		

Fluid Milk

Approximately 92% of the 200 billion pounds of milk produced in the United States is grade A quality milk. Grade A milk in excess of fluid milk demand is processed into other milk products. Figure 5.5 shows how fluid milk typically moves from farm to consumer.

Table 5.4
APPROXIMATE AMOUNT OF MILK USED TO PRODUCE SEVERAL SELECTED DAIRY PRODUCTS

Dairy Product (lb)	Whole Milk (lb)
Butter	21.8
Whole-milk cheese	9.23
Evaporated milk	2.1
Condensed milk	7.4
Ice cream (1 gal)	12.0
Cottage cheese	7.2
Nonfat dry milk	2.15

Source: USDA.

A   B

C   D

Figure 5.5
Milk moves from the farm (A) via tanker trucks (B) to processing facilities where it is packaged (C) into containers (D) or further processed into value-added products. Source: 5.5a: Branex/Fotolia; 5.5b: Blondsteve/Fotolia; Vasily Smirnov/Fotolia; 5.5d: Bramgino/Fotolia.

Depending on the milk fat content and the processing at the plant, the fresh fluid product is labeled whole milk, low-fat milk, or skim milk. In further discussion in this chapter, *milk* refers to whole milk. Whole milk is defined as a lacteal secretion, and when it is packaged for beverage use it must contain not less than 3.25% milk fat and not less than 8.25% milk SNF. Low-fat milk usually has had some or most milk fat removed to have one of the following milk fat levels: 0.5%, 1.0%, 1.5%, or 2.0%, with not less than 8.25% milk SNF. Skim milk has had most of the milk fat removed (it contains less than 0.5% milk fat). It must contain at least 8.25% SNF and may be fortified with nonfat solids to 10.25%.

Most fluid milk consumed in the United States is **homogenized** to prevent milk fat from separating from the liquid portion and rising to the top. Homogenization is a physical process resulting in a stable emulsion of milk fat. A "cream line" does not appear in homogenized milk, nor does it form butter if churned.

Fat globules in raw milk average about 6 μm in diameter. To visualize how small the fat globules in homogenized milk are, note that 25,000 μm equal approximately 1 in. Homogenized milk will likely deteriorate by becoming rancid more rapidly than nonhomogenized milk. Rancidity of homogenized milk is forestalled by pasteurization of milk prior to or immediately following homogenization, thus destroying the action of lipolytic enzymes.

Evaporated and Condensed Milk

The process of preheating to stabilize proteins and then removing about 60% of the water produces **evaporated milk**. It is sealed in the container and then heat-treated to sterilize its contents. Milk fat and SNF of evaporated milk must be at least 7.5% and 25%, respectively. Evaporated skim milk must have at least 20% SNF and not more than 0.5% milk fat. Evaporated milk requires no refrigeration until opened. Once the can is opened, refrigeration is necessary to avoid spoilage.

Concentrated, or **condensed, milk** has milk fat (7.5%) and SNF (25.5%) requirements similar to those for evaporated milk. Concentrated milk also has water removed but, unlike evaporated milk, it is not subjected to further heat treatment to prevent spoilage. Most concentrated milk is sold bulk for industry use. A common form of concentrated milk on grocery shelves is sweetened (sugar added) condensed milk; it is used for candy and other confections.

Dry Milk

Dry milk is prepared by removing water from milk, low-fat milk, or skim milk. All of these products are to contain no more than 5% moisture by weight. Dry milk can be stored for long periods of time if it is sealed in an atmosphere of nonoxidizing gas, such as nitrogen or carbon dioxide. The spray-dried or foam-dried product can be reconstituted easily in warm or cold water with agitation.

Fermented Dairy Products

Buttermilk, yogurt, sour cream, and other **cultured** milk products are produced under rigidly controlled conditions of sanitation, inoculation, incubation, acidification, and/or temperature. The word *cultured* appearing in a product name indicates the addition of appropriate bacteria cultu res to the fluid dairy product and subsequent fermentation. Selected cultures of bacteria convert lactose into lactic acid, which produces a tart flavor. The specific bacterial cultures used and other controls in the fermentation process determine whether buttermilk, sour cream, yogurt, or cheese is the end product.

Today buttermilk is a cultured product rather than the by-product of churning cream into butter as was the procedure in the past. At the correct stage of acid and flavor development, the buttermilk is stirred gently to break the curd that has formed. It is then cooled to stop the fermentation process.

The word **acidified** in the product name or production process indicates that the food was produced by souring milk or cream with or without the addition of microbial organisms.

Cottage cheese dry curd is soft unripened cheese; it is produced by culturing or direct acidification. Finished dry curd contains less than 0.5% milk fat and not more than 80% moisture. Cottage cheese is prepared for market by mixing cottage cheese dry curd with a pasteurized creaming mixture called *dressing*. The finished product contains not less than 4% milk fat and not over 80% moisture. The same process makes low-fat cottage cheese; however, the milk fat range is 0.5–2%.

Sour cream is generally made from lactic acid fermentation, although rennet extract is often added in small quantities to produce a thicker-bodied product. Federal standards provide that cultured sour cream contain not less than 18% milk fat. When stored for more than 3–4 weeks, sour cream may develop a bitter flavor as a result of continued bacterial proteolytic enzyme activity.

Yogurt, as a liquid or gel, can be manufactured from fresh whole, low-fat, or skim milk that is heated before fermentation. Federal standards specify that yogurt contain not less than 3.25% milk fat, low-fat yogurt between 0.5% and 2% milk fat, and nonfat yogurt not more than 0.5% milk fat before any bulky flavors are added.

Today three main types of yogurt are produced: (1) flavored, containing no fruit; (2) flavored, containing fruit (fruit may be at the bottom or blended in); and (3) unflavored.

Cream

Cream is a liquid milk product, high in fat that has been separated from milk. Federal standards require that cream contain not less than 18% milk fat.

Several cream products are marketed (Fig. 5.6). Half-and-half is a mixture of milk and cream containing not less than 10.5% but less than 18% milk fat. Light cream (coffee or table cream) contains not less than 18% but less than 30% milk fat. Light whipping cream or whipping cream contains not less than 30% but less than 36% milk fat. Heavy cream or heavy whipping cream contains not less than 36% milk fat. Dry cream is produced by removal of water only from pasteurized milk and/or cream. It contains not less than 40% but less than 75% milk fat and not more than 5% moisture.

Butter

It is known from food remnants in vessels found in early tombs that the Egyptians cooked with butter and cheese. For many centuries, butter making and cheese making were the only means of preserving milk and cream.

Butter churns are the oldest dairy equipment, dating from prehistoric days when nomads carried milk in a type of pouch made from an animal's stomach. Slung

Figure 5.6
Ice cream and other frozen desserts are high demand examples.
Source: M.u.ozmen/Fotolia.

on the back of a horse or camel, the pouch bounced as the animal moved, churning the cream or milk into butter.

Wooden or crockery dasher-type churns used at the turn of the century are today's attic treasures. Jouncing a long-handled wooden dasher up and down in the deep churn sloshed the cream around until butter particles separated from the remaining liquid, called *buttermilk*.

Butter is made exclusively from milk or cream, or both, and contains not less than 80% milk fat by weight. Federal standards establish U.S. grades for butter based on flavor, color, and salt characteristics.

In modern butter making, fresh sweet milk is weighed, tested for milk fat content, and checked for quality. The cream is then separated by centrifugation to contain 30–35% milk fat for batch-type churning or 40–45% milk fat for continuous churning.

Continuous butter-making operations, which can produce 1,800–11,000 lb an hour, are an industry trend. New continuous butter-making processes employ several different scientific principles to form butter. On a global basis, India is the leading butter-producing nation generating nearly three times more than all of North America.

Cheese

There are more than 400 different kinds of cheese that can be made (Fig. 5.7). Cheeses have more than 2,000 names (the same cheese may have two or more different names). For example, cheddar cheese is one of the American-type cheeses in the United States, along with Colby, washed or stirred curd, and Monterey. Leading cheese types as a percent of production in the United States are cheddar (32%) and Mozzarella (33%).

Most cheeses and cheese products are classified into one of four main groups: (1) soft, (2) semisoft, (3) hard, and (4) very hard (Table 5.5). Classification is based on moisture content in the cheese. Thus the body and texture of cheeses range from soft unripened cheese (such as cottage cheese with 80% moisture) to very hard, grated, shaker cheeses such as Parmesan and Romano. These latter, ripened cheeses have 32–34% moisture.

Cheese flavor varies from bland cottage cheese to pungent Roquefort and Limburger. Modern microbiology makes it possible to add specific species and strains of microorganisms required to produce the desired product. For example, a fungus (*Penicillium roqueforti*) is used to produce Roquefort cheese.

Figure 5.7
A variety of cheeses have been produced to satisfy the varied preferences of consumers.
Source: Dream79/Fotolia.

Table 5.5
CHEESE CLASSIFICATION

Classification	Description	Varieties
Soft (fresh)	Unripened with a fresh, creamy flavor. Typically the most perishable and may be stored in brines to extend shelf life.	Cottage cheese, pot cheese, queso blanco, and cream cheese
Soft (ripened)	Sprayed or dusted with a mold and allowed to ripen. Variable in butterfat ranging from 50 to 70%.	Brie and Camembert
Semisoft	Range from mild and buttery to strong and aromatic, may be smoked or flavored to enhance taste.	Edam, Gouda, Havarti, Muenster, Monterrey Jack
Hard	Cheddar is a style of processing that involves turning and stacking the slabs of young cheese to extract whey to create the characteristic texture. Cheeses are then wrapped in wax-dipped cheesecloth and allowed to ripen. Cheddars are categorized by length of aging process (Current—30 days; Mild—30 to 90 days; Medium—3 to 6 months; Sharp—6 to 9 months; Extra-sharp—9 months to 5 years.	Cheddar, Colby, American, and Swiss
	Swiss cheese varieties are characterized by holes, called eyes, which range in size from tiny to the size of a quarter.	
Very hard	Usually grated or shaved and may be characterized by a distinct grainy texture. Ripening (aging) typically ranges from 14 to 24 months.	Parmesan and Romano
Pasta filata	During processing the curds are dipped into hot water and then stretched to attain the proper consistency and texture before being kneaded and molded into desired shapes.	Mozzarella and Provolone
Blue-veined	Mold is injected into the cheese via needling which also allows fermentation gases to escape and oxygen to enter to sustain mold growth. This process results in the veining effect. The cheese is then salted or brined and allowed to ripen in caves or under similar conditions.	French Roquefort (sheep milk), Italian Gorgonzola, English Stilton, Danish blue, and American Maytag blue.

The most dramatic increase in cheese production in the United States has occurred with Italian varieties. The phenomenal growth of the pizza industry has caused an increased production and importation of Mozzarella cheese.

Cheese making involves a biochemical process called **coagulation** or **curdling**. First the milk is heated, and then a liquid starter culture is added. Bacteria from the culture form acids and turn the milk sour. At a later time, rennet is added to thicken the milk. **Rennet**, obtained from the stomachs of young calves, contains the enzyme rennin. Other enzymes of microbial origin have been used successfully to replace a declining supply of rennin. After this mixture is stirred, a custard-like substance called **curd** is formed. Then the liquid part (**whey**) is removed. The cheese-making process reduces 100 lb of milk to 8–16 lb of cheese.

In cheese plants, disposal of whey is a problem. It is currently estimated that 35 billion pounds of whey containing more than 4 billion pounds of solids are produced. Slightly more than half the whey is used to produce human food or animal feeds. The remaining whey poses waste-disposal problems. There continues to be an increasing industrial use of whey for production of food, feed, fertilizer, alcohol, and insulation.

Table 5.6 identifies the leading countries in yearly cheese production and the most important cheese-producing states in the United States.

Table 5.6
LEADING CHEESE-PRODUCING REGIONS AND STATES

Country	Total Production (bil lb)	State	Total Production[a] (mil lb)
European Union	20.2	Wisconsin	2,635
United States	11.7	California	2,245
Poland	1.5	Idaho	842
Russia	1.3	New Mexico	744
New Zealand	0.6	New York	731
World Total	45.4	U.S. Total	10,597

[a]Does not include cottage cheese.
Source: USDA.

Ice Cream

Although they come in many variations, ice cream and a group of similarly frozen foods are made in a similar way and have many of the same ingredients. Ice cream, frozen custard, French ice cream, and French custard ice cream are frozen dairy products high in milk fat and milk solids. Ice cream may contain egg yolk solids. If egg yolks are in excess of 1.4% by weight, the product is called frozen custard, French ice cream, or French custard ice cream. The leading states for ice cream production (% U.S. production) are California (15%), Indiana (9%), Texas (7%), Minnesota (6%), and Pennsylvania (4%).

Ice milk contains less milk fat, protein, and total solids than ice cream. Ice milk usually has more sugar than does ice cream. Soft ice milk and soft ice cream are soft and ready to eat when drawn from the freezer. About three-fourths of the soft-serve products are ice milks. Some of the soft-serve frozen dairy foods contain vegetable fats that have replaced all or most of the milk fat.

Frozen yogurt has become a popular dairy product. Frozen yogurt has less milk fat and higher acidity than ice cream and less sugar than sherbet.

Sherbet is low in both milk fat and milk solids. It has more sugar than ice cream. The tartness of fruit sherbet comes from the added fruit and fruit acid. Sherbet without fruit is flavored with such ingredients as spices, coffee, or chocolate.

Water ices are nondairy frozen foods. They contain neither milk ingredients nor egg yolk. Ices are made similarly to sherbet.

Mellorine differs from ice cream in that milk fat may be replaced partially by a vegetable fat, another animal fat, or both.

Eggnog

Eggnog contains milk products, egg yolk, egg white, and a nutritive carbohydrate sweetener. In addition, eggnog may contain salt, flavoring, and color additives. Federal standards specify that eggnog shall contain not less than 6% milk fat and 8.25% SNF.

Imitation Dairy Products

The FDA established regulations in 1973 to differentiate between imitation and substitute products. It defined an imitation product as one that looks like, tastes like, and is intended to replace the traditional counterpart but is nutritionally inferior to it.

A substitute product resembles the traditional food and also meets the FDA's definition of nutritional equivalency.

Imitation milks usually contain ingredients such as water, corn syrup solids, sugar, vegetable fats, and protein from soybean, sodium caseinate, or other sources. Although imitation fluid milk may not contain dairy products per se, it may contain derivatives of milk such as whey, lactose, casein, salts of casein, and milk proteins, other than casein.

The dairy industry has developed a program for identification of real dairy foods. A "REAL" seal on a carton or package identifies milk or other dairy foods made from U.S.-produced milk that meets federal and/or state standards. This seal assures consumers that the food is not an imitation or substitute.

HEALTH CONSIDERATIONS

Nutritive Value of Milk

Milk and other dairy products make a significant contribution to the nation's supply of dietary nutrients. Particularly noteworthy are the relatively large percentages of calcium, phosphorus, protein, and B vitamins (Table 5.7).

Human milk is regarded as the best source of nourishment for infants. Cow's milk for infant feeding is modified to meet the nutrient and physical requirements of infants. Cow's milk is heated, homogenized, or acidified so infants can utilize the nutrients efficiently. Sugar is usually added to cow's milk for infant feeding to make the milk more nearly like human milk.

Table 5.7
PERCENTAGE CONTRIBUTION OF DAIRY FOODS (EXCLUDING BUTTER) TO NUTRIENTS CONSUMED IN THE UNITED STATES

Nutrients	1980	1990	2000	2006[a]
Energy	10	9	9	9
Protein	20	20	19	9
Fat	12	12	13	8
Carbohydrate	6	5	5	4
Cholesterol	13	15	16	12
Minerals				
Calcium	75	75	72	70
Phosphorus	34	34	32	31
Magnesium	19	18	16	14
Iron	2	2	2	2
Zinc	18	19	16	16
Vitamins				
Ascorbic acid	3	3	2	2
Thiamin	7	7	5	5
Riboflavin	31	31	26	26
Vitamin B_6	11	10	9	7
Vitamin B_{12}	18	20	22	20
Vitamin A	17	18	15	17

[a]Most recent data available.
Source: USDA.

Milk is low in iron; therefore, young animals consuming nothing but milk may develop anemia. Baby pigs produced in confinement need a supplemental source of iron, usually given as an injection. Babies typically receive supplemental iron, and young children who consume large amounts of milk at the expense of meat should be given supplemental iron.

Milk and milk products are excellent sources of nutrients to meet the dietary requirements of young children, adolescents, and adults. With aging, there is an increased prevalence of osteoporosis, a problem of loss of bone mass. Since osteoporosis involves the loss of bone matrix and minerals, a diet generous in protein, calcium, vitamin D, and fluoride is recommended to overcome the problem. Thus, milk, owing to its content of calcium and other nutrients, is an important food for this segment of the population. Furthermore, for the elderly as well as for the infant and young child, milk is an efficient source of nutrients readily tolerated by a sometimes-weakened digestive system.

Wholesomeness

Milk is among the most perishable of all foods owing to its excellent nutritional composition and its fluid form. As it comes from the cow, milk provides an ideal medium for bacterial growth. Properly processed milk can be kept for 10–14 days under refrigeration. With **ultra-high-temperature (UHT) processing**, milk can be kept for several weeks at room temperature.

Protecting the quality of milk is a responsibility shared by public health officials, the dairy industry, and consumers. The most important safeguards from a health standpoint are requirements that bacterial counts are low and that milk is pasteurized. Milk sold through commercial outlets is certified to be from herds that are tested and found to be free from brucellosis and tuberculosis. The consumer who buys milk and milk products can be assured of obtaining a safe, desirable, wholesome food. There are definite health risks in consuming unpasteurized cow's or goat's milk. This concern is important where families or individuals have purchased an animal with an unknown health background or have purchased milk that has not been processed.

Milk Processing

Milk is taken from the cow by a sanitized milking machine, and then transported through sanitized pipes into holding tanks. When withdrawn from the cow, milk is at the cow's body temperature of about 100°F (38°C). It flows into a refrigerated tank, where it is rapidly cooled to 40–42°F. Cold temperatures maintain the high quality of the milk while it is held for delivery.

Tank truck drivers who pick up milk inspect it to see if it is cold and has an acceptable aroma. They take a sample for testing later at the processing plant. Then the cold milk is pumped from the refrigerated farm tank through a sanitized hose into the insulated tank on the truck. After the milk sample passes several tests at the processing plant, the milk is pumped through sanitized pipes into the processing plant's refrigerated or insulated holding tanks.

Milk is pasteurized at the processing plant. **Pasteurization** is a process of exposing milk to a temperature that destroys all pathogenic bacteria but neither reduces the nutritional value of milk nor causes it to curdle. Milk is most commonly pasteurized at 161 °F (71.5°C) for 15 seconds.

Ultrapasteurized milk and UHT processed milk are heated to 280°F (138°C) for at least 2 seconds; this sterilizes milk and increases shelf life.

As it flows out of the pasteurizer, most milk is homogenized by being pumped through a series of valves under pressure. The milk fat is broken up into particles too small to coalesce. As a result, they remain suspended throughout the milk rather than separating to the top as a layer of cream.

The pasteurized, ultrapasteurized, or UHT processed milk is cooled rapidly to below 45°F (7°C). UHT milk is packaged into presterilized containers and aseptically sealed. Since bacteria cannot enter the UHT milk, it can be kept unrefrigerated for at least 3 months. However, once the container is opened, the UHT milk picks up organisms from the air. Then the UHT milk must be handled and stored like any other fluid milk product.

Milk Intolerance

Milk is the main dietary source of a carbohydrate called *lactose*. In the 1960s, the potential problem of lactose intolerance was emphasized when reports revealed low levels of the enzyme lactase, which aids in the digestion of lactose, in the digestive tracts of 70% of black and 10% of white persons in the United States. Symptoms of lactose intolerance include bloating, abdominal cramps, nausea, and diarrhea. Worldwide, lactose intolerance is relatively high among nonwhite populations. However, results of many studies have disclosed the difference between lactose intolerance and milk intolerance. While a large segment of certain populations may be diagnosed as lactose-intolerant, most of these individuals, once adapted, can tolerate the amount of lactose contained in typical servings of dairy foods.

For the few individuals who are truly milk-intolerant, suitable alternatives are available. These include consumption of recommended amounts of milk in smaller but more frequent servings throughout the day, most cheeses, many fermented and culture-containing dairy products (e.g., yogurt), lactose-hydrolyzed milk, and some dairy products containing up to 75% less lactose.

Milk protein allergy may be considered as another form of milk intolerance. An allergic reaction to the protein component of cow's milk may occur in a few infants who experience allergies early in life. The incidence is probably 1% or less in the infant and child population in industrialized countries, although a range of 0.3–7.5% has been reported. The condition is usually outgrown by 2 years of age, beyond which time true allergic reactions to milk in the general population are rare.

CONSUMPTION

Population growth will no doubt cause an increase in consumption of dairy products; however, per-capita consumption is expected to decrease.

Table 5.8 shows the per-capita sales of milk and milk products in the United States over the past 30 years. In 2005, consumers purchased 205 lb of total dairy products per capita. There is a noticeable consumer preference for low-fat milk. However, sales of butter, cream, and cheese have risen in the past decade.

MARKETING

World

Only a small percentage of the world's production of milk products enters world trade: butter (4%), cheese (4%), skim milk powder (18%), and casein (80%). The European Union (EU) and New Zealand account for about 80% of total world exports of dairy products.

The EU, Canada, and the United States have surpluses of skim milk powder in part because of price stabilization programs and because production costs

Table 5.8
YEARLY PER-CAPITA SALES OF DAIRY PRODUCTS IN THE UNITED STATES (LB)

Product	1985	1995	2000	2005	2010	2013
Plain whole milk	116.7	72.1	71.4	61.4	46.6	44.4
Low-fat milk/skim milk	95.9	125.8	128.9	117.1	112.5	106.8
Buttermilk	4.4	2.8	2.2	1.8	1.5	2.2
Yogurt	4.1	4.8	5.4	9.2	13.6	13.9
Half-and-half, light cream, heavy cream	4.4	3.1	6.3	7.9	12.5	11.1
Butter	3.9	4.2	4.8	4.6	5.0	5.6
Cheese	10.4	15.4	16.8	17.0	32.4	33.5
Cottage cheese	4.1	2.8	2.7	2.6	2.4	2.3
Ice cream	18.1	16.1	16.8	15.4	11.9	11.5
Ice milk	6.9	7.7	7.9	7.8	6.0	6.7

Source: Adapted from USDA.

exceed world market prices. Exports of skim milk powder go principally to countries with a milk deficit and are used in a variety of recombined products. Much of the New Zealand and Australian production of casein is exported to the United States.

Nutritionally, the world may need milk, but there is no real market for its present surpluses of dairy products. Surpluses continue to be a major marketing challenge both in the United States and the world.

United States

Most milk produced on U.S. dairy farms goes to plants and dealers for processing. More than 50% of fluid milk is marketed by supermarkets, primarily in plastic gallon containers. However, some dairy farms have on site processing facilities that allow them the opportunity to direct market to consumers, retail outlets, and food-service enterprises. This value-added approach to marketing milk allows producers more direct control in establishing brand identity and potentially higher profits (Fig. 5.8)

Prices

Cooperative milk marketing associations located near large population centers give producers more bargaining power, as they control approximately 90% of the milk produced. The cooperatives hire professionals to market the milk to prospective purchasers.

There are classes and grades of milk that determine price. Grade A (fluid or market milk) and grade B (manufacturing milk) are determined by the sanitary and microbial quality of the milk. Class I, the highest price class, is grade A milk for fluid use. Classes II and III are milk used in manufactured products. Class II includes milk for cottage cheese, cream, and frozen desserts; class III is milk used for butter and cheese. Class II is grade A milk, and class III is surplus grade A or grade B milk. Producers receive a "blend" price based on the proportion of milk used in each price class.

Forty-four federal "milk-marketing orders" and 14 states establish the prices that processors must pay dairy farmers for about 95% of the fluid milk and fluid milk products consumed in the United States. In some states, milk

A

B

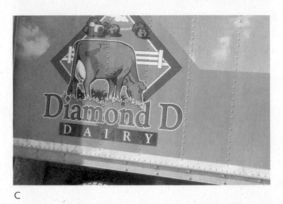
C

Figure 5.8
(A) On farm processing facility. (B) Milk loaded into crates for transport to local homes and stores. (C) Delivery truck. Photos by Justin Field.

control commissions not only determine what farmers are to be paid but the prices stores can charge customers. Federal milk orders are intended to promote orderly marketing conditions for dairy producers and assure consumers an adequate supply of milk.

CHAPTER SUMMARY

- Most of the 1,661 billion lb of world milk production is contributed by cows 1,379 billion lb, water buffalo (215 bil lb), goats (39 bil lb), and sheep (22 bil lb).

- In addition to fluid milk for drinking and cooking, cheese, butter, and dessert dairy products (e.g., ice cream, yogurt, and sherbet) are important consumer dairy products.

- Dairy products contribute significant amounts of protein, calcium, phosphorus, and riboflavin to the human diet.

- Sales of cheese, low-fat skim milk, and yogurt have increased in recent years, while whole milk and cottage cheese have experienced decreased per-capita sales.

KEY WORDS

solids-not-fat (SNF)
colostrum
nutrient density
milk fat
lactose
casein
osteoporosis
homogenized
evaporated milk
condensed milk
dry milk

cultured
acidified
yogurt
cheese classifications
coagulation
curdling
rennet
curd
whey
UHT processing
pasteurization

REVIEW QUESTIONS

1. Describe the contribution of various livestock to the global milk supply.
2. Compare the United States to other leading milk-producing regions.
3. Describe the trends in milk production and cow inventory in the United States.
4. Describe the composition of milk and identify the key carbohydrates, proteins, minerals, and vitamins.
5. List the various primary products and by-products generated by the dairy industry and describe the total production volume of the basic dairy product classes in the United States.
6. Compare the amount of milk required to produce a unit of various processed dairy goods.
7. Describe dairy processing and the steps involved to create evaporated, condensed, dry, and fermented products.
8. Compare the characteristics of various forms of crème and butter.
9. Describe the regional concentration of milk and processed dairy production in the United States.
10. Describe the nutritional contributions of the dairy industry to the human diet.
11. List the various approaches to assuring the wholesomeness and safety of milk.
12. Discuss the consumption trends of milk and dairy products and describe milk marketing in the United States.

SELECTED REFERENCES

Dairy Producer Highlights. 2009. Arlington, VA: National Milk Producers Federation.

Hedrick, T. I., L. G. Harmon, R. C. Chandan, and D. Seiberling. 1981. Dairy products industry in 2006. *J. Dairy Sci.* 64:959.

Lowenstein, M., S. J. Speck, H. M. Barnhart, and J. F. Frank. 1980. Research on goat milk products: A review. *J. Dairy Sci.* 63:1629.

Milk Facts. 2009. Washington, DC: Milk Industry Foundation.

Miller, G. D., et al. 1995. *Handbook of Dairy Foods and Nutrition*. Rosemont, IL: National Dairy Council.

Quality of U.S. Agricultural Products. 1996. Ames, IA: Council for Agricultural Science and Technology.

Sellars, R. L. 1981. Fermented dairy foods. *J. Dairy Sci.* 64:1070.

Speckmann, E. W. L., M. F. Brink, and L. D. McBean. 1981. Dairy foods in nutrition and health. *J. Dairy Sci.* 64:1008.

Tobias, J., and G. A. Muck. 1981. Ice cream and frozen desserts. *J. Dairy Sci.* 64:1077.

Tong, P. 1994. Dairy processing and products. *Encyclopedia of Agricultural Science*. San Diego, CA: Academic Press, Inc.

6
Wool and Mohair

Fibers that grow from the skin of animals give protection from abrasions to their skin and help keep them warm. Skins of common mammals have a covering to which various terms are applied depending on the nature of the growth. For example, cattle, pigs, horses, and dairy goats have hair; sheep have wool; mink and non-Angora rabbits have fur; Angora rabbits have angora; and Angora goats have mohair.

Hair from most mammals has little commercial value (it is used mostly in padding and cushions). Fur of mink and non-Angora rabbits is either naturally beautiful or can be dyed to give attractive colors; therefore, it has considerable value as fur. Fur is composed of fine short fibers and relatively long, coarse guard hairs, in contrast to the skin covering of cattle, horses, and pigs, which is composed entirely of guard hairs.

Furs made from rabbits, mink, foxes, and bison (American buffalo) may be classed as status clothing because they are usually costly and attractive in appearance. White rabbit furs can be dyed any color (including pastels), but colored furs cannot. Because colored furs cannot be dyed, color mutations have been important in the mink business because they provide a wide array of fur colors. At present, the mink industry in the United States is suffering economically because of competition from furs made in other countries and high costs of producing mink.

Hides from young market lambs are often marketed with wool left on the pelt to be processed into heavy winter lambskin coats. Lambs from the Karakul breed of fat-tail sheep are used to produce Persian lambskins. The black, tight curls of fiber are more like fur than wool. Persian lambskins have been produced in the United States, but larger numbers are currently produced in Afghanistan, Iran, and the former Soviet Union.

Hides from sheep with longer wool on the pelts also may be marketed intact for processing to produce ornamental rug pieces. The long wool may be dyed. The long wool may also be loosened and removed from the hides after slaughter and sold separately from the hides.

GROWTH OF HAIR, WOOL, AND MOHAIR

Hair or wool fiber grows from a **follicle** located in the outer layers of the skin. Growth occurs at the base of the follicle, where there is a supply of blood, and cells produced are pushed outward. The cells die after they are removed from the blood supply because they can no longer obtain nutrients or eliminate wastes. A schematic drawing of a wool follicle is presented in Figure 6.1.

The **cuticle** causes the fibers to cling together. The intermingling of wool fibers is known as felting. The felting of wool is advantageous in

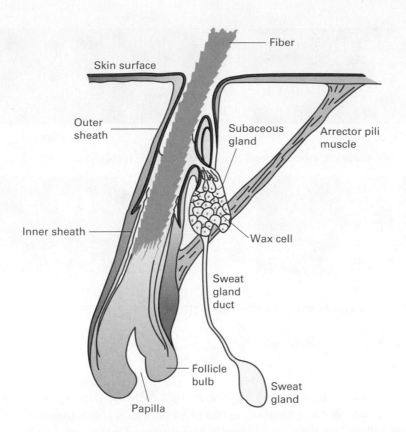

Figure 6.1
Wool follicle structure.
Source: Artist: Sean Field.

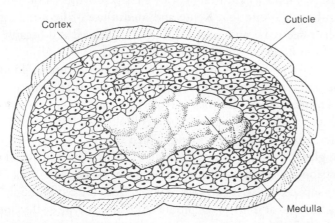

Figure 6.2
Cross-section of a medullated wool fiber.
Source: Adapted from *The Sheep Production Handbook, American Sheep Industry Association.* Volume 7, 2002.

that wool fibers can be entangled to make woolens, but it is also responsible for the shrinkage that occurs when wool becomes wet.

All wool and hair fibers have a similar gross structure, consisting of an outer thin layer (cuticle) and a cortex that surrounds an inner core (**medulla**) in medullated fibers (Fig. 6.2). Only medium and coarse wools have a medulla as it is absent in the structure of fine wools.

Wool fibers have waves called **crimp** (Fig. 6.3). Crimp is caused by the presence of hard and soft cellular material in the cortex. The soft cortex is more elastic and is on the outer side of the crimp.

Hair does not exhibit any crimp because all the cells in the cortex are hard. Also, the inner core of hair is not solid, unlike the inner core of most wool fibers. Some fibers in wool of a low quality are large, lack crimp, and do not have a solid inner core. These fibers are called **kemp**, and they reduce the value of the fleece.

Mohair fiber follicles develop in groups consisting of three primary follicles each. Secondary follicles develop later. Mohair fibers have very little crimp (less than

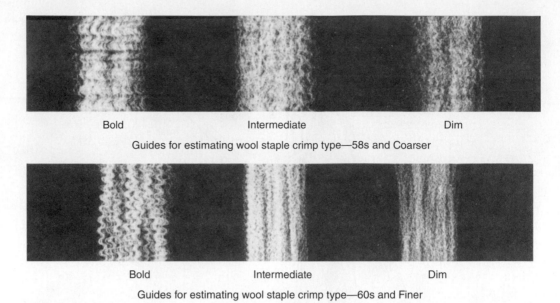

Bold   Intermediate   Dim

Guides for estimating wool staple crimp type—58s and Coarser

Bold   Intermediate   Dim

Guides for estimating wool staple crimp type—60s and Finer

Figure 6.3
Several expressions of crimp in coarser and finer wool. Source: Courtesy of USDA.

one crimp for each inch of length). The mohair cortex is composed largely of so-called *ortho* cells, in contrast to the cortex of wool, which is composed of both *ortho* and *para* cells. Mohair has a different scale structure from wool, but it does have scales and can be made into felt. Kemp and **medullated fibers** reduce the value of wool and mohair fleeces because such fibers do not dye well and because they show in apparel made from fleeces containing the fibers.

The skin of young lambs is made up of two layers: the surface skin (outer skin) is the **epidermis**, and the underlying skin is the **dermis** (Fig. 6.1). Between these two layers is the **basal layer**. Development of wool follicles occurs in skin of the fetus. The basal layer thickens and areas that are to become wool follicles push down into the dermis. Two glands develop: the **sebaceous gland**, which secretes sebum (a greasy substance) and the **sweat gland**, which produces and secretes sweat. The downward growth of the basal layer rests on the papilla, which has a supply of blood. The wool fiber grows toward the skin's surface, and as it becomes removed from the blood supply, the cells die.

Two major types of follicles are produced. The first type to develop is known as the **primary follicle**. These appear on the poll and face of the fetus by 35–50 days' gestation and over all other parts of the body by 60 days. They are arranged in groups of three. All primary follicles are fully developed and are producing fiber at birth.

A second type of follicle, associated with the primary follicles, develops later and is known as the **secondary follicle**. These follicles have an incomplete set of accessory structures—usually the sweat gland and erector muscle are absent. The primary and associated secondary follicles are grouped into follicle bundles.

Since all the primary follicles are formed prior to birth of the lamb, the density of primary follicles per square inch of skin declines rapidly during the first 4 months of postnatal life and then declines gradually until the lamb approaches maturity.

Secondary follicles show little activity during the first week of a lamb's life, after which they undergo a burst of activity. Follicular activity is at maximum between 1 and 3 weeks of age, with marked reduction after the third week.

Adverse prenatal environmental conditions can affect the number of secondary follicles initiated in development. Also, early postnatal influences could affect secondary follicle fiber production.

Growth of the wool fiber takes place in the **root bulb**, which is located in the follicle. Permanent dimensions of the fiber are determined in the area of the bulb, and there are no changes in growth characteristics of the fiber after it is formed. Defective portions of a fiber are caused by a reduction in the size of the fiber that creates a weakened area when sheep are under stressed conditions, such as poor nutrition and high body temperatures. Such fibers are likely to break under pressure.

Two types of undesirable fibers are *kemp* and *medullated* fibers. Both are hollow and brittle, and primary follicles rather than secondary follicles produce both, as is the case with true wool.

FACTORS AFFECTING THE VALUE OF WOOL

Ensuring wool quality is the result of best production practices both before and after shearing (Fig. 6.4). The two most important factors under the control of the producer that affect the amount of wool produced by sheep are nutrition and breeding. The amount of feed available to the sheep (energy intake) and the percentage of protein in the diet influence wool production. Wool production is decreased when sheep are fed diets having less than 8% protein. When the diet contains more than 8% protein, the amount of energy consumed is the determining factor in wool production.

Improving wool production per animal through breeding involves selection for increased clean **fleece weight**, **staple length**, **fineness**, and uniformity of length, and fineness throughout the fleece. Performance testing rams for quantity and quality of wool production can make much progress, but for overall improvement of the flock, both lamb and wool production must be included (Fig. 6.5).

Sheep and Angora goat producers are interested in maximizing net income from their sheep and goat operations. Increasing net income from wool and mohair can be accomplished by selecting to improve wool and mohair production and grade. Giving attention to the following items can enhance the values of wool and mohair that are produced:

1. The sheep or goats should be shorn when the wool or mohair is dry.
2. Inferior portions of the fleece should be removed at the skirting table and sacked separately. **Off sorts** (inferior portions) include parts of the clip that are matted or heavily contaminated with dung or vegetable matter, bellies, topknots, and floor sweepings (Fig. 6.6).
3. Wool or mohair should be sacked by wool grades so that when **core samples** are taken from a sack of otherwise good wool, a few fleeces of low-grade wool (or mohair) will not cause the entire sack of wool or mohair to be placed in a low grade.
4. Wool should be shorn without making many double clips.
5. Fleeces should be tied with paper twine after they are properly folded with the clipped side out.

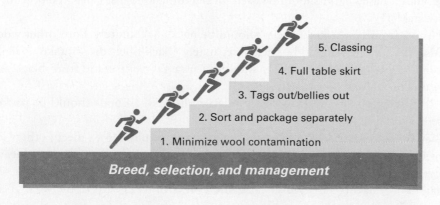

5. Classing
4. Full table skirt
3. Tags out/bellies out
2. Sort and package separately
1. Minimize wool contamination

Breed, selection, and management

Figure 6.4
Preparation steps for wool quality improvement.
Source: Sheep Production Handbook, American Sheep Industry Association, Volume 7, 2002.

Figure 6.5
Depending on the goals of the production flock, lambs and wool are either seen as primary products or by-products. However, balanced selection can assist producers in optimizing production.
Source: 169169/Fotolia.

Figure 6.6
Skirting the fleece.
Source: Sheep Production Handbook, American Sheep Industry Association, Volume 7, 2002.

SKIRTING THE FLEECE

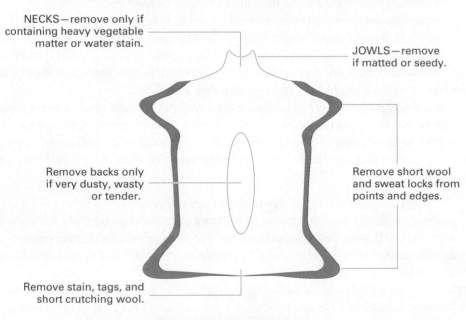

NECKS—remove only if containing heavy vegetable matter or water stain.

JOWLS—remove if matted or seedy.

Remove backs only if very dusty, wasty or tender.

Remove short wool and sweat locks from points and edges.

Remove stain, tags, and short crutching wool.

NOTE—keep all strains separate
—make necessary cast lines
—remove skin pieces

6. A lanolin-based paint should be used for branding animals. This type of paint is scourable.

7. Fleeces from black-faced sheep should be packed separately from other wool. Also, black fleeces should be packed separately. Black fibers do not take on light-colored dyes and consequently stand out in a garment made from black and white fibers dyed a light pastel color.

8. All tags, sweepings, and wool or mohair from dead animals should be packed separately.

9. Hay that has been baled with twine should be carefully fed to sheep; otherwise, pieces of twine will get into the wool or mohair and markedly lower its value.

10. Environmental stresses should be avoided. Being without feed or water for several days or having a high fever can cause a break or weak zone in the wool fibers.

11. Wool containing coarse fibers, or kemp, should be avoided, or kempy fleeces should be packed separately. Cloth and wool containing these coarse fibers are highly objectionable and thus are low in price.

Undesirable fleeces contain off color fibers and are contaminated by foreign material. Specific types of undesirable fleeces include the following:

Burry—wool contains vegetable matter, such as grass seeds and prickly seeds, which adheres tenaciously to wool

Chaffy—wool contains vegetable matter such as hay, straw, and other plant material

Cotted—wool fibers are matted or entangled

CLASSES AND GRADES OF WOOL

The class and grade determine the value of wool. Staple length determines the class, and fineness of fibers determines the grade. The grades of wool are given by any one of three methods, all of which are concerned with a value that indicates the fineness (diameter) of the wool fibers. The three methods of reporting wool grades are as follows:

1. The American grade, or blood (outdated terminology), method is based on the theoretical amount of **fine-wool breeding** (Merino and Rambouillet) represented in the sheep producing the wool (Fig. 6.7). The terms *fine, ½-blood, ⅜-blood, ¼-blood,* and *low-¼ blood* are used to describe typical fiber diameter. Wool classified as *fine* or *⅝-blood* has small fiber diameters, whereas ¼-blood wool is coarse.

2. The spinning count system refers to the number of *hanks* of yarn, each 560 yd long, that can be spun from 1 lb of **wool top**. These grades range from 80 (fine) to 36 (coarse). Thus, a grade of 50 by the spinning count method means that 28,000 yd (560 yd × 50 = 28,000 yd) of yarn can be spun from 1 lb of wool top.

3. The micron diameter method is based on the average of actual measurements of several wool fibers. This is the most accurate method of determining the grade of wool. A micron is $\frac{1}{25,400}$ of an inch.

Wool samples from each of the major grades are shown in Figures 6.7 and 6.8. The systems of grading, along with the breeds from which the grades of wool come, are presented in Table 6.1. The Merino and Rambouillet breeds are considered fine-wool breeds, with the Cotswold, Lincoln, and Romney identified as coarse-wool breeds. In Table 6.1, breeds listed between the fine- and coarse-wool breeds are called **medium-wool breeds**.

Classes of wool are staple, French combing, and clothing; however, fibers of finer grades that do not meet the length requirement may go into a higher class than they would if they were not fine. Much of the staple length wool is combed into **worsted**-type yarns for making garments of a hard finish that hold a press well.

The fineness of wool fibers from a sheep depends on the area of the body from which the fibers came. The body of a sheep can be divided into seven areas. Area 1 is the shoulder and area 2 is the neck, where wool grows longer and finer than average wool from the same sheep. Area 3 is the back. Its wool is long and has average fineness, though it has been exposed to the most weathering. Area 4 is the side. Wool from this area is long, has average fineness, and constitutes most of the wool. Area 5 is the tag, which yields long, coarse, dirty, and stained wool. Area 6 is the britch. Wool from this area is long and coarser than average and tends to be medullated. Area 7 is the belly. Its short, fine, matted wool may be very heavy in vegetable matter. Preference is given to a fleece that has uniformity of length and fineness in all areas of the sheep's body.

Qualitative factors of importance in wool marketing include strength, crimp, softness or handle, color, purity, character, and freedom from contaminants.

Figure 6.7
Wool samples of the major grades of wool based on spinning count and the blood system. Source: Tom Field.

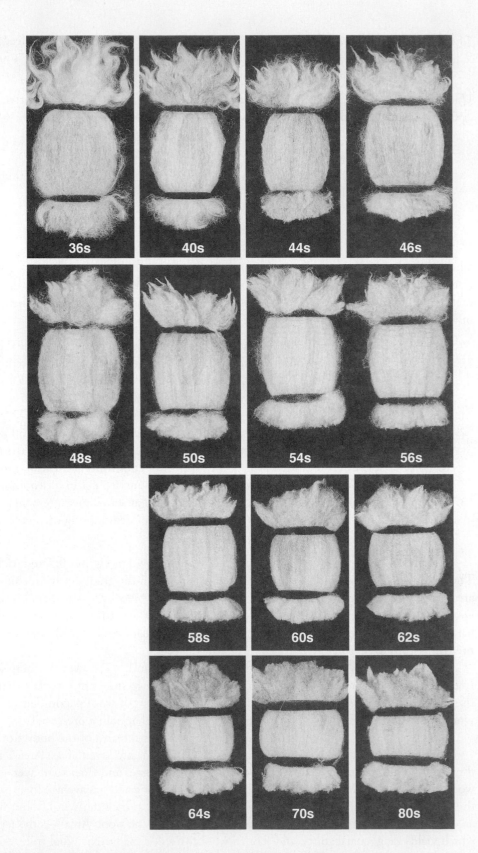

PRODUCTION OF WOOL AND MOHAIR

Production of wool can be expressed in terms of **greasy wool** or **scoured** (clean) **wool**. Greasy wool is wool or the fleece shorn once each year from sheep (Fig. 6.9). Scoured wool originates in the initial wool-processing stage. Scouring is the washing and rinsing of wool to remove grease, dirt, and other impurities.

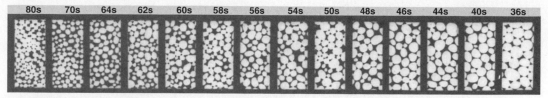

Figure 6.8

Cross-section of magnified wool fibers demonstrating the wool grades based on fiber diameter measured in microns. Source: USDA.

Table 6.1
MARKET GRADES OF WOOL AND BREEDS OF SHEEP THAT PRODUCE THESE GRADES OF WOOL

USDA Grades Based on Spinning Count	USDA Grades Based on American (Blood) System	Grades Based on Micron Diameter	Grades According to Average Grades
80s	Fine	17.70–19.14	Merino
70s	Fine	19.15–20.59	Rambouillet
64s	Fine	20.60–22.04	Targhee, Southdown
62s	½ blood	22.05–23.49	Corriedale, Columbia
50s to 60s	¼ blood to ½ blood	23.50–30.99	Panama, Romeldale
48s to 56s	Low-¼ blood to ½ blood	26.40–32.69	Shropshire, Hampshire, Suffolk, Oxford, Dorset, Cheviot
46s to 48s	Common to low-¼ blood	32.69–38.09	Romney
36s to 40s	Common and braid	36.20–40.20	Cotswold
36s to 46s	Common and braid	32.70–40.20	Lincoln

Source: USDA.

Figure 6.9
Sheep being shorn for their yearly production of wool.
Source: Tom Field.

Table 6.2 shows the production of greasy and scoured wool for several countries of the world. The world greasy wool production is approximately 4.5 billion pounds, with the United States producing approximately 27 million pounds. The U.S. wool production and total yearly supply is shown in Table 6.3. The U.S. wool production by states is shown in Figure 6.10. Scoured or clean wool production represents 50–60% of the greasy wool produced. Total U.S. wool production on a clean basis has been in decline for nearly three decades (Fig. 6.11).

Table 6.2
WORLD PRODUCTION OF GREASY WOOL

Country	Greasy Wool (mil lb)
China	882
Australia	798
European Union	424
New Zealand	364
World total	4,556

Note that Australia, China, the EU, and New Zealand account for 50% of world production.
Source: Adapted from FAO.

Table 6.3
U.S. SHORN WOOL PRODUCTION

Year	No. Head Shorn (mil)	Ave. Fleece Wt. (lb)	Price (cents/lb)	Domestic Production (mil lb)
1990	11.2	7.8	0.80	88.0
1995	8.1	7.8	1.04	63.5
2000	6.1	7.6	0.33	46.4
2005	5.0	7.3	0.80	37.2
2010	4.2	7.3	1.15	35.0
2013	4.0	7.3	1.45	27.0

Source: Adapted from USDA.

Figure 6.10
Top 10 states for wool production, 2013 (1,000 lb). Source: USDA.

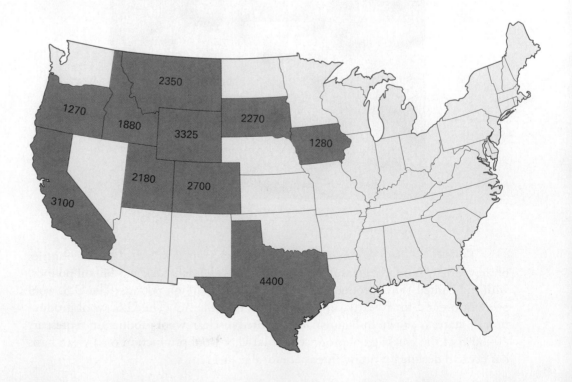

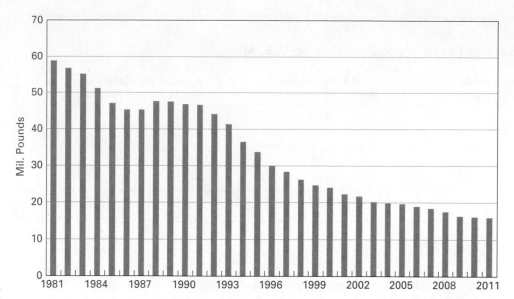

Figure 6.11
Annual U.S. wool production—clean basis.
Data Source: USDA/NASS. Data compiled by Livestock Marketing Information Center. Used by permission of Livestock Marketing Information Center.

Table 6.4
MOHAIR PRODUCTION

Year	No. Goats Clipped (mil)	Avg. Wt. Per Clip (lb)	Total Production (mil lb)	Price per lb	Total Value ($ mil)
1990[a]	1.9	7.8	14.5	0.95	13.8
1995	1.6	7.4	12.0	1.84	22.1
2000	0.3	6.6	1.7	2.20	3.8
2005	0.27	6.7	1.8	2.78	5.1
2010	0.20	6.0	1.1	3.49	3.8
2013	0.14	5.6	0.8	4.25	3.5

[a]Texas production alone.
Source: Adapted from USDA.

Growth of wool varies greatly among sheep breeds. Fine-wool sheep grows 4–6 in. of wool per year, $\frac{1}{4}$- to $\frac{1}{2}$-blood breeds of sheep grow 5–10 in. of wool per year, and common-to-braid-wool breeds grow 9–18 in. of wool per year. Mohair growth is approximately 12 in. per year.

Weights of **fleeces** also vary greatly among the breeds of sheep. Even though fine-wool breeds have fleeces of short length, weight of their fleeces varies from 9–12 lb; $\frac{1}{4}$- to $\frac{1}{2}$-blood breeds of sheep produce fleeces weighing 7–15 lb; and coarse-wool (common and braid) breeds produce fleeces weighing 9–16 lb.

Annual world mohair production is approximately 13.5 million pounds, of which the United States produces about 14% of world production. South Africa leads global production with a 60% share. Texas produces about 97% of the mohair produced in the United States. The pounds and value of mohair production is shown in Table 6.4. Mohair growth is about 1–2 in. per month; therefore, goats are usually shorn twice per year, with length of clip averaging 6–12 in.

Table 6.5
PRICES (DOLLARS PER LB) FOR CLEAN WOOL DELIVERED TO U.S. MILLS

Year	U.S. Origin 64s Grade (20–22 micron)	U.S. Origin 58s Grade (25–28 micron)	Australian Origin 64s Grade (20 and less micron)
1995	2.49	1.70	2.81
2000	1.08	0.61	1.50
2005	1.86	1.26	2.57
2010	3.27	2.03	4.10
2013	4.28	3.81	6.30

Source: Adapted from USDA.

WOOL MARKETING

Wool sales are accomplished via several marketing alternatives including direct sale to warehouses, or via cooperative ventures where local producers pool their wool inventories to gain scale advantages in the marketplace. Worldwide improvement in descriptive technologies has refined the process of selling wool based on objective measurements of quality.

Manufacturers of wool products typically purchase from the central markets. Increasingly, wool is priced on a clean or scoured basis. Prices for raw wool are highly variable (Table 6.3). Clean wool prices for various grades and for U.S. versus Australian sources are provided in Table 6.5. Note that regardless of year, significant price advantages accrue to better grade wool with the fine wool from Australia commanding the highest prices.

USES OF WOOL AND MOHAIR

Sheep, Angora rabbit, and Angora goat fibers (Fig. 6.12) are used in making cloth and carpets. The consumption of wool in the United States is small when compared to the consumption of cotton and numerous manufactured fibers. These trends illustrate the decline in demand for wool. Wool trade from Australia accounts for approximately two-thirds of fine wool imports to the United States, while New Zealand provides in excess of 60% of imported medium-grade wool.

Cloth made from wool has both highly desirable and undesirable qualities. Wool has a pleasant, warming quality and can absorb considerable quantities of moisture while still providing warmth. It is also highly resistant to fire. However, wool has a tendency to shrink when it becomes wet, and cloth made from wool causes some people to itch.

Researchers have contributed greatly to the alteration of wool so that fabrics or garments will not shrink when washed. A process called the WURLAN treatment makes woolen garments machine-washable. In this treatment, the wool fibers are coated with a very thin layer of resin, which adds only 1% to the weight of the wool. The resin does not alter wool in any significant way except to prevent the fibers from absorbing water. There is an increasing trend toward blending wool with other fibers. This improves aesthetics and provides materials with more durability and comfort.

Only fabrics that have passed the quality tests demanded by the International Wool Secretariat and its U.S. branch, the Wool Bureau, Inc., are eligible to carry the Woolmark or Woolblend label.

Woolen garments are as old as the Stone Age, yet as new as today. Wool takes many forms—wovens, knits, piles, and felts—each the result of a different process. It is a long way from a bale of raw wool to a beautiful bolt of fabric and ultimately to beautiful, versatile garments (Fig. 6.13).

Figure 6.12
Angora goats with a full growth of mohair.
Source: Blackcurrent/Fotolia.

A

B

C

Figure 6.13
Natural fiber moves through a process of production, marketing, and processing to generate desirable wool and cashmere products for consumers. (A) Preparing wool for the skirting table. (B) Wool is dyed, carded, and spun before being woven into a desired fabric. (C) High-quality woolen garments are then made available to the public. Source: 6.13a: Tom Field; 6.13b: Rafael Ben-Ari/Fotolia; 6.13c: Silvano Rebai/Fotolia.

CHAPTER SUMMARY

- Wool, mohair, and hair from animals provide useful human products.
- Wool value is determined by fleece weight, cleanliness, staple length, and fineness (fiber diameter).
- Mohair production from Angora goats is concentrated primarily in Texas.
- Cotton, synthetic fibers, and wool imports have provided stiff competition for domestic wool production.

KEY WORDS

follicle
cuticle
medulla
crimp
kemp
medullated fibers
epidermis
dermis
basal layer
sebaceous gland
sweat gland
primary follicle
secondary follicle
root bulb

fleece weight
staple length
fineness
core sample
burry
chaffy
cotted
fine-wool breeds
wool top
medium-wool breeds
worsted
greasy wool
scoured wool

REVIEW QUESTIONS

1. Name the fibers associated with sheep, angora goats, and hogs.
2. Describe the structure of a wool fiber.
3. Compare and contrast the structure of mohair and wool.
4. What are the characteristics of undesirable fleeces?
5. How does diet affect wool production?
6. What are the primary traits affecting wool quality?
7. Describe the types of undesirable fleeces.
8. Compare the three methods of determining wool grade.
9. What are the seven body regions of a sheep and the characteristics of wool from each?
10. Describe the breeds typically affiliated with the various classes of wool.
11. Describe domestic and global wool production.
12. Describe the wool marketing process.

SELECTED REFERENCES

Botkin, M. P., R. A. Field, and C. L. Johnson. 1988. *Sheep and Wool: Science, Production, and Management.* Englewood Cliffs, NJ: Prentice Hall.

Code of Practice for Preparation of Wool Clips in the United States. Englewood, CO: American Sheep Industry Association, Inc.

Hyde, N. 1988. Fabric of history: Wool. *Natl. Geogr.* 173:552.

Lupton, C. J. 1994. Wool and mohair production and processing. *Encyclopedia of Agricultural Science.* San Diego, CA: Academic Press.

Quality of U.S. Agricultural Products. 1996. Ames, IA: Council for Agricultural Science and Technology.

Rogers, G. E., et al. 1989. *The Biology of Wool and Hair.* New York: Chapman and Hall.

7

By-Products of Meat Animals

By-products are products of considerably less value than the major product. However, the value of by-products can account for as much as 8–10% of the total value of a fed steer. In the United States, meat animals produce meat as the major product; hides, fat, bones, and internal organs are considered by-products. In other countries, primary products may be draft (work), milk, hides, and skins, with meat considered a by-product when old, less useful animals are slaughtered.

There are numerous by-products resulting from animal slaughter and the processing of meat, milk, and eggs. Animals dying during their productive lives may also be processed into several by-products; however, these by-products are not used in human consumption.

By-products are commonly classified into two categories based on human consumption: **edible** and **inedible** (Fig. 7.1). Some by-products that are not considered edible by humans are edible to other animals. By-products that are processed into animal feeds, such as feather meal or fish-meal, are good examples. Examples of edible by-products (variety meats) include liver, heart, cheek meat, oxtail, pig's feet, pig's ears, tongue, tripe, and chitlins. Inedible by-products include lung, meat and bone meal, blood meal, hides, and skins. The pharmaceutical industry utilizes fresh bile, bovine fetal blood, pituitary glands, pancreas, and thyroid glands.

Figure 7.2 shows the major by-products obtained from a steer in addition to the retail beef products. The average total value of cattle by-products is highly variable, with the hide comprising the largest part of the total value. Because hides' values have cyclic prices, as is often the case with commodity products, their value as a percent of the total by-product value often range from 45% to 65%.

The use of by-products in a variety of applications is a highly sustainable practice that allows the livestock industries to minimize wastage and lost value.

EDIBLE BY-PRODUCTS

Variety meats are edible products originating from organs and body parts other than the carcass. Liver, heart, tongue, tripe, and sweetbread are among the more typical variety meats. **Tripe** comes from the lining of the stomach; **sweetbread** is the thymus gland.

An average 1,350-lb slaughter steer produces approximately 45 lb of variety meats. Since the U.S. per-capita disappearance is 9 lb, surplus variety meats are exported into countries where consumer demand is more substantial.

Figure 7.1
Edible and inedible by-products from a 1,100-lb steer. Source: USDA.

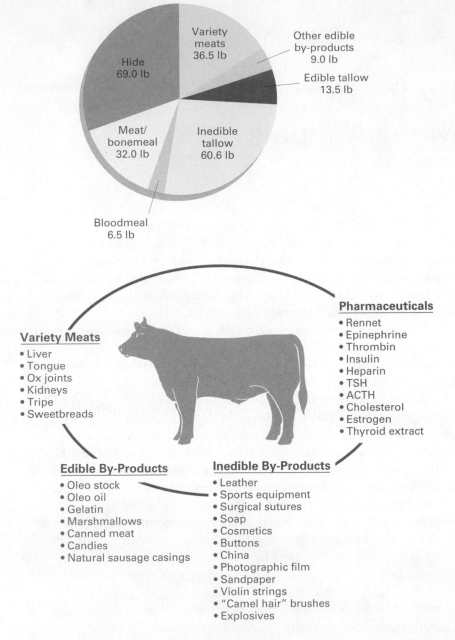

Hide
69.0 lb

Variety meats
36.5 lb

Other edible by-products
9.0 lb

Edible tallow
13.5 lb

Inedible tallow
60.6 lb

Meat/bonemeal
32.0 lb

Bloodmeal
6.5 lb

Figure 7.2
In addition to the retail product of beef are numerous by-products.
Source: Adapted from Field (1996).

Variety Meats
• Liver
• Tongue
• Ox joints
• Kidneys
• Tripe
• Sweetbreads

Pharmaceuticals
• Rennet
• Epinephrine
• Thrombin
• Insulin
• Heparin
• TSH
• ACTH
• Cholesterol
• Estrogen
• Thyroid extract

Edible By-Products
• Oleo stock
• Oleo oil
• Gelatin
• Marshmallows
• Canned meat
• Candies
• Natural sausage casings

Inedible By-Products
• Leather
• Sports equipment
• Surgical sutures
• Soap
• Cosmetics
• Buttons
• China
• Photographic film
• Sandpaper
• Violin strings
• "Camel hair" brushes
• Explosives

Other edible products are fats used to produce **lard** and **tallow**. These products are eventually used in shortenings, margarines, pastries, candies, and other food items. Inedible tallow and greases are used in the manufacture of soap, lubricants, feed, and fatty acids. Nearly 45% of the inedible tallow and nearly 20% of edible tallow and lard are exported. Rendered fats are sold on the basis of percent total fatty acids, unsaponifiable matter, insoluble impurities, free fatty acids, and moisture.

INEDIBLE BY-PRODUCTS

Tallow, hides (skins), and inedible organs are the higher-valued inedible by-products. Not all skins are inedible, as some pork skins are processed into consumable food items. Life-saving and life-supporting pharmaceuticals originate from animal by-products, even though many of the same pharmaceuticals are made synthetically. Table 7.1 shows many of these pharmaceuticals and their animal by-product sources.

Table 7.1
SELECTED PHARMACEUTICALS FROM RED MEAT ANIMALS—THEIR SOURCE AND UTILIZATION

Pharmaceutical	Source	Utilization
Amfetin	Amniotic fluid	Reducing postoperative pain and nausea; enhancing intestinal peristalsis
Catalase	Liver	Preparation of contact lens care products
Cholesterol	Nervous system	Male sex hormone synthesis
Chondroitin sulfate	Trachea	Arthritis treatment
Chymotrypsin	Pancreas	Removing dead tissue; treatment of localized inflammation and swelling
Corticosteroids	Adrenal gland	Treatment for shock, Addison's disease
Corticotrophin (ACTH)	Pituitary gland	Diagnostic assessment of adrenal gland function; treatment of psoriasis, allergies, mononucleosis, and leukemia
Cortisone	Adrenal gland	Relief of hay fever, asthma, and other allergies; heart stimulation
Deoxycholic acid	Bile	Used in synthesis of cortisone for asthma and arthritis
Epinephrine	Adrenal gland	Relief of hay fever, asthma, and other allergies; heart stimulation
Fibrinolysin	Blood	Dead tissue removal; wound-cleansing agent; healing of skin from ulcers or burns
Glucagon	Pancreas	Counteracting insulin shock; treatment of some psychiatric disorders
Glycerin	Tallow	Cough syrups, lozenges, eyewashes, contraceptive creams, gel capsules
Heparin	Intestines, lungs	Anticoagulant used to thin blood; retarding clotting, especially during organ implants; preventing gangrene
Hyaluranidase	Testes	Enzyme that aids drug penetration into cells, cartilage, and joint treatment
Insulin	Pancreas	Treatment of diabetes for 10 million Americans (primarily synthetic insulin is used)
Liver extracts	Liver	Treatment of anemia
Mucin	Stomach	Treating ulcers
Norepinephrine	Adrenal gland	Shrinking blood vessels, reducing blood flow, and slowing heart rate
Ovarian hormone	Ovaries	Treating painful menstruation and preventing abortion
Ox bile extract	Liver	Treatment of indigestion, constipation, and bile tract disorders
Parathyroid hormone	Parathyroid gland	Treatment of human parathyroid deficiency
Plasmin	Blood	Digests fibrin in blood clots, used to treat patients with heart attacks
Rennet	Stomach	Assisting infants in digesting milk; cheese making
Serum	Blood	Variety of macromolecular proteins, buffering agents for cell culture mediums
Thrombin	Blood	Assisting in blood coagulation; treatment of wounds; skin grafting
Thyroid extract	Thyroid gland	Treatment of cretinism
Thyrotropin (TSH)	Pituitary gland	Stimulating functions of thyroid gland
Thyroxin	Thyroid	Treatment of thyroid deficiency
Vasopressin	Pituitary gland	Control of renal function

Source: Adapted from Field et al. (1996).

Table 7.2 lists other inedible by-products and their uses. Animals contribute a large number of useful products. The many uses of inedible fats are discussed later in the chapter.

Hides and skins are valuable as by-products or as major products on a worldwide basis. Countries that are most important in cattle hide production include

Table 7.2
OTHER INEDIBLE BY-PRODUCTS AND THEIR USES

By-Product	Use
Hog heart valves	Replacement of injured or weakened human heart valves; since the first operation in 1971, more than 35,000 hog heart valves have been implanted in humans.
Pig skins	Treating massive human burns; these skins help prepare the patient for permanent skin grafting.
Gelatin (from skin)	Coatings for pills and capsules.
Brains	Cholesterol for an emulsifier in cosmetics.
Blood	Sticking agent for insecticides; a leather finish; plywood adhesive; fabric painting and dyeing.
Hides and skins	Many leather goods from coats, handbags, and shoes to sporting goods.
Bones	Animal feed, glue, buttons, china, and novelties.
Gallstones	Shipped to Southeast Asia for use as ornaments in necklaces and pendants.
Hair	Paint and other brushes; insulation; padding in upholstery, carpet padding, filters.
Meat scraps and blood	Animal feed.
Inedible fats	Animal feed, fatty acids.
Wool (pulled from pelts)	Clothing, blankets, lanolin.
Poultry feathers	Animal feed, arrows, decorations, bedding, brushes.
Glycerin	Adhesives, antifreeze, cleaners, anticorrosive coatings, cosmetics, leather tanning and finishing, resins, and metal processing.

Source: Adapted from Field et al. (1996).

China, the United States, Brazil, and Argentina. Leaders in sheep skin production are China, the EU, Australia, and New Zealand. Goat skin global leaders are China, India, Pakistan, and Bangladesh.

Cattle and buffalo hides comprise 80% of the farm animal hides and skins produced in the world. The United States makes a major contribution in the world hide and skin market, with $1 billion worth of cattle hides produced being exported to other countries (Chapter 2). Pork skin usually remains on the carcass, although some pigs are skinned.

The U.S. hide export market yields approximately $1.3 billion on a yearly basis. Hides, skins, and pelts are made into useful leather products through the tanning process (Table 7.3). One cowhide can yield approximately 144 baseballs, 20 footballs, 18 volleyballs or soccer balls, 12 baseball gloves, or 12 basketballs. The surplus hides are exported primarily to Japan, Korea, Mexico, Taiwan, and Romania. Leather utilization in the United States is categorized as 40% for upholstery, 50% for shoes and shoe leather, and 10% for other uses.

The general term **hide** refers to a beef hide weighing more than 30 lb. Those weighing less than 30 lb are called **skins**. Skins come from smaller animals, such as pigs, sheep, goats, and small wild animals. Those skins from sheep with the wool left on are usually called **sheep pelts**. Hides, skins, and pelts are classified according to (1) species, (2) weight, (3) size and placement of brand, and (4) type of packer producing them.

The value of hides may be reduced by branding, nicking the hide while skinning, warbles (larvae of heel flies), mange, lice, biting and sucking insects, grubs, water, mud, and urine damage. Warbles emerge from the back region of cattle in the spring, making holes in the hide. Sheep hides may be damaged by outgrowths of grasses in the production of seeds (called beards), which can penetrate the skin, as well as by external parasites called **keds**.

A fed steer produces a 65–75-lb hide that is worth nearly $1 per pound. When the hide is converted into the "blue" stage (which means it is treated so it will not

Table 7.3
LEATHER USES RELATED TO TYPES OF HIDES AND SKINS

Skin Origin	Use
Cow and steer	Shoe and boot uppers, soles, insoles, linings; patent leather; garments; work gloves; waist belts; luggage and cases; upholstery; transmission belting; sporting goods, packings
Calf	Shoe uppers; slippers; handbags and billfolds; hat sweatbands; book bindings
Sheep and lamb	Grain and suede garments; shoe linings; slippers; dress and work gloves; hat sweatbands; book bindings; novelties
Goat and kid	Shoe uppers, linings; dress gloves; garments; handbags
Pig	Shoe suede uppers; dress and work gloves; billfolds; fancy leather goods
Horse	Shoe uppers; straps; sporting goods

Source: Adapted from New England Tanners Club, *Leather Facts*.

deteriorate during shipping), it loses about 15 lb. The value, however, has increased to $80–$90. A 60-lb blue hide will produce 40 ft^2 of shoe leather. Shoe leather sells for $2.50 a square foot, so the hide is now worth approximately $100.

Every year, U.S. tanneries convert millions of raw hides and skins into leather. The tanners' value added by manufacture constitutes over $500 million annually. Their product (leather) serves in turn as a raw material for the shoe and leather goods industries that provide jobs for over 200,000 people.

Fewer than half of U.S. cattle hides are converted to leather domestically, most being exported for tanning. Some sheepskins are imported, as the domestic supply does not meet the demand. Pigskins are used primarily as food, but interest in leather production is increasing.

Just as meat is perishable, so too are hides and skins. If not cleaned and treated, they begin to decompose and lose leather-making substance within hours after removal from the carcass. Hides are commonly treated (cured) by adding salt as the principal curing agent. The salt solution penetrates the hides in about 12 hours, and then the hides are bundled for shipment. The hide is the single most valuable by-product of beef cattle. The 2000 Beef Industry Quality Audit found that hide damage due to management-related defects (parasite control, branding, etc.) reduced per-head value by $23.92. One of the largest causes of this loss is rib branding .

Nearly 7 million tons of raw material by-products, at a value of $8 billion, are utilized by the pet food industry annually. Exports of U.S. pet foods have tripled since 1990. Dog food sales in the United States account for more than $9 million, with an additional $1 billion spent on dog treats such as rawhide, cured bovine ears, and hooves. Rendered fats and greases have experienced significant export demand for use in livestock and poultry feed formulations. Asia, South America, and Mexico are the highest-volume markets for these by-products.

New products, such as biodiesel, have been developed as a means to extend demand for rendered fats and oils. Biodiesel is currently in use by some of the public transport system buses in several urban areas. The need for research and development is a critical component of assuring that the products of the livestock industry are not wasted.

THE RENDERING INDUSTRY

A focal point of by-products is the **rendering** industry, which recycles offal, fat, bone, meat scraps, and entire animal carcasses. The sources of these raw materials are packing and processing plants, butcher shops, restaurants, supermarkets, farmers, and ranchers. Some large packing and processing plants have their own rendering plants integrated with other operations.

Renderers have a regular pickup service that amounts to more than 70 million pounds of animal material daily. This amount has been declining in recent years because less fat is shipped out of packing plants. This pickup service is essential to public health, as it reduces a major garbage-disposal problem. Also, since animal by-products have value, consumers can buy meat at a lower price than would otherwise be possible.

Rendering of Red Meat Animal By-Products

Animal fat and animal protein are the primary products of the renderers' art. Originally, animal fats went almost entirely into soap and candles. Today, from the same basic material, renderers produce many grades of tallow and semiliquid fat. The major uses of rendered fat are in animal feeds, fatty-acid production, and soap manufacture.

Fatty acids are used in the manufacture of many products, such as plastic consumer items, cosmetics, lubricants, paints, deodorants, polishes, cleaners, caulking compounds, asphalt tile, printing inks, and others. The fatty-acid industry experienced tremendous growth since 1950.

The rendered animal proteins are processed into several high-protein (more than 50% protein) feed supplements, among which are meat and bone meals and blood meals. The supplements are more commonly fed to young monogastric animals—swine, poultry, and pet animals. These animals require a high-quality protein, particularly the amino acid lysine, which is typical of animal protein supplements.

Blood not processed into blood meal is used to produce products used in the pet food industry. Such products are blood protein (fresh, frozen whole blood) and blood cell protein (frozen, fresh, dewatered blood).

By-products enjoy a relatively strong export market, as indicated by Table 7.4. By-product sales grew in the 1990s, with fats, oils and greases, variety meats, and hides comprising the majority of sales. Gross sales have declined over the last several years. The major customers for U.S. exports of inedible tallow, edible tallow, lard, meat meal and tankage, and feather meal are listed in Table 7.5.

Rendering of Poultry By-Products

Within a few hours after poultry is slaughtered, the offal is at the rendering plant. There it is cooked in steam-jacketed tanks at temperatures sufficient to destroy all pathogenic organisms. After being cooked for several hours, the material is passed through presses

Table 7.4
VALUE OF BY-PRODUCT EXPORTS AND RENDERED ITEMS ($ MILLION)

	1995	2000	2005	2010	2013
Lard, rendered pig fat	29.0	41.6	33,2	30,432	24,837
Tallow, edible	61,204	35,601	62,621	66,678	72,037
Tallow, inedible	335,375	225,292	273,293	610,556	359,940
Beef variety meats	613,320	628,204	447,223	541,063	729,658
Pork variety meats	88,568	98,146	262,603	524,240	751,184
Cattle hides	1,224,803	1,156,687	1,415,091	1,314,845	1,459,656

Source: Adapted from USDA: FAS (2013).

Table 7.5
LEADING CUSTOMERS OF U.S. RENDERED PRODUCT EXPORTS—2013

Animal Fats		Meat Meal and Tankage		Feather Meal	
Country	$1,000	Country	$1,000	Country	$1,000
Mexico	353,571	Indonesia	53,729	Indonesia	37,711
Turkey	42,143	Canada	45,400	Chile	21,462
Canada	40,855	China	21,945	Canada	2,536

Source: USDA, FAS.

Figure 7.3
The export of value-added livestock by-products is economically significant and contributes to higher prices for producers.
Source: Fotos-v/Fotolia.

that remove most of the fat (poultry fat), while the remaining material becomes poultry by-product meal. While it is still hot, the poultry fat is piped to sterile tanks where appropriate antioxidants or "stabilizers" are added, and thence to tank cars or drums for shipment to feed manufacturers or for other uses as shown in Figure 7.3.

Poultry by-product meal (PBPM) is high in protein (approximately 58%) and contains 12–14% fat. PBPM consists of ground dry-rendered clean parts of the carcasses of slaughtered poultry, such as heads, feet, undeveloped eggs, and intestines, exclusive of feathers except in such trace amounts as might occur unavoidably in good factory practice. It should not contain more than 16% ash and not more than 4% acid ash.

DISPOSING OF DEAD LIVESTOCK

Unfortunately not all livestock survive to enter the market place. As such, careful attention should be given to proper disposal of dead animals to avoid cross-contamination of disease to humans or other livestock, to assure environmental quality, and to avoid

Table 7.6
COST ESTIMATES FOR PRIMARY PROTOCOLS OF LIVESTOCK MORTALITY REMOVAL ($ PER MORTALITY)

| Species | Rendering[a] | | Burial | Incineration | Composting |
	MBM[a]: For Feed	MBM[b]: Not for Feed			
Cattle/calves	8.25	24.11	10.63	9.33	30.34
Weaned hogs	7.00	11.53	12.45	4.09	14.04
Preweaned hogs	0.50	0.70	2.01	0.30	1.02
Other	7.00	9.61	1.51	0.29	0.98

[a]Assumes all mortalities are rendered with a charge of $10.00 per mature cattle and $7.00 per calf.
[b]Meat and bone meal.
Source: Adapted from National Renderers Association.

problems resulting from odor. The rendering industry plays a crucial role in this process by providing a safe and resource-efficient means to deal with the problem of dead animal disposal.

The generally approved protocols for disposal of dead livestock are

1. removal by a licensed rendering company,
2. compost the carcass,
3. burn the carcass in an incinerator (approved and permitted), or
4. lury the carcass at least 4 ft. deep.

Each of these options is suitable for certain circumstances, and all are associated with some level of cost (Table 7.6), with larger animals having higher costs of disposal. As state and federal policies are adopted in regards to this issue, arriving at workable solutions is critical from both the perspective of the profitable livestock enterprise owner and the maintenance of environmentally friendly approaches. Currently there are a myriad of state regulations, so producers are advised to contact their local health department or state department of agriculture for guidance in seeking the best solution.

Composting is becoming increasingly recognized as a viable solution in many situations. Composting is the process of accelerating decay processes. There are a variety of factors that affect the success of composting including moisture, type of co-composting material, carbon, oxygen, and nitrogen levels, and level of heat retention in the compost pile. Preferred moisture levels are 40–60% to assure rapid decay without the production of highly offensive odors. Wood chips, sawdust, corncobs, and poultry litter are popular co-composting materials that have proven to be effective. The preferred C:N ratio is 25 to 1 although a wider range is workable. Oxygen levels are most affordably maintained by avoiding overly wet compost, regular churning or turning of the compost pile, and the use of coarse co-composting ingredients. Compost temperatures need to reach the range of 120–150°F to assure optimal decay rates while assuring the destruction of pathogenic microorganisms.

CHAPTER SUMMARY

- Hides, fat, bones, and internal organs are the primary by-products, with hides usually having the highest value.
- Variety meats (liver, heart, tongue, tripe, etc.) are examples of edible by-products, while hides and other internal organs are inedible.
- Numerous inedible by-products produce useful human products—for example, pig heart valves (human heart valves), pig skin (human graft skin), hides and skins (leather, bone buttons, china, etc.), and many pharmaceuticals (e.g., corticosteroids, epinephrine, heparin, insulin, thyrotropin, etc.).
- Hides, fats, and variety meats are the by-products that comprise most of the total value of U.S. livestock exports.

KEY WORDS

edible by-products	hides
inedible by-products	skins
variety meats	sheep pelts
tripe	keds
sweetbread	rendering
lard	poultry by-product meal (PBPM)
tallow	composting

REVIEW QUESTIONS

1. How does by-product value impact the economics of livestock production?
2. Provide examples of edible and inedible by-products.
3. What are the highest valued inedible by-products?
4. What are uses of leather from different sources of hides and skins?
5. What is the volume and value of livestock by-product exports?
6. What is the role of the rendering industry?
7. What countries are the primary buyers of U.S. by-products?
8. Describe the approved protocols for disposal of dead livestock.

SELECTED REFERENCES

Field, T. G., J. Garcia, and J. Ahola. 1996. *Quantification of the Utilization of Edible and Inedible Beef By-products.* Englewood, CO: Colorado State University and The National Cattlemen's Beef Association.

Foreign Agricultural Service. 2009. *Market and Trade Data.* Washington DC: United States Department of Agriculture.

Glanville, T. 1999. *Composting Dead Livestock: A New Solution to an Old Problem.* Ames, IA: Iowa State University.

Kinsman, D. M. 1994. Animal by-products from slaughter. *Encyclopedia of Agricultural Science.* San Diego, CA: Academic Press.

Kirk-Othmer Encyclopedia of Chemical Technology. 2004. New York: John Wiley & Sons.

Lawrence, J. D. 1994. *An Economic Assessment of Trends in Pork By-Product Values.* Ames, IA: Iowa State University Swine Research Report.

North American Rendering. 2009. Alexandria, VA: National Renderers Association, Inc.

Romans, J. R., K. W. Jones, W. S. Costello, C. W. Carlson, and P. T. Ziegler. 1985. Packing house by-products. Chap. 10 in *The Meat We Eat.* Danville, IL: Interstate Printers and Publishers.

8

Market Classes and Grades of Livestock, Poultry, and Eggs

learning objectives

- Describe the market classes and grades of livestock
- Describe the USDA quality and yield grades of beef, pork, and lamb
- Describe the USDA grades for poultry and eggs

The annual production and movement of more than $150 billion worth of livestock, poultry, and their products to U.S. consumers is accomplished through a vast and complex marketing system. **Marketing** is the transformation and pricing of goods and services through which buyers and sellers move livestock and livestock products from the point of production to the point of consumption. Producers need to understand marketing if they are to produce products preferred by consumers, decide intelligently among various marketing alternatives, successfully negotiate prices, and create profitable management systems.

Market classes and grades have been established to segregate animals, carcasses, and products into uniform groups based on preferences of buyers and sellers. The USDA Meat Grading and Certification Branch established extensive classes and grades to simplify the marketing process and to facilitate communication between buyers and sellers. Use of USDA grades is voluntary. Some packers have their own private grades, though these are often used in combination with USDA grades. An understanding of market classes and grades helps producers recognize compositional and quality attributes of products they are supplying to consumers. **Meat inspection**, on the other hand, is administered by the Food Safety and Inspection Service of USDA and is a mandatory process focused on assuring the safety and wholesomeness of products entering the supply chain as well as accurate labeling of products. The inspection process is multifaceted with emphasis on **antemortem** or live animal evaluation prior to harvest, **postmortem** inspection of the carcass and associated tissues after harvest, and reinspection during further processing.

MARKET CLASSES AND GRADES OF RED MEAT ANIMALS

Slaughter Cattle

Slaughter cattle are separated into classes based primarily on age and sex. Age of the animal has a significant effect on tenderness, with younger animals typically producing more tender meat than older animals. Age classifications for meat from cattle are veal, calf, and beef. **Veal** is from young calves, 1–3 months of age, with carcasses weighing less than 150 lb. **Calf** is from animals ranging in age from 3 to 10 months with carcass

Table 8.1
OFFICIAL USDA GRADE STANDARDS FOR LIVE SLAUGHTER CATTLE AND THEIR CARCASSES

Class of Kind	Quality Grades (highest to lowest)	Yield Grades (highest to lowest)
Beef		
Steer and heifer	Prime, choice, select, standard, commercial, utility, cutter, canner	1, 2, 3, 4, 5
Cow	Choice, select, standard, commercial, utility, cutter, canner	1, 2, 3, 4, 5
Bullock	Prime, choice, select, standard, utility	1, 2, 3, 4, 5
Bull	(No designated quality grades)	1, 2, 3, 4, 5
Veal	Prime, choice, select, standard, utility	NA
Calf	Prime, choice, select, standard, utility	NA

Source: USDA.

weights between 150 and 300 lb. **Beef** comes from mature cattle (over 12 months of age) having carcass weights higher than 300 lb. Classes and grades established by the USDA for cattle are based on sex, quality grade, and yield grade, all of which are used in the classification of both live cattle and their carcasses (Table 8.1).

The sex classes for cattle are **heifer**, **cow**, **steer**, **bull**, and **bullock**. Occasionally, the sex class of **stag** is used by the livestock industry to refer to males that have been castrated after their secondary sex characteristics have developed. Sex classes separate cattle and carcasses into more uniform carcass weights, tenderness groups, and processing methods. **Quality grades** are intended to measure certain consumer palatability characteristics, and **yield grades** measure amounts of fat, lean, and bone in the carcass. Slaughter steers representing some of the eight quality grades and five yield grades are shown in Figures 8.1 and 8.2.

Quality grades are based primarily on two factors: (1) **maturity** (physiological age) of the carcass and (2) amount of **marbling**. Maturity is determined primarily by observing bone and cartilage structures. For example, soft, porous, red bones with a maximum amount of pearly white cartilage characterize A maturity, whereas very little cartilage and hard, flinty bones characterize C, D, and E maturities. As maturity in carcasses increases from A to E, the meat becomes less tender.

Marbling is intramuscular fat or flecks of fat within the lean, and it is evaluated at the exposed ribeye muscle between the 12th and 13th ribs. Ten degrees of marbling, ranging from abundant to devoid, are designated in the USDA marbling standards. Various combinations of marbling and maturity that identify the carcass quality grades are shown in combinations of marbling and maturity in Figure 8.3. Figure 8.4 shows ribeye sections representing several different quality grades and several different degrees of marbling. Note that C maturity starts at 42 months of age. Slaughter cows more than 42 months of age will grade commercial, utility, cutter, or canner regardless of amount of marbling.

Yield grades, sometimes referred to as **cutability grades**, measure the quantity of **boneless, closely trimmed retail cuts (BCTRC)** from the major wholesale cuts of beef (round, loin, rib, and chuck). BCTRC should not be confused with the total percentage of retail cuts (including hamburger) from a beef carcass. Calculated yield grades range from 1 to 5, with 1 denoting the highest percentage of BCTRC. Although marketing communications generally quote yield grades in whole numbers, these yield grades are often shown in tenths in research data and information provided in genetic evaluations.

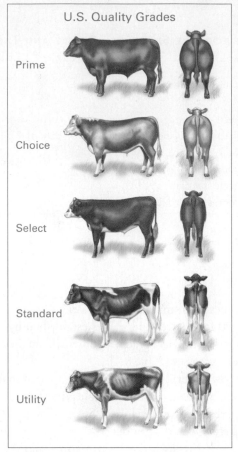

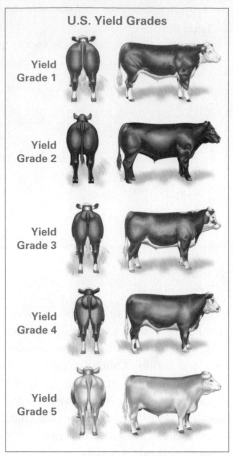

Figure 8.1
USDA quality grades (Commercial, Cutter, and canner omitted). Source: USDA.

Figure 8.2
USDA yield grades for market cattle. Source: USDA.

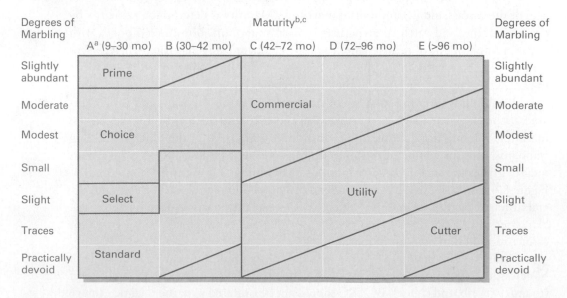

[a]Assumes that firmness of lean is comparably developed with the degree of marbling and that the carcass is not a "dark cutter."
[b]Maturity increases from left to right (A through E).
[c]The A maturity portion of the figure is the only portion applicable to bullock carcasses.

Figure 8.3
Relationship between marbling, maturity, and carcass quality grade. Source: USDA.

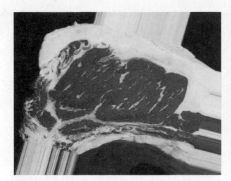

USDA Prime Typical – Moderately Abundant Marbling – A Maturity

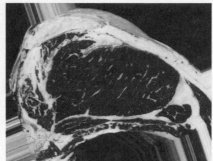

USDA Choice Typical – Modest Marbling – A Maturity

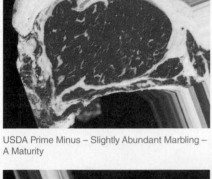

USDA Prime Minus – Slightly Abundant Marbling – A Maturity

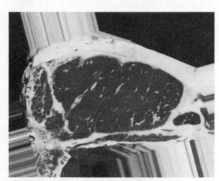

USDA Choice Plus – Moderate Marbling – A Maturity

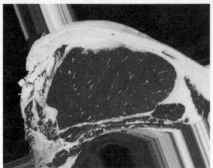

USDA Choice Minus – Small Marbling – A Maturity

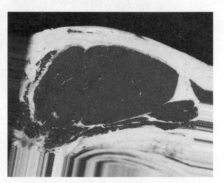

USDA Select – Slight Marbling – A Maturity

Figure 8.4
Exposed ribeye muscles (between the 12th and 13th ribs) showing various degrees of marbling associated with several beef carcass quality grades.
Source: USDA.

Table 8.2 shows the yield grades and their respective percentage of BCTRC. As an example, a carcass with a yield grade of 3.0 has a BCTRC of 50%; a 700-lb carcass with this yield grade would produce 350 lb of boneless, closely trimmed retail cuts from the round, loin, rib, and chuck.

Yield grades are determined from the following four carcass characteristics: (1) amount of fat measured in tenths of an inch over the **ribeye** muscle, also known as the **longissimus dorsi** (Fig. 8.5); (2) kidney, pelvic, and heart fat (usually estimated as percent of carcass weight); (3) area of the ribeye muscle, measured in square inches (Fig. 8.6); and (4) hot carcass weight. The last measurement reflects amount of intermuscular fat. As carcass weights increase, the amount of fat between muscles tends to increase as well. Thus carcass weight serves as an indicator of the **intermuscular** fat depot.

Measures of fatness in beef carcasses have the greatest effect in determining yield grade. Preliminary yield grade (PYG) is determined by estimating or measuring the outside fat of the ribeye muscle. Figure 8.7 shows the five yield grades with varying amounts of fat over the ribeye muscle and the area of the ribeye.

Table 8.2

BEEF CARCASS YIELD GRADES AND THE YIELD OF BCTRC (BONELESS, CLOSELY TRIMMED RETAIL CUTS) FROM THE ROUND, LOIN, RIB, AND CHUCK

Yield Grade	BCTRC (%)	Yield Grade	BCTRC (%)	Yield Grade	BCTRC (%)
1.0	54.6	2.8	50.5	4.6	46.4
1.2	54.2	3.0	50.0	4.8	45.9
1.4	53.7	3.2	49.6	5.0	45.4
1.6	53.3	3.4	49.1	5.2	45.0
1.8	52.8	3.6	48.7	5.4	44.5
2.0	52.3	3.8	48.2	5.6	44.1
2.2	51.9	4.0	47.7	5.8	43.6
2.4	51.4	4.2	47.3	—	—
2.6	51.0	4.4	46.8	—	—

Source: Adapted from USDA.

Figure 8.5
Location of the fat measurement over the ribeye (longissimus dorsi) muscle. Source: Colorado State University.

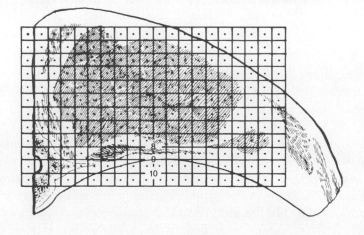

¾ the length of the longissimus dorsi muscle

Fat thickness measurement

Figure 8.6
Plastic grid is placed over the ribeye muscle to measure the area. Each square represents 0.1 in. Source: Colorado State University.

Quality grading and yield grading of beef carcasses by packers are voluntary. Approximately 93% of the 25 billion pounds of federally inspected carcass beef produced in 2012 was quality-graded and nearly 100% was yield-graded. Most of the graded beef originates from market steers and heifers fed in feedlots. The distribution of carcass quality and yield grades from fed beef is shown in Figure 8.8. Imaging

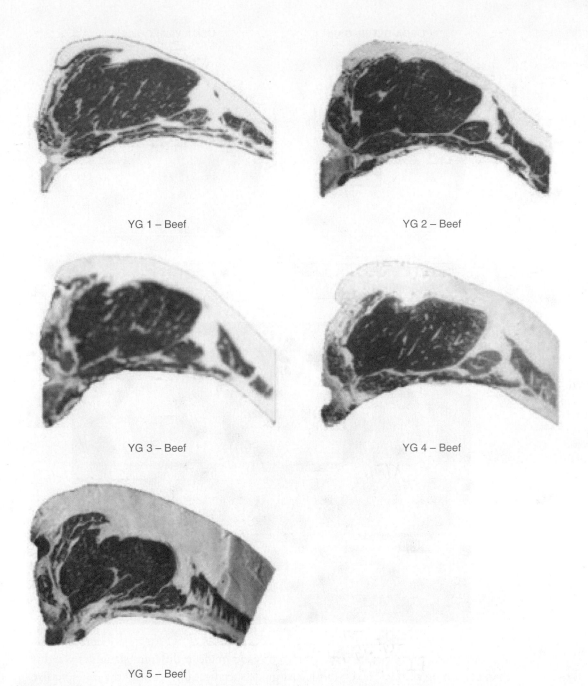

YG 1 – Beef

YG 2 – Beef

YG 3 – Beef

YG 4 – Beef

YG 5 – Beef

Figure 8.7
The five yield grades of beef shown at the juncture of the 12th and 13th ribs. Source: USDA.

technology is being incorporated to enhance the accuracy of traditional grading systems (Fig 8.9).

Feeder Cattle

The revised 2000 USDA **feeder grades** for cattle are intended to predict feedlot weight gain and the slaughter weight end point of cattle fed to a desirable fat-to-lean composition. The two criteria used to determine feeder grade are **frame size** and **thickness**. The four thickness descriptions are shown in Figure 8.10. Some examples of feeder cattle grade terminology are "large no. 1," "medium no. 2," and "small no. 2." Feeder cattle are given a USDA grade of "inferior" if the cattle are unhealthy or double-muscled.

Although frame size and ability to gain weight in the feedlot are apparently related in the sense that large-framed cattle usually gain fastest, frame size appears to be a more

Figure 8.8
Distribution of quality grades and yield grades of fed beef. Source: Adapted from USDA.

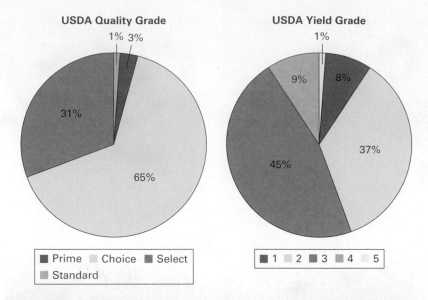

Figure 8.9
Scanning technology is being utilized to augment quality grade determination and to assign USDA yield grades. The advantage of these systems is that variation due to human grading error is limited and thus more accurate assessments are made. Source: USDA.

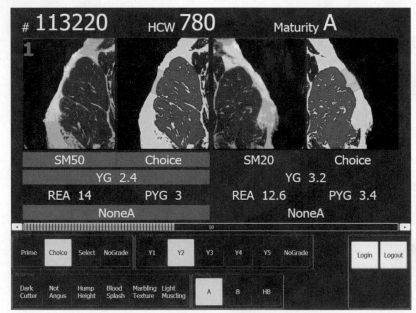

accurate predictor of carcass composition or yield grade at different slaughter weights than of gaining ability. The USDA feeder grade specifications identify the different live weights from the three frame sizes when they reach the choice grade (Table 8.3).

Slaughter Swine

Sex classes of swine are **barrow**, **gilt**, **sow**, **boar**, and stag. Boars and sows are older breeding animals, whereas gilts are younger females that have not produced any young. The barrow is the male pig castrated early in life, and the stag is the male pig castrated after having attained puberty. Pork from stags often has a strong, undesirable flavor referred to as **boar-taint**. Because of relationships between sex and sex condition and the acceptability of prepared meats to the consumer, separate grade standards have been developed for barrow and gilt carcasses and for sow carcasses. There are no official grade standards for boar and stag carcasses.

The traditional grades for barrow and gilt carcasses were based on two general criteria: (1) quality characteristics of the lean and (2) expected combined yields of the four lean cuts (ham, loin, blade [Boston] shoulder, and picnic shoulder). There are only two quality grades for lean in a pork carcass: acceptable and unacceptable.

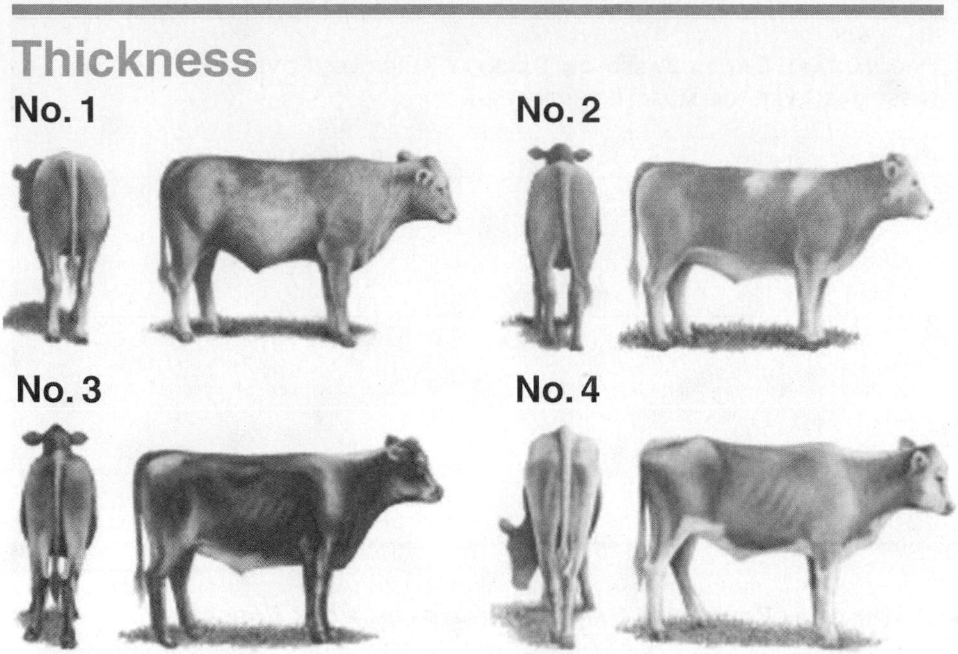

Thickness

No. 1

No. 2

No. 3

No. 4

Figure 8.10
The four thickness standards of the USDA feeder cattle grade system. Source: USDA.

Table 8.3

ESTIMATED SLAUGHTER WEIGHTS OF LARGE-, MEDIUM-, AND SMALL-FRAME SLAUGHTER CATTLE WHEN THEY REACH THE CHOICE GRADE

	Slaughter Weight	
Frame Size	Steers (lb)	Heifers (lb)
Large	>1,350	>1,250
Medium	1,200–1,350	1,100–1,250
Small	<1,200	<1,100

Table 8.4

EXPECTED PORK YIELDS OF THE FOUR LEAN CUTS, BASED ON PERCENT OF CHILLED CARCASS WEIGHT

Grade	Four Lean Cuts (%)
U.S. no 1	>60.4
U.S. no 2	57.4–60.3
U.S. no 3	54.4–57.3
U.S. no. 4	<54.4

Source: Adapted from USDA.

Observing the exposed surface of a cut muscle, usually between the 10th and 11th ribs, assesses quality of the lean. Acceptable lean is gray-pink in color, has fine muscle fibers, and has fine marbling. Carcasses that have unacceptable lean quality (too dark or too pale, soft, or watery) or bellies too thin for suitable bacon production are graded U.S. utility, as they are soft and oily. Carcasses with acceptable lean quality are graded U.S. no. 1, U.S. no. 2, U.S. no. 3, or U.S. no. 4. These grades are based on expected yields of the four lean cuts, as shown in Table 8.4. Yield differences of the four wholesale cuts exist because of variation in the amount of muscling and fatness

Table 8.5

PRELIMINARY GRADE BASED ON BACKFAT THICKNESS OVER THE LAST RIB (ASSUMES AVERAGE MUSCLE THICKNESS)

Preliminary Grade[a]	Backfat Thickness (in.)
U.S. no. 1	<1.00
U.S. no 2	1.00–1.24
U.S. no 3	1.25–1.49
U.S. no 4	>1.50[b]

[a]Swine with thick muscling qualify for next higher grade; those with thin muscling are downgraded to next lower grade.
[b]Animals with an estimated last-rib backfat thickness of 1.75 in. or over have to be U.S. no. 4 and cannot be graded U.S. no. 3, even with thick muscling.
Source: USDA.

Table 8.6

PREDICTION EQUATIONS FOR ESTIMATING POUNDS OF FAT-FREE LEAN FROM PORK CARCASSES

Fat Measurement Protocol	Equation for lb of Fat-Free Lean
Stainless steel ruler to measure last-rib backfat	$23.568 - (21.348 \times \text{last-rib backfat}) + (0.503 \times \text{hot carcass weight})$[a]
Fat-O-Meter optical grading probe to measure fat and muscle depth between the 3rd and 4th from the last rib	$15.31 - (31.277 \times \text{fat depth}) + (3.813 \times \text{loin muscle depth}) + (0.51 \times \text{hot carcass weight})$[a]
Ultrasound measurement of the average fat and muscle depth spanning the last rib to the 10th rib	$6.783 - (15.745 \times \text{avg. fat depth}) + (4.007 \times \text{avg. loin muscle depth}) + (0.47 \times \text{hot carcass weight})$[a]

[a]To convert to a percent fat-free lean, divide by hot carcass weight and multiply by 100.
Source: Adapted from USDA and American Meat Science Association.

in relation to skeletal size. Measurements of backfat and estimates of muscling are the factors used to determine numerical yield grade (Table 8.5). There are three muscle scores—thick, average, and thin.

Currently, pork packers are providing information that describes the pounds of lean produced per carcass (Table 8.6).

Assuming a 190-lb carcass with 0.8 in. of back fat measured by handheld ruler, then the percent lean would be calculated as: *23.568 − (21.348 × 0.8) + (0.503 × 190) = 102.06 lb of lean ÷ 190 lb carcass = 53.72% lean.*

The Fat-O-Meter is an electronic scanning technology that objectively measures fat thickness and loin eye depth with a line speed of 1,300 carcasses per hour. The development of electronic scanning technologies for application in the beef, pork, and lamb grading processes is a major focus and will likely replace the current yield estimation systems. Assuming the use of the Fat-O-Meter on a 190-lb carcass with an average fat depth of 0.9 and a loin depth of 2.4, the percent lean would be calculated as follows: *15.31 − (31.277 × 9) + (3.813 × 2.4) + (0.51 × 190) = 93.21 lb of lean ÷ 190 lb carcass = 49.06% lean.*

Slaughter Sheep

Slaughter sheep are classified by their sex and maturity. Live sheep and their carcasses are also graded for quality grades and yield grades (Table 8.7).

Table 8.7
USDA MATURITY GROUPS, SEX CLASSES, AND GRADES OF SLAUGHTER SHEEP

Maturity Group	Sex Class	Quality Grade (highest to lowest)	Yield Grade (highest to lowest)
Lamb	Ewe, wether, or ram	Prime, choice, good, utility	1, 2, 3, 4, 5
Yearling mutton	Ewe, wether, or ram	Prime, choice, good, utility	1, 2, 3, 4, 5
Mutton	Ewe, wether, or ram	Choice, good, utility, cull	1, 2, 3, 4, 5

Source: USDA.

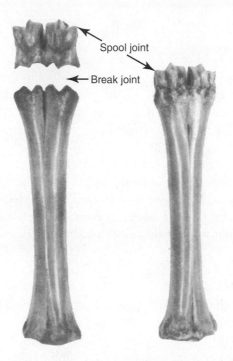

Spool joint

Break joint

Figure 8.11
Break and spool joints. The cannon bone on the left exhibits the typical break joint in which the foot and pastern are removed at the cartilaginous junction. In the cannon bone on the right, the cartilaginous junction has ossified, making it necessary for the foot and pastern to be removed at the spool joint. Source: USDA.

Lamb carcasses ranging from approximately 2 to 14 months of age always have the characteristic break joint on one of their shanks following the removal of their front legs (Fig. 8.11). The **break joint** is the growth plate of the long bone that has not yet ossified. **Mutton** carcasses are distinguished from lamb carcasses by the appearance of the spool joint instead of the break joint. The break joint ossifies as the sheep matures. Yearling mutton carcasses ranging from 12 to 15 months of age usually have the spool joint present but may occasionally have a break joint. Yearling mutton is also distinguished from lamb and mutton by color of the lean (intermediate between the pinkish red of lamb and the dark red of mutton) and shape of rib bones. Most U.S. consumers prefer lamb to mutton because it has a milder flavor and has superior tenderness.

Quality grades are determined from a composite evaluation of conformation, maturity, and **flank streaking**. There are also minimum standards for **flank firmness and fullness**; however, most lambs meet these standards. Conformation is an assessment of overall thickness of muscling in the lamb carcass. Maturity of lamb carcasses is determined by bone color and shape and muscle color. Flank streaking (streaks of fat within the flank muscle) predicts marbling because lamb carcasses are not usually ribbed to expose marbling in the ribeye muscle. Flank firmness and fullness are

Table 8.8
LAMB CARCASS YIELD GRADES AND PERCENTAGE OF RETAIL CUTS

Yield Grade	Percent Retail Cuts	Yield Grade	Percent Retail Cuts
1.0	49.0	3.5	44.6
1.5	48.2	4.0	43.6
2.0	47.2	4.5	42.8
2.5	46.3	5.0	41.8
3.0	45.4	5.5	41.0

Source: USDA.

determined by taking hold of the flank muscle with the hand. Most lamb carcasses grade either prime or choice, with few carcasses grading in the lower grades.

Lamb carcasses must be both quality-graded and yield-graded. Yield grades estimate boneless, closely trimmed retail cuts from the leg, loin, rack, and shoulder. The approximate percentage of retail cuts for selected yield grades are shown in Table 8.8.

Yield grades are determined by fat thickness over the loin eye muscle. Figure 8.12 shows cross-sections of lamb carcasses representing the five yield grades and the amount of fat thickness over the loin eye for each yield grade. Multiplying the adjusted fat thickness by 10 and adding 0.4 estimates the yield grade. For example, a lamb with 0.25 in. of fat would have an estimated yield grade of 2.9 (e.g., [0.25 × 10] + 0.4 = 2.9).

Feeder Lamb Grades

Lambs weighing less than 100 lb at weaning are considered feeder lambs. They require additional feeding to produce a more desirable carcass.

There are no official USDA grades for feeder lambs. Some promising research has been done to evaluate some possible feeder lamb grades. These grades, based on frame size, project the slaughter weight of the lambs when they would reach 0.25 in. of fat thickness over the loin eye muscle. The frame sizes and slaughter weights are: *large* (over 120 lb), *medium* (100 lb–120 lb), and *small* (less than 100 lb). There are also body-thickness categories based on conformation of slaughter lambs: no. 1 (prime conformation), no. 2 (choice conformation), and no. 3 (less-than-choice conformation).

Goat Grades

USDA has not created a uniform set of grading standards for live goats or carcasses. They have adopted voluntary standards for live meat goats and Meat Purchase Specifications for Fresh Goat. The live standards are categorized as Selection #1 (thickly muscled), Selection #2 (moderately muscled), and Selection #3 (poor muscle with inferior meat type conformation).

Goat carcasses are categorized by the same muscle scoring system used in live goats as well as an estimate of lean color and of fat cover. Lean color ranges from light pink (A) to dark red (C) with light pink considered more desirable by consumers. A score of B indicates an intermediate color. Furthermore, scores of 1 (least fat), 2 or 3 (most fat) are assigned based on the degree of subcutaneous fat cover. Fat deposition in goats tends to be highest just behind the shoulder extending toward the breast with very little cover over the back and rear flank.

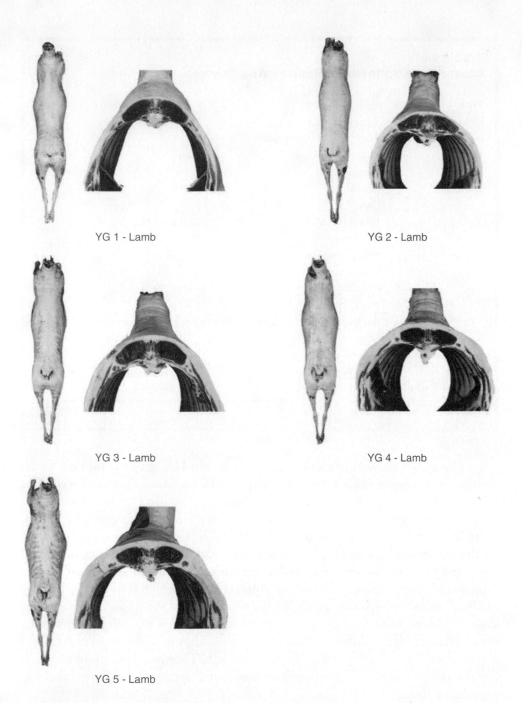

Figure 8.12
*The five yield grades
of lamb showing the
progressive increases in the
amount of fat.* Source: USDA.

YG 1 - Lamb

YG 2 - Lamb

YG 3 - Lamb

YG 4 - Lamb

YG 5 - Lamb

MARKET CLASSES AND GRADES OF POULTRY PRODUCTS

Poultry Meat

The class of poultry must be displayed on the package label or a tag on the wing of the bird. The age class indicates tenderness, as meat from younger birds is more tender than that from older birds. Table 8.9 shows the age-classification labels for poultry meat.

Broiling, frying, roasting, or barbecuing usually requires more tender meat from the young age classes. The mature, less tender poultry meat is best prepared by baking, stewing, or fabricating or by including it in other prepared dishes.

Table 8.9
LABELS USED TO IDENTIFY POULTRY AGE CLASSES

Type	Young	Mature
Chicken	Young chicken	Mature/old chicken
	Rock Cornish game hen	
	Broiler	Hen
	Fryer	Stewing or baking chicken
	Roaster	Fowl
	Capon	
Turkey	Young turkey	Mature/old turkey
	Fryer-roaster	Yearling turkey
	Young hen	
	Young tom	
Duck	Duckling	Mature/old
	Young duckling	
	Broiler duckling	
	Fryer duckling	
	Roaster duckling	
Goose	Young goose	Mature/old goose
Guinea	Young guinea	Mature/old guinea

Source: USDA.

The grades are U.S. grade A, U.S. grade B, and U.S. grade C for each of the classes, with A being the highest grade (Fig. 8.13). Carcasses of A quality are free of deformities that detract from their appearance or that affect normal distribution of flesh. They have a well-developed covering of flesh and a well-developed layer of fat in the skin. They are free of pinfeathers and diminutive feathers, exposed flesh on the breast and legs, and broken bones. They have no more than one disjointed bone, and they are practically free of discolorations of the skin and flesh and defects resulting from handling, freezing, or storage. Carcasses of B quality may have moderate deformities. They have a moderate covering of flesh, sufficient fat in the skin to prevent a distinct appearance of the flesh through the skin, and no more than an occasional pinfeather or diminutive feather. They may have moderate areas of exposed flesh and discoloration of the skin and flesh. They may have disjointed parts but no broken bones, and they may have moderate defects resulting from handling, freezing, or storage.

Eggs

The grading of shell eggs involves classifying individual eggs according to established standards. Eggs are graded by sorting them into groups, each group having similar weight and quality characteristics. Table 8.10 shows how eggs are classified according to a size-and-weight relationship.

The USDA quality standards used to grade individual shell eggs are as follows:

Exterior Quality Factors	*Interior Quality Factors*
Cleanliness of shell	Albumen thickness
Soundness of shell	Condition of yolk
(cracks, and texture)	Size and condition of air cell
Shape	Abnormalities (e.g., blood spots, meat spots)

Figure 8.13
Classes and grades of ready-to-cook poultry.
Source: USDA.

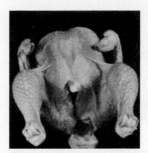

Broiler – A Quality

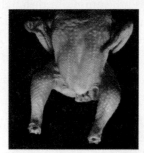
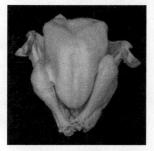

Broiler – B Quality

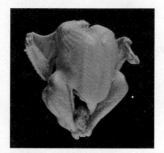

Turkey – A Quality

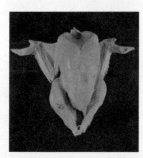

Turkey – B Quality

Table 8.10
CONSUMER WEIGHT CLASSES OF EGGS, MINIMUM NET WEIGHT PER DOZEN

Size	Ounces (per doz.)
Jumbo	30
Extra large	27
Large	24
Medium	21
Small	18
Peewee	15

Source: USDA.

Exterior quality factors are apparent from external observation; **interior quality factors** involve an assessment of egg content. The latter is accomplished through a process called **candling** (visually appraising the eggs while light is shone through them).

Although shell color is not a factor in the U.S. standards and grades, eggs are usually sorted by color and sold as either "whites" or "browns." Eggs sell better when sorted by color and packed separately. Contrary to popular opinion, there are no differences other than color between similar-quality brown- and white-shelled eggs.

The U.S. standards for quality of individual shell eggs are applicable to eggs from domestic chickens only; these standards are summarized in Table 8.11. Consumer egg grades are U.S. grade AA, U.S. grade A, and U.S. grade B. Figure 8.14 shows selected shields for communicating the quality grade.

Table 8.11
SUMMARY OF U.S. STANDARDS FOR QUALITY OF INDIVIDUAL SHELL EGG

Quality Factor	AA Quality	A Quality	B Quality	Dirty	Check
Shell	Clean, unbroken; practically normal	Clean, unbroken; practically normal	Clean to slightly stained,[a] unbroken; abnormal	Unbroken; adhering dirt or foreign material, prominent stains, moderate stained areas in excess of B quality	Broken or cracked shell but membranes intact, not leaking[c]
Air cell	$\frac{1}{8}$ in. or less in depth; unlimited movement and free or bubbly	$\frac{3}{16}$ in. or less in depth; unlimited movement and free or bubbly	Over $\frac{3}{16}$ in. in depth; unlimited movement and free or bubbly		
White	Clear, firm	Clear, reasonably firm	Weak and watery; small blood and meat spots present[b]		
Yolk	Outline slightly defined; practically free from defects	Outline fairly well defined; practically free from defects	Outline plainly visible; enlarged and flattened; clearly visible germ development, but no blood; other serious defects		

[a]Moderately stained areas permitted ($\frac{1}{32}$ of surface if localized, or $\frac{1}{16}$ if scattered).
[b]If they are small (aggregating not more than $\frac{1}{8}$ in. in diameter).
[c]Leaker has broken or cracked shell and membranes, and contents are leaking or free to leak.
Source: USDA, Egg Grading Standards.

Figure 8.14
USDA grade shields showing two egg quality grades. Source: USDA.

CHAPTER SUMMARY

- Market classes and grades have been established to segregate animals, carcasses, and products into uniform groups based on preferences of buyers and sellers. However, movement toward branded value-added products reduces the need for commodity grading programs.

- The United States Department of Agriculture (USDA) classes and grades make the marketing process simpler and more easily communicated.

- Red meat animals are separated into classes based primarily on age and sex. Age of the animal significantly affects tenderness. Sex class separates animals and carcasses into more uniform carcass weights, tenderness groups, and processing methods.

- Feeder and slaughter grades exist for most red meat animals. For slaughter animals, quality grades are used to measure consumer palatability characteristics, while yield grades measure the amount of fat, lean, and bone in the carcass.

- Poultry meat classes are based on age (affects tenderness), and the grades evaluate thickness and fullness of meat in the breast and thigh.

- Eggs are graded for size (weight), shell condition, and interior quality (yolk, white, and air cell).

KEY WORDS

marketing
market classes and grades
meat inspection
antemortem
postmortem
veal
calf
beef
heifer
cow
steer
bull
bullock
stag
quality grades
yield grades
maturity
marbling
cutability grades

boneless, closely trimmed retail cuts (BCTRC)
ribeye
longissimus dorsi
intermuscular
feeder grades
frame size
thickness
barrow
gilt
sow
boar
boar-taint
break joint
mutton
flank streaking
flank firmness and fullness
exterior quality factors
interior quality factors
candling

REVIEW QUESTIONS

1. Define marketing and describe the role of market classification systems and grade standards.
2. Compare and contrast USDA grading and inspection.
3. Discuss the age and sex classes of livestock.
4. Compare and contrast quality and yield grading.
5. Describe the quality and yield grades of the primary livestock species and the factors used to determine them.
6. How does grade affect pricing of carcasses and live animals?
7. What is the role of a live animal grading system?
8. Describe poultry age, classes, and grades.
9. Describe the interior and exterior characteristics used in egg grading.
10. Describe the egg size classes and associated weight per dozen standards.

SELECTED REFERENCES

Boggs, D. L., R. A. Merkel, M. E. Doumit, and K. Bruns. 2006. *Livestock and Carcass: An Integrated Approach to Evaluation, Grading and Selection.* Dubuque, IA: Kendall/Hunt.

Meat Buyer's Guide. 2006. Chicago, IL: North American Meat Processor's Association.

Meat Evaluation Handbook. 2001. Chicago, IL: American Meat Science Association.

USDA. *Egg Grading Manual.* USDA Agriculture Handbook no. 75.

USDA. *Facts about U.S. Standards for Grades of Feeder Cattle.* USDA Agricultural Marketing Service AMS-586.

USDA. *Official United States Standards for Grades of Feeder Pigs; Grades of Slaughter Swine; Grades of Pork Carcasses; Grades of Veal and Calf Carcasses; Grades of Lamb, Yearling Mutton and Mutton Carcasses; Grades of Carcass Beef and Slaughter Cattle;* and *Grades of Poultry.* 2006. USDA: Agricultural Marketing Service.

Visual Evaluation of Market Animals

Most red meat animals ready for harvest, if purchased live, are evaluated based on visual appraisal of their apparent carcass merit. Combining meaningful performance records and effective visual appraisal best identifies productivity of breeding and slaughter meat animals. Opinions regarding the relationship of form and function in red meat animals (cattle, sheep, and swine) differ widely among producers. Opinions range from a nearly complete dependence on visual appraisal to determine the relative merit of breeding, feeding, and finished animals to those who are solely focused on objective quantitative measures in making selection and marketing decisions. Historically, livestock producers have focused attention on defining the so-called ideal animal prototype that expresses value-determining characteristics. Determining, producing, and replicating the ideal has fascinated livestock breeders and producers for centuries. However, it is important to separate true relationships from opinion in the area of animal form and function.

Type is defined as an ideal or standard of perfection combining all the characteristics that contribute to an animal's usefulness for a specific purpose. **Conformation** implies the same general meaning as *type* and refers to form and shape of the animal. Both *type* and *conformation* describe an animal according to its external form and shape that can be evaluated visually or measured more objectively with a tape, ultrasound machine, or other device.

Red meat animals have three productive stages: (1) breeding (reproduction), (2) feeder (growth), and (3) slaughter (carcass or product). It has been well demonstrated that performance records are much more effective than visual appraisal in improving performance in terms of reproduction and growth. Keen observation allows for the early identification of problems stemming from injury, disease, poor body condition, physical defects, and other factors that can be assessed visually. Visual appraisal of slaughter red meat animals can be used effectively to predict carcass composition (fat and muscling), primarily when relatively large differences exist.

An important training experience for people desiring to work in the livestock industry is to participate in judging teams. These programs have the potential to sharpen observational powers that improve management, enhance selection of breeding stock that are free from significant physical defects, better understand compositional differences to assist in making marketing decisions, and to heighten the ability of people to make defensible and logical choices.

learning objectives

- Identify the external parts of cattle, swine, and sheep
- Describe the conformational characteristics of market cattle
- Describe the location of wholesale cuts on live cattle, hogs, and sheep
- Describe the role of visual appraisal in assessing composition

EXTERNAL BODY PARTS

Effective communication in many phases of the livestock industry requires knowledge of the external body parts of the animal. The locations of the major body parts for swine, cattle, and sheep are shown in Figures 9.1, 9.2, and 9.3, respectively.

LOCATION OF THE WHOLESALE CUTS IN THE LIVE ANIMAL

The next step for effective visual appraisal, after becoming familiar with the external parts of the animal, is to understand where major meat cuts are located in the live animal. Wholesale and retail cuts of the carcasses of beef, sheep, and swine are identified in Chapter 3. A carcass is evaluated after the animal has been slaughtered, eviscerated, and, in the case of beef and swine, split into two halves. The lamb carcass remains as a whole carcass. Figure 9.4, which shows the relationship of the wholesale cuts to the live animal, assists in correlating carcass assessment to live-animal evaluation.

Figure 9.1
The external parts of swine.
Source: Tom Field.

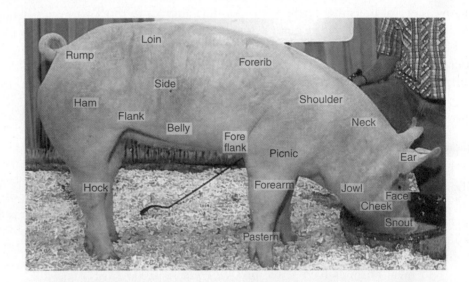

Figure 9.2
The external parts of cattle.
Source: Tom Field.

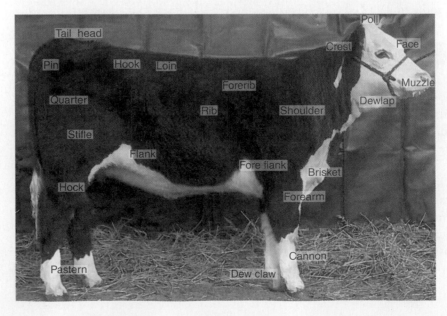

Figure 9.3
The external parts of sheep.
Source: Tom Field.

VISUAL ASSESSMENT OF LIVESTOCK CARCASS COMPOSITION

The carcass is composed of fat, lean (red meat), and bone. The meat industry's goal is to produce relatively large amounts of highly palatable lean and minimal amounts of fat and bone. These composition differences are reflected in the yield grades of beef and lamb and the lean percentage of swine discussed in Chapter 8. Effective visual appraisal of carcass composition requires knowledge of fat deposition, muscle mass growth, and the appropriate anatomical sites for effective determination of compositional differences.

The conformation characteristics of two steers, which are primarily influenced by fat deposits and some muscling differences, are contrasted in Figures 9.5 and 9.6. The conformation of these two steers is also contrasted to the conformation of a slaughter steer that is underfinished and thinly muscled (Fig. 9.7). All body parts referred to in these figures can be identified in Figures 9.2 and 9.4, with the exception of the twist—the distance from the top of the tail to where the hind legs separate as observed from a rear view.

Innovative demonstration techniques can help illustrate compositional differences. Figures 9.8 to 9.14 contain images from livestock that were harvested and then frozen in a standing position. This approach allowed for the removal of different tissue layers to illustrate compositional variation and to create cross-sections that demonstrate lean and fat variation at unique anatomical locations on the animals.

Figures 9.8 and 9.9 show the fat and lean composition at several cross-sections of a yield grade 2 steer and a yield grade 5 steer. The ribbons on the frozen carcasses show the location of the cross-sections. Cross-section 3 removes the bulge from the round, cross-section 4 is in front of the hip bone down through the flank, cross-section 5 is at the 12th and 13th rib, and cross-section 6 is at the point of the shoulder down through the brisket. Note the contrasts in percentages of fat (18% versus 44%) and lean (66% versus 43%). Both steers graded choice and were slaughtered at approximately 1,100 lb.

Figure 9.4

Location of the wholesale cuts on the live steer, pig, and lamb. (A) live steer, (B) pig, and (C) lamb.

Source: Figure by Kevin Pond from Colorado State University. Used by permission of Kevin Pond.

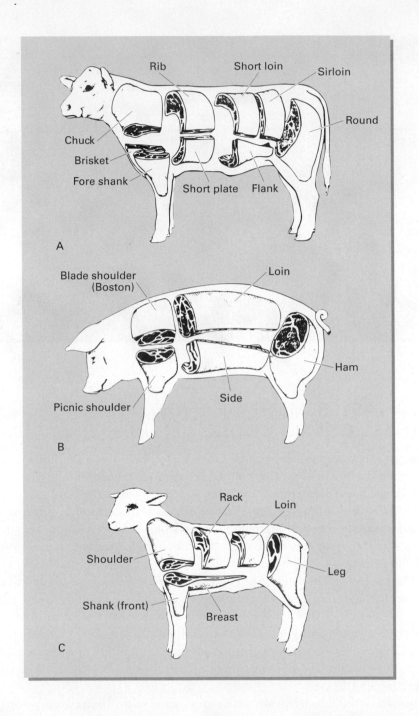

Figures 9.10 and 9.11 illustrate the cross sections of a yield grade 1 lamb and a yield grade 5 lamb. Cross-section 4 is cut through the shoulder area, cross section 5 is at the back, and cross-section 6 is through the leg area.

Figures 9.12 and 9.13 show the cross-section of a U.S. no. 1 hog and a U.S. no. 4 hog. Percent lean cuts are used in pork carcass evaluation instead of yield grades. Although the terminology is different, both systems measure fat-to-lean composition. Percent lean cuts are measured by obtaining the total weight of the trimmed Boston shoulder, picnic shoulder, loin, and ham (Fig. 9.4), and then dividing the total by carcass weight. The lean cuts percentage for the U.S. no. 1 would be 60% or more, as contrasted with less than 54% for the U.S. no. 4. The ribbon on the frozen carcass marks the location of the cross-sections. Cross-section 4

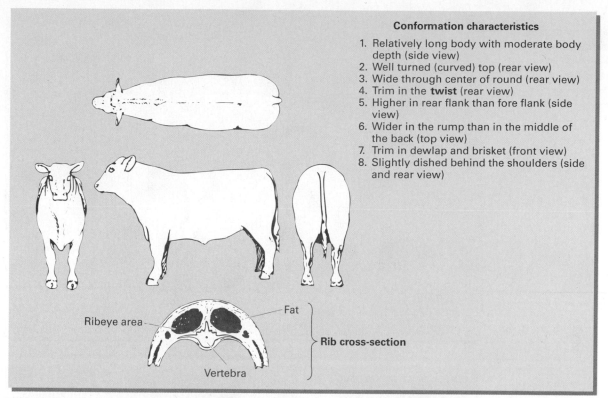

Conformation characteristics

1. Relatively long body with moderate body depth (side view)
2. Well turned (curved) top (rear view)
3. Wide through center of round (rear view)
4. Trim in the **twist** (rear view)
5. Higher in rear flank than fore flank (side view)
6. Wider in the rump than in the middle of the back (top view)
7. Trim in dewlap and brisket (front view)
8. Slightly dished behind the shoulders (side and rear view)

Figure 9.5

Conformation characteristics of slaughter steer typical of yield grade 2. Compare with Figure 9.8. Source: Figure by Kevin Pond from Colorado State University. Used by permission of Kevin Pond.

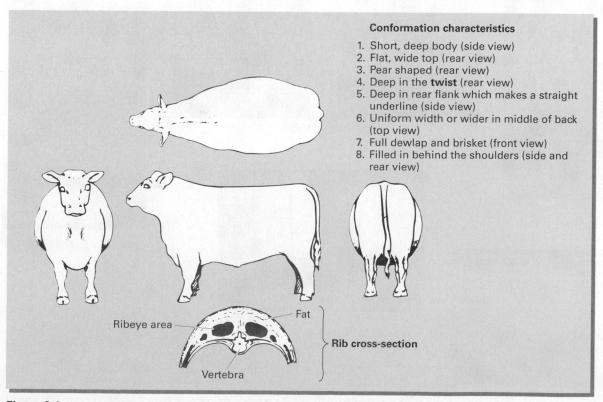

Conformation characteristics

1. Short, deep body (side view)
2. Flat, wide top (rear view)
3. Pear shaped (rear view)
4. Deep in the **twist** (rear view)
5. Deep in rear flank which makes a straight underline (side view)
6. Uniform width or wider in middle of back (top view)
7. Full dewlap and brisket (front view)
8. Filled in behind the shoulders (side and rear view)

Figure 9.6

Conformation characteristics of slaughter steer typical of yield grade 5. Compare with Figure 9.9. Source: Figure by Kevin Pond from Colorado State University. Used by permission of Kevin Pond.

Figure 9.7

Conformation characteristics of slaughter steer that is underfinished and thinly muscled. Source: Figure by Kevin Pond from Colorado State University. Used by permission of Kevin Pond.

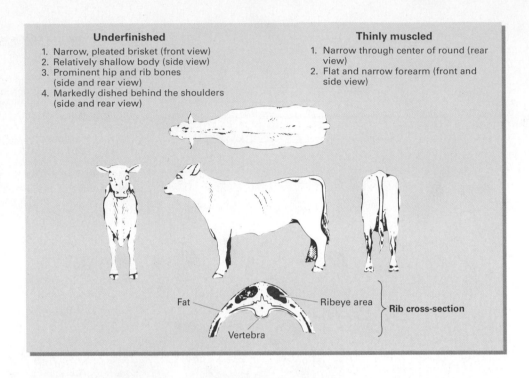

Underfinished

1. Narrow, pleated brisket (front view)
2. Relatively shallow body (side view)
3. Prominent hip and rib bones (side and rear view)
4. Markedly dished behind the shoulders (side and rear view)

Thinly muscled

1. Narrow through center of round (rear view)
2. Flat and narrow forearm (front and side view)

Fat — Ribeye area — } **Rib cross-section**

Vertebra

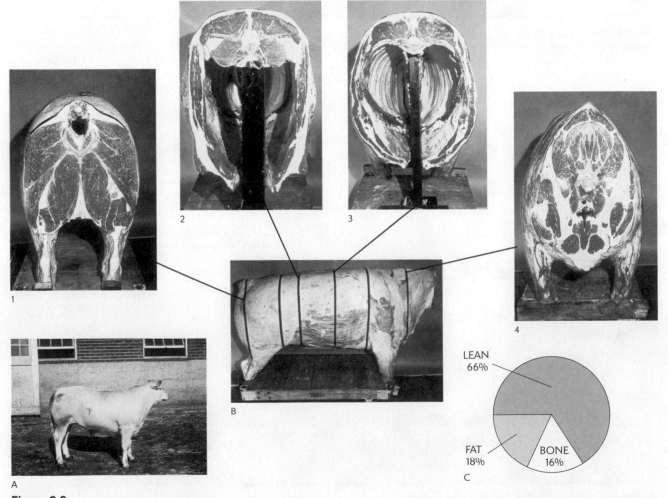

LEAN 66%

FAT 18%   BONE 16%

Figure 9.8

Composition of a yield grade 2 steer. A. Side view of live cattle. B. Black ribbons delineate cross-sections from the rump, hip, mid-carcass, and shoulder. C. Percentages of lean, fat and bone from this steer. Source: Iowa State University.

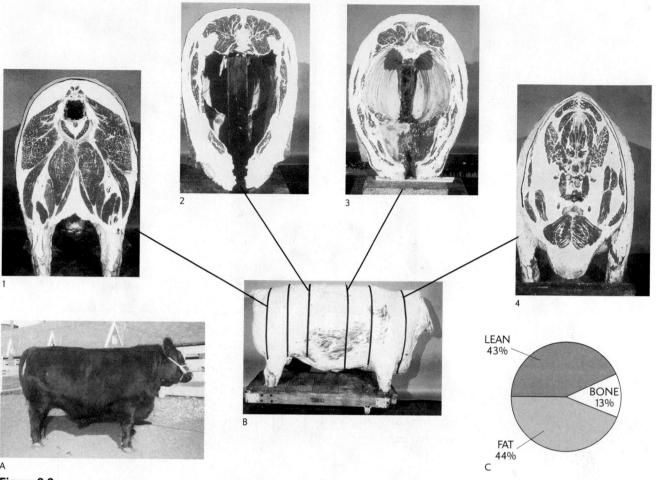

Figure 9.9

Composition of a yield grade 5 steer. A. Side view of live cattle. B. Black ribbons delineate cross-sections from the rump, hip, mid-carcass, and shoulder. C. Percentages of lean, fat and bone from this steer. Source: Iowa State University.

is through the shoulder area, 5 through the middle of the back, and 6 is through the ham area.

Even though the sizes and shapes of slaughter cattle, sheep, and swine are different, these three species are remarkably similar in muscle structure and fat-deposit areas. Therefore, what is learned from visual evaluation of one species can be applied to another species. Regardless of species, an animal that shows a square appearance over the top of its back and appears blocky and deep from a side view usually has a large accumulation of fat. **Fat** accumulates first in flank areas, **brisket**, **dewlap**, and throat (**jowl**); between the hind legs; and over the edge of the loin. Fat also fills in behind the shoulders and gives the animal a smooth appearance. Movement of shoulder blade can be observed when leaner cattle and swine walk. A slaughter red meat animal that has an oval turn to the top of its back and thickness through the center part of its hind legs (as viewed from the rear) has a high proportion of lean to fat. It is important that slaughter animals have an adequate amount of fat, because thin animals typically do not produce a highly palatable consumer product.

The wool covering of sheep can easily camouflage fat depots. The amount of fat in sheep can be determined by pressing the fingers of the closed hand lightly over the

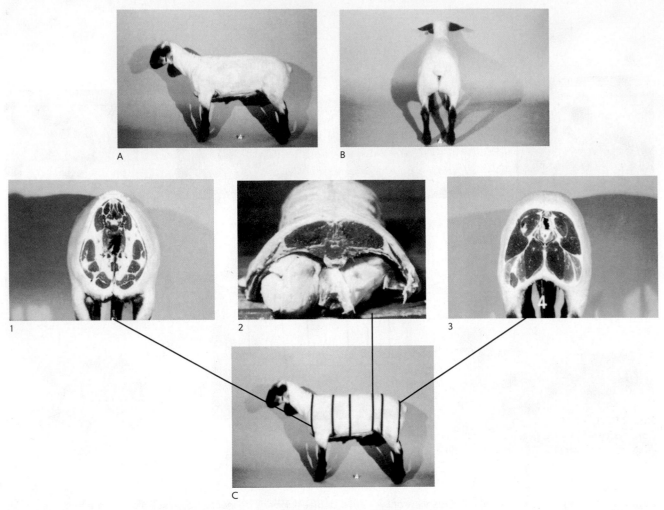

Figure 9.10

Composition of a yield grade 1 lamb. A. Side view. B. Rear view. C. Black ribbons delineate where cross-sections were made from the shoulder, mid-carcass, and rump. Source: Iowa State University.

last two ribs and over the spinal processes of the vertebrae. Sheep that have a thick padding of fat in these areas will produce poor yield-grading carcasses.

Accuracy in visually appraising slaughter red meat animals is obtained by making visual estimates of yield grades and percent lean cuts and their component parts, and then comparing the visual estimates with the carcass measurements. Accurate visual appraisal can be used as one tool in producing red meat animals with a more desirable carcass composition of lean to fat.

The proper ratio of fat to lean has been debated since the first organized system of marketing livestock and meat began. In the 1940s, fat-type animals, such as the lard-type hog, baby beef, and fat lamb, were in vogue. Over time, vegetable oils largely replaced lard and other animal fats. As consumers began to favor leaner meats, the market place and the grading system evolved in accordance with these demand shifts. Since the late 1980s and early 1990s the amount of fat on retail cuts has been significantly reduced—for example, down to ¼ in., ⅛ in., or no fat on many cuts. This fat reduction has occurred primarily due to trimming of excess fat at the packing plant. Certain breeding and feeding practices can reduce the trimmable fat more economically.

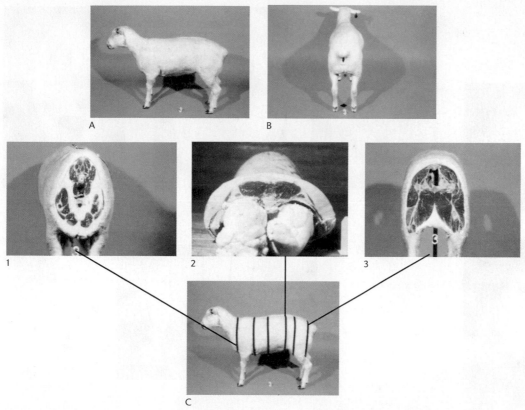

Figure 9.11
Composition of a yield grade 5 lamb. A. Side view. B. Rear view. C. Black ribbons delineate where cross-sections were made from the shoulder, mid-carcass, and rump. Source: Iowa State University.

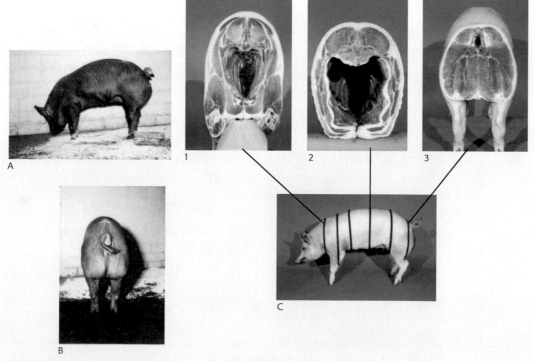

Figure 9.12
Composition of a U.S. number 1 hog. A. Side view. B. Rear view. C. Black ribbons delineate where cross-sections were made from the shoulder, mid-carcass and ham. Source: Iowa State University.

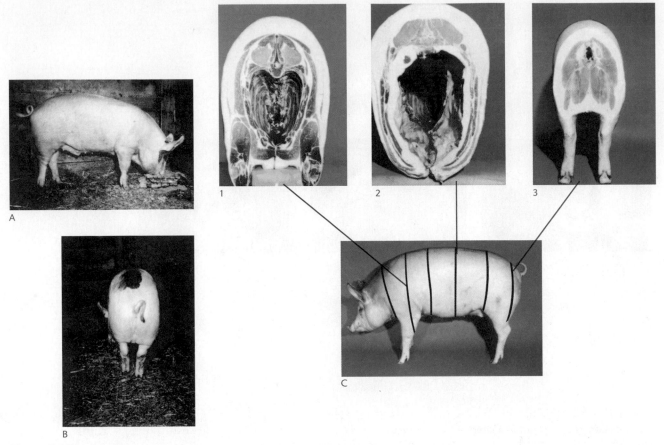

Figure 9.13
Composition of a U.S. number 4 hog. A. Side view. B. Rear view. C. Black ribbons delineate where cross-sections were made from the shoulder, mid-carcass, and ham. Source: Iowa State University.

Figure 9.14A illustrates muscling differences between two yearling stocker steers of similar size. The steer on the left has considerably more thickness than the steer on the right and thus would command a higher value from a feeder buyer. Figure 9.14B and 9.14C illustrates that two finished steers of comparable weights and USDA quality grades can have significantly different levels of carcass fat with the steer in Figure 9.14B having about twice the amount of trimmable fat waste as compared to the steer in Figure 9.14C. Again these differences would be reflected in the total value received for each. Figure 9.14D illustrates how fat depositions increase from the middle of the back to the edge of the loin. Finally, Figure 9.14E illustrates the potential effect of genetics on carcass composition and value. These two Hereford bulls are of comparable weight but differing lean to fat ratios. The bull on the left would be expected to sire calves that would be more likely to produce yield grade 4 or 5 carcasses while the bull on the right would be expected to sire progeny that would have yield grades of 1 and 2. This comparison assumes that the bulls were mated to cows comparable to their own type and that the steer progeny were harvested at approximately 1,150 lb.

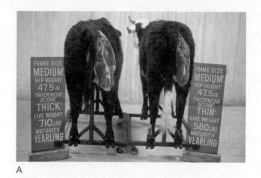

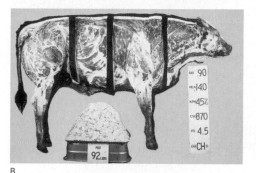

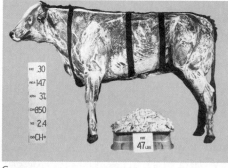

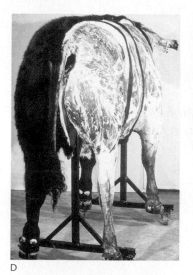

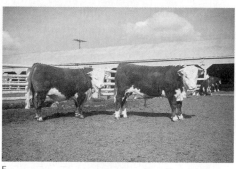

Figure 9.14

Compositional differences that affect value in the beef industry. A. Muscularity differences are shown in two medium-framed steers. The steer on the left is more muscular than the steer on the right. B and C. Pounds of fat trimmed from one side of a yield grade 4 steer (92 lb) and a yield grade 2 steer (47 lb). Black stripes are hide and fat remaining on key reference points on each steer. D. Rear view illustrating how fat deposition increases from the back to the edge of the loin. E. Two yearling Hereford bulls of approximately the same weight. The bull on the left would be expected to sire steers having yield grades of 4 or 5. The bull on the right would be expected to sire steers that would be yield grade 1 and 2.

Source: Colorado State University.

The role of managers throughout the livestock supply chain should understand compositional and value differences and then employ protocols and techniques to positively affect positive change.

CHAPTER SUMMARY

- Numerous red meat slaughter animals are evaluated and priced on the visual appraisal of the animal's carcass merit.

- Carcass composition (fat, lean, and bone) can be determined quite accurately by visually appraising live animals.

- An understanding of the anatomical structure, especially the location of major muscle and patterns of fat deposition, is needed for accurate visual appraisal.
- Thickness through the center of the round (cattle), ham (pig), and leg (lamb) gives a visual estimate of muscling. Squareness over the animal's top and rectangular appearance from a side view usually gives a visual assessment that the animal is excessively fat.

KEY WORDS

type
conformation
fat

brisket
dewlap
jowl

REVIEW QUESTIONS

1. What are the three productive stages of red meat-producing animals?
2. Why are type and conformation important?
3. Describe the external body parts of cattle, swine, and sheep.
4. Locate the wholesale cuts on live cattle, hogs, and lambs.
5. How can visual appraisal aid in determining compositional differences?
6. Describe the anatomical reference points to determine the degree of fatness in livestock.

SELECTED REFERENCES

Boggs, D. L., R. A. Merkel, M. E. Doumit, and K. Bruns. 2006. *Livestock and Carcass: An Integrated Approach to Evaluation, Grading and Selection.* Dubuque, IA: Kendall/Hunt.

10
Reproduction

Reproductive efficiency in farm animals, as measured by number of calves or lambs per 100 breeding females or number of pigs per litter for example, is a trait of significant economic importance in farm animal production. Reproduction is typically considered at least twice as important to economic returns of livestock production as growth or carcass performance. It is essential to understand the reproductive process in the creation of new animal life because it is a focal point of overall animal productivity. Producers who manage animals for optimal reproductive rates must understand the production of viable sex cells, estrous cycles, mating, pregnancy, and birth. (Some aspects of reproductive behavior are presented in Chapter 22.)

TERMINOLOGY FOR ANATOMICAL POSITIONING

A variety of terms are utilized to describe the position of organs and tissues. A partial list is provided below to serve as a reference for the reader. These terms will be useful in a number of chapters in the text and as terminology frequently encountered in the animal sciences.

- Anterior—toward the front
- Posterior—toward the back
- Cranial—toward the anterior or front (four-legged animals)
- Caudal—toward the posterior or tail of four-legged animals
- Ventral—toward the belly
- Dorsal—toward the back
- Proximal—close to the midline
- Distal—away from the midline
- Lateral—toward the side or away from the midline
- Medial—toward the midline
- Superior—over, higher
- Inferior—under, lower

FEMALE ORGANS OF REPRODUCTION AND THEIR FUNCTIONS

Figures 10.1 and 10.2 show the reproductive organs of the cow and sow. Figure 10.3 illustrates the reproductive tract of the mare.

The organs of reproduction of the typical female farm mammal include a pair of ovaries, which are suspended by ligaments just caudal of the kidneys, and a pair of open-ended tubes, the **oviducts** (also called the **Fallopian tubes**), which lead directly into the uterus (womb).

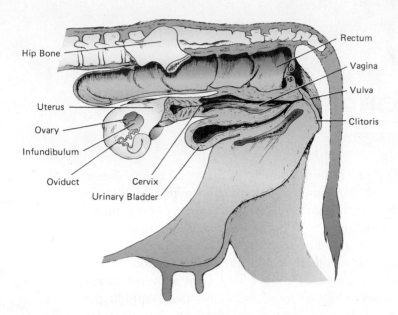

Figure 10.1

Reproductive organs of the cow. Source: Figure by Kevin Pond from Colorado State University. Used by permission of Kevin Pond.

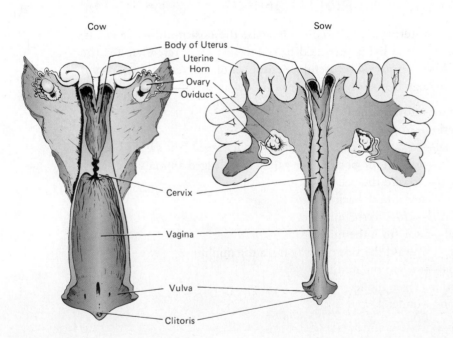

Figure 10.2

A dorsal view of the reproductive organs of the cow and the sow. The most noticeable difference is the longer uterine horns of the sow compared to the cow.

Source: Figure by Kevin Pond from Colorado State University. Used by permission of Kevin Pond.

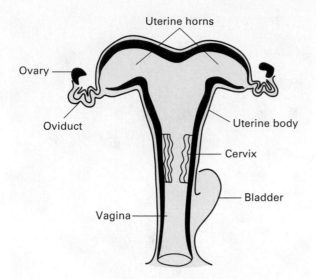

Figure 10.3
Reproductive tract of the mare. Source: Sean Field.

Table 10.1
SUMMARY OF THE PRIMARY FUNCTIONS OF THE ORGANS OF THE FEMALE REPRODUCTIVE TRACT

Organ	Function(s)
Oviduct	Transport of ova and sperm Site of fertilization and early cleavage
Ovary	Production of oocytes Production of estrogen (Graafian follicle) Production of progesterone (corpus luteum)
Uterus	Assisting in sperm transport Regulation of the corpus luteum Site of implantation and pregnancy Expulsion of fetus and fetal membranes
Cervix	Facilitating sperm transport (sow and mare) Prevention of uterine contamination
Vagina	Copulatory organ Birth canal

The **uterus** itself has two horns, or branches, that in farm mammals merge together at the lower part into the uterine body. The uterine body is protected by the **cervix**, which serves as a pathway to and from the uterus. Its surface is fairly smooth in the mare and the sow, but is folded in the cow and ewe. The cervix opens into the vagina, a relatively large canal or passageway that leads posterior to the external structures—the vulva and clitoris. The urinary bladder empties into the vagina through the urethral opening. Table 10.1 summarizes the primary functions of the organs of the female reproduction system.

Ovaries

Ovaries produce ova (female sex cells, also called *eggs*) and the hormones estrogen and progesterone. Ova (Fig. 10.4) are the largest single cells in the body; each ovum develops inside a recently formed **follicle** within the ovary (Fig. 10.5). Some tiny follicles develop and ultimately attain maximum size (about 0.8–1.5 in. in diameter)

Figure 10.4
Bull sperm and cow egg magnified 300×. The ovum is about $\frac{1}{200}$ in. in diameter, while the sperm is $\frac{1}{6000}$ in. in diameter. Each is a single cell and contains half the chromosome number typical of other body cells. Source: Figure by George Seidel from Colorado State University. Used by permission of George Seidel.

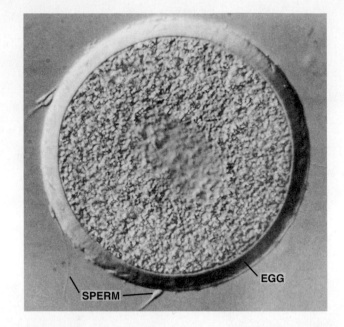

SPERM EGG

Figure 10.5
The large structure outlined with a circle of dark cells, is a follicle located on a cow's ovary (magnified 265×). The smaller circle, near the center, is the egg. The large light gray is the fluid that fills the follicle. When the follicle ruptures, the egg will move into the oviduct by anatomical action of the infundibulum. Source: Figure by George Seidel from Colorado State University. Used by permission of George Seidel.

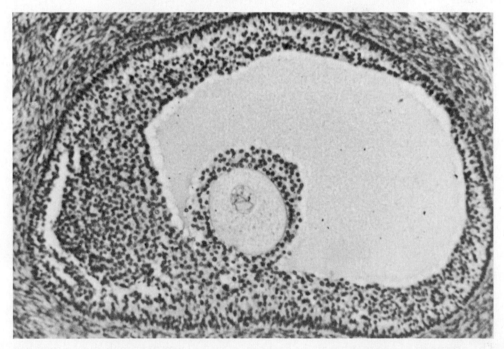

after having migrated from deep in the ovary to the surface of the ovary. These growing follicles produce estrogens. The mature (Graafian) follicles rupture, thus freeing the ovum in a process called ovulation.

Follicles can be classified as primordial (least mature), primary, secondary, tertiary, and Graafian (most mature). In addition to changes in size and complexity, follicles vary in their hormonal responsiveness as they mature. Follicles undergo a multistage process of development that may result in the formation of a **Graafian follicle** capable of ovulation. This developmental process can be defined as the steps of recruitment, selection, and dominance (Fig. 10.6). If the follicle fails to reach dominance, it will undergo a degenerative process known as atresia.

After the ovum is released from the mature follicle, cells of the follicle change into a **corpus luteum**, or "yellow body." The corpus luteum produces progesterone, which becomes a vitally important hormone for maintaining pregnancy.

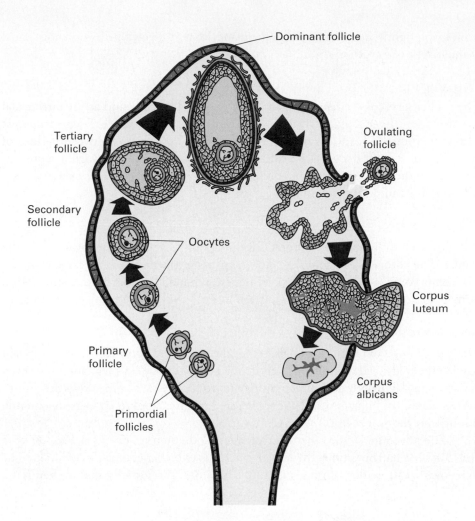

Figure 10.6
The progression of follicular development from primordial stage to ovulation, formation of the corpus luteum, and the formation of the corpus albicans once the CL regresses. Source: Sean Field.

The Oviducts

Immediately after ovulation, the ova are caught by the fingerlike projections of the **infundibulum**, which guide the ova into the tubular portion of the oviduct. The ova are tiny (approximately $\frac{1}{200}$ in. in diameter, about the size of a dot made by a sharp pencil). Sperms are transported through the uterus into the oviduct after the female is inseminated (naturally or artificially). Therefore, the oviducts are where ova and sperm meet to initiate fertilization. After fertilization, 3–5 days are required in cows and ewes (about the same amount of time in other farm animals) for the ova to travel down the remaining two-thirds of the oviduct. From the oviduct, the newly developing embryos pass to the uterus and soon attach to its lining.

The Uterus

The uterus varies in shape from the type that has long, slender left and right horns, as in the sow, to the type that is primarily a fused body with short horns, as in the mare. In the sow, the embryos develop in the uterine horn; in the mare, the embryo develops in the body of the uterus. Each surviving embryo develops into a fetus and remains in the uterus until **parturition** (birth).

The posterior outlet of the uterus is the cervix, an organ composed primarily of connective tissue that constitutes a formidable gateway between the uterus and the vagina. Like the rest of the reproductive tract, the cervix is lined with mucosal cells. These cells undergo significant changes as the animal experiences the various phases of the estrous cycle as well as the process of pregnancy. The cervical passage changes from being tightly closed or sealed in pregnancy to a relatively open, very moist canal at the height of estrus.

The Vagina

The vagina serves as the female organ of copulation at mating and as the birth canal at parturition. Its mucosal surface changes during the estrous cycle from very moist when the animal is ready for mating to almost dry, even sticky, between periods of heat. The tract from the urinary bladder joins the posterior ventral vagina; from this juncture to the exterior vulva, the vagina serves the dual role of a passageway for the reproductive and urinary systems.

The Clitoris

A highly sensitive organ, the clitoris is located ventrally and at the lower tip of the vagina. The clitoris is the homologue of the penis in the male (i.e., it came from the same embryonic source as the penis). Some research indicates that clitoral stimulation or massage following artificial insemination in cattle will increase conception rates.

Reproduction in Poultry Females

The hen differs from farm mammals in that the young are not suckled, the egg is laid outside the body, and there are no well-defined estrous cycles or pregnancy. However, variation in day length does affect poultry reproductive rates. Since eggs are an important source of human food, hens are selected and managed to lay eggs consistently throughout the year.

The anatomy of the reproductive tract of the hen is shown in Figures 10.7 and 10.8. At hatching time, the female chick has two ovaries and two oviducts. The right ovary and oviduct do not develop. Therefore, the sexually mature hen has a

Figure 10.7

Reproductive organs of the hen in relation to other body organs. The single ovary and oviduct are on the hen's left side; an underdeveloped ovary and oviduct are sometimes found on the right side, having degenerated in the developing embryo.

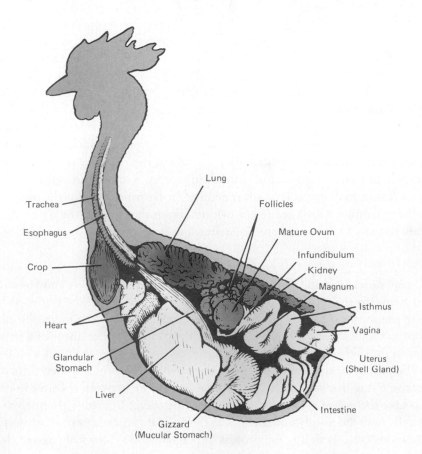

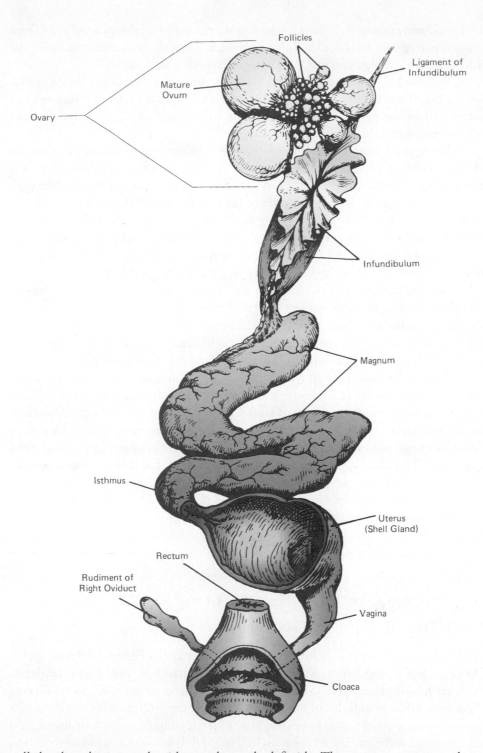

Figure 10.8
Reproductive organs of the hen. Sections of the uterus and cloaca are cut away to better view internal structure.

well-developed ovary and oviduct only on the left side. The ovary appears as a cluster of tiny gray eggs or yolks in front of the left kidney and attached to the back of the hen. The ovary is fully formed, although very small, when the chick is hatched. It contains approximately 3,600–4,000 miniature ova. As the hen reaches sexual maturity, some of the ova develop into mature yolks (yellow part of the laid egg). The remaining ova are variable in size from the nearly mature to microscopic.

The oviduct is a long, glandular tube leading from the ovary to the **cloaca** (common opening for reproductive and digestive tracts). The oviduct is divided into five parts: (1) the infundibulum (3–4 in. long), which receives the yolk; (2) the **magnum** (approximately 15 in. long), which secretes the thick albumen, or white, of the egg;

(3) the **isthmus** (about 4 in. long), which adds the shell membranes; (4) the uterus (approximately 4 in. long), or shell gland, which secretes the thin white albumen, the shell, and the shell pigment; and (5) the vagina (about 2 in. long).

Ovulation is the release of a mature yolk (ovum) from the ovary. When ovulation occurs, the infundibulum engulfs the yolk and starts it on its way through the 25 to 27-in. oviduct. The yolk moves by peristaltic action through the infundibulum into the magnum area in about 15 minutes.

During the 3-hour passage through the magnum, more than 50% of the albumen is added to the yolk. The developing egg passes through the isthmus in about 1¼ hours. Here water and mineral salts and the two shell membranes are added. During the egg's 21-hour stay in the uterus, the remainder of the albumen is added, followed by the addition of shell and shell pigment. Moving finally into the vagina, the fully formed egg enters the cloaca and is laid. The entire time from ovulation to laying is usually slightly more than 24 hours. About 30 minutes after a hen has laid an egg, she releases another yolk into the infundibulum, and it will likewise travel the length of the oviduct.

After the fertilized egg is incubated for 21 days, the chick is hatched. The egg is biologically structured to support the growth and life processes of the developing chick embryo during incubation and for 3–4 days after the chick is hatched.

There are several egg abnormalities that occur because of factors affecting ovulation and the developmental process. Double-yolked eggs result when two yolks are released about the same time or when one yolk is lost into the body cavity for a day and is picked up by the infundibulum when the next day's yolk is released. Yolkless eggs are usually formed when a bit of tissue that is sloughed off the ovary or oviduct stimulates the secreting glands of the oviduct and a yolkless egg results. The abnormality of an egg within an egg is due to reversal of direction of an egg by the wall of the oviduct. One day's egg is added to the next day's egg, and shell is formed around both. Soft-shelled eggs generally occur when an egg is laid prematurely and insufficient time in the uterus prevents the deposit of the shell. Thin-shelled eggs may be caused by dietary deficiencies, heredity, or disease. Glassy- and chalky-shelled eggs are caused by malfunctions of the uterus of the laying bird. Glassy eggs are less porous and will not hatch but may retain their quality.

MALE ORGANS OF REPRODUCTION AND THEIR FUNCTIONS

The reproductive organs of the bull and boar are illustrated in Figures 10.9 and 10.10. The organs of reproduction of a typical male farm mammal include two **testicles**, which are held in the scrotum. Male sex cells (called **sperm** or *spermatozoa*) are formed in the **seminiferous tubules** of the testicles. The sperm from each testicle then pass through a ductal system into the **epididymis**, which is a highly coiled tube that is held in a covering on the exterior of the testicle. The epididymis functions to concentrate, store, transport, and facilitate maturation of sperm cells. The epididymis empties into a larger tube, the **vas deferens** (also called the **ductus deferens**). The two vasa deferentia converge at the upper end of the urethral canal, where the urinary bladder opens into the urethra. In some species, the wall of the upper end of the vas deferens is thickened and forms a secretory gland called the **ampulla**. The urethra is the large canal that leads through the penis to the outside of the body. The penis has a triple role: It serves as a passageway for semen and urine and it is the male organ of copulation.

The left and right parts of the seminal vesicles, which lie against the urinary bladder, consist of glandular tissue that secretes into the urethra a substance that supplies nutrients for the sperm. The prostate gland contains 12 or more tubes, each of

Figure 10.9
Reproductive organs of the bull.

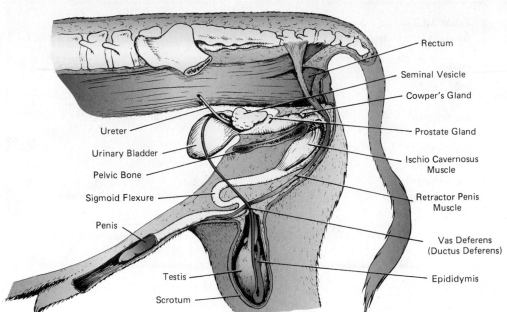

Rectum

Seminal Vesicle

Cowper's Gland

Prostate Gland

Ischio Cavernosus Muscle

Retractor Penis Muscle

Vas Deferens (Ductus Deferens)

Epididymis

Ureter

Urinary Bladder

Pelvic Bone

Sigmoid Flexure

Penis

Testis

Scrotum

Figure 10.10
Reproductive organs of the boar.

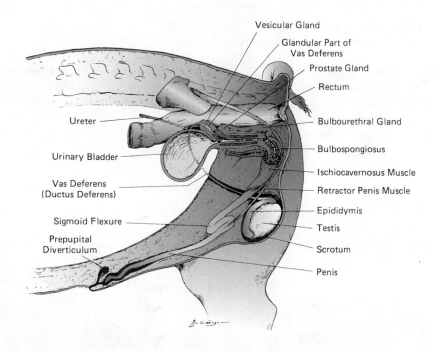

Vesicular Gland

Glandular Part of Vas Deferens

Prostate Gland

Rectum

Bulbourethral Gland

Bulbospongiosus

Ischiocavernosus Muscle

Retractor Penis Muscle

Epididymis

Testis

Scrotum

Penis

Ureter

Urinary Bladder

Vas Deferens (Ductus Deferens)

Sigmoid Flexure

Prepupital Diverticulum

which empties into the urethra. Another gland, the bulbourethral (Cowper's) gland, which also empties its secretions into the urethral canal, is posterior to (behind) the prostate. Table 10.2 outlines the primary activities of the male reproductive organs and tissues. Table 10.3 summarizes the comparative anatomical and seminal differences of males for several species.

Testicles

The testicles are suspended from the body cavity by the spermatic cord and produce (1) sperm cells that fertilize the ova of the female, and (2) a hormone called **testosterone** that conditions the male so that his appearance and behavior are masculine. Details of the structure of the spermatozoa of the bull are shown in Figure 10.11. The diameter of sperm is approximately 1/6000 in. Figure 10.12 illustrates several

Table 10.2
SUMMARY OF THE PRIMARY FUNCTIONS OF THE MALE REPRODUCTIVE TRACT

Organ	Primary Function(s)
Testicles	Testosterone production (interstitial cells)
	Spermatozoa production (seminiferous tubules)
Epididymis	Concentration, storage, maturation, and transport of spermatozoa
Scrotum	Support of testicles
	Temperature control (tunica dartos)[a]
Vas deferens	Sperm transport
Accessory glands	Addition of fluid volume, nutrients, and buffers to semen
Penis	Copulatory organ

[a]The cremaster and pampiniform plexus of the spermatic cord also assist with temperature regulation of the testes.

Table 10.3
COMPARATIVE ANATOMY AND SEMINAL CHARACTERISTICS OF MALES

Animal	Weight of Paired Testes (g)	Volume per Ejaculate (ml)	Type of Ejaculate	Sperm Concentration (mil per ml)	Total Sperm per Ejaculate (bil)
Bull (cattle)	650	3–8	Rapid	800–2,000	5–15
Ram (sheep)	550	0.8–2.0	Rapid	2,000–3,000	2–4
Boar (swine)	750	150–350	Prolonged	200–300	30–60
Stallion (horse)	160	60–100	Rapid	150–300	5–15
Buck (goat)	—	0.6–1.0	—	2,000–3,500	1–8
Tom (turkey)	—	0.2–0.8	—	8,000–30,000	1–20
Cock (chicken)	25	0.2–1.5	—	3,000–7,000	0.6–3.5

Source: Adapted from Hafez (1993), McDonald (1989), and others.

Figure 10.11
A diagrammatic sketch of the structure of bull sperm.
Artist: Sean Field.

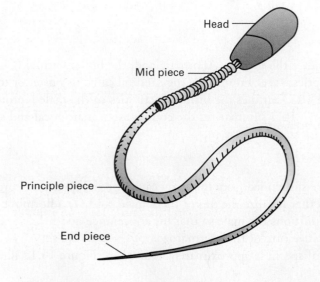

Head

Mid piece

Principle piece

End piece

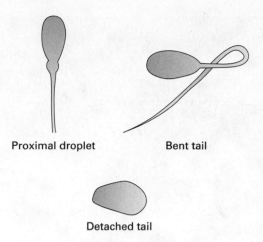

Proximal droplet

Bent tail

Detached tail

Figure 10.12
Examples of sperm abnormalities including twisted tail, detached head, and proximal droplets.
Artist: Sean Field.

incidences of spermatozoa abnormalities that impair fertility. During a breeding soundness exam a semen sample is collected and the percent of normal cell is evaluated to assist in predicting fertility. In the case of proximal droplets where a small amount of fluid attaches to the spermatozoa, this condition is most often observed in young males and will improve with maturity.

If both testicles are removed (as is done in castration), the individual loses sperm production capability and is sterile. Also, without testosterone a masculine appearance is not apparent, and the castrate approaches the status of a neuter—an individual whose appearance is somewhere between that of a male and that of a female. A steer does not have the crest, or powerful neck, of the bull. The bull has heavier, more muscular shoulders and a deeper voice than a counterpart steer. If a bull calf is castrated, the reproductive organs, such as the vas deferens, seminal vesicles, and prostate and bulbourethral glands, all but cease further development. If castration is done in a mature bull, the remaining genital organs shrink in size and in function.

Within each testicle, sperm cells are generated in the seminiferous tubules, and testosterone is produced in the cells between the tubules, called **Leydig cells or interstitial cells** (Fig. 10.13).

The Epididymis

The epididymis is the storage site for sperm cells, which enter it from the testicle to mature. In passing through this very long tube (95–115 ft in the bull, longer in the boar and stallion), the sperm acquire the potential to fertilize ova. Sperm taken from the part of the epididymis nearest the testicle and inseminated into females are not likely to be able to fertilize ova, whereas sperm taken near the vas deferens have the potential to fertilize.

In the sexually mature male animal, sperm reside in the epididymis in large numbers. In time, the sperm mature, then degenerate, and are absorbed in the part of the epididymis farthest from the testicle, unless they have been moved on into the vas deferens to be ejaculated.

The Scrotum

The scrotum is a two-lobed sac that contains and protects the two testicles. It also regulates temperature of the testicles, maintaining them at a temperature lower than body temperature (3–7°F lower in the bull and 9–13°F lower in the ram and goat). When the environmental temperature is low, the tunica dartos muscle of the scrotum contracts, pulling the testicles toward the body and its warmth; when the

Figure 10.13

A cross-section through the seminiferous tubules of the bovine testis (magnified 240×). The tubule in the lower right-hand corner demonstrates the more advanced stages of spermatogenesis as the spermatids are formed near the lumen (opening) of the tubule. Source: Figure by George Seidel from Colorado State University. Used by permission of George Seidel.

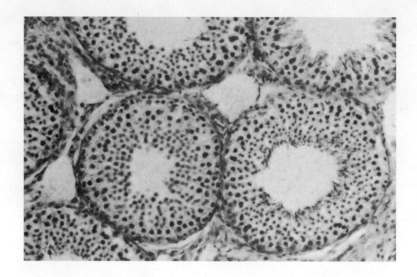

environmental temperature is high, this muscle relaxes, permitting the testicles to drop away from the body and its warmth. The cremaster muscle that is aligned with the spermatic cord functions in a similar manner to the tunica dartos as an aid to temperature control. Furthermore, the pampiniform plexus acts as a heat exchanger to cool blood as it enters the testes. This heat-regulating mechanism of the scrotum begins at about the time of puberty. Hormone function precedes puberty by 40–60 days.

When the environmental temperature is elevated such that the testicles cannot cool sufficiently, the formation of sperm is impeded, and a temporary condition of lowered fertility results. Providing shade, keeping the males in the shade during the heat of the day, even providing air conditioning, are ways to manage and prevent this temporary sterility.

The Vas Deferens

The vas deferens is essentially a transportation tube that carries the sperm-containing fluid from each epididymis to the urethra. The vasa deferentia join the urethra near its origin as the urethra leaves the urinary bladder. In the mature bull, the vas deferens is about 0.1 in. in diameter except at its upper end, where it widens into a reservoir, or ampulla, about 4–7 in. long and 0.4 in. wide.

Under the excitement of anticipated mating, the secretion loaded with spermatozoa from each epididymis is propelled into each vas deferens and accumulates in the ampulla of the deferent duct. This brief accumulation of sperm in the ampulla is an essential part of sexual arousal. The sperm reside briefly in the ampulla until the moment of ejaculation, when the contents of each ampulla are pressed out into the urethra, and then through the urethra and the penis en route to their deposition in the female tract.

The ampulla is found in the bull, stallion, goat, and ram—species that ejaculate rapidly. It is not present in the boar or dog, animals in which ejaculation normally takes several minutes (8–12 minutes is typical in swine). In such animals, large numbers of sperm travel all the way from the epididymis through the entire length of the vas deferens and the urethra. On close observation of the boar at the time of mating, the muscles over the scrotum can be seen quivering rhythmically as some of the contents of each epididymis are propelled into the vasa deferentia and on into the urethra. This slow ejaculation of the boar contrasts to the sudden expulsion of the contents of the ampulla of the vas deferens at the height of the mating reaction, or orgasm, in the bull, stallion, ram, and goat.

The Urethra

The urethra is a large, muscular canal extending from the urinary bladder. The urethra runs posteriorly through the pelvic girdle and curves downward and forward through the full length of the penis. Very near the junction of the bladder and urethra, tubes from the seminal vesicles and tubes from the prostate gland join this large canal. The bulbourethral gland joins the urethra at the posterior floor of the penis.

Accessory Sex Glands

The ampullae, seminal vesicles, prostate, and bulbourethral glands are known as the **accessory sex glands**. Their primary functions are to add volume and nutrition to the sperm-rich fluid coming from the epididymis. Semen consists of two components: the sperm and the fluids secreted by the accessory sex glands. The semen characteristics of some farm animals are shown in Table 10.3.

The Penis

The penis is the organ of copulation. It provides a passageway for semen and urine. It is an organ characterized especially by its spongy, erectile tissue that fills with blood under considerable pressure during periods of sexual arousal, making the penis rigid and erect. The bull, ram, and boar have fibroelastic penile structure. The penis in the stallion, dog, and cat is vascular in structure.

The penis of the bull is about 3 ft in length and 1 in. in diameter, tapering to the free end, or glans penis. In the bull, boar, and ram, the penis is S-shaped when relaxed. This S curve, or sigmoid flexure, becomes straight when the penis is erect. The sigmoid flexure is restored after copulation, when the relaxing penis is drawn back into its sheath by a pair of retractor penis muscles. The stallion penis has no sigmoid flexure; it is enlarged by engorgement of blood in the erectile tissues.

The free end of the penis is termed the **glans penis**. The opening in the ram penis is at the end of a hair-like appendage that extends about 0.8–1.2 in. beyond the larger penis proper. This appendage also becomes erect and, during ejaculation, whirls in a circular fashion, depositing semen in the anterior vagina. It does not regularly penetrate the ewe's cervix, as some investigators claim it does. Only a small portion of the penis of the bull, boar, ram, and goat extends beyond its sheath during erection. The full extension awaits the thrust after entry into the vagina has been made. The stallion and ass usually extend the penis completely before entry into the vagina.

All the accessory male sex organs depend on testosterone for their tone and normal function. This dependence is especially apparent when the testicles are removed (as in castration); the usefulness of the accessory sex organs is then diminished or even terminated.

Reproduction in Male Poultry

The reproductive tract of male poultry is shown in Figure 10.14. There are several differences when compared to the reproductive tracts of the farm mammals previously described. The testes of male poultry are contained in the body cavity. Each vas deferens opens into small papillae, which are located in the cloacal wall. The male fowl has no penis but does have a rudimentary organ of copulation. The sperm are transferred from the papillae to the rudimentary copulatory organ, which transfers the sperm to the oviduct of the hen during the mating process. The sperm are stored in primary sperm-host glands located in the oviduct. These sperm are then released on a daily basis and transported to secondary storage glands in

Figure 10.14
Male poultry reproductive tract (ventral view).

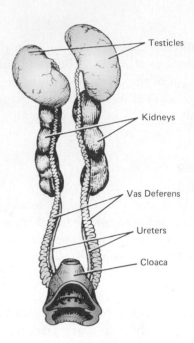

Testicles

Kidneys

Vas Deferens

Ureters

Cloaca

the infundibulum. Fertilization occurs in the infundibulum. Sperm stored in the oviduct are capable of fertilizing the eggs for 30 days in turkeys and 10 days in chickens.

HORMONAL CONTROL OF TESTICULAR AND OVARIAN FUNCTION

Testicular Function

Reproduction is controlled via a series of hormonal signals from the endocrine system (Table 10.4). Testicles produce their hormones under stimuli coming to them from the anterior pituitary (AP) gland situated at the base of the brain. The AP produces and secretes two hormones important to male reproductive performance. **Luteinizing hormone (LH)** and **follicle-stimulating hormone (FSH)** are known as **gonadotropic hormones** because they stimulate the gonads (ovary and testicle). LH produces its effect on the interstitial tissue (Leydig cells) of the testicle, causing the tissue to produce the male hormone, testosterone. FSH stimulates cells in the seminiferous tubules to nourish the developing spermatozoa.

Some species that respond to change in length of daylight exhibit more seasonal fluctuation than others in reproductive activities. These influences of daylight are exerted on the neurophysiological mechanism in the brain. The hypothalamus makes up the floor and part of the wall of the third ventricle of the brain and secretes through blood vessels releasing factors that affect the AP and its production of FSH and LH.

Ovarian Function

Ovarian hormones in the sexually mature female owe the cyclicity of their production to hormones that originate in the hypothalamus and AP (Fig. 10.15). The hypothalamus produces **gonadotropin-releasing hormone (GnRH)** that in turn stimulates the AP to release FSH and LH. GnRH is released in the presence of estrogen when progesterone levels are low.

Table 10.4

THE MAJOR REPRODUCTION HORMONES: THEIR SOURCE, TARGET ORGANS, AND FUNCTIONS

Hormone	Source	Target	Function(s)
Gonadotropin-releasing hormone (GnRH)	Hypothalamus	Anterior pituitary	Release of LH and FSH
Luteinizing hormone (LH)	Anterior pituitary	Ovary (luteal cells)	Ovulation
			CL formation
			Progesterone production
		Testis (interstitial cells)	Testosterone production
Follicle-stimulating hormone (FSH)	Anterior pituitary	Ovary (granulosa cells)	Development of follicle
		Testis (sertoli cells)	Estrogen synthesis
			Sperm production
Prolactin	Anterior pituitary	Mammary tissue	Milk synthesis
			Maternal behavior
Oxytocin	Posterior pituitary	Uterus	Gamete transport
		Mammary tissue	Uterine contraction
			Milk let down
Estrogen	Follicle	Uterus	Mating behavior
	Placenta	Hypothalamus	Uterine growth
		Mammary tissue	Secondary sex characteristics
			Promotes GnRH release
Progesterone	Corpus luteum	Uterus	Maintains pregnancy
	Placenta	Mammary tissue	Mammary development
		Hypothalamus	Inhibits GnRH release
Testosterone	Interstitial cells of testis	Skeletal muscle	Anabolic growth
		Seminiferous tubules	Sperm production
		Male reproductive tract	Secondary sex characteristics
Prostaglandin F$_{2\alpha}$	Uterus (endometrium)	Corpus luteum	Regression of CL
		Uterus (myometrium)	Uterine contraction
		Graafian follicles	Ovulation
Placental lactogen	Placenta	Mammary gland	Promotes lactation
Relaxin	Ovary	Cervix, pelvis	Cervical dilation
	Placenta		Pelvic expansion
Cortisol (fetal)	Adrenal gland (fetus)	Dam's uterus	Initiation of parturition
Melatonin	Pineal gland	Hypothalamus	Partial control of seasonal reproductive patterns

Source: Adapted from Hafez (1993), McDonald (1989), Senger (2003), and others.

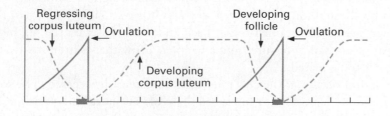

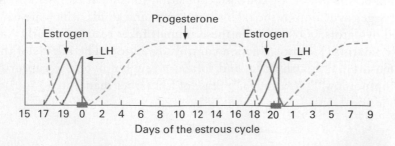

Days of the estrous cycle

Figure 10.15

Events and hormone changes during the estrous cycle of the cow.

FSH and LH work in concert to stimulate follicular development and then a surge of LH initiates ovulation. As the follicle matures, **estrogen** is released, which stimulates not only the release of GnRH but also stimulates sexual behavior of the female.

Following ovulation, the luteal cells form into the corpus luteum that produces progesterone. Progesterone then acts to inhibit the release of LH and FSH. When the influence of progesterone is removed from the hypothalamus–AP axis then cyclicity is reinitiated.

Estrous Cycle

Female farm animals exhibit a predictable pattern of reproductive events referred to as the **estrous cycle**. The estrous cycle is initiated by the expression of heat (estrus) and concludes at the following estrus. This pattern is typically ongoing in mature females until disrupted by other physiological events such as pregnancy or lactation; or as the results of environmental effects such as poor nutrition, disease, or seasonality in some species such as sheep and horses.

These patterns are of three types based on the timing and frequency throughout the year—polyestrus (cattle, swine), seasonally polyestrus (sheep, goats, horses), and monoestrus (dogs). Regardless of the cycle category, each estrous cycle is defined by the dominant or controlling structure present on the ovary. Reproductive physiologists refer to the two portions of the estrous cycle as the luteal phase (approximately 80% of the cycle) and the follicular phase (20% of the cycle). The luteal phase extends from ovulation until regression of the corpus luteum (CL) during which time progesterone is the dominant hormone. In the follicular phase (CL regression to ovulation), estrogen produced by the follicles is in control.

The estrous cycle can be further broken down into four stages—proestrus and estrus (follicular phase) and metestrus and diestrus (luteal phase). During proestrus, the follicles form and produce estrogen, which peaks during estrus and brings the female into heat during which time she exhibits sexual receptivity to the male. Following ovulation, metestrus occurs as the CL is formed and progesterone secretion is initiated. The final stage, diestrus, is defined by the sustained release of progesterone by the CL. Understanding and manipulating the estrous cycle is the foundation of most reproductive management strategies.

Estrus or **heat** is the period of time when the female will accept the male for breeding purposes. The female of each species exhibits some behavior patterns that demonstrate she is in heat (see Chapter 22). For example, a mare that is in estrus, when "teased" with the presence of a stallion will not avoid or kick him. The mare in heat will stand solidly, sometimes squatting and urinating when approached by the stallion. Similarly, a sow will brace or "lock up" when she is in heat. When pressure is applied to her back and sides either by the mating action of the boar or via hand pressure from a handler, a sow that is in heat will stop and brace herself to be mated.

Ovulation occurs in the cow after estrus behavior has ceased. The sow, ewe, goat, and mare ovulate toward the latter phase of, but nevertheless during, estrus. These species all ovulate spontaneously; that is, ovulation takes place whether copulation occurs or not. By contrast, copulation (or similar stimulation) is necessary to trigger ovulation in the rabbit, cat, ferret, and mink, which are considered "induced ovulators." Ovulation in these animals takes place at a fairly consistent time after mating. Hormonal action controls ovulation. The follicle of the ovary grows, matures, fills with fluid, and softens a few hours before rupturing. The follicle ruptures owing to a sudden release of LH rather than bursting as a result of pressure inside.

Table 10.5
AGE AT PUBERTY, LENGTH OF ESTROUS CYCLE, DURATION OF HEAT, AND TIME OF OVULATION

Animal	Puberty (mo)	Estrous Cycle Length (days)		Duration of Estrus		Approx. Time from Onset of Ovulation
		Average	Range	Average	Range	
Heifer, cow (cattle)	6–18	21	14–29	18 hrs	12–30 hrs	18–48 hrs
Ewe (sheep)[a]	6–12	17	14–19	30 hrs	24–36 hrs	24–27 hrs
Mare (horse)[b]	20	22	19–24	7 days	4–8 days	24–48 hrs prior to end of heat
Gilt, sow (swine)	5–10	21	19–23	60 hrs	2–3 days	38–42 hrs
Doe (goat)[a]	4–8	21	18–22	39 hrs	1–3 days	24–36 hrs
Heifer, cow (bison)	24	21	18–22	2 days	—	18–45 hrs

[a]Typically fall breeders in Northern Hemisphere.
[b]Typically spring breeders in Northern Hemisphere.
Source: Adapted from Hafez (1993), McDonald (1989), and others.

The FSH produced by the AP accounts for the increase in size of the ovarian follicle and for the increased amount of estradiol and estrone, which are products of the follicle cells. LH, another hormone from the AP, alters the follicle cells and granulosa cells of the ovary, changing them into luteal cells. LH from the AP in turn stimulates the luteal cells to produce progesterone.

In the course of 7–10 days, what was formerly an egg-containing follicle develops into the corpus luteum, a luteal body of about the same size and shape as the mature follicle. The luteal cells of the corpus luteum produce a sufficient quantity of progesterone to depress FSH secretion from the AP until the luteal cells reach maximum development (in nonpregnant animals), cease their development, and (in 3 weeks) lose their potency and disappear.

If pregnancy occurs, the corpus luteum continues to function, persisting in its progesterone production and preventing further estrous cycles. Thus, no more heat occurs until after pregnancy has terminated. If pregnancy does not occur, the uterus releases the hormone $PGF_{2\alpha}$ (**prostaglandin**). $PGF_{2\alpha}$ causes the regression of the corpus luteum and thus declining levels of progesterone. As previously stated, when the influence of progesterone is removed, GnRH secretion increases and another follicle matures to ovulatory status. The process of follicle development, ovulation, and corpus luteum development and regression is shown in Figure 10.6. Table 10.5 shows the length of estrus, estrous cycles, and time of ovulation for the different farm animals.

Seasonal Effects on Reproduction

Polyestrous females (cows and sows) have regular, uniform estrous cycles that occur throughout the year. Seasonal polyestrous females (ewes, does, and mares) display a concentrated or clustered pattern of estrous cycle into specific seasons of the year. Sheep and goats are short-day breeders that are stimulated to begin cycling as day length declines. Females of these species are most fertile in the late summer and fall months of the year. Some breeds are less affected by day length. For example, Dorset, Merino, Rambouillet, and Finnsheep are breeds that exhibit a more extended pattern of estrous cycles throughout the year. Seasonality is absent in sheep and goats raised in close proximity to the equator.

Mares are induced to cyclicity as day length extends in the spring months and are thus considered long-day breeders. Mares enter a phase of anestrous or reproductive dormancy in the seasons of the year characterized by declining levels of daylight. Male livestock are less affected by changes in day length but may experience reduced fertility especially in periods of extreme heat.

The amount of daylight acts both directly and indirectly on the animal. It acts directly on the central nervous system by influencing the secretion of hormones and indirectly by affecting plant growth, thus altering the level and quality of nutrition available. When selection has resulted in improvements in the traits associated with reproduction, individuals exhibit higher levels of fertility (i.e., more intense expression of estrus, occurrence of estrus over more months of the year, or occurrence of spermatogenesis at a high level over more months of the year) than unselected individuals.

Seasonal species are responsive to melatonin, a hormone produced by the pineal gland in response to declining periods of light. Thus, the mare's estrous cycle is stimulated by decreasing periods of dark that results in declining levels of melatonin secretion resulting in increased GnRH secretion and thus initiation of the ovulatory cascade. Conversely, increased levels of melatonin resulting from the increasing levels of darkness in the fall months stimulate the ewe.

The use of artificial lighting systems and other day-length manipulation strategies can be used to influence the cyclicity of these species.

PREGNANCY

When the sperm and the egg unite (fertilization), conception occurs, which is the beginning of the gestation period. The fertilized egg begins a series of cell divisions (Fig. 10.16). About every 20 hours, embryonic cells duplicate their genes and divide, progressing through the 2-, 4-, 8-, and 16-cell stages, and so on. In farm mammals, the embryo migrates through the oviduct to the uterus in 3–4 days, by which time it has developed to the 16- or 32-cell stage. The chorionic, allanotic, and amniotic membranes develop around the new embryo and together are referred to as the **conceptus**. A process called maternal recognition of pregnancy signals the presence of the embryo and allows attachment to the uterus. The signal is sent via a system of recognition factors and if not complete within approximately 2 weeks for farm animals, termination of the pregnancy will result.

However, if maternal recognition is achieved, then implantation will occur via formation of the placenta. The **placenta** is a temporary organ that facilitates the metabolic connection between the dam and fetus as well as producing hormones that maintain pregnancy (progesterone), stimulate development of the mammary glands to prepare for lactation, and enhance growth of the fetus. This period of attachment or **placentation** (20–30 days in cows, 14–21 days in sows, 36–38 days in mares, and 13–14 days in ewes) is critical. Unless the environment is sufficiently favorable, the embryo dies. Embryonic mortality causes a significant economic loss in farm animals, especially swine, where multiple ovulations and embryos are typical of the species. High temperatures are an example of a contributor to increased embryonic death.

The chorion is the fetal component of the placenta and has numerous protrusions or projections along its surface that protrude toward the uterus. These chorionic villi can be characterized into four types: diffuse, cotyledonary, zonary, and discoid. **Diffuse** (sows and mares) chorionic villi uniformly cover the surface of the chorion. **Cotyledonary** attachment (cows, ewes and does) is accomplished via a series of approximately 100 structures called **placentomes**. The maternal side of the

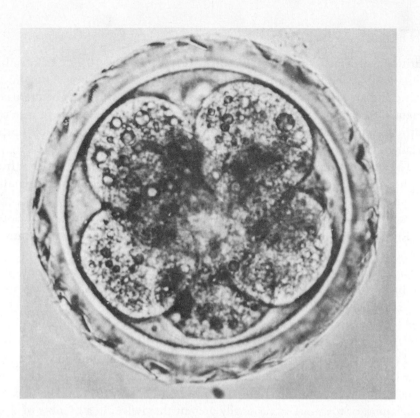

Figure 10.16
A bovine embryo in the six-cell stage of development (magnified 620×). Source: Figure by George Seidel from Colorado State University. Used by permission of George Seidel.

Table 10.6
GESTATION LENGTH, TYPE OF PLACENTATION, AND NUMBER OF OFFSPRING BORN

Animal	Gestation Length (days)	Type of Placentation	Usual Number of Offspring Born
Cow (cattle)	285	Cotyledonary	1
Ewe (sheep)	147	Cotyledonary	1–3
Mare (horse)	336	Diffuse	1
Sow (swine)	114	Diffuse	6–14
Doe (goat)	150	Cotyledonary	2–3
Cow (bison)	270	Cotyledonary	1

Source: Adapted from multiple sources.

placentome is called the **caruncle** while the fetal side is called the **cotyledon**. Placentomes in cattle is convex in shape and concave in the ewe. Zonary attachment (dogs and cats) involves the congregation of the villi into a specific zone or band on the surface of the chorion. Discoid (primates and rats) involves one or two distinct disc-like structures that provide the site of exchange.

The actual exchange of materials between the fetus and its mother occurs via the following mechanisms:

• Simple diffusion—water and gases move from high to low concentrations.
• Carrier diffusion—glucose the main source of fetal energy and other nutrients are transported via carrier molecules.
• Active transport—a system of pumps moved sodium, potassium, and calcium.

The duration of pregnancy in various species is included in Table 10.6. Length of pregnancy varies chiefly with the breed and age of the mother.

PARTURITION

Parturition (birth) marks the termination of pregnancy (Fig. 10.17). Extraembryonic membranes that had been formed around the embryo in early pregnancy are shed at this time and are known as **afterbirth**. These membranes attach to the uterus during pregnancy and become known as the placenta, which is responsible for the transfer of nutrients and wastes between mother and fetus. The placenta produces hormones, especially estrogens and progesterone, in some farm animal species.

The **parturition** process is initiated by release of the hormone cortisol from the fetal adrenal cortex. Progesterone levels decline, whereas estrogen, prostaglandin $F_{2\alpha}$, and oxytocin levels rise, resulting in uterine contractions.

Parturition is a synchronized process. The cervix, until now tightly closed, relaxes. Relaxation of the cervix, along with pressure generated by uterine muscles on the contents of the uterus, permits the passage of the fetus into the vagina and on to the exterior. Another hormone, relaxin, is thought to aid in parturition. Relaxin, which originates in the corpus luteum or placenta, helps to relax cartilage and ligaments in the pelvic region.

At the beginning of parturition, the offspring typically assumes a position that will offer the least resistance as it passes into the pelvic area and through the birth canal. Fetuses of the cow, mare, and ewe assume similar positions in which the front feet are extended with the head between them (Fig. 10.18). Fetal piglets do not orient themselves in any one direction, which does not appear to affect the ease of birth. Calves, lambs, or foals may occasionally present themselves in a number of abnormal positions (Fig. 10.18). In many of these situations, assistance needs to be given at parturition. Otherwise, the offspring may die or be born dead, and in some instances, the mother may die as well. An abnormally small pelvic opening or an abnormally large fetus can cause some mild to severe parturition problems.

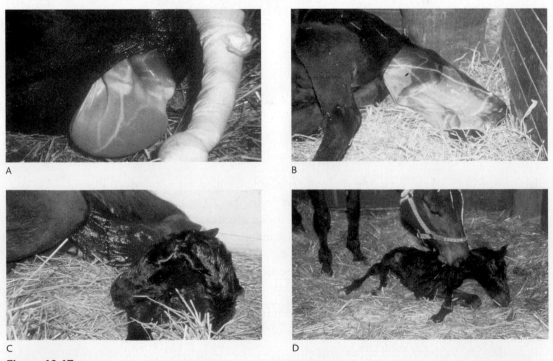

Figure 10.17

Parturition in the mare. (A) The water bag (a fluid-filled bag) becomes visible. (B) The head and front legs of the foal appear. (C) The foal is forced out by the uterine contractions of the mare. (D) In a few moments, the mare is licking the foal to increase circulation and stimulate the newborn. Source: Colorado State University.

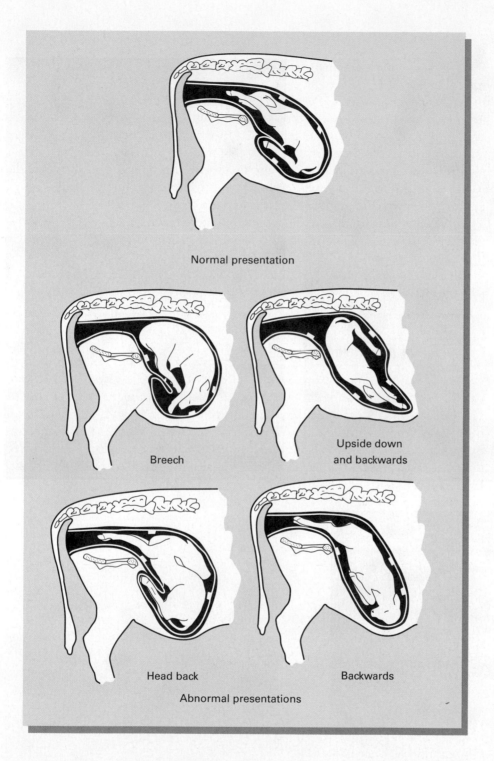

Normal presentation

Breech

Upside down
and backwards

Head back

Backwards

Abnormal presentations

Figure 10.18
Normal and some abnormal presentations of the calf at parturition. Source: Figure from *Handbook of Livestock Management Techniques* by Richard Battaglia, 4E, pp. 146, 149, 150. Copyright © 2007 by Pearson Education, Inc. Reprinted and Electronically reproduced by permission of Pearson Education Inc., Upper Saddle River, New Jersey.

Producers can manage their herds and flocks for high reproductive rates. Management decisions are most critical that cause males and females selected for breeding to reach puberty at early ages, have high conception rates, and have minimum difficulty at parturition. Keeping animals healthy, providing adequate levels of nutrition, selecting genetically superior animals, and paying attention to parturition are some of the more critical management efforts required to minimize reproductive loss. The ability of every female in the herd to give birth to live offspring is the key to successful farm animal reproduction (Fig. 10.19).

Figure 10.19
Live offspring are born to each breeding female in the herd or flock when the intricate mechanisms of reproduction function properly. (A) Beef cow and calf. (B) Ewe and lambs. (C) Nursing pigs. (D) Chicks starting on feed. Source: 10.19a: Coleman Locke; 10.19b: Tom Field; 10.19c: Juhanson/Fotolia; 10.19d: Willee Cole/Fotolia.

CHAPTER SUMMARY

- Reproductive efficiency, as measured by number of offspring born in a herd or flock of breeding females, is a trait of high economic importance.

- Knowledge of the anatomy and physiology of the male and female reproductive organs is essential in understanding the production of ova and sperm, estrous cycles, mating, pregnancy, and parturition.

- Follicle-stimulating hormone, luteinizing hormone, estrogen, progesterone, and testosterone are important hormones in the reproductive process.

- Reproduction in poultry is unique in that ova (eggs) are incubated outside the body, produced from only one ovary and there are no well-defined estrous cycles or pregnancy.

KEY WORDS

oviducts (Fallopian tubes)
uterus
cervix
ovaries
follicle
Graafian follicle
corpus luteum
infundibulum
parturition
cloaca
magnum
isthmus
testicles
sperm
seminiferous tubules
epididymis
vas deferens
ductus deferens
ampulla
Leydig cells or interstitial cells
testosterone

accessory sex glands
glans penis
luteinizing hormone (LH)
follicle-stimulating hormone (FSH)
gonadotropic hormone
gonadotropin-releasing hormone (GnRH)
estrogen
estrous cycle
estrus (heat)
prostaglandin
conceptus
placenta
placentation
Diffuese
Cotyledonary
placentomes
caruncle
cotyledon
afterbirth
parturition

REVIEW QUESTIONS

1. Define the terms associated with anatomical positioning.
2. Describe the primary reproductive organs and their functions in male and female livestock and poultry.
3. Describe several abnormal egg formation conditions.
4. Compare and contrast the seminal characteristics of male livestock and poultry.
5. List the hormones and their roles in regulating reproductive function.
6. Describe the estrous cycle and define its phases/stages.
7. Identify which species demonstrate polyestrus, seasonal polyestrus and monoestrus reproductive performance.
8. Compare and contrast age at puberty, length of estrous cycles, duration of estrus, and timing of ovulation for the primary livestock species.
9. Describe the effects of seasonal factors on reproduction.
10. Describe the structures that establish and maintain pregnancy.
11. Compare and contrast gestation length, placentation, and normal fetal number for livestock.
12. Describe normal parturition.

SELECTED REFERENCES

Battaglia, R. A. 2007. *Handbook of Livestock Management.* Upper Saddle River, NJ: Pearson-Prentice Hall.

Bearden, H. J., J. W. Fuquay, and S. T. Willard. 2004. *Applied Animal Reproduction.* Upper Saddle River, NJ: Prentice Hall.

Bone, J. F. 1988. *Animal Anatomy and Physiology.* Reston, VA: Reston Publishing.

Frandson, R. D. 1986. *Anatomy and Physiology of Farm Animals.* Philadelphia, PA: Lea & Febiger.

Hafez, E. S. E. (ed.). 1993. *Reproduction in Farm Animals.* Philadelphia, PA: Lea & Febiger.

King, G. J. (ed.). 1993. *Reproduction in Domesticated Animals.* Amsterdam, the Netherlands: Elsevier Science Publishers.

McDonald, L. E. 1989. *Veterinary Endocrinology and Reproduction.* Philadelphia, PA: Lea & Febiger.

Pickett, B. W., J. L. Voss, E. L. Squires, and R. P. Amann. 1981. *Management of the Stallion for Maximum Reproductive Efficiency.* Fort Collins: Colorado State University Press, Animal Reprod. Lab. Gen. Series 1005.

Senger, P. L. 2003. *Pathways to Pregnancy and Parturition.* Pullman, WA: Current Conceptions, Inc.

11

Artificial Insemination, Estrous Synchronization, and Embryo Transfer

Manipulation of the reproductive patterns of farm animals has long been of interest to livestock managers seeking to improve efficiency and profitability. The science of reproductive physiology has yielded tremendous technological breakthroughs over the past several centuries (Table 11.1).

In the process of **artificial insemination (AI)**, semen is deposited in the female reproductive tract by artificial techniques rather than by natural mating. AI was first documented as successfully accomplished in the dog in 1780 and in horses and cattle in the early 1900s. AI techniques are also available for use in sheep, goats, swine, poultry, and a number of other species.

The primary advantage of AI is that it permits extensive use of genetically superior sires to maximize genetic improvement. The adoption of AI by the livestock industry has allowed for the development of extensive progeny testing programs to identify outstanding sires that can then be utilized widely in the industry. For example, a bull may sire 30–50 calves naturally per year over a productive lifetime of 3–8 years. In an AI program, a bull can produce 200–400 units of semen per ejaculate, with four ejaculates typically collected per week. If the semen is frozen and stored for later use, hundreds of thousands of calves can be produced by a single sire (one calf per 1.5 units of semen), and many of these offspring can be produced long after the sire is dead. Additional advantages of AI include:

- Disease control—venereal disease can substantially lower reproductive rates. Artificial breeding allows the testing of sires to determine if they are carrying pathogens, avoids the transmission of disease that may occur in natural mating systems, and via the use of antimicrobials in semen extenders, the presence of disease-causing organisms in the semen can be substantially reduced.
- Reduces potential for injury to the male and female. Breeding injuries to males are relatively common in natural mating. Particularly in the case of horses, natural mating may be somewhat violent with mares kicking or striking the stallion as well as the sire biting and raking the mare with his hooves prior to and during mating. The use of AI minimizes the opportunity for injuries that may reduce or eliminate reproductive potential.
- Allows use of sires that have been injured or have poor breeding habits such as low **libido (sexual drive)**. Sires that have chronic lameness, other musculoskeletal injuries, or breeding habits that limit their

learning objectives

- Describe the advantages and disadvantages of various reproductive technologies
- Describe the process of semen collection and processing
- Describe the extent of AI utilization in the livestock industry
- Explain the process of artificial insemination
- Define the role of estrous synchronization in reproductive management
- Compare and contrast several synchronization protocols
- Describe the procedure to transfer embryos
- Explain the value of sexed semen to the livestock industry

Table 11.1
HISTORY OF REPRODUCTIVE TECHNOLOGY DEVELOPMENT

Date	Technology	Animal Model
1780	Successful artificial insemination	Dog
1890	First successful embryo transfer	Rabbit
1949	First successful farm animal ET	Sheep
1951	Successful ET	Pigs/cattle
1973	Offspring produced from frozen embryos	Cattle

effectiveness in natural mating. For example, Tom turkeys tend to have low libido and have so much breast muscle that they are incapable of natural mating. Thus AI is the norm in the turkey industry.

• Extends the productive life of older sires. Fertility is diminished as males age due to reduced production and viability of sperm. With a carefully planned collection and seminal evaluation schedule, a mature and highly desirable sire can produce progeny well beyond normal expectations. This is of particular value in the horse industry where most breeds do not allow registration of progeny from matings using frozen semen.

• Provides early detection of reduced semen quality. Seminal characteristics such as normal morphology, motility, and other quality factors are evaluated at each collection and thus problems can be detected quickly and remedial action taken.

There are disadvantages to AI that should be considered before adopting the technology. These include increased intensity of management; increased costs resulting from higher labor requirements, animal handling facilities, equipment, acquisition of semen; and the potential for low reproductive rates if AI protocols are not implemented correctly.

Furthermore, high-quality management of disease prevention and nutrition coupled with access to trained technicians are required for success.

The success of an AI program is dependent on superior estrous detection, high levels of fertility, high quality of semen, and skilled semen handling and insemination technique.

SEMEN COLLECTING AND PROCESSING

There are several different methods of collecting semen. The most common method is the **artificial vagina** (Fig. 11.1), which is constructed to be similar to an actual vagina. The artificial vagina is commonly used to collect semen from bulls, stallions, rams, buck goats, and rabbits. The semen is collected by having the male mount an estrous female or training him to mount another animal or object. When the male

Figure 11.1
Longitudinal section of an artificial vagina. Source: Figure by Kevin Pond from Colorado State University. Used by permission of Kevin Pond.

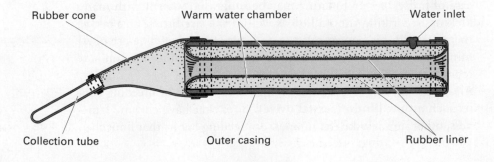

Rubber cone   Warm water chamber   Water inlet

Collection tube   Outer casing   Rubber liner

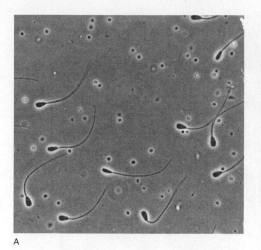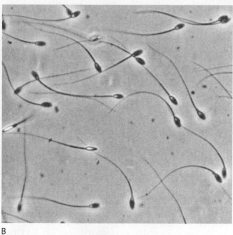

A   B

Figure 11.2
(A) Normal bull semen. (B) Normal stallion semen.
Source: Figure by George Seidel from Colorado State University. Used by permission of George Seidel.

mounts, the penis is directed into the artificial vagina by the person collecting the semen, and the ejaculated semen accumulates in the collection tube. Semen from the boar and dog is not typically collected with an artificial vagina. It is collected by applying pressure with a gloved hand, after grasping the extended penis, as the animal mounts another animal or object.

Semen can also be collected via an **electroejaculator**, in which a probe is inserted into the rectum and through a wave of electrical pulses ejaculation results. This is used most commonly in bulls and rams that are not easily trained to use the artificial vagina and from which semen is collected infrequently.

If semen is collected too frequently, the numbers of sperm per ejaculate declines. Semen from a bull is typically collected twice a day for 2 days a week. Semen from rams can be collected several times a day for several weeks, but bucks (goats) must be ejaculated less frequently. Boars and stallions give large numbers of sperm per ejaculate, so semen from them is usually collected every other day at most.

Semen is collected from tom turkeys two to three times per week. The simplest and most common technique for collecting sperm from toms is the abdominal massage. This technique usually requires two persons: the "collector" helps hold the bird and operates the semen collection apparatus, while the "milker" stimulates a flow of semen by massaging the male's abdomen.

After semen is collected, it is evaluated for volume, sperm concentration, motility of the sperm, and sperm abnormalities (Fig. 11.2). The semen is usually mixed with an extender that dilutes the ejaculate to a greater volume. This greater volume allows a single ejaculate to be processed into several units of semen, where one unit of semen is used for each female inseminated. The extender is usually composed of nutrients such as milk and egg yolk, a citrate buffer, and antibiotics. The amount of extender used is based on the projected number of viable sperm available in each unit of extended semen. For example, each unit of semen for insemination in cattle should contain 10 million motile, normal spermatozoa.

Some semen is used fresh; however, it can be stored this way only for a day or two. Most semen has glycerol added as a cryoprotectant prior to freezing in liquid nitrogen and storage in plastic straws (Fig. 11.3A) or pellets. Bull semen can be stored in this manner for an indefinite period of time and retain its fertilization capacity. Most cattle are inseminated with frozen semen that has been thawed (Fig. 11.3B). The fertilizing capacity of frozen semen from boars, stallions, and rams is only modestly satisfactory; however, improvement has been noted in recent years. Turkey semen cannot be frozen satisfactorily. For maximum fertilization capacity, it should be used within 30 minutes of collection.

Figure 11.3
(A) Bovine semen is typically packaged in plastic straws and then stored in liquid nitrogen. The bull's name, registration number, and inventory code are usually printed on each straw to ensure accurate identification of semen during breeding season.
(B) Semen is then thawed in a water bath and the straw is loaded into a warmed insemination tube prior to placement of semen into the female's reproductive tract.
Source: 11.3a and b: Tom Field.

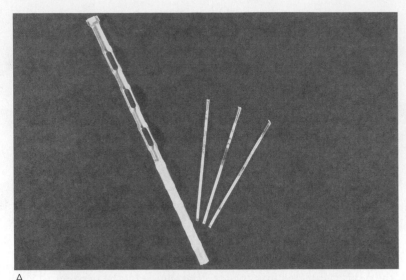

A

B

INSEMINATION OF THE FEMALE

Prior to insemination, the frozen semen is thawed. Thawed semen cannot be refrozen and used again because refreezing will kill the sperm cells.

High conception rates using AI depend on the female's cycling and ovulating; detecting estrus; using semen that has been properly collected, extended, and frozen; thawing and handling the semen satisfactorily at the time of insemination; insemination techniques; and avoiding extremes in stress and excitement to the animal being inseminated. Percent of total matings achieved by AI are summarized in Table 11.2. The turkey industry has the highest use (100%) with the sheep industry least likely to use the technology.

Semen collection, processing, and sales is a robust business that involves sales not only domestically but also in the global marketplace. The primary genetic providers in the swine industry, for example, are companies with international ownership and market reach. Table 11.3 illustrates the number of units of dairy and beef cattle semen sold in the U.S. market, exported abroad, and custom collected typically for within herd use. The use and sale of semen in the dairy industry is considerably larger scale than in the beef industry. The importance of the global market for some breeds is substantial. Note that in the beef breeds listed in Table 11.3, export sales eclipse marketings within the United States. For example, the sale of Brahman and Brahman influenced genetics domestically is relatively small. However, the export market for these genetics is 12 times higher.

Table 11.2
PERCENT OF AI MATINGS IN VARIOUS ANIMAL INDUSTRIES

Animal	% AI
Beef cattle	8–10
Dairy cattle	80–90
Horses	40–50
Swine	80–85
Sheep	<5
Turkeys	100

Table 11.3
UNITS OF SEMEN SOLD FOR DOMESTIC USE, CUSTOM FROZEN, AND EXPORTED (2012)

Breed	Domestic Sales (1,000 units)	Custom Frozen (1,000 units)	Export Sales (1,000 units)
Holstein	19,965	2,287	14,515
Jersey	2,479	290	1,268
Total dairy	23,024	2,782	16,583
Angus	1,238	1,049	2,007
Red Angus	100	150	298
Hereford	18	200	100
Brahman and Brahman influenced	14	220	175
Total beef	1,669	2,854	2,695

Source: Adapted from the National Association of Animal Breeders and USDA.

Detecting Estrus

Estrus must be detected accurately because it signals time of ovulation and determines proper timing of insemination. The best indication of estrus is the condition called **standing heat**, in which the female stands still when mounted by a male or another female.

Cows are typically checked for estrus twice daily, in the morning and evening. They are usually observed for 30 minutes to detect standing heat. Other observable signs are restlessness, attempting to mount other cows, and a clear, mucous discharge from the vagina. Some producers use sterilized bulls or hormone-treated cows as "heat checkers" in the herd. These animals are sometimes equipped with a chin marker that greases or paints marks on the back of the cow when she is mounted. Electronic systems are available but may be cost-prohibitive. Alternatives include the use of chalk on the tail heads of cows or patches that are secured to the tail head and change color under the pressure of mounting behavior (Fig. 11.4).

Estrus in sheep or goats is checked using sterilized males equipped with a brisket-marking harness. Does and ewes exhibit swollen vulvas, male-seeking behavior, tail flagging, frequent urination, and increased vocalization during estrus. Gilts and sows in heat assume a rigid stance with ears erect when pressure is applied to their back and sides. Heat detection in swine can be accomplished by handlers pressing down on the sow's back. The presence of a boar and the resulting sounds and odors can stimulate heat in swine. The females in estrus will also indicate they are in heat

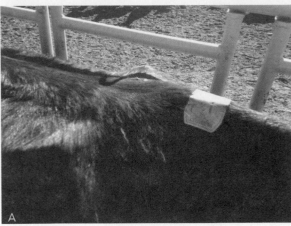

Figure 11.4

Heat detection patches can be used to detect estrus activity in cattle. (A) The patch is silver when placed onto the tailhead of the female. (B) If she is mounted either by a teaser bull or other females, the patch will change to a bright orange color. Source: 11.4a and b: Tom Field.

as they become more active in attempting to locate a boar. Signs of estrus in the mare are elevation of the tail, contractions of the vulva (**winking**), spreading of the legs, and frequent urination. Mares will frequently be exposed to a stallion in a process called **teasing** to stimulate estrus.

Proper Insemination

The duration of estrus and ovulation time are quite variable in farm animals. This variability poses difficulty in determining the best time for insemination. An additional challenge is that sperm are short-lived when put into the female reproductive tract. Also, estrus is sometimes expressed without ovulation occurring; sows, for example, typically show estrus 3–5 days after farrowing, but ovulation does not occur. Sows should not be bred at this time either artificially or naturally.

Insemination time should be as close to ovulation time as possible; otherwise sperm stay in the female reproductive tract too long and lose their fertilizing capacity. Cows found in estrus in the morning are usually inseminated the evening of the same day, and cows in heat in the evening are inseminated the following morning. Insemination, therefore, should occur toward the end or after estrus has been expressed in the cow. Ewes are usually inseminated in the second half of estrus, and goats are inseminated 10–12 hours after the beginning of estrus. Sows ovulate from 30 to 38 hours after the beginning of estrus, so insemination is recommended at the end of the first day and at the beginning of the second day of estrus. When sows are inseminated both days conception rate is improved compared to insemination on only one day.

Insemination of dairy cows occurs while the cow is standing in a stall or stanchion. Beef cows are penned and inseminated in a chute that restrains the animal. The most common insemination technique in cattle involves the inseminator having one arm in the rectum to manipulate the insemination tube through the cervix (Fig. 11.5). The insemination tube is passed just through the cervix, and the semen is deposited into the body of the uterus.

The insemination procedure for sheep and goats is similar to that for cattle; however, a **speculum** (a tube approximately 1.5 in. in diameter and 6 in. long) allows the inseminator to observe the cervix in sheep and goats. The inseminating tube is passed through the speculum into the cervix, where the semen is deposited into the

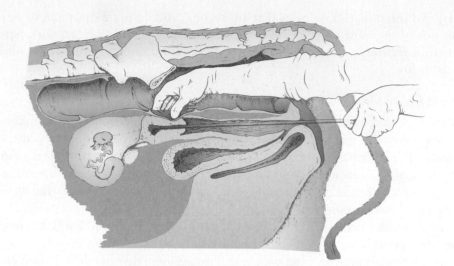

Figure 11.5
Artificial insemination of the cow. Note that the insemination tube has been manipulated through the cervix. The inseminator's forefinger is used to determine when the insemination rod has entered into the uterus.
Source: Robert Taylor & Tom Field.

uterus or cervix. Typically, the female will be placed in a breeding stanchion that elevates her rear quarters for easier access. AI is more effective in goats than sheep.

The sow is usually inseminated without being restrained. The inseminating tube is easily directed into the cervix because the vagina tapers into the cervix. The semen is expelled into the body of the uterus. The mare is hobbled or adequately restrained prior to insemination. The vulva area is washed, and the tail is wrapped or put into a plastic bag. The plastic-covered arm of the inseminator is inserted into the vagina, and the index finger is inserted into the cervix. The insemination tube is passed through the cervix, and the semen is deposited into the uterus.

The turkey hen is inseminated by first applying pressure to the abdominal area to cause version of the oviduct. The insemination tube is inserted into the oviduct approximately 2 in., and the semen is released. The first insemination is made when 5–10% of the flock has started laying eggs. A second insemination a week later ensures a high level of fertility. Thereafter, insemination is done at 2-week intervals. Fertility in the turkey usually persists at a high level for 2–3 weeks after insemination. This is possible because sperm are stored in special glands of the hen, where they are nourished and retain their fertilizing capacity. Most of these glands are located near the junction of the uterus and vagina.

ESTROUS SYNCHRONIZATION

Estrous synchronization involves controlling or manipulating the estrous cycle so that females in a herd express estrus at approximately the same time. It is a useful part of an AI program because checking heat and breeding animals, particularly in large pasture areas, is time consuming and expensive. Estrous synchronization is a tool used in successful embryo transfer programs and it can be used with natural service where bulls are used intensively in breeding cows for a few days. Synchronization of estrus is used in approximately 8% of U.S. cowherds but in herds with more than 200 head of cows, the rate increases to nearly 20% (NAHMS, 2008).

Success of a synchronization program depends on many factors, including facilities, available labor, the body condition, postcalving interval, and fertility level of the cows and heifers. Also important are herd health, high-quality semen, qualified inseminators, females that are cycling, and accurate detection of estrus. But one of the most serious considerations affecting a herd's success with estrous synchronization is selecting the proper method. Several example protocols are listed below.

Prostaglandin

In 1979, prostaglandin was cleared for use in cattle. **Prostaglandins** are naturally occurring fatty acids that have important functions in several of the body systems. The prostaglandin that has a marked effect on the reproductive system is prostaglandin F_2 alpha. Lutalyse®, Estrumate®, and Prostamate® are commonly used prostaglandins. It is important to handle all hormonal synchronization products carefully and all label warnings should be observed.

Earlier in Chapter 10, it was pointed out that the corpus luteum (CL) controls the estrous cycle in the cow by secreting the hormone progesterone. Progesterone prevents the expression of heat and ovulation. Prostaglandin destroys the CL, thus eliminating the source of progesterone. About 3 days after the injection of prostaglandin, the cow will be in heat. For prostaglandin to be effective, the cow must have a functional CL. It is ineffective in heifers that have not reached puberty or in noncycling mature cows. In addition, prostaglandin is ineffective if the CL is immature or has already started to regress. Prostaglandin is, then, only effective in heifers and cows that are in days 5–18 of their estrous cycle. Because of this relationship, prostaglandin is given in either a one- or two-injection system separately or in combination with other products that influence the estrous cycle.

One-Injection System with 10 Days of Breeding

Days 1–5:   The first 5 days are a conventional AI program of heat detection and insemination.

Day 6:   At this point, the herd owner must decide whether to proceed with the prostaglandin injection. If the percent cycling is satisfactory and drug costs are considered reasonable, the breeder injects the females not already detected in heat with prostaglandin.

Days 6–11:   The synchronized AI breeding season continues with 5 more days of heat detection and insemination (Fig. 11.6).

Days 27–33:   Cattle that synchronize but do not settle on the first service will return to heat. Repeat services can be performed at this time to produce additional AI calves.

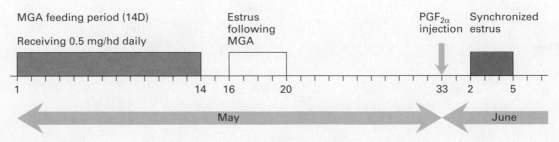

June 2 is the first day of the breeding season.

Figure 11.6
The MGA-prostaglandin (PGF$_{2\alpha}$) system for synchronizing estrus. Source: Figure by Kevin Pond from Colorado State University. Used by permission of Kevin Pond.

This is the most popular protocol that uses only prostaglandin to synchronize estrus and can result in more than 90% of cyclic cows being bred the first 10 days of the breeding season. This system can be shortened by not inseminating during days 1–5. However, this will lower heat detection to 65–75%. In any event, if less than 20% of the females express heat in the first 5 days, success will be limited. Timed inseminating should not be used with this system.

MGA and Prostaglandin

The use of **MGA (melengestrol acetate)** in conjunction with prostaglandin results in an inexpensive, easily administered synchronization system. This system best fits the breeding of heifers being raised in semiconfinement. The MGA–prostaglandin system showing the most promise is the following 31- to 36-day system:

Days 1–14: Feed 0.5 mg MGA per head per day. The MGA prevents heifers from coming into heat during this period. MGA can be fed as topdressing or can be mixed in with the feed. It is important to provide enough bunk space for all heifers to have an opportunity to eat all the required amount of MGA.

Day 15: Stop feeding MGA. Heifers should come into heat within the next 7 days. The fertility at this heat is very low, so do not inseminate.

Day 31: Inject prostaglandin. Heifers and cows respond to the prostaglandin by coming into heat within the next 5 days.

Days 32–38: Inseminate according to estrus. While the synchronized estrus will be relatively tight, some females may not show heat until approximately 4 days after injection. Some heifers may not initiate a cycle, so semen is wasted if timed insemination is used. For natural service, the prostaglandin shot can be waived with bull turnout occurring on that day.

Research studies suggest that tighter estrus synchronization is achieved if the prostaglandin is given on day 33, 19 days following the end of MGA feeding.

Prostaglandin is not a wonder drug; it will work only in well-managed herds where a high percentage of the females are cycling. Biologically, it has been well demonstrated that prostaglandin can synchronize estrus. Producers, however, must weigh the cost against the economic benefit. Caution should be exercised in administrating prostaglandin to pregnant cows as it may cause abortion. Estrous synchronization in cattle may not be advisable in areas where inadequate protection can be given to young calves during severe blizzards. Also, herd health programs must be excellent to prevent high losses from calf scours and other diseases that become more serious where large numbers of newborn calves are grouped together.

Controlled Internal Drug Release (CIDR)

The FDA approved the use of **CIDR** technology in synchronization of beef cows and heifers in 2002, thus providing the ability to use intravaginal release of progesterone. Marketed as the CIDR® Cattle Insert, the device is a T-shaped nylon structure coated with a layer of silicone containing 10% progesterone by weight (Fig. 11.7). The device is inserted into the vagina with a lubricated applicator and it can be removed by grasping and pulling the flexible tail of the CIDR. Minimal levels of vaginitis may be observed but do not interfere with fertility. Retention rates typically exceed 95% when correct insertion protocols are followed.

The CIDR is inserted on day 0, females are administered an injection of prostaglandin on day 6, and the CIDR is removed on day 7. An alternate protocol calls for the injection on the same day as CIDR removal. Standing heat should be observed

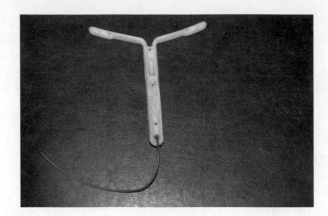

Figure 11.7
CIDR (controlled internal drug release) is a T-shaped device with flexible wings that collapse to form a rod for insertion into the vagina. The CIDR's nylon core is covered with a progesterone-coated silicone layer. A vinyl cord is attached to the tail of the device to aid in removal so that the handler can grasp the blue string that extends external to the vulva and with gentle pressure can remove the device as the wings fold upward allowing it to slip easily from the reproductive tract. A CIDR can be used as a component of estrous synchronization in cattle, sheep and goats.
Source: Tom Field.

within 1 to 3 days following removal of the CIDR. This protocol has been effective in both noncycling and cycling heifers and cows in terms of percentage of animals expressing estrus and conceiving.

Synchronization with Natural Service

Some producers cannot economically use AI but would like the benefits of estrous synchronization. Research has shown that natural service can be used with estrous synchronization if managed properly. One bull per 15 to 20 females in a small pasture (or drylot), and rotated every 24 hours with a rested bull, is recommended during the synchronization period of 4–5 days. Observations have shown that one bull may service 5 to 20 females in a 24-hour period. Pregnancy rates during the synchronized period have ranged from 60–80% and 75–95% during a 30-day breeding season.

Bulls can be used in all synchronization programs. The most popular program is feed MGA for 14 days, wait 17 days, then place the bulls with the females. The advantages of this program are low drug costs, no heat detection, and less demand on the bulls in a short time period.

P.G. 600

P.G. 600 (pregnant mare serum gonadotropin) is utilized to enhance fertility efficiency in the swine industry. This product is used to induce fertile estrus in noncycling gilts over 5.5 months of age and weighing a minimum of 185 lb as well as weaned sows. The advantage of synchronizing gilts lies in the ability to more efficiently schedule breeding and farrowing facilities as well as lowering the days between weaning and return to estrus. However, as is the case with all synchronization schemes, the cost and benefit relationship must be carefully evaluated prior to adoption.

In a sense, there are some natural occurrences of estrous synchronization. The weaning process in swine is an example; the sow will typically show heat 3–8 days after the pigs are weaned. Estrus is suppressed through the suckling influence, so when the pigs are removed, the sow will show heat. Several beef cattle synchronization protocols call for 48-hour calf removal to enhance response rates.

EMBRYO TRANSFER

The primary function of **embryo transfer** is to increase the reproductive rate of valuable females by tenfold or more in a given year and fivefold or more in a cow's lifetime. Even greater increases in reproductive rates can be expected as new technologies are improved. Other uses of embryo transfer are to obtain offspring from infertile cows and to export or import breeding stock to reduce disease risk.

Embryo transfer is sometimes referred to as *ova transplant* or *embryo transplant*. In this procedure, an embryo in its early stage of development is removed from its own mother's (the donor's) reproductive tract and transferred to another female's (the recipient's) reproductive tract. The first successful embryo transfers were accomplished in rabbits in 1890 and in cattle in 1951. Commercial embryo transfer companies have been established in the United States and several foreign countries. More than 400,000 embryos are transferred annually throughout the world with about one-half of the embryos utilized in the United States. Approximately one-half of the embryos are fresh, with the remaining half frozen.

Superovulation is the production of a greater-than-normal number of eggs. Females that are donors for embryo transfer are injected with fertility drugs, which usually cause several follicles to mature and ovulate. The two most common methods of superovulation are using pregnant mare serum gonadotropin (PMSG) or follicle-stimulating hormone (FSH), with the latter usually producing more usable embryos. FSH injections are given over 3–4 days with prostaglandin $F_{2\alpha}$ (e.g., Lutalyse®) administered usually on the third day.

A nonsurgical embryo collection procedure occurs when a flexible rubber tube (Foley catheter) with three passageways is passed through the cervix and into the uterus. A rubber balloon, which is built into the anterior end of the tube, is then inflated to about half the size of a golf ball so that it expands to fill the uterine lumen and to prevent fluid from escaping around the edges. There are two holes in the tube anterior to the balloon that lead into separate passageways, one for fluid entering the uterus and the other for fluid draining from the uterus. A balanced salt solution (to which antibiotics and heat-treated serum are added) is placed in a container and held about 3 ft above the cow. The container is connected to the Foley catheter by an inflow tube. A second tube is connected to the other passageway of the catheter to drain off the medium that has washed (flushed) the ova and embryos out of the uterus. The solution is then collected in tall cylinders holding about 2 pints of fluid. Filtering the collected fluid through a cup-sized container with a fine stainless-steel filter in the bottom and then searching with a microscope can isolate the embryos.

At approximately 12 hours and 24 hours after first standing estrus, the donor is bred artificially. Usually frozen-thawed semen is used and 2–4 straws of high-quality semen is recommended for each insemination. The recipients are usually synchronized with a prostaglandin. The recipient cow must be in estrus within 1.5 days of the donor's estrus for best results. Recipients are selected for calving ease, high milk production, and in excellent health status.

An ideal response is 5 to 12 usable embryos per donor per superovulation treatment; however, a range of 0 to more than 20 embryos can be expected . On average, two to four calves will result per superovulated donor if fertile donors are utilized in a well-managed embryo transfer program. A potential donor cow usually produces more embryos if she is 3–10 years of age, has calved regularly each year of her productive life, usually conceives in two services or less, has exhibited regular estrous cycles, comes from fertile bloodlines, did not retain her placenta, and has no history of calving difficulties. The bull, to be bred to the donor cow, should have a history of excellent semen production and a high conception rate determined by his semen having been used in an AI program.

The key to the justification of embryo transfer is identifying genetically superior cows and bulls. Making repeat matings where genetically superior calves have previously been produced can eliminate much of the guesswork. Embryo transfer is usually confined to seedstock herds, where genetically superior females can be more easily identified and high costs can be justified. Embryo transfer calves have an estimated average break-even price of $1,500–$2,500 at 1 year of age (costs for embryo transfer are $500–$1,000 above the usual production costs). Seedstock producers must evaluate the marketability of the embryo transfer calves. Some genetically superior embryo transfer calves may not sell for a sufficient amount to cover costs. Some embryo transfer calves that are not genetically superior may be merchandised for high prices because of a demand that has been previously created.

Donor selection that combines expected high levels of financial return, genetic superiority, and high reproductive potential indicates that the numbers of donors worthy of an embryo transfer program are relatively few. This "relatively few" number could be justified into several thousand head, which is less than 1% of the total U.S. cowherd.

Although embryo transfer was done surgically in the past, today new nonsurgical techniques are used. The embryos are transferred by way of an artificial insemination gun shortly after being collected. The recipient females need to be in the same stage (within 36 hours) of the estrous cycle for highly successful transfers to occur. Large numbers of females must be kept for this purpose, or estrous synchronization of a smaller number of females will be necessary. The embryos can be frozen in liquid nitrogen and remain dormant for years or decades.

Although the conception rate is lower for frozen embryos and higher for fresh embryos, ongoing research is narrowing the difference in conception rate. In cattle, approximately 85% of frozen embryos are normal after thawing; however, 40–55% of those normal embryos result in confirmed pregnancies at 60–90 days. In contrast, fresh embryos transferred the same day of collection have a pregnancy rate of 55–65%.

While there is no decline in the embryo recovery rate the first three times a donor is superovulated, additional superovulation treatments result in a reduced number of embryos for some donors. Most donors can be superovulated three or four times a year. Season, breed, and lactating versus dry donors appear to have little effect on the success of embryo transfer. Surveys have shown that 2% of the recipients abort between 3 and 9 months of gestation, 4% of the embryo transfer calves die at birth, and another 4% die between birth and weaning—representing a total loss of 10% between 3 months gestation and weaning. These losses may be similar to those experienced in a typical natural service program.

Recent advances in reproduction technology include the birth of the first test-tube calf (fertilization occurred outside the cow's body) in 1981 and the first identical twin calves resulting from embryo splitting. Splitting one embryo (Fig. 11.8) to produce two identical offspring is one option in the cloning of livestock.

Cloning

Cloning is the process of creating a biological copy of an organism. Cattle, sheep, pigs, goats, mules and horses have been successful cloned using a complex process called somatic cell nuclear transfer (SCNT). SCNT involves the transfer of genetic material from the nucleus of one cell into a recipient, unfertilized egg that has had its nucleus removed prior to the exchange. Cell division is then activated and the egg is implanted into a recipient dam.

The first clone cloned from an adult cell was Dolly the sheep at Scotland's Roslin Institute in 1996. To date, cloning remains an inefficient method of reproduction given the low success rate of embryo creation from the transfer of nuclear material from a parent cell into enucleated (nuclear material removed) eggs. It is quite doubtful that cloning of livestock to produce food and fiber will be commercialized.

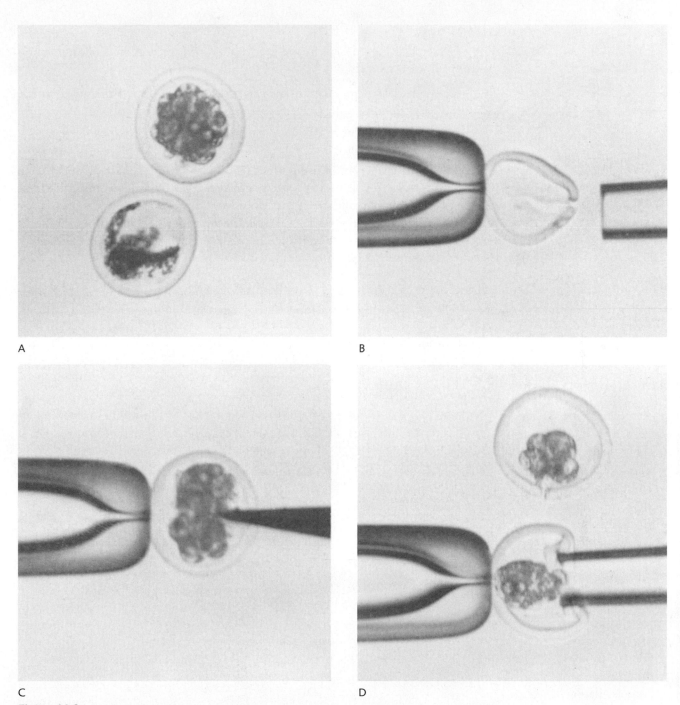

A

B

C

D

Figure 11.8

The process of embryo splitting. (A) A morula (40–70 cells of the fertilized egg) and an unfertilized egg. (B) The unfertilized egg with the cellular contents aspirated from it. (C) The morula is divided into two groups of cells with the microsurgical blade. (D) One-half of the cells are left in the morula while the other one-half of the cells are placed inside the other unfertilized egg. The result is two genetically identical embryos ready for transfer. Source: Courtesy of Williams et al., 1983 NAAB Conference, Beef AI and Embryo Transfer.

However, there may be medical applications that involve the creation of transgenic clones designed to produce compounds in their milk that may be useful in treating diseases such as hemophilia and cystic fibrosis. There may also be applications in regards to efforts to preserve genetic diversity by propagating rare species of both wild and domestic origin.

Sexed Semen

The commercialization of sexed semen technology offers a multitude of opportunities to livestock producers:

1. Initiating female pregnancies in first-calf heifers to help control rates of dystocia.
2. Creation of target-specific progeny. For example, production of only male calves from terminal sire AI matings.
3. Creating only female progeny from high-performance maternal genetics.

One of the challenges in using sexed semen as a genetic tool is the reality that because sorting protocols using flow cytometry are relatively slow, inseminations must be made with low-dose sperm concentrations. The normal sperm concentration for AI in sheep and cattle is 60 and 20 million, respectively. Thus, the use of highly fertile sires coupled with superior female management protocols, superior estrus detection, and insemination is required to achieve favorable results.

The use of sexed semen results in predicting fetal sex rates of greater than 85–90% for cattle and sheep. The technology is being applied to cattle, sheep, swine, horses, elk, yaks, gorillas, sea mammals, dogs, and cats.

CHAPTER SUMMARY

- Artificial insemination (AI) is the deposition of semen in the female reproductive tract by methods other than natural mating.

- The advantages of AI are the extended use of sires, controlling reproductive diseases, and using sires that have been injured or are dangerous when used naturally.

- AI is widely used in dairy cattle and turkeys and to a lesser extent in other farm animal species.

- Estrous synchronization is controlling or manipulating the estrous cycle so that females express estrus at approximately the same time.

- Prostaglandin F_2 alpha, melengestrol acetate (MGA), and GnRH are commonly used products to synchronize estrus in cattle.

- Embryo transfer involves removing embryos from females (donors) and transferring them into the reproductive tracts of other females (recipients).

KEY WORDS

artificial insemination (AI)
libido (sexual drive)
artificial vagina
electroejaculator
standing heat
winking
teasing
speculum

estrous synchronization
prostaglandin
MGA (melengestrol acetate)
CIDR
P.G. 600
embryo transfer
superovulation
cloning

REVIEW QUESTIONS

1. What are the advantages and disadvantages of AI?
2. List the protocols for collected semen.
3. Describe semen processing following collection.
4. Compare and contrast the use of AI across farm animal enterprises.
5. Describe procedures to detect estrus and the importance of timing of insemination to achieve high conception rates.
6. Compare AI protocol differences in various livestock species.

7. What is the motivation to utilize estrous synchronization?

8. Compare various protocols to achieve synchronization of estrous.

9. What is the rationale for the use of ET?

10. Describe the basic steps of ET.

11. Explain cloning and sexed semed; define the value of these technologies to livestock production.

SELECTED REFERENCES

American Breeder's Service. 1991. *A.I. Management Manual.* DeForest, WI.

Beal, W. E. 1998. Making Embryo Transfer Easier. Proc. Virginia Beef Genetics Management Conference. Roanoke, VA.

Beal, W. E. and J. M. DeJarnette. 2002. Understanding the Estrous Cycle of the Cow. Proc. Cattlemen's College, NCBA Convention. Denver, CO.

Ernst, R. A., F. X. Ogasawara, W. F. Rooney, J. P. Schroeder, and D. C. Ferebee. 1970. *Artificial Insemination of Turkeys.* University of California Agric. Ext. Pub. AXT-338.

Estienne, M. J. and A. F. Harper. 2001. *Uses of P.G. 600 in Swine Breeding Herd Management.* Virginia Cooperative Extension Livestock Update.

Foote, R. H. 1994. Embryo transfer in domestic animals. *Encyclopedia of Agricultural Science.* San Diego, CA: Academic Press, Inc.

Geary, T. 1997. Synchronization programs update. Proc. of the Range Beef Cow Symposium XV. Rapid City, SD.

Hafez, S. E. 1987. *Reproduction in Farm Animals.* Philadelphia, PA: Lea & Febiger.

Herman, H. A. 1981. *Improving Cattle by the Millions. NAAB and the Development and Worldwide Application of Artificial Insemination.* Columbia, MO: University of Missouri Press.

Kiessling, A. A., W. H. Hughes, and M. R. Blankevoort. 1986. Superovulation and embryo transfer in the dairy goat. *J. Am. Vet. Med. Assoc.* 188:829.

Lake, P. E, and J. M. Steward. 1978. *Artificial Insemination in Poultry.* Scotland Ministry of Agriculture, Fisheries, and Food-N. 213.

National Animal Health Monitoring System (NAHMS). 2008. Part 1: Reference of beef cow-calf management practices, 2007-08. Washington, DC: USDA, APHIS, V.S.

Seidel, G. E., Jr. 1981. Superovulation and embryo transfer in cattle. *Science* 211:351.

Seidel, G. E. Jr. 1995. Reproductive Technologies for Profitable Beef Production. Proc. Natl. Assn. of Animal Breeders Conference. Sheridan, WY.

Seidel, G. E., Jr. S. M. Seidel, and R. A. Bowen. 1978. *Bovine Embryo Transfer Procedures.* Fort Collins: Colorado State University Exp. Sta., Animal Repro. Lab., Gen. Series 975.

12

Genetics

Gregor Mendel, an Augustinian monk, began working with an experimental garden in 1856 and for nearly a decade worked diligently to observe and record the outcome of crossing lines of garden peas from which emerged the mathematical rules governing genetic inheritance. Mendel observed seven characteristics or traits in the pea population—seed shape, seed color, flower color, pod shape, pod color, positioning of the pod and flower, and stem length. While his original manuscript was largely ignored for more than three decades, Mendel is recognized as the founder of modern genetics.

The keen observations made by Mendel were the basis for scientific exploration into the mechanisms by which genetic information was passed from generation to generation. Cambridge University scientists James Watson and Francis Crick in 1953 suggested a three-dimensional model of the basic structure of DNA (deoxyribonucleic acid) that explained the process of inheritance. The findings of these and many other geneticists have revolutionized the livestock and poultry industries. Harnessing the power of genetics has played a significant role in the productivity gains attained in domesticated livestock and poultry species across the globe.

MITOSIS AND MEIOSIS

Each species has a characteristic number of **chromosomes**, the rodlike bodies contained within the nucleus that carries the genetic code. In body cells or **somatic cells**, chromosomes occur in pairs with each parent contributing half of the genetic material to their respective offspring. For example, turkeys have 41 chromosome pairs while swine carry only 19 pairs of chromosomes (Table 12.1). Comparatively, humans have 23 pairs or 46 individual chromosomes. Somatic cells replicate via a process called **mitosis** (Fig. 12.1) during which each parent cell divides into two identical daughter cells each carrying a full chromosome complement. However, during the formation of **gametes** (spermatozoa in the male and ova in the female), the cell division process is termed **meiosis** and results in the creation of daughter cells that carry one-half the normal number of chromosomes so that during fertilization each parent has contributed a sample one-half of their genetic material to the offspring.

PRODUCTION OF GAMETES

The testicles of the male and the ovaries of the female produce sex cells (gametes) by a process called **gametogenesis**. The gametes produced by the testicles are called *sperm;* the gametes produced by the ovaries are

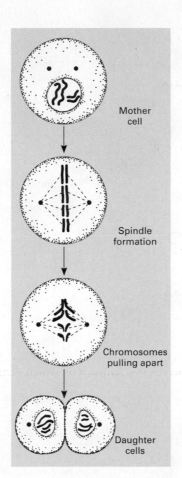

Figure 12.1
Mitosis. Source: Figure by Kevin Pond from Colorado State University. Used by permission of Kevin Pond.

Mother cell

Spindle formation

Chromosomes pulling apart

Daughter cells

Table 12.1
PAIRS OF CHROMOSOMES IN LIVESTOCK AND POULTRY

Species	Number of Pairs
Turkeys	41
Chickens	39
Horses	32
Cattle	30
Goats	30
Sheep	27
Swine	19

called *eggs* or *ova*. Specifically, the production of sex cells that become sperm is called **spermatogenesis**; the production of ova is known as **oogenesis**. The unique type of cell division in which gametes (sperm or ova) are formed is called meiosis. Each newly formed gamete contains only one member of each of the chromosome pairs present in the body cells.

Consider gametogenesis in a theoretical species in which there are only two pairs of chromosomes. Meiosis occurs in the primordial germ cells (cells capable of undergoing meiosis) located near the outer wall of the **seminiferous tubules** of each testicle and near the surface of each ovary. The initial steps in meiosis are similar for the male and female. The chromosomes replicate themselves so that each

Figure 12.2

Meiosis or reduction cell division in the testicle and ovary (example with two pairs of chromosomes).

Source: Figure by Kevin Pond from Colorado State University. Used by permission of Kevin Pond.

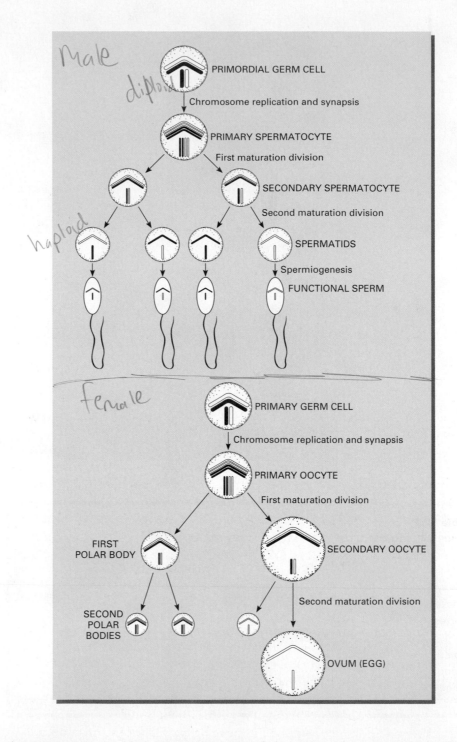

chromosome is doubled. Then each pair of chromosomes comes together in extremely accurate pairing called **synapsis**. After chromosome replication and synapsis, the cell is called a **primary spermatocyte** in the male and a **primary oocyte** in the female. Because subsequent differences exist between spermatogenesis and oogenesis, the two are described here separately.

SPERMATOGENESIS

The process of spermatogenesis in the theoretical species with only two chromosome pairs is shown in Figure 12.2. The primary spermatocyte contains two pairs of chromosomes in synapsis. Each chromosome is doubled following replication. Thus, the

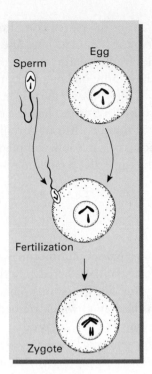

Figure 12.3
Combining of chromosomes through fertilization (two pairs of genes used for simplification of example). Source: Figure by Kevin Pond from Colorado State University. Used by permission of Kevin Pond.

primary spermatocyte contains two bodies or structures formed by four parts. By two rapid cell divisions, in which no further replication of the chromosomes occurs, four cells, each of which contains two chromosomes, are produced. The spermatid cells each lose much of their cytoplasm and develop a tail. This process of spermatogenesis results in formation of sperm. Four sperm are produced from each primary spermatocyte. Whereas four chromosomes in two pairs are present in the primordial germ cell, only two chromosomes (one-half of each chromosome pair) are present in each sperm. Thus, the number of chromosomes in the sperm has been reduced to half the number as compared to the primordial germ cell.

OOGENESIS

The process of oogenesis is shown in Figure 12.2. Like the primary spermatocyte of the male, the primary oocyte of the female contains a tetrad. The first maturation division produces one relatively large nutrient-containing cell (the secondary oocyte) and a smaller cell (the first polar body). Each of these two cells contains a dyad. The unequal distribution of nutrients that results from the first maturation division in the female serves to maximize the quantity of nutrients in one of the cells (the secondary oocyte) at the expense of the other (the first polar body). The second maturation division produces the egg (ovum) and the second polar body, each of which contains two chromosomes. The first polar body may also divide, but all polar bodies soon die and are reabsorbed. Note that the egg, like the sperm, contains only one chromosome of each pair that was present in the primordial germ cell. The gametes each contain one member of each pair of chromosomes that existed in each primordial germ cell.

FERTILIZATION

When a sperm and an egg of our theoretical species unite to start a new life, each contributes one chromosome to each pair of chromosomes in the fertilized egg, now called a **zygote** (Fig. 12.3). **Fertilization** is defined as the union of the sperm and the egg along with the establishment of the paired condition of the chromosomes. The zygote is termed **diploid** (*diplo* means "double") because it has chromosomes in pairs; one member of each pair comes from the **sire** (father) and one member comes from the **dam** (mother). Gametes have only one member of each pair of chromosomes; therefore, gametes are termed **haploid** (*haplo* means "half"). Gametogenesis thus reduces the number of chromosomes in a cell to half the diploid number. Fertilization reestablishes the normal diploid number. It is important to remember that gametogenesis produces gametes that carry a random sample half of each parent's genetic code. Which sperm fertilizes the egg is also a random event. As such, genetic diversity in the population is maintained.

DNA AND RNA

The two members of each typical pair of chromosomes in a cell are alike in size and shape and carry **genes** that affect the same hereditary characteristics. Such chromosomes are said to be **homologous**. The genes are points of activity found in each of the chromosomes that govern the way in which traits develop. The genes form the coding system that directs enzyme and protein production and thus controls the expression of a specific trait or characteristic.

Chromosomes are composed of **deoxyribonucleic acid (DNA)** that is structured as two strands of nucleotides wrapped around each other and connected at the base. The structural configuration is termed "double-helix." Segments of DNA are genes. Each gene is a specific DNA sequence that codes for a particular protein. The location of a gene on the chromosome is called the **locus**. DNA is composed of deoxyribose sugar, phosphate, and four nitrogenous bases. The combination of deoxyribose, phosphate, and one of the four bases is called a **nucleotide**; when many nucleotides are chemically bonded to one another, they form a strand that composes one-half of the DNA molecules. (A molecule formed by many repeating sections is called a **polymer**.)

The bases of DNA are the parts that hold the key to inheritance. The four bases are adenine (A), thymine (T), guanine (G), and cytosine (C). In the two strands of DNA, A is always complementary to (pairs with) T and G is always complementary to C (Fig. 12.4). During meiosis and mitosis, the unwinding and pulling apart of the DNA strands replicate the chromosomes and a new strand is formed alongside the old. The old strand serves as a template, so that wherever an A occurs on the old strand, a T will be directly opposite it on the new, and wherever a C occurs on the old strand, a G will be placed on the new. Complementary bases pair with each other until two entire double-stranded molecules are formed where originally there was one.

The genetic code carries information to trigger formation of specific proteins. For each of the 26 amino acids of which proteins are made, there is at least one "triplet" sequence of three nucleotides. For example, two DNA triplets, TTC and TTT, code for the amino acid lysine; four triplets, CGT, CGA, CGG, and CGC, all code for the amino acid alanine. If we think of a protein molecule as a word, and amino acids as the letters of the word, each triplet sequence of DNA can be said to code for a letter of the word, and the entire encoded message, the series of base triplets, is the gene.

The processes by which the code is "read" and protein is synthesized are called **transcription** and **translation**. To understand these processes, another group of molecules, the **ribonucleic acids (RNAs)**, must be introduced. There are three types of RNA; **transfer RNA** (tRNA), which identifies both an amino acid and a base triplet in mRNA; **messenger RNA** (mRNA), which carries the information codes for a particular protein, and **ribosomal RNA** (rRNA), which is essential for ribosome structure and function. The DNA template codes all three RNA forms.

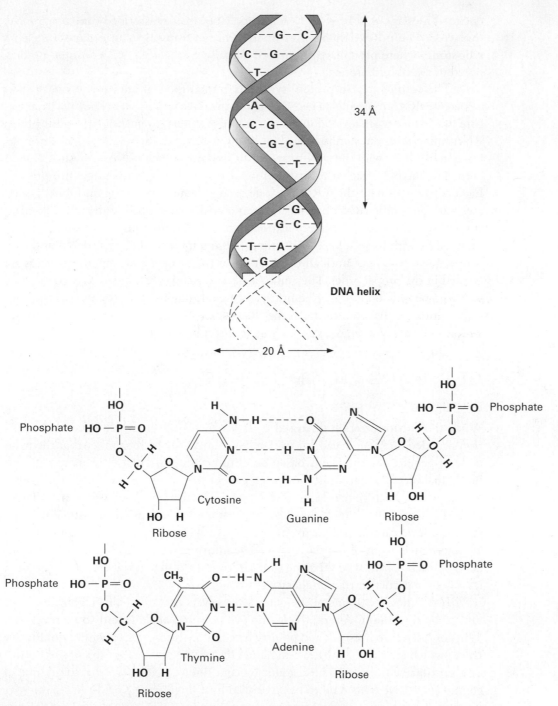

Figure 12.4
DNA helix and structure of nucleotides.

The first step in protein synthesis is transcription. Just as the DNA molecule serves as a template for self-replication using the pairing of specific bases, it can also serve as a template for the mRNA molecule. Messenger RNA is similar to DNA but is single-stranded, coding for only one or a few proteins. The triplet sequence that codes for one amino acid in mRNA is called a **codon**. There are also three "stop" codons that serve as punctuation in the genetic code and don't carry code that triggers specific protein synthesis (UAA, UAG, and UGA). Through transcription, the encoded message held by the DNA molecule becomes transcribed onto the mRNA molecule. The mRNA then leaves the nucleus and travels to an organelle called a **ribosome**, where protein synthesis will actually take place. The ribosome is composed of rRNA and protein.

The second step in protein synthesis is the union of amino acids with their respective tRNA molecules. The DNA codes the tRNA molecules. They have a structure and contain an anticodon that is complementary to an mRNA codon. Each RNA unites with one amino acid. This union is very specific such that, for example, lysine never links with the tRNA for alanine but only to the tRNA for lysine.

The mRNA attaches to a ribosome for translation of its message into protein. Each triplet codon on the mRNA (which is complementary to one on DNA) associates with a specific tRNA bearing its amino acid, using a base-pairing mechanism similar to that found in DNA replication and mRNA transcription. This matching of each tRNA with its specific mRNA triplets begins at one end of the mRNA and continues down its length until all the codons for the protein-forming amino acids are aligned in the proper order. The amino acids are chemically bonded to each other by so-called peptide bonding as the mRNA moves through the ribosome, and the fully formed protein disassociates from the tRNA–mRNA complex and is ready to fulfill its role as a part of a cell or as an enzyme to direct metabolic processes.

GENES AND CHROMOSOMES

Principles of Inheritance

While the previous section discussed in detail the process by which genetic information is coded and then "read" to initiate protein synthesis, remember that the principle of genetic inheritance is based on chromosomes occurring in pairs and that individual pieces of chromosomes called genes carry the genetic code that influence specific traits. Each gene is located at a specific address on the chromosome referred to as the locus (plural, loci). Each member of a homologous chromosome pair has genetic information at each locus that will be passed from parent to offspring. **Alleles** are alternative forms of a specific gene. The information carried by the homologous genes in each pair may be the same (homozygous) or different (heterozygous) in their impact on a particular trait or characteristic.

If the message of one of the genes overpowers that of its allelic pair, then it is said to be **dominant**. Alternatively, the overpowered allele is said to be **recessive**. Traits of interest to livestock and poultry breeders are categorized as either **qualitative traits**, which are typically discrete in nature (color, the presence or absence of horns), or **quantitative traits**, which exist along a continuum (height, weight, speed, rate of gain). Qualitative traits tend to be controlled by a single gene pair while quantitative traits tend to be influenced by a number of gene pairs.

A special pair of chromosomes, the so-called sex chromosomes (X and Y), exist as a pair in which one of the chromosomes does not correspond entirely to the other in terms of what loci are present. The Y chromosome is much shorter in length than the X chromosome.

The X and Y chromosomes determine the sex of mammals. A female has two X chromosomes, and a male has an X and a Y chromosome. The female, being XX, can contribute only an X chromosome to the offspring. The male contributes either an X or a Y, thus determining the sex of the newborn.

In all bird species, including chickens and turkeys, the female determines the sex of the offspring. The sex chromosomes are identical in the sperm *(XX),* but they are different in the egg *(XY).*

SIMPLE INHERITANCE

The fundamental importance of genetics is the ability to predict the outcomes resulting from specific matings. In the case of coat color in beef cattle, black pigmentation is dominant to red (Fig. 12.5). It is common for the allele coding for black color (the dominant form of the gene) to be represented as *B* while the recessive form coding for red is represented as *b*. In this case, neither the dominant nor the recessive condition is inherently better or worse than the other. Because the allele (*B*) is completely dominant (the various forms of dominance are discussed later in the chapter), any animal carrying at least one copy of B will be black. Thus, the homozygous dominant (*BB*) and the heterozygote (*Bb*) will express black coat color. In the case of the heterozygote, the recessive gene is carried but not expressed. On the other hand, the homozygous recessive *(bb)* will always be red.

Consider what happens when the three combinations (*BB, Bb, bb*) are cross-mated. Using the genes designated as *B* = black and *b* = red in cattle, the six mating possibilities are *BB* × *BB*, *BB* × *Bb*, *BB* × *bb*, *Bb* × *Bb*, *Bb* × *bb*, and *bb* × *bb*. Keep in mind that our discussion of these crosses is applicable to any other single pair crosses in other species. To illustrate the predicted outcomes of these matings relative to coat color, the **Punnett square** methodology developed in the early twentieth century by Ronald Punnett, an English geneticist, will be employed. A grid of perpendicular lines is drawn with the genotype of one parent listed on the top row and the genotype of the other parent listed down the left column. The final step is to then copy the row and column information down or across the empty squares to show the predicted frequency of the possible outcomes originating from the cross.

	Parent A—first allele	Parent A— second allele
Parent B— first allele		
Parent B—second allele		

Homozygous-Dominant × Homozygous-Dominant *(BB × BB)*. When two homozygous-dominant individuals are mated, each can produce only one kind of gamete—the gamete carrying the dominant gene. In this particular example, each gamete carries gene *B*. The union of gametes from two homozygous-dominant parents results in a zygote that is homozygous-dominant *(BB)*. Thus, homozygous-dominant parents produce only homozygous-dominant offspring.

Figure 12.5
Black coat color is dominant (B) to red color (b). Black hided cattle may be BB or Bb while red cattle are bb.
Source: Tom Field.

	B	B
B	**BB**	**BB**
B	**BB**	**BB**

Homozygous-Dominant × Heterozygous *(BB × Bb)*. The mating of a homozygous-dominant with a heterozygous individual results in an expected ratio of one homozygote-dominant: one heterozygote. The homozygous-dominant parent produces only one kind of gamete *(B)*. The heterozygous parent produces in equal proportion two kinds of gametes, one carrying the dominant gene *(B)* and one carrying the recessive gene *(b)*. The chances are equal that a gamete from the parent producing only the one kind of gamete (the one having the dominant gene) will unite with each of the two kinds of gametes produced by the heterozygous parent; therefore, the number of homozygous-dominant offspring *(BB)* and heterozygous offspring *(Bb)* produced should be approximately half and half.

	B	B
B	**BB**	**BB**
b	**Bb**	**Bb**

Homozygous-Dominant × Homozygous-Recessive *(BB × bb)*. The homozygous-dominant individual can produce only gametes carrying the dominant gene *(B)*. The recessive individual must be homozygous-recessive to express the recessive trait; therefore, the individual produces only gametes carrying the recessive gene *(b)*. When the two types of gametes unite, all offspring produced receive both the dominant and the recessive gene and are thus all heterozygous *(Bb)*.

	B	B
b	**Bb**	**Bb**
b	**Bb**	**Bb**

Heterozygous × Heterozygous *(Bb × Bb)*. Each of the two heterozygous parents produces two types of gametes in equal ratios: one carries the dominant gene *(B)*; the other carries the recessive gene *(b)*. The two types of gametes produced by one parent each have an equal chance of uniting with each of the two types of gametes produced by the other parent. Thus, four equal-chance unions of gametes are possible. If the gamete carrying the *B* gene from one parent unites with the gamete carrying the *B* gene from the other parent, the offspring produced are homozygous-dominant *(BB)*. If the gamete carrying the *B* gene from one parent unites with the gamete carrying the *b* gene from the other parent, the offspring are heterozygous *(Bb)*. The latter combination of genes can occur in two ways: the *B* gene can come from the male parent and the *b* gene from the female parent, or the *B* gene can come from the female parent and the *b* gene from the male parent. When the gametes carrying the *b* gene from both parents unite, the offspring produced are recessive *(bb)*. The total expected ratio among the offspring of two heterozygous parents is 1*BB*: 2*Bb*: 1*bb*. This 1:2:1

ratio is the **genotypic ratio**. The appearance of the offspring is in the ratio of three dominant: one recessive because the *BB* and *Bb* animals are all black (dominant) and cannot be genetically distinguished from one another visually. This 3:1 ratio, based on external appearance, is called the **phenotypic ratio**.

	B	*b*
B	**BB**	**Bb**
b	**Bb**	**bb**

Heterozygous ✕ Homozygous-Recessive *(Bb ✕ bb)*. The heterozygous individual produces two kinds of gametes, one carrying the dominant gene *(B)* and the other carrying the recessive gene *(b)*, in approximately equal numbers. The recessive individual produces only the gametes carrying the recessive gene *(b)*. There is an equal chance that the two kinds of gametes produced by the heterozygous parent will unite with the one kind of gamete produced by the recessive parent. The offspring produced when these gametes unite thus occur in the expected ratio of 1 heterozygous *(Bb)*:1 homozygous-recessive *(bb)*.

	B	*b*
b	**Bb**	**bb**
b	**Bb**	**bb**

Homozygous-Recessive ✕ Homozygous-Recessive *(bb ✕ bb)*. The recessive individuals are homozygous; therefore, they can produce gametes carrying only the recessive gene *(b)*. When these gametes unite, all offspring produced will be recessive *(bb)*. This example illustrates the principle that recessives, when mated together, breed true and express red color.

	b	*b*
b	**bb**	**bb**
b	**bb**	**bb**

Knowledge of the combinations resulting from each of the six fundamental types of matings provides a background for understanding complex crosses. When more than one pair of genes is considered in a mating, one can understand the expected results; that is, one can combine each of the combinations of one pair of genes with each of the other combinations of one pair of genes to obtain the expected ratios.

MULTIPLE GENE PAIRS

Suppose that there are two pairs of genes to be considered, each pair independently affecting a particular trait—one determines coat color in cattle while the other determines whether the animal is **polled** (hornless) or horned (Figure 12.6). The genes are designated as follows: *B* = black (dominant), *b* = red (recessive), *P* = polled (dominant), and *p* = horned (recessive).

Figure 12.6
Hereford cattle may be either Horned (A) or Polled (B) (without horns). This trait is controlled by a single gene pair where the polled gene is dominant. Polled cattle are either PP or Pp while horned cattle are homozygous recessive (pp). Source: American Hereford Association.

If a bull that is heterozygous for both traits *(BbPp)* is mated to cows that are also heterozygous for both traits *(BbPp),* the expected phenotypic and genotypic ratios can be estimated. The results of crossing *Bb* × *Bb* (ratio of three black : one red in the offspring, or 3/4 black and 1/4 red) have previously been shown. Similarly, a cross of *Pp* × *Pp* gives a ratio of three polled: one horned. The *BbPp* × *BbPp* combination produces the following phenotypic ratios:

9 black polled, 3 black horned, 3 red polled, and 1 red horned (9: 3: 3: 1)

Expressed as percentages:

- 12 of 16 are black—75%
- 4 of 16 are red—25%
- 12 of 16 are polled—75%
- 4 of 16 are horned—25%
- 9 of 16 are black polled—56.25%
- 3 of 16 are black horned—18.75%
- 3 of 16 are red polled—18.75%
- 1 of 16 are red horned—6.25%

Table 12.2 demonstrates the use of a Punnett square to predict the different phenotypes and genotypes that can be produced from a *BbBp* (bull) × *BbPp* (cow) mating. The four different sperm genotypes have equal opportunity to combine with each of the four different forms of eggs. Nine different genotypes and four different phenotypes are possible from a heterozygote × heterozygote mating when two genes are involved.

A related condition to horns is the presence of a loose bony growth called a *scur*. However, two different gene pairs control the scurred condition versus the horned/ polled condition.

Table 12.2
GENOTYPES AND PHENOTYPES WITH TWO HETEROZYGOUS GENE PAIRS

Sperm	Eggs			
	BP	*Bp*	*bP*	*bp*
BP	BBPP[a]	BBPp	BbPP	BbPp
	black, polled[b]	black, polled	black, polled	black, polled
Bp	BBPp	BBpp	BbPp	Bbpp
	black, polled	black, horned	black, polled	black, horned
bP	BbPP	BbPp	bbPP	bbPp
	black, polled	black, polled	red, polled	red, polled
bp	BbPp	Bbpp	bbPp	bbpp
	black, polled	black, horned	red, polled	red, horned

[a]Genotype. Number of different genotypes: 1 (*BBPP*), 2 (*BBPp*), 2 (*BbPp*), 4 (*BbPp*), 1 (*BBpp*), 1 (*bbPP*), 2 (*Bbpp*), 2 (*bbPp*), 1 (*bbpp*).
[b]Phenotype. Number of different phenotypes: 9 black, polled; 3 black, horned; 3 red, polled; 1 red, horned.

GENE INTERACTIONS

A gene may interact with other genes in the same chromosome (**linear interaction**), with its corresponding gene in a homologous chromosome (*allelic interaction*), or with genes in nonhomologous chromosomes (*epistatic interaction*). In addition, genes interact with the cytoplasm and with the environment. The environmental factors with which genes interact are internal, such as hormones, and external, such as nutrition, temperature, and amount of light.

Linear interactions are known to exist in lower animals *(Drosophila),* but have not been demonstrated in farm animals.

Allelic Interactions

When contrasting genes occupy corresponding loci in the same (homologous) pair of chromosomes, each gene exerts its influence on the trait, but the effects of each of the genes depend on the relationship of dominance and recessiveness. Allelic interactions are also called *dominance interactions*. When dissimilar genes occupy corresponding loci, complete dominance might exist. In this situation, only the effect of the dominant gene is expressed. A good example, as previously discussed, is provided by cattle, in which the hornless (polled) condition is dominant to the horned condition. The heterozygous animal is polled and is phenotypically indistinguishable from the homozygous polled animal.

There may be lack of dominance, in which heterozygous animals show a phenotype that is different from either homozygous phenotype and is usually intermediate between them. A good example is observed in sheep, in which two alleles for ear length lack dominance to each other. Sheep that are *LL* have long ears, *Ll* have short ears, and *ll* are earless. Thus the heterozygous *Ll* sheep are different from both homozygotes and are intermediate between them in phenotype.

Lack of dominance can also be considered additive gene action. Additive gene action occurs when each gene has an expressed phenotypic effect. For example, if *D* gene and *d* gene influence rate of gain (where $D = 0.10$ lb/day and $d = 0.05$ lb/day), then *DD* = increase in gain of 0.20 lb, *Dd* = increase in gain of 0.15 lb, and *dd* = increase of gain of 0.10 lb.

There is evidence that many pairs of genes affect production traits (like rate of gain) in an additive manner. For example, consider two pair of genes that influence daily gain, where $D = 0.10$ lb, $d = 0.05$ lb, $N = 0.10$ lb, and $n = 0.05$ lb:

$$DDNN = 0.40 \text{ lb}$$
$$DDNn = 0.35 \text{ lb}$$
$$DdNN$$
$$DdNn$$
$$DDnn = 0.30 \text{ lb}$$
$$ddNN$$
$$Ddnn = 0.25 \text{ lb}$$
$$ddNn$$
$$ddnn = 0.20 \text{ lb}$$

Sometimes heterozygous individuals are superior to either of the homozygotes. This condition, in which the heterozygotes show overdominance, is an example of a selective advantage for the heterozygous condition. Overdominance means that heterozygotes possess greater vigor or are more desirable in other ways (such as producing more milk or being more fertile) than either of the two homozygotes that produced the heterozygote. Because the heterozygotes of breed crosses are more vigorous than the straightbred parents, they are said to possess **heterosis**. This greater vigor or productivity of crossbreds is also said to be an expression of **hybrid vigor**.

The effects of dominance on the expression of traits that are important in livestock production (production traits like fertility, milk production, growth rate, feed conversion efficiency, carcass merit, and freedom from inherited defects) can be illustrated by bar graphs (Fig. 12.7). Figure 12.7A, which illustrates complete dominance, shows equal performance of the homozygous-dominant and heterozygous individuals, both of which are quite superior to the performance of the recessive individual.

With lack of dominance, one homozygote is superior, the heterozygote is intermediate, and the other homozygote is inferior (Fig. 12.7B). When genes that show the effects of overdominance control traits, the heterozygote is superior to either of the homozygous types (Fig. 12.7C).

Another form of gene dominance is referred to as **partial dominance**. The heterozygote expresses a phenotype that is intermediate to either homozygote but more closely resembles the expression of the homozygous-dominant pairing. Hyperkalemic

Figure 12.7

Bar graphs illustrating (A) complete dominance; (B) lack of dominance; (C) and overdominance.

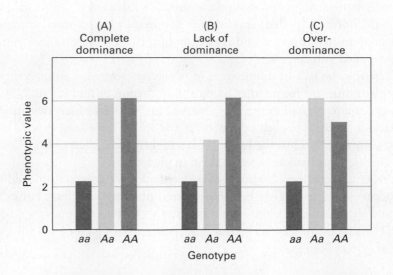

periodic paralysis (HYPP) in horses is representative of partial gene dominance. HYPP causes erratic episodes of muscular tremor that ranges from trembling to acute paralysis.

The gene responsible for HYPP was spread when selection for dramatic levels of muscularity was in vogue in the halter and stock horse industry and a highly favored bloodline proved to be a carrier. The gene has been difficult to remove from the population because it is not completely lethal and because carriers of the gene tend to be heavily muscled—a trait favored by breeders of halter and stock horses.

A gene or a pair of genes in one pair of chromosomes may alter or mask the expression of genes in other chromosomes. A gene that interacts with other genes that are not allelic to it is said to be **epistatic** to them. In other cases, a gene or one pair of genes in one pair of chromosomes may influence many other genes in many pairs of chromosomes.

Several coat colors in horses are due to **epistatic interactions**. For example, two basic coat colors in horses are black and chestnut. The recessive gene *e* produces the chestnut color; thus, *EE* and *Ee* are black, while *ee* is chestnut. Some horses possess a gene for white color that masks all other genes for color that exist in the genotype. This gene is called the dominant white gene *W*. The recessive gene *w* allows other color genes in the genotype to be expressed. Thus, the genotypes and phenotypes are:

EEWW or *EEWw*: white *EEww*: black

EeWW or *EeWw*: white *Eeww*: black

eeWW or *eeWw*: white *eeww*: chestnut

Epistasis can result from dominant or recessive genes or a combination of both. For example, in White Rock chickens, *C* = color, *c* = albino, *W* = color, and *w* = white. Homozygous *cc* prevents *W* from being expressed, and *ww* prevents *CC* from expression. In this example, the mating *CcWw* × *CcWw* would result in only two genotypes (colored or white). This contrasts with four phenotypes if *C* and *W* showed complete dominance, or nine phenotypes if there were no dominance expressed in the two pairs of genes. Thus, epistasis changes the phenotypic ratio, but the number of different genotypes remains the same.

INTERACTIONS BETWEEN GENES AND THE ENVIRONMENT

Genes interact with both external and internal environments. The external environment includes temperature, light, altitude, humidity, disease, and feed supply. Some breeds of cattle (Brahman) can withstand high temperatures and humidity better than others. Some breeds (Scottish Highland) can withstand the rigors of extreme cold better than others.

Perhaps the most important external environmental factor is feed supply. Some breeds or types of cattle can survive when feed is in short supply for considerable periods of time, and they may consume almost anything that can be eaten. Other breeds of cattle select only highly palatable feeds, and these animals have poor production when good feed is not available.

Allelic, epistatic, and environmental interactions all influence the degree to which genetic improvement can be made through selection. When the external environment has a large effect on production traits, genetic improvement is quite low. For example, if animals in a population are maintained at different nutritional levels, those that are best fed obviously grow faster. In such a case, much of the difference in growth may be due to the nutritional status of the animals rather than to differences in their genetic composition. The producer has two alternatives: (1) standardize the environment so that it causes less variation among the animals, or (2) maximize the expression of the production trait by improving the environment. The first approach is geared to increasing genetic improvement so that selection is more effective. Improvements made via this approach are permanent. The latter approach does not improve the genetic qualities of animals, but it allows animals to express their genetic potentials. The most sensible approach is to expose the breeding animals to an environment similar to one in which commercial animals are expected to perform economically.

BIOTECHNOLOGY

Biotechnology is defined as the use of living organisms to improve, modify, or produce industrially important products or processes. Microorganisms, for instance, have been used for centuries in the production of food and fermented substances, such as some dairy products. **Genetic engineering** of animals by selection and hybridization was first implemented at the turn of the century. Superovulation, sexing of semen, embryo splitting (cloning), embryo transfer (discussed in Chapter 11), identification of genetic markers, and genetic therapy are some advances in the field of biotechnology.

Research has provided the tools to identify and manipulate genetic material at the molecular level. These genetic engineering methods, which are used to alter hereditary traits, have created a renewed excitement in biotechnology as well as a flurry of critics who voice concern about the long term effects of such applications. The American Association for the Advancement of Science ranks genetic engineering as among the four major scientific advancements of the twentieth century, similar in significance to unlocking the atom, escaping the earth's gravity, and the computer revolution.

Applications

Enzymes are used as "genetic scissors" to cut the DNA at designated places. These genetic components can then be reconstructed into unique combinations not easily achieved with natural selection. In addition to altering existing genes, synthetic genes can be constructed.

Genes can be injected from other animals of the same or different species (Fig. 12.8). The technology mimics nature in that genes can be moved around within and occasionally between species by some types of viral infections. The manipulation of the genetic material does not occur at the whole animal level, but is performed

Figure 12.8
A fertilized swine egg photographed at the moment it is microinjected with new genetic material. The vacuum in the large pipette at the bottom anchors the cell while a mixture containing the genetic material is forced through the smaller pipette into one of the egg's pronuclei. Source: Courtesy of R. E. Hammer and R. L. Brinster, University of Pennsylvania School of Veterinary Medicine.

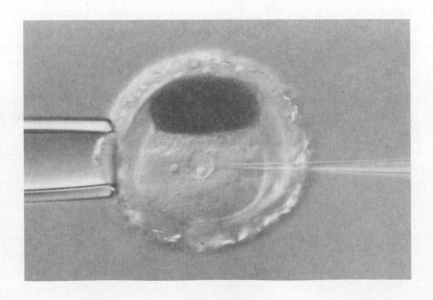

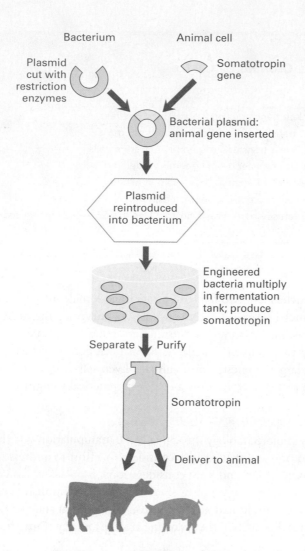

Figure 12.9
*Somatotropin production
for use in cattle and swine.*

in vitro. The genes are then either inserted into embryos that are transferred to recipient animals or inserted into cells that are grafted into living animals.

Bovine somatotropin (BST), a bovine growth hormone, is a natural protein produced in the pituitary glands of cattle. It helps direct the energy in feed to meet the animal's needs for growth and milk production. BST can be produced outside the animal's body by the genetic engineering process illustrated in Figure 12.9. Providing supplemental BST to the dairy cow increases feed consumption and milk production. BST directs more of the additional feed toward increasing milk production than to body maintenance.

Porcine somatotropin (PST) has also been produced by a process similar to that used for BST (Fig. 12.9). Research has shown that PST given as a supplement increases litter size, reduces the time needed for pigs to reach market weight, improves feed efficiency, and reduces carcass fat production.

Table 12.3 illustrates the advancement of cloning technologies applied to cattle. The application of cloning and other biotechniques raises practical, legal, and ethical questions. Cloning is not likely to gain widespread utilization due to the low efficiency of gaining a pregnancy (<25%), increased birth weights of clones, and a cost of $15,000 or more per clone.

Nuclear fusion is the union of nuclei from two sex cells (sperm and eggs). In terms of biotechnology, this means not only can nuclei from a sperm and egg be united but also that the union of nuclei from two eggs or two sperm can occur.

Table 12.3
PROGRESSION OF EVENTS INVOLVING THE CLONING OF CATTLE

Date	Event
1986	First bovine produced via nuclear transfer using embryo blastomeres.
1993	First bovine cloned by nuclear transfer using cells from inner cell mass.
1993	Bovine fetuses produced by nuclear transfer using embryonic stem cells.
1997	First bovine clone born by nuclear transfer using nonembryo-derived cells.
2000	A clone of the Holstein cow—Lauduc Broker Mandy—sold at auction for $82,000.
2001	Cloned Holstein 2-year-old heifers successfully deliver offspring.
2002	First successful clone of the Guernsey cow—Westlynn Tom Dee 'EX-96'—the highest scored living female in the breed.

The union of nuclei from two females results in all female offspring, while the union of two sperm nuclei produces ½ females *(XX)*, ¼ surviving males *(XY)*, and ¼ lethal males *(YY)*. Thus, nuclei from two outstanding males (or females) can be united and then transferred to a recipient female. This has been accomplished in laboratory animals but not in larger, domestic farm animals. Even self-fertilization, once considered possible only in plants, may become a choice in genetically engineered farm animals.

Future Expectations and Concerns

Future advances in biotechnology through genetic manipulation offer both great promise and a myriad of issues that require considerable effort to address in terms of environmental impact, ethical and legal considerations, food safety concerns, and public acceptance. Many of these changes will take years to accomplish because genetic engineering tasks are complicated and expensive. The present state of the art has a high number of failures. Research in the following areas gives insight into future possibilities.

1. Identification of the **genome** (all the genes) of domestic animals creates unique opportunities in terms of better selection of breeding stock, improvement of animal health, and improving the nutritional value of animal products to name a few. This process allows the identification of genetic markers or regions of the chromosome that are responsible for significant levels of genetic variation within a trait.

2. Animal and plant productivity (i.e., growth, changing the fat-to-lean ratio, milk production) could be significantly increased and the cost of animal production could be reduced. The National Center for Food and Agriculture Policy evaluated six biotech crops in the United States and found that they yielded an additional 4 billion pounds from the same acreage, increased farm income by $1.5 billion, and cut pesticide use by 46 million pounds as compared to conventional plants. Transgenic crops approved for cultivation in the United States include alfalfa, canola, chicory, corn, cotton, flax, papaya, plum, potato, rice, soybean, squash, sugar beet, and tomato. Approximately one-third of the U.S. corn crop and three-quarters of the cotton and soybean harvested in the United States originate from genetic engineering. However, no transgenic food animal product has been introduced to the marketplace. In the United States, the sale of milk or meat from cloned animals or their offspring is subject to a voluntary moratorium.

3. Animals with resistance to specific disease can be propagated. Feedstuffs can also be developed with the capability to provide vaccines against disease.

4. Parasites could be controlled by genetic interference with their immune systems.

5. Domestic and international markets could be expanded through new genetically engineered products. Nutritionally enhanced foods including improvements in flavor and palatability characteristics can be produced. Rice with the ability to produce beta-carotene—the precursor to vitamin A—is responsible for blindness prevention in many parts of the world. Furthermore, biotechnology may hold the key to preventing widespread hunger in the world. Norman Borlaug, the father of the green revolution, advocated for the use of technology and science in pursuit of a solution to malnutrition. He stated that "we have the capacity to sustain 10 billion people on the earth, the question is whether or not farmers and ranchers will be allowed to use the technologies to do so."

6. Development of rapid, accurate tests for identifying food contaminants could potentially allow food inspection to move from visual assessment to a sensitive system able to detect specific pathogens or substances. The use of DNA probes and biosensors such as enzymes, antibodies, or microbial cells is currently under evaluation.

7. Diseases may be controlled via the ability to turn associated genes off and on. "Smart" drugs may be developed that can distinguish between healthy and cancerous cells and target only those that are diseased.

Future advances in biotechnology will likely bring many significant genetic changes and challenges that will benefit both humans and domestic animals. However, genetic engineering of animals through the use of recombinant DNA methodologies that combine genetic fragments from different sources raises questions of ethics, food product safety, and environmental safety.

CHAPTER SUMMARY

- Cells contain several pairs of chromosomes that are composed of deoxyribonucleic acid (DNA). Genes are segments of DNA that are the basic units of inheritance.

- Two chromosomes (*X* and *Y*) determine the sex of animals. In livestock, *XX* is female while *XY* is male. In poultry, *XX* is male and *XY* is female. The parent carrying the *XY* chromosome determines the sex of the offspring.

- Genetic variation and genetic change occur through chromosome (gene) segregation and chromosome (gene) recombination resulting from sex cell formation and fertilization.

- Genes produce their effects through dominance, overdominance (heterosis), lack of dominance (additive), or epistasis (interactions between pairs of genes).

- Biotechnology and genetic engineering are bringing changes and challenges to both humans and domestic animals.

KEY WORDS

chromosomes
somatic cells
mitosis
gametes
meiosis
gametogenesis
spermatogenesis
oogenesis
seminiferous tubules
synapsis
primary spermatocyte

primary oocyte
zygote
fertilization
diploid
sire
dam
haploid
genes
homologous
deoxyribonucleic acid (DNA)
locus

nucleotide
polymer
transcription
translation
ribonucleic acid (RNA)
transfer RNA
messenger RNA
ribosomal RNA
codon
ribosome
homozygous
heterozygous
alleles
dominant
recessive
qualitative traits
quantitative traits

Punnett square
genotypic ratio
phenotypic ratio
polled
interaction (linear, allelic, epistatic)
heterosis
hybrid vigor
partial dominance
epistatic
epistatic interactions
biotechnology
genetic engineering
in vitro
transgenic
nuclear fusion
genome

REVIEW QUESTIONS

1. Compare the number of chromosome pairs across livestock species.
2. Describe mitosis and meiosis.
3. Explain the process of spermatogenesis and oogenesis.
4. What is the genetic outcome of fertilization?
5. Describe the structure and role of DNA.
6. What is the role of RNA in animal genetics?
7. Explain simple inheritance using coat color and horned vs polled conditions in cattle.
8. Demonstrate the use of the Punnett Square to determine phenotypic and genotypic ratios.
9. Describe allelic, epistatic, and environmental interactions.
10. Discuss the role of biotechnology in livestock and poultry production.

SELECTED REFERENCES

Acquaah, G. 2004. *Understanding Biotechnology—An Integrated and Cyber-Based Approach*. Upper Saddle River, NJ: Pearson Prentice Hall.

American Health Institute. 1988. *Bovine Somatotropin (BST)*. Alexandria, VA: AHI.

Bailey, E. 2001. The Equine Genome. Equus 260:51.

Berkowitz, D. B. 1994. Transgenic animals. *Encyclopedia of Agricultural Science*. San Diego, CA: Academic Press.

Bourdon, R. M. 2000. *Understanding Animal Breeding*. Upper Saddle River, NJ: Prentice Hall, Inc.

Cundiff, L. V., L. D. Van Vleck, L. D. Young, and G. D. Dickerson. 1994. Animal breeding and genetics. *Encyclopedia of Agricultural Science*. San Diego, CA: Academic Press.

Genetic Engineering: A Natural Science. St. Louis, MO: Monsanto Company.

Genetically modified livestock: Progress, prospects, and issues. 1993. *J. Anim. Sci.* 71 (Supplement 3).

Green, R. D. 1999. DNA + EPDs = Market Assisted Selection. Proceedings of Cattlemen's College. Charlotte, NC.

Hartl, D. L., and E. W. Jones. 2001. *Genetics—Analysis of Genes and Genomes*. Sudbury, MA: Jones and Bartlett Publishers.

Legates, J. E. 1990. *Breeding and Improvement of Farm Animals*. New York: McGraw-Hill.

McGarvey, R. 2002. Biotech breakthroughs. *Harvard Business Review*, (August), 51–59.

Petters, R. M. 1986. Recombinant DNA, gene transfer and the future of animal agriculture. *J. Anim. Sci.* 62:1759.

13
Genetic Change Through Selection

IMPORTANCE OF GENETICS TO THE LIVESTOCK INDUSTRY

Traits of economic importance in livestock and poultry vary in terms of the level of genetic influence on expressed differences observed within groups or populations. In most cases, the level of genetic control over the expression of these traits is sufficient to allow for improvements to be made through disciplined application of breeding principles. Robert Bakewell, an English farmer, is often credited as the innovator of selecting sires and dams and using disciplined mating systems to create desired levels of performance in specific traits. In the mid-1760s, Bakewell initiated the application of genetic knowledge to the business of livestock production through his approach of deliberate and selective breeding decisions. The effort to improve the characteristics of livestock and poultry has been ongoing ever since.

STRUCTURE OF THE BREEDING INDUSTRY

While each industry has its own unique approach to creating and distributing improved genetics (see Chapters 24, 26, 28, 30, 32, and 34), the basic model is described in Figure 13.1. Over time, specialized breeding farms called **elite seedstock producers** have focused on creating breeding animals as their primary business objective. Through the use of genetic information originating from specie and breed specific performance programs, technologies such as artificial insemination and embryo transfer that allowed for rapid propagation of highly desired sires and dams, and a host of other information and technology innovations; these breeding enterprises were able to carve out a market niche as suppliers of improved and predictable breeding stock. The elite seedstock producer sells primarily to **multiplier seedstock producers** who work to create a higher volume of improved sires and dams for use by the **commercial breeder**. Both elite and multiplier seedstock enterprises focus on revenue streams originating from the sale of breeding animals, semen, and embryos.

The commercial breeder may produce their own replacement breeding females but almost always depend on the seedstock sector for production of breeding males. The commercial breeder then sells offspring (progeny, grandprogeny, and great-grandprogeny of the animals produced by the seedstock sector) as feeder or finished animals that will be harvested and processed, or eggs from laying operations, wool or fiber, or milk.

learning objectives
- Explain the concept of genetic variation
- Compare and contrast qualitative and quantitative traits
- Describe the genetic model and its components
- Calculate adjusted 205-day weights
- Calculate the rate of genetic change
- Outline evidence of genetic change in the livestock industry
- Describe the various approaches to selection and the role of selection tools in genetic improvement

Figure 13.1
Structure and flow of genetics.

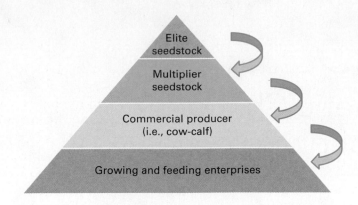

Genetic decisions are driven by a variety of forces:

- Market signals driven by pricing differences favoring one type over another. Examples include demand for leaner products, improved palatability, improved growth rate, higher yields, etc.
- Environmental constraints associated with a particular production setting that favor one genotype over another
- Trends driven by preferences for particular colors or aesthetic characteristics
- Natural selection where certain genotypes are propagated by virtue of having higher reproductive and survival rates under specific conditions

BREEDS

Bakewell and subsequent livestock breeders created unique populations known as breeds. A **breed** is a population of animals that share a distinctive set of characteristics that have been established via a process of deliberate selection to "fix" these traits so that they are passed from generation to generation. Breeds have been created to match specific functions related to both meeting market requirements and the unique environmental constraints of specific climates and regions. Breeds are often categorized into groups based on either region of origin or function.

For example, beef cattle breeds are often categorized as British (Angus, Hereford, Shorthorn), European (Charolais, Gelbvieh, Limousin, Simmental), or Zebu (Brahman and Brahman derivatives such as Brangus and Braford). Zebu refers to an origin of eastern Asia. Goats are categorized as meat (Boer, Spanish), dairy (Alpine, Nubian) or fiber (Angora) breeds while sheep breeds are classified as fine wool (Merino, Rambouillet), meat-type (Hampshire, Dorset, Suffolk), and dual-purpose (Columbia, Targhee). Horse breeds are labeled as light horse, draft (Belgian, Clydesdale, Percheron), and ponys (Shetland, Welsh). The light horse breeds are further categorized as:

- Hunter—Thoroughbred, Warmbloods
- Saddle—Arabian, Morgan, Saddlebred
- Stock—Quarter Horse, Appaloosa, Paint

While individuals within a breed share common characteristics, there is tremendous potential for genetic variation within a breed allowing for wide-ranging performance in a host of traits. It is often said that there is as much variation within a breed as there is between breed averages. Specific breeds will be discussed in more detail in Chapters 24, 26, 28, 30, 32, and 34.

CONTINUOUS VARIATION AND MANY PAIRS OF GENES

Mammalian genomes are composed of 30,000 to 40,000 genes. Most economically important traits in farm animals, such as milk production, egg production, growth rate, and carcass composition, are controlled by very large numbers of gene pairs; therefore, it is necessary to expand one's thinking beyond inheritance involving one and two pairs of genes.

Consider even a simple hypothetical example of 20 pairs of heterozygous genes (one gene pair on each pair of 20 chromosomes) affecting yearling weight in sheep, cattle, or horses. The estimated numbers of genetically different gametes (sperm or eggs) and genetic combinations are shown in Table 13.1. Remember that for one pair of heterozygous genes, there are three different genetic combinations (i.e., *AA, Aa,* and *aa*), and for two pairs of heterozygous genes, there are nine different genetic combinations (Chapter 12, Table 12.2).

Most farm animals are likely to have some heterozygous and some homozygous gene pairs, depending on the mating system being utilized. Table 13.2 shows the number of gametes and genotypes where eight pairs of genes are either heterozygous or homozygous and each gene pair is located on a different pair of chromosomes.

Many economically important traits in farm animals show continuous variation primarily because many pairs of genes control them. As these many genes express them, and the environment also influences these traits, producers usually observe and

---

Table 13.1

NUMBER OF GAMETES AND GENETIC COMBINATIONS WITH VARYING NUMBERS OF HETEROZYGOUS GENE PAIRS

No. of Pairs of Heterozygous Genes	No. of Genetically Different Sperm or Eggs	No. of Different Genetic Combinations (genotypes)
1	2	3
2	4	9
n	2^n	3^n
20	$2^n = 2^{20} = $ ~1 million	$3^n = 3^{20} = $ ~3.5 billion

---

Table 13.2

NUMBER OF GAMETES AND GENETIC COMBINATIONS WITH EIGHT PAIRS OF GENES WITH VARYING AMOUNTS OF HETEROZYGOSITY AND HOMOZYGOSITY

Genes in Sire:	Aa	Bb	Cc	Dd	Ee	FF	GG	Hh	
Genes in Dam:	Aa	Bb	CC	Dd	Ee	FF	gg	Hh	Total
No. of different sperm for sire	2 ×	2 ×	2 ×	2 ×	2 ×	1 ×	1 ×	2 =	64
No. of different eggs for dam	1 ×	2 ×	1 ×	2 ×	2 ×	1 ×	1 ×	2 =	16
No. of different genetic combinations possible in offspring	2 ×	3 ×	2 ×	3 ×	3 ×	1 ×	1 ×	3 =	324

measure large differences in the performance of animals for any one trait. For example, if a large number of calves were weighed at weaning (~205 days of age) in a single herd, there would be considerable variation in the calves' weights. Distribution of weaning weights of the calves would be similar to the examples shown in Figures 13.2 or 13.3. The bell-shaped curve distribution demonstrates that most of the calves are near the average for all calves, with relatively few calves having extremely high or low weaning weights when compared at the same age.

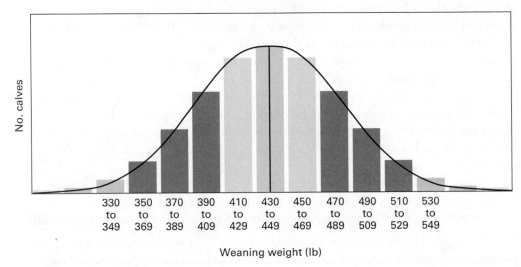

Figure 13.2

Variation or difference in weaning weight in beef cattle. The variation shown by the bell-shaped curve could be representative of a breed or a large herd. The dark vertical line in the center is the average or the mean—in this example, 440 lb.

Figure 13.3

A normal bell-shaped curve for weaning weight showing the number of calves in the area under the curve (400 calves in the herd).

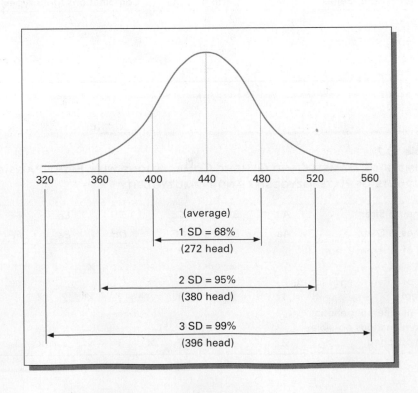

Figure 13.3 shows how the statistical measurement of standard deviation (SD) is used to describe the variation of differences in a herd where the average weaning weight is 440 lb and the calculated standard deviation is 40 lb. Using herd average and standard deviation, the variation in weaning weight is shown in Figure 13.2 and can be described as follows:

1 SD: 440 lb ± 1 SD (40 lb) = 400 – 480 lb
         (68% of the calves are in this range)
2 SD: 440 lb ± 2 SD (80 lb) = 360 – 520 lb
         (95% of the calves are in this range)
3 SD: 440 lb ± 3 SD (120 lb) = 320 – 560 lb
         (99% of the calves are in this range)

One percent of the calves (four calves in a herd of 400) would be on either side of the 320- to 560-lb range. Most likely, two calves would be below 320 lb and two calves would weigh more than 560 lb.

Animals have multiple traits that can be measured or described. **Quantitative traits** are those that can be objectively measured, and the observations typically exist along a continuum. Examples include growth traits, skeletal size, speed, and others. **Qualitative traits** are descriptively or subjectively measured and would include hair color, horned versus polled, and so forth. Many gene pairs control quantitative traits, while few, if not just one, gene pairs control qualitative traits.

The observation or measurement of each trait is referred to as the **phenotype**. Phenotypic variation exists within the trait due to two primary sources of influence—**genotype** and **environment** (Fig. 13.4). Table 13.3 shows the typical phenotypic means and standard deviations for a sample of traits from the primary livestock species.

Weaning weight is a phenotype, since the expression of this characteristic is determined by the genotype (genes received from the sire and the dam) and the environment to which the calf is exposed. The genetic part of the expression of weaning weight is obviously not simply inherited. There are many pairs of genes involved, and, at the present time, the individual pairs of genes cannot be identified similarly to traits controlled by one to two pair of genes.

Genotype is the result of both the cumulative effects of the animal's individual genes for the trait and the effect of the gene combinations. **Environmental effects** can be thought of as the summation of all nongenetic influences.

Breeding value or parental worth for a trait can be defined as that portion of genotype that can be transferred from parent to offspring. **Breeding value** in mathematical terms is the total of all the independent genetic effects on a given trait of an individual. **Nonadditive value** is that portion of genotype that is attributed to the gene combinations unique to a particular animal. Because genetic combinations are reestablished in each successive graduation, nonadditive value

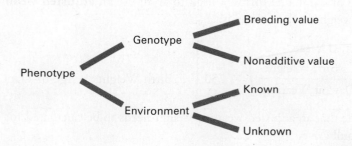

Figure 13.4
The genetic model.

Table 13.3
TYPICAL PHENOTYPIC MEANS AND STANDARD DEVIATIONS FOR A SAMPLE OF TRAITS FROM FARM ANIMALS

Species	Trait	Mean	Standard Deviation
Cattle (beef)	Birth weight	80 lb	10 lb
	Yearling weight (bulls)	950 lb	60 lb
	Mature weight	1,100 lb	85 lb
	Backfat thickness (steers)	0.4 in.	0.1 in.
Cattle (dairy)	Calving interval	404 days	75 days
	Milk yield	13,000 lb	560 lb
Horses	Wither height (mature)	60 in.	1.8 in.
	Time to run $\frac{1}{4}$ mile	20 seconds	0.6 seconds
	Time to run 1 mile	96 seconds	1.3 seconds
	Cutting score	209 points	10.3 points
Swine	Litter size (# born alive)	9.8 pigs	2.8 pigs
	Days to 230 lb	175 days	12 days
	Loineye area	4.3 in.^{2}	0.25 in.^{2}
Poultry	Hatchability (chickens)	90%	2.2%
	Egg weight (layers)	0.13 lb	0.01 lb
	Feed conversion ratio (broilers)	5.40 lb/lb	0.88 lb
	Breast weight (broilers)	0.64 lb	0.007 lb
Sheep	60-day weaning weight	45 lb	8 lb
	Grease fleece weight	8 lb	1.1 lb
	Staple length	2.5 in.	0.5 lb

Source: Adapted from Bourdon, 2000.

does not pass from generation to generation. Therefore, breeding value responds to selection while nonadditive value is accessed via choice of mating system (i.e., crossbreeding).

There are two basic types of environmental effects—known and unknown. **Known effects** have an average effect on individuals in a specific category. Examples include age, age of dam, and gender. Calves born earlier in the calving season and thus older at time of weaning typically weigh more than their younger counterparts. **Unknown effects** are random in nature and are specific to an individual phenotype. Known effects can be quantified and used to adjust phenotypic measures to allow more accurate selection. Unknown effects are more difficult to account for, but breeders can use management to minimize their impacts.

One attempt to remove the effects of environmental influence is the use of adjusted records. For example, weaning weight can be adjusted to account for age of calf and age of dam. The formula used to compute an **adjusted weaning weight** or 205-day weight for beef cattle is:

$$\left[\left(\frac{\text{Actual Wean Weight} - \text{Birth Weight}}{\text{Age in Days at Weaning}}\right) \times 250\right] + \text{Birth Weight} + \text{Age-of-Dam Adjustment}$$

Assume that an adjusted weaning weight needs to be calculated for the following pair of bull calves:

	Bull A	Bull B
Actual weaning weight	550 lb	520 lb
Age at weaning	230 days	190 days
Birth weight	82 lb	75 lb
Age of dam	3 years	6 years

Use adjustments listed below.

Age of Dam (years)	Bull Calf (lb)	Heifer Calf (lb)
2	+60	+54
3	+40	+36
4	+20	+18
5–10	+0	+0
>10	+20	+18

Adjusted weaning weight for Bull A:

$$\left[\frac{(550 - 82)}{230} \right] + 82 + 40 = 439 \text{ lb}$$

Adjusted weaning weight for Bull B:

$$\left[\frac{(520 - 75)}{190} \right] + 75 + 0 = 555 \text{ lb}$$

The interpretation of these solutions is that if bulls A and B had been weaned at 205 days of age, and if their dams had been mature cows between the ages of 5 and 10, then their weights would be estimated at 539 and 555 pounds, respectively.

The phenotype will more closely predict the genotype if producers expose their animals to a similar environment; however, the latter must be within economic reason. This resemblance between phenotype and genotype, where many pairs of genes are involved, is predicted with the estimate of heritability. Use of **heritability**, along with selecting animals with superior phenotypes, is the primary method of making genetic improvement in traits controlled by many pairs of genes. The application of this method is discussed in more detail later in the chapter.

Traits influenced little by the environment can also show considerable variation. An example is the white belt, a breed-identifying characteristic in Hampshire swine (Fig. 13.5). Note that the white belt can be nonexistent in some Hampshires, whereas others can be almost completely white. To be eligible for pedigree registration, Hampshire pigs must be black with the white belt entirely circling the body, including both front legs and feet. Too much white can also limit registration. It is difficult to select a herd of purebred Hampshires that will breed true for desired belt pattern. Apparently, some of the genes for this trait exist in heterozygous or epistatic combinations.

Selection

Selection is differential reproduction—preventing some animals from reproducing while allowing other animals to become parents of numerous offspring. In the latter situation, the selected parents should be genetically superior for the economically important traits. Factors affecting the rate of genetic improvement from selection include selection differential, heritability, and generation interval.

Figure 13.5
Variation in belt pattern in Hampshire swine. Source: Courtesy of National Swine Registry.

Selection Differential

Selection differential, sometimes called **reach**, is the superiority (or inferiority) of the selected animals compared to the herd average. To improve weaning weights in beef cattle, producers cull as many below-average-producing cows as economically feasible. Then replacement heifers and bulls are selected that are above the herd average for weaning weight. For example, if the average weaning weight of the selected replacement heifers is 480 lb in a herd averaging 440 lb, then the selection differential for the heifers would be 40 lb. Part of this 40-lb difference is due to genetic differences, and the remaining part is due to differences caused by the environment.

Heritability

A **heritability** estimate describes the percent of total phenotypic variation (phenotypic differences) that is due to breeding value. Heritability can also be defined as that portion of the selection differential that is passed from parent to offspring. If the parents' performance is a good estimate of progeny performance for that trait, then the heritability is said to be high.

Realized heritability is the portion actually obtained compared to what was attempted in selection. To illustrate realized heritability, let us suppose a farmer has a herd of pigs whose average postweaning gain is 1.80 lb/day. If the farmer selects from this original herd a breeding herd whose members have an average gain of 2.3 lb/day, the farmer is selecting for an increased daily gain of 0.5 lb/day.

If the offspring of the selected animals gain 1.95 lb, then an increase of 0.15 lb (1.95 – 1.80) instead of 0.5 lb (2.3 – 1.8) has been obtained. To find the portion obtained of what was reached for in the selection, 0.15 is divided by 0.5, which gives 0.3. This figure is the realized heritability. If 0.3 is multiplied by 100%, the result is 30%, which is the percentage obtained of what was selected for in this generation.

Table 13.4 shows heritability estimates for several species of livestock. Traits having heritability estimates of 40% and higher are considered highly heritable. Those with estimates 20–39% are classified as having medium heritability; and low-heritability traits have heritability estimates below 20%. Heritability is a measure determined by studying populations and, as such, is not specific to individuals. Furthermore, heritability may differ from breed to breed and environment to environment.

Table 13.4
HERITABILITY ESTIMATES FOR SELECTED TRAITS IN SEVERAL SPECIES OF FARM ANIMALS

Species Trait	Percent Heritability	Species Trait	Percent Heritability
Beef Cattle		Riding performance	
Age at puberty	40%	Jumping (earnings)	20
Weight at puberty	50	Dressage (earnings)	20%
Scrotal circumference	50	Cutting ability	5
Birth weight	40	Thoroughbred racing	
Gestation length	40	Log earnings	50
Body condition score	40	Time	15
Calving interval	10	Pacer: best time	15
Percent calf crop	10	Trotter	
Weaning weight	30	Log earnings	40
Postweaning gain	45	Time	30
Yearling weight	40	**Poultry**	
Yearling hip (frame size)	40	Age at sexual maturity	35
Mature weight	50	Total egg production	25
Carcass quality grade	40	Egg weight	50
Yield grade	30	Broiler weight	40
Tenderness of meat	50	Mature weight	50
Longevity	20	Egg hatchability	10
Dairy Cattle		Livability	10
Services per conception	5	**Sheep**	
Birth weight	40	Multiple births	20
Milk production	25	Birth weight	30
Fat production	25	Weaning weight	30
Protein	25	Postweaning gain	40
Solids-not-fat	25	Mature weight	50
Type score	30	Fleece weight	40
Feet and leg score	10	Face covering	50
Teat placement	30	Loineye area	40
Udder score	20	Carcass fat thickness	50
Mastitis (susceptibility)	10	Weight of retail cuts	40
Milking speed	20	**Swine**	
Mature weight	50	Litter size	10
Excitability	25	Birth weight	10
Goats		Litter weaning weight	15
Milk production	30	Postweaning gain	30
Mohair production	20	Feed efficiency	30
Horses		Backfat (live animal)	40
Withers height	45	Carcass fat thickness	50
Pulling power	25	Loineye area	50
		Percent lean cuts	45

Generation Interval

Generation interval is defined as the average age of the parents when the offspring are born. **Generation interval** is calculated by adding the average age of all breeding females to the average age of all breeding males and dividing by 2. The generation interval is approximately 2 years in swine, 3–4 years in dairy cattle, and 5–6 years in beef cattle. When the rapid speed of genetic change in the poultry industry is evaluated, the rapid generation turnover is a major factor.

PREDICTING GENETIC CHANGE

The rate of **genetic change** that can be made through selection can be estimated using the following equation:

$$\text{Genetic change per year} = \frac{(\text{Heritability} \times \text{Selection Differential})}{\text{Generation Interval}}$$

Consider the following example of selecting for weaning weight in beef cattle and predicting the genetic change. The herd average is 440 lb and bulls and heifers are selected from within the herd. The selection differential measures the phenotypic superiority for inferiority of the selected animals compared to the herd average—for example, bulls (535 lb – 440 lb = 95 lb); heifers (480 lb – 440 lb = 40 lb). Heritability times selection differential measures the genetic superiority of the selected animals compared to the herd average.

Bulls	*Weight*
Average of selected bulls	535 lb
Average of all bulls in herd	440 lb
Selection differential	95 lb
Heritability	0.30
Total genetic superiority	28 lb
Only half passed on (28 ÷ 2)	14 lb

Heifers	*Weight*
Average of selected heifers	480 lb
Average of all heifers in herd	440 lb
Selection differential	40 lb
Heritability	0.30
Total genetic superiority	12 lb
Only half passed on (12 ÷ 2)	6 lb

Fertilization combines the genetic superiority of both parents, which in this example is 20 lb (14 lb + 6 lb). The calves obtain half of their genes from each parent; therefore ½(28) + ½(12) = 20 lb.

The selected heifers represent only approximately 20% of the total cowherd, so the 20 lb is for one generation. Since the heifer replacement rate in the cowherd is 20%, it would take 5 years to replace the cowherd with selected heifers. Because of this generation interval, the genetic change per year from selection would be 20 lb ÷ 5 = 4 lb/year.

Genetic Change for Multiple Trait Selection

The previous example of 4 lb/year shows genetic change if selection was for only one trait. If selection is practiced for more than one trait, genetic change is $\frac{1}{\sqrt{n}}$, where n is the number of traits in the selection program. If four traits were in the selection program, the genetic change per trait would be $\frac{1}{\sqrt{n}} = \frac{1}{2}$. This means that only half the progress would be made for any one trait compared to giving all the selection to one trait. This reduction in genetic change per trait should not discourage producers from multiple-trait selection, as herd income is dependent on several traits. Maximizing genetic progress in a single trait may not be economically feasible, and it may lower productivity in other economically important traits.

Sometimes traits are genetically correlated, meaning that some of the same genes affect both traits. In swine, the genetic correlation between rate of gain and feed per pound of gain (from similar beginning weights to similar end weights) is negative. This is desirable from a genetic improvement standpoint because animals that gain faster require less feed (primarily less feed for maintenance). This relationship is also desirable because rate of gain is easily measured, whereas feed efficiency is expensive to measure.

Yearling weight or mature weight in cattle is positively correlated with birth weight, which means that as yearling or mature weight increases, birth weight also increases. This weight increase may pose a potential problem, since birth weight to a large extent reflects calving difficulty and calf death loss.

EVIDENCE OF GENETIC CHANGE

The previously shown examples of selection are theoretical. Does selection really work, or is the observed improvement in farm animals a result of improving only the environment? Let's evaluate several examples. Recognize that single trait selection strategies focused on dramatic, ongoing increased performance may not be sustainable. Because some genes affect more than one trait, some traits are negatively correlated (improvement in one trait results in less desirable performance in another trait), and increases in productivity may also be associated with increased costs of nutritional inputs, the law of unintended consequence must be at the forefront of the decision-making process. Animals selected to be too large, too small, produce too much milk, or any other extreme may be suboptimized in other traits such as structural correctness, fitness, or survivability. Thus selection for extremes is rarely desirable.

There have been marked, visual meat-to-bone ratio changes in the thick-breasted modern turkey selected from the narrow-breasted wild turkey. Estimates for genetic change for meat-to-bone ratio have been 0.5% per year for the modern turkey. Genetic selection and improving the environment, particularly through improved nutrition and health, have produced the modern turkey. Toms turkeys can produce 27 lb of live weight (22 lb dressed weight) in 5 months; the wild turkey weighs 10 lb in 6 months.

The modern turkey needs an environment with intensive management and is not likely to survive in the same environment as the wild turkey. For example, turkeys are mated artificially because the heavy muscling in the breast prevents them from mating naturally. Selection has been effective in changing rate of growth and meatiness in turkeys, but it has necessitated a change in the environment for these birds to be productive. Therefore, the improvement in turkeys has resulted from improvement in both the genetics and the environment. However, extended selection pressure for muscle mass has resulted in toms (male turkeys) that are incapable of natural service of a female due to the extended size of the breast muscle.

There are tremendous size differences in horses. Draft horses have been selected for large body size and heavy muscling to perform work. The light horse is more moderate in size, and ponies are small. Miniature horses have genetic combinations that result in a very small size. The miniature horse is considered a novelty and a pet. Table 13.5 shows the height and weight variations in different types of horses. Apparently these different horse types have been produced from horses that originally were quite similar in size and weight. The extreme size differences in horses result primarily from genetic differences, as the size differences are apparent when the horses are given similar environmental opportunities.

Table 13.5
HEIGHT AND WEIGHT DIFFERENCES IN MATURE DRAFT HORSES, LIGHT HORSES, PONIES, AND MINIATURE HORSES

| | Approximate Height at Withers | | Approximate Mature Weight |
Horse Type	Hands	Inches	(lb)
Draft	17	68	1,600
Light	15	60	1,100
Pony	13	52	700
Miniature	8	32	250

Table 13.6
TRENDS IN U.S. PORK PRODUCTION ON A PER ANIMAL BASIS

Year	Pork Production (lb per breeding animal)	Live (lb per hog)	Retail Meat Yield (lb per hog)
1955	NA	237	123
1960	NA	236	124
1965	1,315	238	127
1970	1,442	240	129
1975	1,583	238	131
1980	1,766	242	133
1985	2,140	245	136
1990	2,209	249	141
1995	2,554	256	144
2000	3,037	262	151
2005	3,59	269	156
2010	4,003	270	157

Source: Adapted from National Pork Producers Council, National Pork Board, and USDA.

The effectiveness of selection for increased growth and retail yield in swine is shown in Table 13.6. These data illustrate that the average harvest weight in swine has increased 34 pounds since 1960 while markedly improving both individual animal retail yield and total production per breeding sow.

Selection for milk production in dairy cattle poses an additional challenge in a genetic improvement program because the bull does not express the trait. This is an example of a **sex-limited trait**.

Genetic evaluation of a bull is based primarily on how his daughters' milk production compares to that of their **contemporaries**. Milk production is a moderately heritable trait (25%), and the average milk cow today produces nearly five times as much milk as the average cow in 1940; this is a most noteworthy example of selection pressure, even though the trait is not expressed in the bull (Table 13.7). Even when multiple traits receive the attention of selection pressure, progress can be made. Figure 13.6 documents the genetic trends for the Red Angus cattle breed over a period of nearly 50 years for growth traits, milk production, and longevity (stayability). In the poultry industry, disciplined selection over a period of 40 years enabled producers to reduce the age of marketing from 12 weeks to less than 6 weeks. Over the same time period, the feed required to raise a broiler was cut in half.

However, even in traits of moderate heritability, selection may not yield progressive improvements in phenotype. In the All-American Quarter Horse Futurity (440 yd), 5-year average winning times during the 26-year period from 1986 to 2012 did not show appreciable and sustained improvement (Table 13.8). In this case, biomechanical limitations make it very difficult to make horses progressively faster.

A project conducted at Colorado State University involved mating a group of commercial Hereford cows to Hereford bulls representative of the breeding cattle population in the 1950s, 1970s, and 1990s. The performance comparison of the resulting progeny illustrated that breeders had been very successful in changing the growth rates of their cattle (Table 13.9). Note that birth weights increased along with growth rate and frame size. Intense selection for increased growth may lead to undesirable levels of birth weight and mature size.

Table 13.7
CHANGES IN MILK PRODUCTION IN THE UNITED STATES, 1940–2011

Year	No. Cows (mil)	Average Milk Per Cow (lb)	Total Milk (bil lb)
1940	23.7	4,622	109.4
1950	21.9	5,314	116.6
1960	17.5	7,029	123.1
1970	12.0	9,751	117.0
1980	10.8	11,875	128.4
1990	10.1	14,645	148.3
2000	9.2	18,204	167.6
2005	9.0	21,854	170.0
2010	9.1	22,526	191.9
2011	9.2	22,684	195.3

Source: USDA.

Table 13.8
AVERAGE WINNING TIMES IN THE ALL-AMERICAN QUARTER HORSE FUTURITY (440 YARDS)

Years	Average 5-year Winning Times
1986–1990	21.16
1991–1995	21.47
1996–2000	21.45
2001–2005	21.29
2006–2010	21.15
2011–2012	21.14

Figure 13.6
Genetic trends since 1954 for the six traits presented in a national sire evaluation.

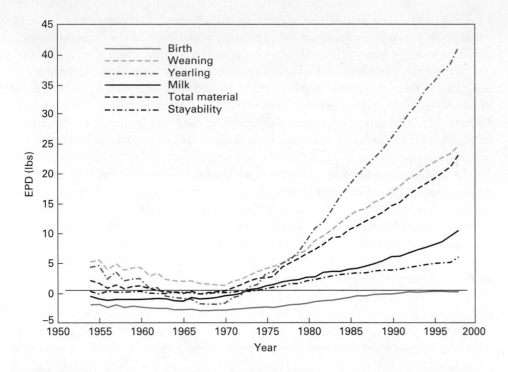

Table 13.9
GROWTH OF HEREFORD CATTLE SIRED BY DIFFERENT GENERATIONS OF BULLS

Sire Generation	Birth Wt. (lb)	On-Test Wt. (lb)	Off-Test Wt. (lb)	Frame Size
1950	82.5	665	1,083	3.7
1970	85.9	717	1,158	4.9
1990	91.4	791	1,261	5.5

Source: Adapted from Tatum and Field, 1996.

SELECTION METHODS

The three typical methods of selection are (1) tandem, (2) independent culling level, and (3) selection index. **Tandem** is selection for one trait at a time. This method can be effective if the situation calls for rapid change in a single, highly heritable trait. When the desired level is achieved in one trait, then selection is practiced for the second trait. However, the tandem method is rather ineffective if selection is for more than two traits or if the desirable aspect of one trait is associated with the undesirable aspect of another trait.

For example, in the case of dairy cattle, milk production and milk fat percentage are negatively correlated. If a producer were interested in increasing milk yield, selection pressure solely focused on that trait would eventually result in decreases in percent milk fat. If the producer then decided to practice single trait selection to increase percent milk fat, progress could be made in that trait. However, such an approach would lead to a decline in milk yield over time due to the negative correlation between the traits. Tandem selection makes it difficult to sustain progress when more than one trait is of importance. Thus, this approach is typically not recommended.

Independent culling level establishes minimum culling levels for each trait in the selection program. Even though it is the second most effective type of selection, it is the most prevalent because it is fairly simple to implement. Independent culling level is most useful when the number of traits being considered is small and when only a small percentage of offspring is needed to replace the parents. Table 13.10 shows an example of using independent culling levels in selecting yearling bulls. Birth weight is correlated with calving ease. Yearling weight indicates rate of growth. Puberty and semen production are estimated with scrotal circumference. Any bull that does not meet the minimum or maximum level is culled. Birth weights are evaluated on a maximum level (upper limit) since high birth weights result in increased calving difficulty.

The boxed records in Table 13.10 indicate bulls that did not meet the independent culling levels. Bulls A, B, and D would be culled. A disadvantage of this method is that it may cull a relatively superior animal for only slightly missing a target criteria in a single trait.

The **selection index** method recognizes the value of multiple traits and places an economic weighting on the traits of importance. Such a calculation allows an overall ranking of the animals from best to worst utilizing a highly objective approach. The selection index is the most effective system but the most difficult to develop. The disadvantages of this system include the possibility of shifts in economic value of traits over time and potential failure to identify functional defects or weaknesses.

A comparison of Tables 13.10 and 13.11 reveals several differences resulting from the application of independent culling level versus selection index. In both cases, bull E would be identified as the most desirable. Bull D is culled using independent culling level due to inadequate scrotal circumference. However, in the selection index, bull D is ranked second because scrotal measurement wasn't included. On the other hand, bull C barely escapes culling under the independent culling level system with just acceptable performance in all three traits. The selection index ranks bull C next to last.

An advantage of independent culling levels is that selection can occur during different productive stages during the animal's lifetime (e.g., at weaning).

Table 13.10
INDEPENDENT CULLING LEVEL SELECTION IN YEARLING BULLS

		Bull				
Trait	Culling Level	A	B	C	D	E
Birth weight (lb)	85 (max)	105	82	85	93	76
Yearling weight (lb)	1,000 (min)	1,142	980	1,001	1,098	1,160
Scrotal circumference (cm)	30 (min)	34	37	31	29	35

Table 13.11
SELECTION INDEX METHOD OF RANKING YEARLING BULLS

Bull ID	Birth Weight (lb)	Yearling Weight (lb)	Index Value = YW − 5.8 (BW)	Ranking
A	105	1,142	533	3
B	82	980	504	5
C	85	1,001	508	4
D	93	1,098	559	2
E	76	1,160	719	1

This is more cost-effective than the index method, where no culling would occur until records are recorded for all traits. For example, bull A would have been quickly culled by the independent culling level method as opposed to waiting until yearling weight and thus incurring additional cost, as would be the case with the selection index. A combination of the two methods may be most useful and cost-effective.

BASIS FOR SELECTION

Effective selection requires that the traits in question be heritable, relatively easy to measure, and associated with economic value; that genetic estimates or predictions be accurate; and that genetic variation be available. The notion of making sustained genetic progress in a herd is the basis for development of breed associations and utilization of performance data.

In the not-too-distant past, breeders depended almost entirely on visual appraisal as the basis for selection. Modern breeders have access to a functional array of selection tools. The basis for modern selection is the availability of breeding value estimates often referred to as predicted differences or expected progeny differences.

Predicted Differences or Expected Progeny Differences

Expected progeny differences (EPDs) are calculated for a variety of traits by utilizing information on the individual, on siblings (half and full), on ancestors, and, best of all, on progeny. As more information is utilized in the calculation, the accuracy of the estimate improves.

The dairy industry has been the leader of the modern performance movement. By 1929, all 48 states had dairy cow record associations and in the mid-1930s a progeny test program for dairy sires had been initiated. Breed associations have maintained accurate pedigree records outlining the parentage of seedstock animals for several hundred years. However, only in the past 30–50 years have breed associations focused on developing performance databases for a multitude of traits.

The earliest efforts at objective across-herd comparisons utilized central tests. In these tests, young bulls, rams, or boars were brought together in a common environment to be evaluated, primarily for growth traits. These tests yield information on only a few traits and all the information is obtained from the individual. Because of these limitations, designed progeny tests were initiated to compare sires via information collected on their respective progeny. While an improvement, these designed progeny tests were relatively limited in scope and expensive to conduct.

The advent of **best linear unbiased prediction (BLUP)** techniques has allowed field data to be utilized in computation of valid breeding values that could be used to compare animals across herds. Breed associations or large seedstock companies sponsor most of these national sire or animal evaluation systems.

The poultry and dairy industries have made the best use of sophisticated genetic information. Genetic information is also widely available in the beef cattle industry. The swine industry initiated a national genetic evaluation system to evaluate maternal sow lines in 1997—a follow-up to the 1995 terminal sire line genetic evaluation program. The sheep industry genetic evaluation programs have been most extensive in Australia and New Zealand, with a focus on wool traits such as fiber diameter and fleece weight. Genetic prediction estimates for equines have received more attention in Europe than in the United States.

The utilization of performance data is one of the most profit-oriented decisions a livestock producer can make. Table 13.12 illustrates the progress of the dairy herds

Table 13.12

COMPARISON OF AVERAGE PER COW MILK PRODUCTION FROM DHIA VS NON-DHIA HERDS

Year	DHIA Herds	Non-DHIA Herds
1906	5,034	3,600
1950	9,000	5,300
1970	13,000	9,747
1990	18,031	14,782
1995	19,005	16,405
2000	20,462	17,771
2005	21,854	19,443
2007	22,282	19,951

Table 13.13

IMPORTANCE OF FACTORS IN PURCHASING BREEDING BULLS

Factor	Percent of Respondents by Level of Importance			
	Not	Moderate	Very	Extreme
Birth weight	20.3	20.0	38.0	21.7
Weaning weight/yearling weight	20.2	15.7	42.9	21.2
Hip height/frame score	14.2	27.0	42.6	16.2
Expected progeny differences	30.5	25.3	31.5	12.7
Appearance/structural soundness	2.5	3.0	43.3	51.2
Price	8.1	23.7	37.9	30.3

Source: National Animal Health Monitoring System, USDA, 1994.

utilizing the Dairy Herd Improvement Association (DHIA) system versus those who did not. DHIA herds have had, on average, a clear productivity advantage.

Unfortunately, producers do not always accept or utilize the newest generation of genetic prediction tools. Table 13.13 illustrates that producers often tend to rely on visual appraisal or raw data rather than the more accurate EPDs that are available.

CHAPTER SUMMARY

- Phenotype (what is seen or measured) is determined by genotype (genetic makeup and the environment to which the animal is exposed).

- Heritability measures the proportion (0–100%) of the total phenotypic variation that is due to genetics. Traits high in heritability are ≥40%, while low-heritability traits are >20%.

- Selection differential is the superiority (or inferiority) of the selected animals compared to the average of the group from which they came. Generation interval is the average age of the parents when the offspring are born.

- Genetic change per year =

$$\frac{(\text{Heritability} \times \text{Selection Differential})}{\text{Generation Interval}}$$

- Independent culling level is the most common method of selection.

- Expected progeny differences ought to be the basis for an effective selection program.

KEY WORDS

elite seedstock producers	adjusted weaning weight
multiplier seedstock producers	heritability
commercial breeder	selection
breed	selection differential
quantitative traits	reach
qualitative traits	realized heritability
phenotype	generation interval
genotype	genetic change
environment	sex-limited traits
environmental effects	contemporaries
breeding value	tandem
nonadditive value	independent culling level
known effects	selection index
unknown effects	best linear unbiased prediction (BLUP)

REVIEW QUESTIONS

1. Compare the role of elite seedstock, multiplier seedstock, and commercial breeders.
2. What are the forces that drive selection?
3. Describe the importance of the development of breeds to the livestock industry.
4. Demonstrate the calculation of the number of gametes and genotypes arising from various numbers of heterozygous gene pairs.
5. Use the bell curve to describe genetic variation.
6. Describe the genetic model and its components.
7. Demonstrate the calculation of 205-day adjusted weaning weights.
8. Explain the importance of selection differential, generation interval, and heritability to generating change in a specific trait.
9. Compare the heritability of various traits.
10. Demonstrate the genetic change formula.
11. Explain the impact of simultaneous selection for multiple traits on the speed of genetic change.
12. Discuss examples of change in livestock species resulting from selection.
13. Compare the various selection methods.
14. Discuss the advantage of using quantitative approaches to selection such as EPDs.

SELECTED REFERENCES

Bourdon, R. M. 2000. *Understanding Animal Breeding.* Upper Saddle River, NJ: Prentice Hall.

Bowling, A. T. 1996. *Horse Genetics.* Center for Agriculture and Biosciences International. Cambridge, UK: Cambridge University Press.

Cundiff, L. V., L. D. Van Vleck, L. D. Young, and G. D. Dickerson. 1994. Animal breeding and genetics. *Encyclopedia of Agricultural Science.* San Diego, CA: Academic Press.

Freeman, A. E. and G. L. Lindberg. 1993. Challenges to dairy management: Genetic considerations. *J. Dairy Sci.* 76:3143.

Genetics and Goat Breeding. Proceedings of the Third International Conference on Goat Production and Disease. 1982. *Dairy Goat Journal.* Scottsdale, AZ.

Hetzer, H. O. and W. R. Harvey. 1967. Selection for high and low fatness in swine. *J. Anim. Sci.* 26:1244.

Hintz, R. L. 1980. Genetics of performance in the horse. *J. Anim. Sci.* 51:582.

Legates, J. E. 1990. *Breeding and Improvement of Farm Animals.* New York: McGraw-Hill.

National Animal Health Monitoring System. 1994. Beef Cow/Calf Reproductive and Nutritional Management Practices. USDA: APHIS.

Van Vleck, L. D., E. Oltenacu, and J. Pollack. 1986. *Genetics for the Animal Sciences.* New York: W. H. Freeman.

14
Mating Systems

Seedstock breeders (sometimes called **purebred producers**) and commercial breeders (producers) are the two general classifications of animal breeders. **Seedstock** livestock historically are considered purebreds for which ancestry is recorded on a pedigree by a breed association (Fig. 14.1). Most commercial slaughter livestock are **crossbreds**, resulting from crossing two or more breeds or lines of breeding.

Animal breeders make three critical decisions—choosing the individuals that become parents, determining the rate of reproduction from each individual (especially sires), and deciding which mating system is most likely to yield beneficial results given the constraints of the operation. Genetic improvement can be optimized in most herds and flocks by utilizing a combination of selection and mating systems.

Mating systems are identified primarily by genetic relationship of the animals being mated. Two major systems of mating are **inbreeding** and **outbreeding**. Inbreeding is the mating of animals more closely related than the average of the breed or population. Outbreeding is the mating of animals not as closely related as the average of the population.

Since mating systems are based on the **relationship** of the animals being mated, it is important to understand more detail about genetic relationship. Proper pedigree evaluation also involves understanding the genetic relationships between animals. Relationship is best described as knowing which genes two animals have in common and whether the genes in an animal or animals exist primarily in a heterozygous or homozygous condition. Figure 14.2 shows the mating systems and their relationship to homozygosity and heterozygosity.

INBREEDING

Because inbreeding is the mating of related animals, the resulting inbred offspring have an increased homozygosity of gene pairs compared to noninbred animals in the same population (breed or herd). An example of inbreeding is shown in Figure 14.3, where animal A has resulted from mating a sire B to his full sister C. B and C are highly genetically related because they are full siblings. However, they do not have the same identical genetic makeup because each has received a sample half of genes from each parent. In the arrow pedigree the arrow represents a sample half of genes from the parent to the offspring.

Animal A is inbred because it has resulted from the mating of related animals. Although the calculation is not shown in the figure, animal A has 25% more homozygous gene pairs as compared with a noninbred animal in the same population. Keep in mind that breeders

learning objectives
- Describe the roles of the animal breeders
- Describe the advantages and disadvantages of inbreeding, linebreeding, outcrossing, and crossbreeding
- Describe heritability and heterosis
- Calculate percent heterosis
- Discuss the role of composite and hybrid breed formation

American **Simmental** Association
1 Simmental Way
Bozeman MT 59718

406.587.4531/fax: 406.587.9301
simmental@simmgene.com

Performance Data Sheet

ASR H3 W9139

2496847
Registration Number

SIMANGUS BULL

W9139 LE
Tatto-Loc

	Name	Breeds	Reg Number	YR Born	H/P/S	Type	Cow Awd
	HC HUMMER 12M	PB SM	2,174,450	2002	P	PCB	
Sire	**TNT HUMMER H3 R360**	PB SM	2,321,773	2005	P	PCB	
	TNT MISS DEATRA K96	PB SM	2,092,410	2000	P	PCB	
ASR H3 W9139		1/2 AN 1/2 SM	2,496,847	02-28-09	P		
	HUNTS CALCULATOR 2720	AN0011871751		1992	P	BTF	
Dam	**ASR MS CALCULATOR 451**	AN0014909774		2004	P		
	FF SENSATIONAL LADY M214	AN0014379108		2002	P		

Expected Progeny Difference and Progeny Counts
for the latest epds go to www.simmental.org

	Individual				Sire					Dam				Maternal Grand Sire				
	EPD	Ac	Rnk		EPD	Ac	Rnk	Hds	Prog	EPD	Ac	Rnk	Prog	EPD	Ac	Rnk	Hds	Prog
CE	0.0	0			9.2	36												
BW	0.6	31			3.2	71				-1.3	14			-4.0	81			
WW	34.4	26			34.9	67				21.1	4			17.0	79			
YW	67.8	31			65.6	61				41.9	5			39.5	79			
MCE	0.0	0			3.0	18												
MM	6.9	9			7.9	29				7.1	9			3.2	77			
MWW	24.1	10			25.4	30				17.6	13			11.7	77			
Stay					25.0	15												
CW	4.0	25			6.4	49				-14.2	6			-19.4	57			
YG	0.15	17			-0.09	32				0.39	12			0.45	56			
Marb	0.48	17			0.30	31				0.69	16			0.86	74			
B Fat	0.05	19			0.01	42				0.12	17			0.17	75			
REA	-0.01	17			0.47	29				-0.47	16			-0.32	73			
WBSF					-0.20	9												
Dau Ht																		
Dau Wt																		
API	108.0				127.9					99.0				119.5				
TI	67.0				66.9					66.0				69.8				

Within Herd Performance

Birth Adj Wt	Birth Ratio	Birth Rank	Wean Adj Wt	Wean Ratio	Wean Rank	Year Adj Wt	Year Ratio	Year Rank
89	101	25/54	662	110	6/39	1306	108	7/32

Registration and Transfer of Calf at Side

Calf Tattoo/Loc: _____   Birth Date: _____   Sex: _____   Birth Wt: _____   Birth Type: _____

Calf Name: _____   Sire Reg. Nbr and Name: _____

If calf is sold to party other than buyer of dam, indicate name and address of buyer who purchased the calf.

Name: _____   Member No: _____

Address: _____   City: _____   State: _____   Zip: _____

This calf at side will be registered and transferred. Signature with transfer certifies the accuracy of this information.

Figure 14.1
Pedigree of a SimAngus (50% Simmental : 50% Angus) bull (ASR H3 W9139). Both pedigree and performance data on the animal, parents, and maternal grandsire are provided.

have little control over assuring that only desired or beneficial genes are paired in the homozygous condition.

Figure 14.4 shows another example of inbreeding, where animal X has resulted from a sire–daughter mating. Sire A was mated to dam D, with D being sired by A. The increased homozygosity of animal X is 25%, which is the same as for animal A in Figure 14.3.

There are two different forms of inbreeding:

1. *Intensive inbreeding*—mating of closely related animals whose ancestors have been inbred for several generations.
2. *Linebreeding*—a mild form of inbreeding where inbreeding is kept relatively low while maintaining a high genetic relationship to an ancestor or line of ancestors.

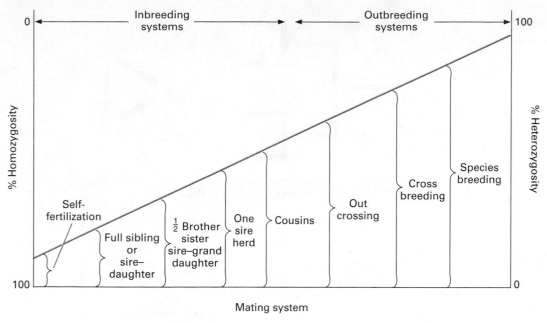

Figure 14.2

Relationship of the mating system to the amount of heterozygosity or homozygosity. Self-fertilization is currently not an available mating system in animals.

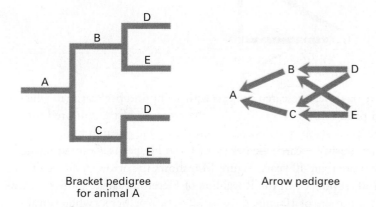

Bracket pedigree for animal A   Arrow pedigree

Figure 14.3

Bracket pedigree and arrow pedigree showing animal A resulting from a full brother–sister mating.

Intensive Inbreeding

Intensive inbreeding occurs when closely related animals are mated for several generations. Figure 14.5 shows an example of intensive inbreeding. The increase in the homozygosity of animal H's genes is higher than 25% because both the sire and grandsire of animal H were inbred (compare with Fig. 14.4).

There are numerous genetically different inbred lines that can be produced in a given population such as a breed. The number of different, completely homozygous inbred lines is $2n$, where n is the number of heterozygous gene pairs. Thus, with two heterozygous gene pairs, 22 or four different inbred lines are possible. For example, with *BbPp* genes in a completely heterozygous herd (see Chapter 12), the different completely homozygous lines that can result from inbreeding are *BBPP, BBpp, bbPP,* and *bbpp*. Each of these four inbred lines is genetically pure (homozygous) for these two pair of genes and each will breed true when animals are mated within the same homozygous line.

Figure 14.4
Bracket pedigree and arrow pedigree showing animal X resulting from a sire–daughter mating.

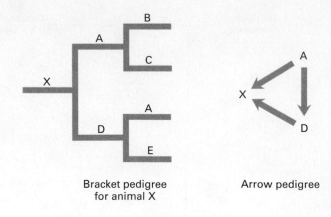

Bracket pedigree
for animal X

Arrow pedigree

Figure 14.5
An arrow pedigree showing animal H resulting from three generations of sire–daughter matings. Note that J is the sire of H and also the sire of K, the latter being the dam of H. Animal J has resulted from two previous, successive generations of sire–daughter matings.

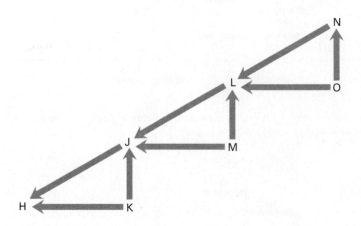

There have been research projects with swine and beef cattle in which inbred lines have been produced, anticipating that the crossing of these inbred lines would produce highly predictable and desired phenotypes. One research project was conducted at the San Juan Basin Research Center (SJBRC) in Hesperus, Colorado, where cattle were inbred for more than 40 years. Figure 14.6 shows the arrow pedigree of an inbred bull (Royal 4160) produced in the Royal line of Hereford cattle. Note that College Royal Domino 3 is the sire of 10 animals in Royal 4160's pedigree, while Royal 3016 sired six animals in the same pedigree. The increased homozygosity of this bull is 58%.

Information obtained from inbreeding studies with livestock demonstrates the following results and observations:

1. Increased inbreeding is usually detrimental to reproductive performance and preweaning and postweaning growth. Also, inbred animals are more susceptible to environmental stresses. Whereas 60–70% of the inbred lines show the detrimental effects of increased inbreeding, 30–40% of the lines show no detrimental effect, with some lines demonstrating improved productivity.

2. In a Colorado research beef herd, the inbred lines showed a yearly genetic increase of 2.6 lb in weaning weight over a 26-year period, while the crosses of inbred lines made a 4.6 lb increase over the same time period. Heterosis is demonstrated in the line crosses, and the 4.6 lb increase is typical of what breeders might expect from using intense selection in an outbred herd.

3. Inbreeding quickly identifies some desirable genes and also undesirable genes, particularly the serious recessive genes that are hidden when left in the heterozygous state.

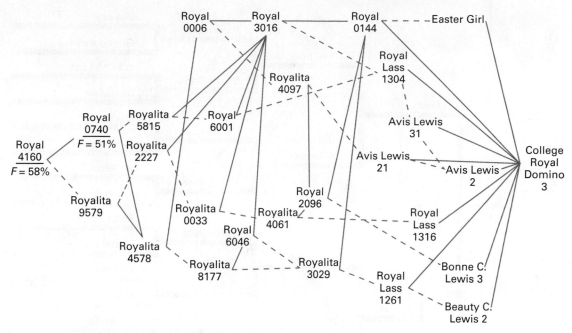

Figure 14.6

Arrow pedigree of Royal 4160, which has an inbreeding coefficient of 58%. Solid lines represent the genetic contribution of the bulls, whereas the cow's contribution is represented by a broken line. Source: Figure by Kevin Pond from Colorado State University. Used by permission of Kevin Pond.

4. Inbred animals with superior performance are the most likely to have superior breeding values, which result in more uniform progeny with high levels of genetically influenced productivity.

5. Crossing of inbred lines results in heterosis; however, in most cases heterosis compensates for inbreeding depression.

6. Crossing of inbred lines of animals has not yielded the same results as crossing inbred lines of corn. The reasons appear to be: (1) inbreeding animals is slower (they cannot self-fertilize as corn does); (2) it is easier and less costly to produce more inbred lines of corn; (3) inbred lines of animals are eliminated because of extremely poor reproductive performance and being less adapted to environmental stress.

7. There is merit in using some inbreeding in developing new lines of poultry and swine. (This is discussed in more detail later in the chapter.)

It is not logical for breeders to develop their own lines of highly inbred beef cattle. Inbreeding depression usually affects the economics of the operation. Both seedstock and commercial producers can take advantage of highly productive inbred lines by crossing these inbred bulls with unrelated cows.

Inbreeding such as sire–daughter matings, are logical ways to test for undesirable recessive genes. Also, seedstock producers use inbreeding in well-planned linebreeding programs. Inbreeding should be considered when breeders have difficulty introducing sires from other herds that are genetically superior to those they are producing.

Linebreeding

Linebreeding is a low-risk form of inbreeding used to maintain a high genetic relationship to an outstanding ancestor, usually a sire. Seedstock producers who have high levels of genetic superiority in their herds and find it difficult to locate sires that are superior to the ones they are raising in their herds best use this mating system.

228 CHAPTER FOURTEEN • MATING SYSTEMS

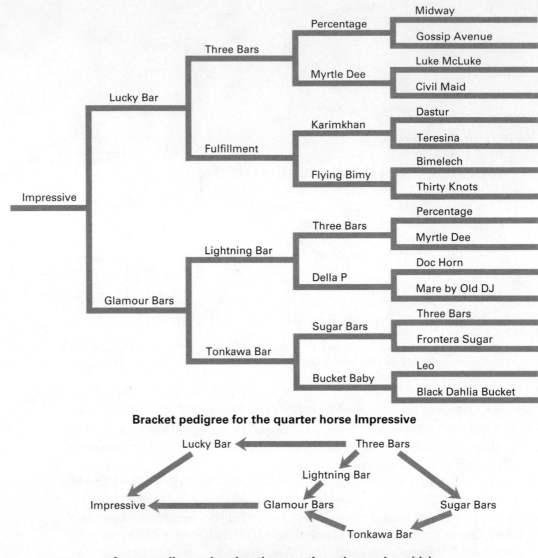

Bracket pedigree for the quarter horse Impressive

**Arrow pedigree showing the genetic pathways by which
Three Bars contributes to the inbreeding and linebreeding of Impressive**

Figure 14.7
Horse pedigree illustrating linebreeding.

Occasionally a breeder may produce a sire with a superior combination of genes that consistently produces high-producing offspring. Younger sires may not outproduce some of these sires. This is observed in some dairy bulls that remain competitively superior as long as they produce semen. These sires warrant use in a linebreeding program.

Figure 14.7 gives an example of linebreeding. Impressive, an outstanding quarter horse stallion is linebred to his ancestor, Three Bars, by three separate pathways. The inbreeding of Impressive is approximately 9%, whereas the genetic relationship of Impressive to Three Bars is approximately 44%. Inbreeding below 20% is considered low, whereas a genetic relationship is high when it approaches 50%.

Impressive has nearly the same genetic relationship as if Three Bars had been his sire (44 versus 50%). Progeny of Impressive have produced outstanding records, particularly as halter-point and working-point winners in the show ring. Unfortunately,

Impressive has also been identified as a carrier of the gene that leads to hyperkalemic periodic paralysis, or HYPP. HYPP is the result of a genetic mutation, but it has not been confirmed if this mutation originated with Impressive.

OUTBREEDING

The four types of outbreeding are as follows:

1. *Species cross*—crossing of animals of different species (e.g., horse to donkey or cattle to bison).
2. *Crossbreeding*—mating of animals of different established breeds.
3. *Outcrossing*—mating of unrelated animals within the same breed.
4. *Grading up*—mating of purebred sires to commercial-grade females and their female offspring for several generations. Grading up can involve some crossbreeding or it can be a type of outcrossing system.

Species Cross

A species designation is part of the zoological classification used in taxonomy (a branch of zoology) to classify animals on the basis of similarities in body structure. All livestock are classified as the phylum **Chordata** and the subphylum **Vertebrata**. Cattle, sheep, hogs, horses, and goats are categorized into the class **Mammalia** while poultry are in the class **Aves**. Order, family, genus, and species designations are listed in Table 14.1.

Some animals of different species but the same genus can be crossed to produce viable offspring. Animals of different genus cannot be successfully crossed because chromosome number and other genetic differences interfere with normal fertilization. Therefore, a **species cross** is the most divergent form of outbreeding that can be attained.

A common species cross is the **mule**, resulting from crossing a male or **jack** donkey and a female or mare horse *(Equus asinus × Equus caballus)*. Mules existed in large numbers as work animals before the advent of the tractor. The **hinny** is the reciprocal cross of the mule *(Equus caballus* stallion × *Equus asinus* jennet). The hinny never achieved the level of popularity as the mule.

Mare mules are usually sterile, which gives verification to genetic differences between the ass and horse. There have been a few reports of fertile mare mules.

Table 14.1
ORDER, FAMILY, GENUS, AND SPECIES OF COMMON FARM ANIMALS

Common Name	Order	Family	Genus	Species
Bison	Artiodactyla	Bovidae	Bison	Bison
Cattle	Artiodactyla	Bovidae	Bos	taurus or indicus
Chicken	Galliformes	Phasianidae	Gallus	domesticus
Donkey	Perissodactyla	Equidae	Equus	asinus
Duck	Anseriformes	Anatidae	Anas	platyrhyncha
Goat	Artiodactyla	Bovidae	Capra	hircus
Goose	Anseriformes	Anatidae	Anser	anser
Horse	Perissodactyla	Equidae	Equus	caballus
Sheep	Artiodactyla	Bovidae	Ovis	aries
Swine	Artiodactyla	Suidae	Sus	scrofa
Turkey	Galliformes	Meleagrididae	Meleagris	gallopavo

Crossing of the zebu or humped cattle with European-type cattle *(Bos indicus × Bos taurus)* is common in the southeastern part of the United States. These crosses are more adaptable and productive in hot, humid environments than either of the straight species. Some authorities question if *Bos indicus* and *Bos taurus* are separate species, and their crosses are usually referred to as *crossbreds* rather than species crosses.

Numerous crosses of American bison and cattle have been made. Some of these crosses have been designated as separate breeds called **Cattalo** or **Beefalo**. These crosses are intended to be more adaptable to harsh environments (cold temperatures and limited forage). Fertility problems have existed in these crosses, and their numbers are limited.

Sheep and goats have been crossed even though they have different genus classifications. Fertilization occurs but embryos die in early gestation.

There are other species crosses that have occurred. Most species crosses have little commercial value. Recent advances in genetic engineering might make some genetic combinations between species more feasible. Gene splicing (inserting a gene or genes from one animal to another) has occurred between species. In the future, opportunity to combine desirable genes both within a species and between species could become a reality.

Crossbreeding

There are two primary reasons for using **crossbreeding**: To take advantage of (1) breed complementation and (2) heterosis (hybrid vigor). Breed complementation is the resulting benefit of crossing breeds so their strengths and weaknesses complement one another. Contrary to the high level of marketing and promotion that occurs in the purebred industry, there is no one breed that is superior in all desired production characteristics; therefore, planned crossbreeding programs that use breed complementation can significantly increase herd productivity.

Crossbreeding, if properly managed, allows for tapping into the genetic power of heterosis, which has a marked effect on productivity in swine, poultry, and beef cattle. Heterosis tends to be most impactful on survival rates of newborns, longevity, and reproductive merit. **Heterosis** is defined as the increase in productivity in the crossbred progeny above the average of breeds or lines that are crossed. An example of calculating heterosis is shown in Table 14.2. The calculated heterosis for calf-crop percent in this example is 5%, whereas heterosis for weaning weight is 4%.

Crossbreeding is sometimes questioned when the performance of crossbred animals is less than that of one of the parental breeds. In Table 14.2, for example, the

Table 14.2
COMPUTATION OF HETEROSIS FOR PERCENT CALF CROP AND WEANING WEIGHTS

	Calf Crop (%)	Weaning Weight (lb)	Lb Calf Weaned per Cow Exposed
Breed A	82	460	377
Breed B	78	540	421
Average of the two breeds (without heterosis)	80	500	399
Average of crossbreds (with heterosis)	84	520	447
Superiority of crossbreds over average of two breeds	4	20	48
Percent heterosis	5% (4 ÷ 80)	4% (20 ÷ 500)	12% (48 ÷ 399)

Table 14.3
RELATIONSHIP OF HERITABILITY AND HETEROSIS FOR MOST TRAITS

Traits	Heritability	Heterosis
Reproduction	Low	High
Growth	Medium	Medium
Carcass	High	Low

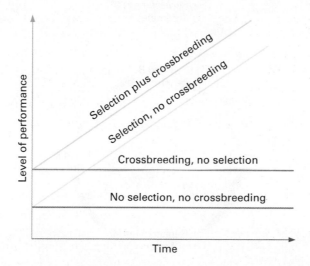

Figure 14.8
Improvement in performance with various combinations of selection and crossbreeding.

weaning weight of the crossbreds is 520 lb, while one parent (Breed B) is 540 lb. However, for calf-crop percent, the crossbreds performed at 84%, which is higher than the levels attained by either Breed A or Breed B. The value of crossbreeding in this example is best demonstrated by combining calf-crop percent and weaning weight (calf-crop percent × weaning weight = lb calf weaned per cow exposed). Note in Table 14.2 that the value for lb of calf weaned per cow exposed is 447 lb for the crossbreds, whereas it is 377 lb for Breed A and 421 lb for Breed B.

The amount of heterosis expressed is related to the heritability of the trait. Table 14.3 shows that heterosis is highest for low-heritability traits and lowest for high-heritability traits. These relationships are helpful to commercial producers in selecting and crossbreeding to enhance genetic improvement. Figure 14.8 illustrates the relative importance of selection and crossbreeding in an improvement program. This figure demonstrates that selecting genetically superior animals is more important than crossbreeding. However, using the two methods in combination gives the highest level of performance.

Crossbreeding is commonly used in swine, beef cattle, and sheep. Little cross-breeding is done in dairy cattle because of the primary emphasis on one trait (milk production) and the superiority of the Holstein breed for that trait. However, in some cases where single-trait selection for milk yield has produced lines that have structural problems, low milk fat, or poor reproductive rates the use of crossbreeding has been incorporated. Poultry breeders utilize heterosis primarily through crossing lines that have been developed from separate and distinct lines created via intensive inbreeding and breed crossing strategies.

Figure 14.9

Two-breed rotational cross. Females sired by breed A are mated to breed B sires, and females sired by breed B are mated to breed A sires.

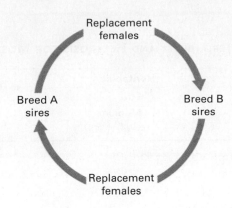

Figure 14.10

Three-breed rotational cross. Females sired by a specific breed are bred to the breed of sire next in rotation.

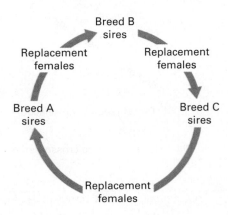

Figure 14.11

Terminal (Static) or modified-terminal crossbreeding system. It is terminal or static if all females in herd (A × B) are then crossed to breed C Sires. All male and female offspring are sold. It is a modified-terminal system if part of females are bred to A and B sires to produce replacement females. The remainder of the females are terminally crossed to breed C sires.

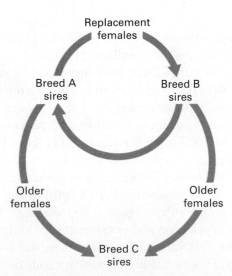

Figures 14.9 through 14.11 show the crossbreeding systems most frequently used in swine, beef cattle, and sheep. More specific detail in using these crossbreeding systems for these species is given in Chapters 25, 29, and 31.

Outcrossing

Outcrossing is the process of mating unrelated animals within the same breed in an attempt to maximize heterozygosity. The motivation to increase the level of heterozygous gene pairings is to mask the expression of deleterious recessive alleles. However, there are limits to this strategy as members of the same breed share a degree of homozygosity estimated to be between 10% and 20%. Perhaps a more accurate description of outcrossing is the disciplined crossing of animals that are more distantly related than the average of the population. In truth, the best approach to outcrossing with a goal of achieving very high levels of heterozygosity is to cross animals from different breeds altogether.

Grading Up

The continuous use of purebred sires of the same breed in a grade herd or flock is called **grading up**. In this situation, grading up is similar to outcrossing. The accumulated percentage of inheritance of the desired purebred is 50% (1/2), 75% (3/4), 84.5% (7/8), and 94% (15/16) for four generations when grading up is practiced. The fourth generation resembles the purebred sires so closely in genetic composition that it approximates the purebred level.

The grading-up system is useful in the breeding of cattle and horses, but it has little value in breeding sheep, swine, or poultry. High-producing purebred sheep, swine, and poultry breeding stock are available at reasonable prices; therefore, the breeder can buy them for less than he or she can produce them by grading up. The use of production-tested males that are above average in performance in a commercial herd can grade up the herd not only to a general purebred level but also to a high level of production.

A use of grading up on a large scale occurred with the introduction of many European beef cattle breeds to the United States. Most of the introduction was accomplished with males (bulls or semen), as females were less available and more expensive because of the numbers needed; grading up allowed relatively rapid propagation of these imported breeds.

Imported bulls (or their semen) have been used on commercial cows or purebred cows of other breeds. Grading up is a type of crossbreeding, although the intent is not to maintain heterosis but to increase the frequency of genes from the introduced breed.

After several successive generations of mating the new breed to cows carrying a certain percentage of the new breed, the resulting offspring have been designated purebreds. In most breeds, this designation has been given when the calves had 7/8 or 15/16 of the genetic composition of the new breed. Figure 14.12 shows how these matings are made. It would require a minimum of 7 years to produce the first 15/16 calves of the new breed.

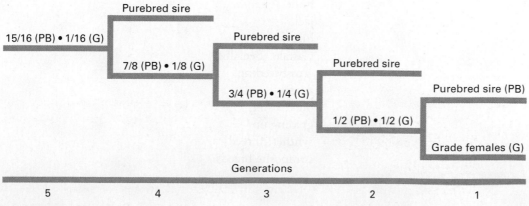

Figure 14.12
Utilizing grading up to produce purebred offspring from a grade herd.

FORMING NEW LINES OR BREEDS

As breeders of livestock and poultry endeavor to create populations of animals particularly well suited to specific sets of environmental or climatic conditions while maintaining characteristics desired in the marketplace, there may be motivation to create "new" breeds through the systematic combination of animals from existing breeds. These are sometimes given a general classification of **synthetic breeds** and **composite breeds**. In beef cattle, the Brangus, Barzona, Beefmaster, and Santa Gertrudis breeds are composite (synthetic) breeds formed several decades ago, whereas MARC I (crosses of Charolais, Brown Swiss, Limousin, Hereford, and Angus breeds) and RX3 (crosses of Red Angus, Hereford, and Red Holstein breeds) are examples of composites more recently developed. Columbia, Targhee, and Polypay are examples of synthetic breeds of sheep.

Crossing several swine breeds, a practice associated with some inbreeding, has been used to develop the hybrid boars now being merchandised by several companies. Hybrid boars are used extensively in the swine industry.

In poultry, breeding for egg production differs from breeding for broiler production. Different traits are emphasized in the production of these two products. Both inbreeding and heterosis are utilized in the production of specific lines and strains of birds that are highly productive in the production of either eggs or broilers.

CHAPTER SUMMARY

- Genetic relationship estimates the genes two animals have in common because the same ancestors appear in the first six generations of their pedigrees.

- Mating systems are identified by the genetic relationship of the animals being mated.

- Inbreeding is the mating of animals more closely related than the average of the population, while outbreeding is the mating of animals not as closely related as the average of the population.

- Inbreeding increases genetic homozygosity, while outbreeding increases genetic heterozygosity.

- Linebreeding is a mild form of inbreeding while maintaining a high genetic relationship to an outstanding ancestor.

- Outcrossing is the mating of unrelated animals within the same breed.

- Crossbreeding is the mating of animals from different breeds, resulting in heterosis (hybrid vigor).

KEY WORDS

purebred producers
seedstock
crossbreds
inbreeding
outbreeding
relationship
intensive inbreeding
linebreeding
Chordata
Vertebrata
Mammalia
Aves

species cross
mule
jack
hinny
Cattalo (Beefalo)
crossbreeding
heterosis
outcrossing
grading up
synthetic breeds
composite breeds

REVIEW QUESTIONS

1. What are the three critical decisions make by breeders?
2. Describe inbreeding and outbreeding.
3. Discuss the advantages and disadvantages of the two forms of inbreeding.
4. Compare and contrast the four approaches to outbreeding.
5. Demonstrate the calculation of percent heterosis.
6. Discuss the impact of crossbreeding on the livestock industry.
7. Provide examples of the formation of new breeds or lines in the livestock and poultry business.

SELECTED REFERENCES

Bourdon, R. M. 2000. *Understanding Animal Breeding.* Upper Saddle River, NJ: Prentice Hall.

Cundiff, L. V., L. D. Van Vleck, L. D. Young, and C. D. Dickerson. 1994. Animal breeding and genetics. *Encyclopedia of Agricultural Science.* San Diego, CA: Academic Press.

Legates, J. E. 1990. *Breeding and Improvement of Farm Animals.* New York: McGraw-Hill.

15
Nutrients and Their Functions

learning objectives

- Discuss the role of water, carbohydrates, fats, proteins, minerals, and vitamins in nutrition
- Illustrate proximate analysis
- Calculate percent crude protein, percent digestibility, and percent dry matter
- Discuss the process of energy utilization
- Classify the basic feedstuffs

A **nutrient** is any feed constituent that functions in the support of life. There are many different feeds available to animals to provide nutrients.

Most animal feeds are classified as **concentrates** or **roughages**. Concentrates include cereal grains (e.g., corn, wheat, barley, oats, and milo), oil meals (e.g., soybean meal, linseed meal, and cottonseed meal), molasses, and dried milk products. Concentrates are high in energy, low in fiber, and highly (80–90%) digestible. Roughages (Fig. 15.1) include legumes such as alfalfa (pasture and hays), grasses (pasture and hays), and straws, the latter being by-products from the production of grass, seed, and grain. Native and improved pastures are vital sources of roughage utilized by cattle, sheep, goats, and horses. Additional roughages are silage, stovers (dried corn, cane, or milo stalks and leaves with the grain portion removed), and soilage (cut green feeds). Roughages are less digestible than concentrates. Roughages are typically 50–65% digestible, but the digestibility of some straws is significantly lower.

NUTRIENTS

The six basic classes of nutrients—water, carbohydrates, fats, proteins, vitamins, and minerals—are found in varying amounts in animal feeds. Nutrients are composed of at least 20 of the more than 100 known chemical elements. These 20 elements and their chemical symbols are: calcium (Ca), carbon (C), chlorine (Cl), cobalt (Co), copper (Cu), fluorine (F), hydrogen (H), iodine (I), iron (Fe), magnesium (Mg), manganese (Mn), molybdenum (Mo), nitrogen (N), oxygen (O), phosphorus (P), potassium (K), selenium (Se), sodium (Na), sulfur (S), and zinc (Zn).

Water

The terms **water** and **moisture** are used interchangeably. Typically, water refers to drinking water, whereas moisture is used in reference to the amount of water in a given feed or ration. The remainder of the feed, after accounting for moisture, is referred to as **dry matter**. Moisture is found in all feeds, ranging from 10% in air-dry feeds to more than 80% in fresh green forage. For example, a typical grass pasture would contain a mix of vegetation that would be roughly 25% dry matter. Thus for every pound of vegetation consumed approximately ¾ of a pound would be in the form of water. Assuming that a mature cow consumed 25 pounds of green forage per day, her water intake from the feed source would be nearly 19 pounds (25 lb × .75 % moisture = 18.75 lb) or 2.25 gallons.

Figure 15.1

Roughage feeds include (A) hay (grass or alfalfa) baled for use in times of low availability of grazing, (B) silage created by chopping whole plants such as corn, packing silage into a pit where fermentation breaks down the plant making nutrients more available to the animal, (C) grazing improved forage such as winter rye, and (D) grazing crop aftermath following mechanical harvesting of the crop. Source: 15.1a, b, c, and d: Tom Field.

Water has important physiological and metabolic functions. It enters into most of the metabolic reactions, assists in transporting other nutrients, helps maintain normal body temperature, and gives the body its physical shape (water is the major component within cells).

Carbohydrates

Carbohydrates contain carbon, hydrogen, and oxygen. Carbohydrates can be classified based on the complexity of their chemical structure. Sugars or monosaccharides are the least complex and contain five or six carbons. Fructose, glucose, and galactose are examples of simple sugars. When two sugar molecules are linked together, the compound is referred to as a disaccharide; examples include lactose, maltose, and sucrose. Polysaccharides are formed when more than two sugars are linked together; examples include cellulose, glycogen, starch, and hemicellulose. Carbohydrates are found in both plant and animal tissues with most common forms being disaccharides and polysaccharides.

Monogastric or nonruminant animals must convert carbohydrates to glucose to assure absorption. Starch is the primary source of carbohydrate in the diets of swine and poultry and so most diets for these species are based on concentrate feeds such as grains and grain by-products. However, ruminant animals (cattle, sheep and goats) are best equipped to efficiently utilize cellulose, which is found in abundance

Table 15.1
STRUCTURE OF SATURATED AND UNSATURATED FATTY ACIDS

Saturated Fatty Acids	Structure[a]	Unsaturated Fatty Acids	Structure[a]
Acetic	C2:0	Palmitoleic	C16:1
Proprionic	C3:0	Oleic	C18:1
Butyric	C4:0	Linoleic	C18:2
Lauric	C12:0	Linolenic	C18:3
Myristic	C14:0	Arachidonic	C20:4
Palmitic	C16:0		
Stearic	C18:0		
Arachidic	C20:0		
Lignoceric	C24:0		

[a]The number following C is the number of carbon atoms, the second number represents the number of double bonds.

in forages. Cellulose is most effectively digested by the microbial organisms present in the rumen. Horses, often referred to as hindgut fermenters, are capable of digesting high-fiber plants containing cellulose and hemicellulose.

Livestock and poultry store minimal amounts of energy in the form of carbohydrates. However, minor amounts of glucose are converted into glycogen, which is an energy form that can be rapidly converted to energy during times of peak activity.

Fats

Fats and oils, also referred to as **lipids**, contain carbon, hydrogen, and oxygen, although there is more carbon and hydrogen in proportion to oxygen than with carbohydrates. Fats are solid and oils are liquid at room temperature. Fats contain 2.25 times more energy per pound than carbohydrates. Lipids provide both energy and essential fatty acids.

Most fats are composed of three fatty acids attached to a glycerol backbone. For example:

Glycerol + three fatty acids ⟶ a triglyceride + water

Certain fatty acids are saturated or unsaturated, depending on their particular chemical composition. Saturated fatty acids have single bonds tying the carbon atoms together (e.g., — C—C—C—C—), whereas unsaturated fatty acids have one or more double bonds (e.g., =C=C—C=C—). Saturated fatty acids range in complexity from a two-carbon form to a 24-carbon form (Table 15.1). The majority of fatty acids in animal tissue are straight chained and composed of an even number of carbons.

The term **polyunsaturated fatty acids** are applied to those having more than one double bond. Although more than 100 fatty acids have been identified, linoleic and α-linolenic acids have been determined to be a dietary essential for livestock. Two functions of the essential fatty acids are (1) as precursors of prostaglandins and (2) as structural components of cells.

Proteins

Proteins are polypeptides of high molecular weight that contain hydrogen, oxygen, carbon, and nitrogen. If a protein contains only amino acids, it is said to be a **simple protein**. A **complex protein** contains additional non–amino acid substances, such as heme (hemoglobin), carbohydrate (glycoproteins), or lipid (lipoproteins).

Protein is the only nutrient class that contains nitrogen. Proteins in feeds contain 16% nitrogen on average. This is why feeds are analyzed for the percent nitrogen in the feed, with the percent multiplied by 6.25 (100% ÷ 16% = 6.25) to convert it to percent protein. If, for example, a feed is 3% nitrogen, 100 g of the feed contains 3 g nitrogen. Multiplying 6.25 × 3 gives 18.75%, meaning that 100 g of this feed contains 18.75 g protein.

Protein synthesis is controlled by the genetic code contained within DNA. Formation of proteins is dependent on the availability of amino acids, often referred to as the building blocks of the animal's body. Proteins are composed of various combinations of some 25 amino acids. **Limiting amino acids** are those not provided in sufficient quantity to allow for the normal synthesis of a protein. Because some feeds are deficient in certain amino acids, diet formulations must assure that limiting amino acids are sufficiently abundant to assure normal protein formation. The building blocks for growth (including growth of muscle, bone, and connective tissue), milk production, and cellular and tissue repair are amino acids that come from proteins in feed. The interstitial (between cells) fluid, blood, and lymph require amino acids to regulate body water and to transport oxygen and carbon dioxide. All enzymes are proteins, so amino acids are also required for enzyme production. Amino acids have an amino group (NH_2) in each of their chemical structures. There are many different combinations of amino acids that can be structured together. The chemical, or peptide, bonding of amino acids is illustrated using alanine and serine, which results in the formation of a dipeptide:

It can be seen that amino acids have a basic portion, NH_2, and an acid portion, and it is because of these that they can combine into long chains to make proteins. When digestion occurs, the action is at the peptide linkage to free amino acids from one another.

Amino acids may be classified as either **essential** or **nonessential** (Table 15.2). Essential amino acids must be provided via dietary means, or in the case of ruminants, via microbial action. The amino acids most commonly deficient are lysine, methionine, and tryptophan. Nonessential amino acids are required by the animal for normal growth but can be synthesized by the animal.

Protein can also be evaluated for quality by evaluating the amount and ratio of essential amino acids present. Proteins vary in terms of absorption rate or biological value. Egg protein is considered the highest in value (>90%). Proteins from animal sources (60–80%) are greater than plant proteins (40–60%).

Table 15.2
ESSENTIAL AND NONESSENTIAL AMINO ACIDS

Essential	Nonessential
Arginine	Alanine
Histidine	Aspartic acid
Isoleucine	Aspartic acid
Leucine	Citrulline
Lysine	Cystine
Methionine	Glutamic acid[a]
Phenylalanine	Glycine[a]
Taurine[b]	Hydroxyproline
Threonine	Proline[a]
Tryptophan	Serine
Valine	Tyrosine

[a]Under some scenarios, these may be essential to assure adequate growth in chicks.
[b]Essential amino acid required by cats.

Table 15.3
MICRO AND MACRO MINERALS

Macro	Micro
Calcium	Chromium
Chlorine	Cobalt
Magnesium	Copper
Phosphorus	Fluorine[a]
Potassium	Iodine
Sodium	Iron
Sulfur	Manganese
	Molybdenum[a]
	Selenium[a]
	Zinc

[a]Beneficial in some regions but toxic if fed in excess.

Minerals

Chemical elements other than carbon, hydrogen, oxygen, and nitrogen are called **minerals**. They are inorganic because they contain no carbon; organic nutrients do contain carbon. Some minerals are referred to as **macro** (required in larger amounts) and others are **micro** or trace minerals (required in smaller amounts) (Table 15.3).

Calcium and phosphorus are required in specified amounts and in a specified ratio to each other for bone growth and repair and for other body functions. The blood plasma contains sodium chloride; the red blood cells contain potassium chloride. The osmotic relations between the plasma and the red blood cells are maintained by proper concentrations of sodium chloride and potassium chloride. Excessive sweating that results from heavy physical work in hot weather may deplete sodium chloride. It is essential that salt and plenty of water be available under such conditions. The acid–base balance of the body is maintained at the proper level by minerals.

Microminerals may become a part of the molecule of a vitamin (e.g., cobalt is a part of vitamin B_{12}) and they may become a part of a hormone (e.g., thyroxin, a hormone made by the thyroid gland, requires iodine for its synthesis).

Certain important metabolic reactions in the body require the presence of minerals. Selenium and vitamin E both appear to work together to help prevent white muscle disease, which is a calcification of the striated muscles, the smooth muscles, and the cardiac muscles. Both vitamin E and selenium are more effective if the other is present. Excesses of certain minerals may be quite harmful. For example, excess amounts of fluorine, molybdenum, and selenium are highly toxic.

Vitamins

Vitamins are organic nutrients needed in very small amounts to provide for specific body functions in the animal. There are 16 known vitamins that function in animal nutrition. Vitamins may be classed as either **fat-soluble** or **water-soluble**.

The fat-soluble vitamins are vitamins A, D, E, and K. Vitamin A helps maintain proper repair of internal and external body linings. Because the eyes have linings, lack of vitamin A adversely affects the eyes. Vitamin A is also a part of the visual pigments of the eyes. Vitamin D is required for proper use of calcium and phosphorus in bone growth and repair. A major function of vitamin D is to regulate the absorption of calcium and phosphorus from the intestine. Vitamin D is produced by the action of sunlight on sterols of the skin; therefore, animals that are exposed to sufficient sunlight make all the vitamin D they need. Vitamin K is important in blood clotting; hemorrhage might occur if the body is deficient in vitamin K. Vitamin E is an excellent antioxidant and has a role in preventing the breakdown of cell membranes by free radicals. Vitamin E exists in several forms classified as tocopherols.

The water-soluble vitamins are ascorbic acid (vitamin C), biotin, choline, cyanocobalamin (vitamin B_{12}), folic acid, niacin, pantothenic acid, pyridoxine (vitamin B_6), riboflavin (B_2), and thiamin (vitamin B_1). More diseases caused by inadequate nutrition have been described in the human than in any other animal, and among the best known are those caused by a lack of certain vitamins: beriberi (lack of thiamin); pellagra (lack of niacin); pernicious anemia (lack of vitamin B_{12}); rickets (lack of vitamin D); and scurvy (lack of vitamin C).

In ruminant animals, microorganisms in the rumen make all of the water-soluble vitamins. Water-soluble vitamins also appear to be readily available to horses; perhaps some are made by fermentation in the cecum. Water-soluble vitamins cannot be synthesized by monogastric animals and must therefore be in their feed. Most fat-soluble vitamins are not synthesized by either ruminants or monogastrics and must be supplied in the diets of both groups (an exception is vitamin K, which is synthesized by rumen bacteria in ruminants). Many vitamins are supplied through feeds normally given to animals.

PROXIMATE ANALYSIS OF FEEDS

The nutrient composition of a feed cannot be determined accurately by visual inspection. A system has been devised by which the value of a feed can be approximated. **Proximate analysis** separates feed components into groups according to their feeding value. This analysis is based on a feed sample and analysis, and therefore is no more accurate than how representative the sample is of the entire feed source.

The inorganic and organic components resulting from a proximate analysis are water, crude protein, crude fat (sometimes referred to as *ether extract*), crude fiber, nitrogen-free extract, and ash (minerals). Figure 15.2 shows these components resulting from a feed that would have a laboratory analysis of 88% dry matter,

Figure 15.2
Proximate analysis showing the inorganic and organic components of a feed (similar to wheat) on a natural or air-dry basis.

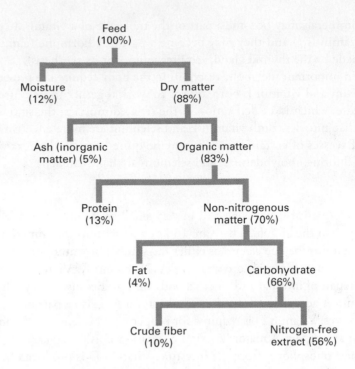

13% protein, 4% fat, 10% crude fiber, and 56% nitrogen-free extract (NFE) on a natural or air-dry basis. The analysis might be reported on a dry-matter basis (no moisture) as 14.8% protein, 4.5% fat, 11.4% crude fiber, and 63.6% nitrogen-free extract. Therefore, caution needs to be exercised in interpreting the proximate analysis results because different laboratories may report their analytical values on either an air-dry (or as-fed) basis or a dry-matter basis.

Values reported on a dry-matter basis can be converted to an as-is basis by multiplying the value, crude protein for example, by the dry-matter percentage and dividing by 100. For example, a hay sample measuring 15% crude protein (CP) and a dry matter of 90% would have a CP on an as-fed basis of:

$$15 \times \frac{90}{100} = 13.5\% \text{ CP}$$

Conversely, values can be converted for an as-is or wet basis to a dry-matter basis. For example, an oat sample has a crude protein measurement of 13% on an as-fed basis and contains 89% dry matter. The crude protein content on an as-fed basis is determined by:

$$13 \times \frac{100}{8.9} = 14.6\% \text{ CP}$$

The proximate analysis for the six basic nutrients does not distinguish the various components of a nutrient. For example, ash content of a feed does not tell the amount of calcium, phosphorus, or other specific minerals. Figure 15.3 gives the chemical analysis for organic and inorganic nutrients. There are specific chemical analyses for each of these nutrients in a feed if such an analysis is needed.

DIGESTIBILITY OF FEEDS

Digestibility refers to the amount of various nutrients in a feed that are absorbed from the digestive tract. Different feeds and nutrients vary greatly in their digestibility. Many feeds have been subjected to digestion trials, in which feeds of known

Figure 15.3
Chemical analysis scheme of inorganic and organic nutrients.

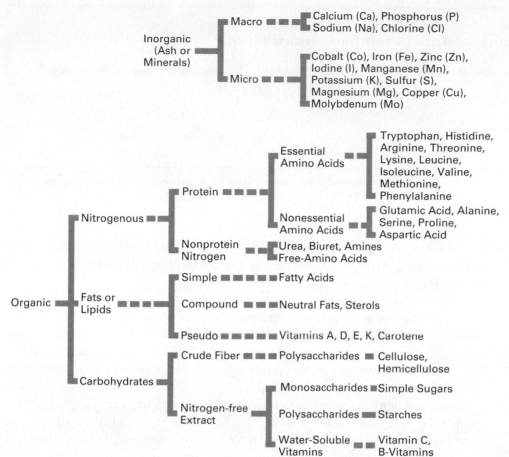

nutrient composition have been fed to livestock and poultry. Feces have been collected and the nutrients in the feces analyzed. The difference between nutrients fed and nutrients excreted in the feces is the apparent digestibility of the feed.

For example, the digestibility of protein is obtained by determining the digestibility of nitrogen in a feed. Digestibility is expressed as a percentage of nitrogen, for example, as follows:

$$\frac{\text{Nitrogen in feed} - \text{Nitrogen in feces}}{\text{Nitrogen in feed}} \times 100 = \text{Percentage digestibility}$$

As an example, if 25 lb of feed contains 3.2 g of nitrogen and 100 g of feces contains 0.8 g of nitrogen, the percent digestibility of nitrogen is:

$$\frac{3.2 - 0.8}{3.2} \times 100 = 75\%$$

Note that the determination of 3.2 g of nitrogen in 100 g of feed enables the percentage of protein in the feed to be estimated as 20% ($3.20 \times 6.25 = 20$).

ENERGY EVALUATION OF FEEDS

Energy is the capacity to accomplish work on the macro scale and is defined as the amount of heat produced when a nutrient is completely oxidized during digestion. Carbohydrates, fats, and proteins can all be used to provide energy; however, carbohydrates supply most of the energy, as they typically have a per unit energy cost advantage compared to other nutrients.

Table 15.4
AN EXAMPLE OF CALCULATING TOTAL DIGESTIBLE NUTRIENTS (TDN)

Nutrient	Amount of Nutrient (g)	Digestibility (%)	Amount of Digestible Nutrient (g)
Protein	20	75	15.00
Carbohydrates	55	85	46.75
Soluble (NFE)	10	20	2.0
Insoluble (fiber)	11.15	85	9.50
			TDN = 73.25

Energy can be used to power movement of the animal, but most of it is used as chemical energy to drive reactions necessary to convert feed into animal products and to keep the body warm or cool.

Energy needs of animals generally account for the largest portion of feed consumed. Several systems have been devised to evaluate feedstuffs for their energy content. **Total digestible nutrients (TDN)** estimates of feeds were historically the most commonly used energy estimation system. TDN is typically expressed in pounds, kilograms, or percentages after obtaining the proximate analysis and digestibility figures for a feed. The formula for calculating TDN is TDN = (digestible crude protein) + (digestible crude fiber) + (digestible nitrogen-free extract) + (digestible crude fat × 2.25). The factor of 2.25 is used to equate fat to a carbohydrate basis, since fat has 2.25 times as much energy as an equivalent amount of carbohydrate.

An example of calculating TDN in 100 g of feed is shown in Table 15.4.

TDN is roughly comparable to **digestible energy (DE)** but it is expressed in different units. TDN and DE both tend to overvalue roughages.

Even though there are some apparent shortcomings in using TDN as an energy measurement of feeds, it works well in balancing rations for cows and growing cattle. However, TDN is being replaced by estimates of **net energy (NE)** in many ration formulation systems. The net energy system (NE) is a more precise energy measurement of feeds. This system usually measures energy values in megacalories per pound or kilogram of feed. The calorie basis, which measures the heat content of feed, is as follows:

Calorie (cal)—amount of energy or heat required to raise the temperature of 1 g of water 1°C.

Kilocalorie (kcal)—amount of energy or heat required to raise the temperature of 1 kg of water 1°C.

Megacalorie (Mcal)—equal to 1,000 kilocalories or 1 million calories.

Figure 15.4 shows various ways in which animals utilize the energy of feeds and the various energy measurements of feeds. Some energy from the feed is lost in feces (not digested), urine (digested but not used by the body cells), and gases from microbial fermentation of the feed, and heat loss resulting from digestion and metabolism of the feed.

Maintenance and Production

Feeds provide energy that the animal uses to supply two basic functions: (1) maintenance—a steady state in which the animal is neither gaining nor losing energy; and (2) production. Maintenance energy is used to maintain basal metabolism, to provide

Figure 15.4
Measures of energy and energy utilization.

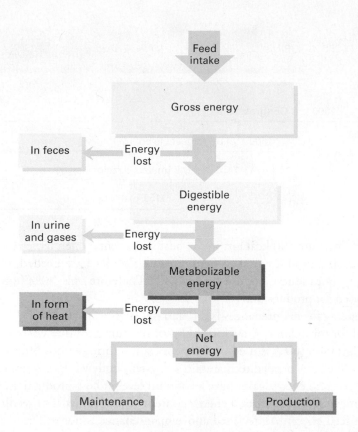

for the voluntary activity of the animal, to generate heat to keep the body warm, and to provide energy to cool the body.

Production activities include fetal development, semen production, growth, fat deposition, and production of milk, eggs, and wool. These functions are only sustained if either stored or consumed energy is in excess of that required for maintenance. Successful livestock management depends on the ability to anticipate the nutritional needs of animals at different stages of the lifecycle and to utilize available feed resources to effectively meet nutritional requirements of the animal. For example, a heifer that is still growing but also nursing her first calf has a different set of nutritional needs than a mature cow that is dry and in the third trimester of pregnancy.

Measurement of Energy

Gross energy (GE) is the quantity of heat (calories) released from the complete burning of the feed sample in an apparatus called a *bomb calorimeter*. GE has little practical value in evaluating feeds for animals because the animal does not metabolize feeds in the same manner as a bomb calorimeter. For example, oat straw has the same GE value as corn grain. Digestible energy (DE) is GE of feed minus fecal energy. **Metabolizable energy (ME)** is DE of feed minus energy in urine and gaseous products of digestion. Net energy (NE) is the ME of feed minus the energy used in the consumption, digestion, and metabolism of the feed. This energy lost between ME and NE is called **heat increment**.

Another way to illustrate the several measures of feed energy and how they are utilized is shown in Figure 15.5. In this example, 2,000 kcal of GE (in approximately 1 lb of feed) is fed to a laying hen. The DE shows that 400 kcal were lost in the feces. The ME (1,450 kcal) is used for heat increment, maintenance, and production (eggs

Figure 15.5

Energy utilization by a laying hen. Data represent approximately 1 lb of feed containing 2,000 kcal of gross energy.

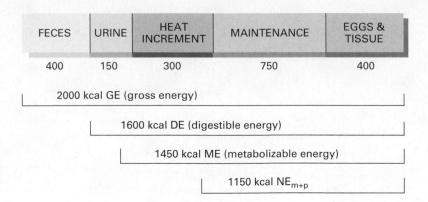

and tissue). There are 300 kcal lost in the heat increment, which leaves 1,150 kcal for maintenance and production. For maintenance, 750 kcal are needed, which leaves 400 kcal for egg production and tissue growth. Therefore only 20% (440/2,000) of the GE is used for production.

Net energy for maintenance (NE_m) and net energy for gain (NE_g) is more commonly used for formulating diets for feedlot cattle than any other energy system. Net energy for lactation (NE_l) is used in dairy cow ration formulation. NE_m in animals is the amount of energy needed to maintain a constant body weight. Animals of known weight, fed for zero energy gain, have a constant level of heat production.

NE_g measures the increased energy content of the carcass after feeding a known quantity of feed energy. All feed fed above maintenance is not utilized at a constant level of efficiency. Higher rates of gain require more feed per unit of gain as composition of gain varies with rate of gain.

FEEDS AND FEED COMPOSITION

Classification of Feeds

Feeds are naturally occurring ingredients in diets of farm animals used to sustain life. The terms *feed* and *feedstuffs* are generally used interchangeably; however, **feedstuffs** are a more inclusive term. Feedstuffs can include certain nonnutritive products such as additives to promote growth and reduce stress, to give flavor and palatability, to add bulk, or to preserve other feeds in the ration.

The National Research Council (NRC) classification of feedstuffs is as follows:

1. Dry roughages and forages
   Hay (legume and nonlegume)
   Straw
   Fodder
   Stover
   Other feeds with greater than 18% fiber (hulls and shells)
2. Range, pasture plants, and green forages
3. Silages (corn, legume, and grass)
4. Energy feeds (cereal grains, mill by-products, fruits, nuts, and roots)
5. Protein supplements (animal, marine, avian, and plant)
6. Mineral supplements
7. Vitamin supplements
8. Nonnutritive additives (antibiotics, coloring materials, flavors, hormones, preservatives, and medicants)

Roughages and forages are used interchangeably, although roughage usually implies a bulkier, coarser feed. In the dry state, roughages have more than 18% crude fiber. The crude fiber is primarily a component of cell walls that is not highly digestible. Roughages are also relatively low in TDN, although there are exceptions; for example, corn silage has over 18% crude fiber and approximately 70% TDN.

Feedstuffs that contain 20% or more protein, such as soybean meal and cottonseed meal, are classified as protein supplements. Feedstuffs with less than 18% crude fiber and less than 20% proteins are classified as energy feeds or concentrates. Cereal grains are typical energy feeds, which are reflected by their high TDN values.

Nutrient Composition of Feeds

Feeds are analyzed for their nutrient composition, as discussed earlier. The ultimate goal of nutrient analysis of feeds is to predict the productive response of animals when they are fed rations of a given composition.

The nutrient compositions of some of the more common feeds are shown in Tables 15.5 (ruminants) and 15.6 (nonruminant animals). The information in these tables represents averages of numerous feed samples. Feeds are not constant in composition, and an actual analysis should be obtained whenever economically feasible. The actual analysis is not always feasible or possible because of lack of available laboratories and insufficient time to obtain the analysis. Therefore, feed-analysis tables become the next-best source of reliable information on nutrient composition of feeds. It is not uncommon to expect the following deviations of actual feed analysis from the table values for several feed constituents: crude protein ($\pm15\%$), energy values ($\pm10\%$), and minerals ($\pm30\%$).

Digestible protein is included in some feed-composition tables when amino acid composition has been adequately determined. This is most typical for swine and poultry diet formulation. Crude protein is more commonly found in feed-composition tables and used in formulating diets for ruminants.

Digestible protein (DP) can be calculated from crude protein (CP) content by using the following equation (%DP and %CP are on a dry-matter basis):

$$\%DP = 0.9\ (\%CP) - 3$$

Five measures of energy values—TDN, ME, NE_m, NE_g, and NE_l)—are shown in Table 15.5. TDN is shown because there are more TDN values for feeds and because TDN has been a standard system of expressing the energy value of feeds. Some individuals seek ME (metabolizable energy) values for feed because these values are in calories rather than pounds. NE_m and NE_g values are used primarily to formulate feedlot diets and diets for growing replacement heifers, as these values offset the major problem associated with the TDN energy system. NE_l is used in formulating diets for dairy cows.

Historically, crude fiber (CF) was estimated using a prolonged protocol of alternatively boiling and filtrating feed samples in acid and then alkali solutions. The process is laborious and not repeatable in accurately determining the carbohydrate components of various feeds. Increasingly CF analysis is being replaced with neutral-detergent fiber (NDF) and acid-detergent fiber (ADF) protocols. These procedures provide a more accurate assessment of feed digestibility.

By-Product Feeds

The livestock industry has historically sought feed ingredients from a variety of sources. Key among them has been feedstuffs that are by-products of agricultural and fishery production, food processing, brewing, and production of alternative fuels such as ethanol. By-product feeds can be very useful both in terms of their nutritional composition but also in terms of cost competitiveness. By-product feeds listed in Tables 15.5 and 15.6 are italicized. Certainly, some of these by-products tend to be

Table 15.5
NUTRIENT COMPOSITION OF SELECTED FEEDS COMMONLY USED IN DIETS OF RUMINANTS

Feed	On a Dry-Matter Basis (moisture-free)									
	Dry Matter (%)	TDN[a] (%)	NE$_m$[b] (Mcal/lb)	NE$_g$[c] (Mcal/lb)	NE$_l$[d] (Mcal/lb)	Crude Protein (%)	Crude Fiber (%)	ADF[e] (%)	Calcium (%)	Phosphorus (%)
Alfalfa hay (early bloom)	90	60	0.52	0.26	0.61	18	23	31	1.41	0.22
Almond hull	90	52	.49	.19	.52	2	17	32	.23	.11
Bakery waste—dried	92	89	1.00	.69	.94	1.3	11	13	.14	.26
Barley (grain)	89	84	0.92	0.63	0.87	13	6	7	0.06	0.38
Beet pulp—dried	91	78	.86	.57	.81	10	16	33	.69	.10
Bermuda grass (hay)	90	49	0.42	0.18	0.53	6	30	39	0.46	0.20
Bluegrass (grazed)	31	72	0.77	0.49	0.74	17	25	29	0.37	0.30
Bone meal (steamed)	95	16	0.27	—	0.11	13	—	—	27.0	12.74
Brewers grains—dry	92	66	.69	.41	.68	25	15	23	.33	.15
Brewers grains—wet	21	66	.69	.41	.68	25	15	23	.33	.15
Brome (grazed early)	34	74	0.80	0.52	0.76	18	24	31	0.45	0.34
Citrus pulp	18	78	.86	.57	.81	10	16	20	—	—
Corn (whole)	88	90	1.01	0.70	0.83	9	2	3	0.02	0.30
Corn (flaked)	86	95	1.08	0.76	0.93	9	1	3	0.02	0.28
Corn silage (mature)	34	70	0.75	0.47	0.73	8	24	51	0.28	0.23
Corn (high moisture)	72	93	1.06	0.73	0.94	10	3	3	0.32	0.02
Cottonseed meal	92	78	0.86	0.56	0.81	44	13	20	0.21	1.19
Dicalcium phosphate	96	0	0	0	0	0	0	0	22.0	18.65
Distillers grain—dehydrated	94	86	.96	.66	.90	23	12	17	.43	.11
Orchard grass (hay)	91	54	0.50	0.25	0.54	8	37	45	0.30	0.26
Limestone (ground)	98	0	0	0	0	0	0	0	34.0	0.02
Meadow hay (native)	92	50	0.45	0.20	0.49	6	33	44	0.43	0.15
Milk dried	99	119	1.39	1.01	1.27	26	0	0	1.36	1.09
Milo (sorghum) (ground)	87	84	0.93	0.64	0.84	10	3	9	0.04	0.32
Molasses (cane)	76	75	0.79	0.50	0.77	5	0	0	1.00	0.10
Oats (grain)	89	76	0.81	0.52	0.78	13	11	15	0.05	0.41
Soybean hulls	91	77	.85	.55	.80	12	40	50	.49	.21
Soybean meal (solvent)	89	84	0.92	0.64	0.88	49	7	10	0.38	0.71
Sudan grass (grazed)	18	70	0.73	0.44	0.71	17	23	29	0.46	0.36
Timothy (grazed)	26	72	0.65	0.36	0.65	18	32	36	0.40	0.28
Wheat (grazed early)	22	73	0.79	0.50	0.76	28	17	30	0.40	0.42
Wheat (straw)	89	42	0.29	.05	0.44	3	43	55	0.16	0.05
Wheatgrass, crested (early)	28	75	0.81	0.53	0.60	21	22	28	0.46	0.32

[a]Total digestible nutrients.
[b]Net energy-maintenance.
[c]Net energy-gain.
[d]Net energy-lactation.
[e]Acid-detergent fiber.
Source: Adapted from Preston, 20, various NRC Publications. Feeds in italics are by-products.

Table 15.6
Nutrient Composition of Selected Feeds Commonly Used in Rations of Nonruminant Animals (Air-Dry Basis)

Feed	Poultry ME[a] (Kcal/lb)	Swine ME[a] (Kcal/lb)	Digestible Protein (%)	Minerals					Vitamins					Amino Acids		
				Calcium (%)	Phosphorus (%)	Iron (ppm)	Manganese (ppm)	Zinc ppm	Niacin (mg/lb)	Pantothenic Acid (mg/lb)	Riboflavin (mg/lb)	Choline (mg/lb)	B$_{12}$ (mg/lb)	Lysine (%)	Methionine (%)	Tryptophan (%)
Alfalfa meal (dehydrated)	672	1,020	8.3	1.30	0.23	400	35	21	19	14	6	583	—	0.73	0.28	0.45
Barley (grain)	1,250	1,305	8.2	0.08	0.42	80	16	30	35	3.6	0.7	461	—	0.53	0.18	0.17
Blood meal	1,465	875	62.3	0.28	0.22	2,500	5	22	10	1.4	1.2	244	0.20	6.9	1.0	1.0
Corn (grain)	1,540	1,520	7.0	0.01	0.25	23	6	15	10	2.3	0.5	222	—	.24	0.18	0.7
Feather meal	1,310	1,030	70.1	0.20	0.70	70	9	55	8.9	3.8	0.9	379	0.27	.66	0.55	0.4
Fish meal (whole)	1,180	1,120	60.7	7.00	3.50	80	10	80	23.5	3.7	2.1	1,338	0.07	4.3	1.65	0.7
Meat and bone meal (45%)	1,080	1,090	45.0	11.0	5.9	500	10.1	9.0	22	2.3	7.9	864	0.03	2.2	0.53	0.18
Milo (sorghum)	1,505	1,470	7.8	0.04	0.29	52	13	14	16	5	0.5	288	—	0.27	0.1	0.1
Oats	1,160	1,215	9.9	0.01	0.35	70	38	38	6.2	4.3	0.6	427	—	0.4	0.2	0.18
Soybean meal (solvent)	1,020	1,405	41.7	0.25	0.60	120	27	60	12.1	7.2	1.3	1,151	—	2.9	0.65	0.6
Wheat, hard (red winter)	1,440	1,465	11.7	0.05	0.41	50	40	34	23.5	4.5	0.6	446	—	0.40	0.25	0.18
Whey (dried)	860	1,450	12.6	0.87	0.79	160	7	3	4.5	19.5	11.6	756	—	1.1	0.2	0.2

[a]Metabolizable energy.
Source: Adapted from *Feedstuffs*, 2012, various NRC publications. Feeds in italics are by-products.

limited to certain regions. For example, citrus by-products, almond hulls, cotton gin remnants, and soybean hulls may be very cost competitive ingredients in regions where production of the primary product is prevalent. However, transportation of these ingredients to other regions may not be cost effective due to transport costs, shrink, or storage issues. By-product origin feeds will become increasingly important to global livestock production as a means to enhance sustainability.

CHAPTER SUMMARY

- Nutrients are feed constituents that support or sustain life.
- The six classes of nutrients are water, carbohydrates, fats, proteins, vitamins, and minerals.
- Carbohydrates and fats are the primary energy sources in feed.
- Proteins are composed of some 25 amino acids; the latter are known as the building blocks of the animal's body.
- Proteins contain 16% nitrogen and compose most of the muscle mass.

- Vitamins and minerals are required in smaller amounts than the other nutrient classes; however, they are necessary for certain metabolic reactions to occur.
- The proximate analysis of a feed sample identifies components that reflect the feeding value—that is, water, crude protein, fat (ether extract), crude fiber, nitrogen-free extract, and ash (minerals).
- Energy evaluation of feeds is measured by total digestible nutrients (TDN), digestible energy (DE), metabolizable energy (ME), and net energy (NE).

KEY WORDS

nutrient
concentrates
roughages
water
moisture
dry matter
carbohydrates
lipids (fats)
polyunsaturated fatty acids
proteins (simple and complex)
limiting amino acids
amino acids (essential and nonessential)
minerals (macro and micro)

vitamins (water- and fat-soluble)
proximate analysis
digestibility
total digestible nutrient (TDN)
digestible energy (DE)
net energy (NE)
calorie
kilocalorie
megacalorie
metabolizable energy (ME)
heat increment
feedstuffs
digestible protein

REVIEW QUESTIONS

1. Compare and contrast roughage and concentrate feeds.
2. Explain the importance of quantifying dry matter.
3. How are fats classified?
4. Compare the energy value of carbohydrates versus fats.
5. Discuss the importance of limiting amino acids to ration formulation.
6. List the primary micro and macro minerals.
7. On what basis are vitamins classified?
8. Diagram proximate analysis.

9. Calculate percent crude protein and percent digestibility.
10. Discuss the importance of quantifying TDN.
11. Describe how the energy content of feeds can be measured and which measures are most useful.
12. What are the NRC feedstuff calculations.
13. Calculate digestible protein.
14. Discuss the importance of by-product feeds to the livestock and poultry industries.
15. Use nutrient composition tables to compare and contrast various feeds.

SELECTED REFERENCES

Cheeke, P. J. 2005. *Applied Animal Nutrition.* Upper Saddle River, NJ: Pearson Prentice Hall.

Church, D. C. 1991. *Livestock Feeds and Feeding.* Englewood Cliffs, NJ: Prentice Hall.

Ensminger, M. E., J. E. Oldfield, and W. W. Heinemann. 1990. *Feeds and Nutrition.* Clovis, CA: Ensminger Publishing Co.

Jurgens, M. H. 1993. *Animal Feeding and Nutrition.* 5th ed. Dubuque, IA: Kendall-Hunt.

Martin, D. W., P. A. Mayes, V. W. Rodwell, and D. K. Granner. 1985. *Harper's Review of Biochemistry.* Los Altos, CA: Lange Medical Publications.

National Research Council. *Nutrient Requirements of Beef Cattle* (2000); *Dairy Cattle* (2001); *Goats* (2007); *Horses* (2007); *Poultry* (1994); *Sheep* (2007); and *Swine* (2012). Washington, DC: National Academy Press.

Preston, R. L. 2013. Feed Composition Guide. *Beef.* St. Paul, MN.

🏠16
Digestion and Absorption of Feed

Animals obtain substances needed for all body functions from the feeds they eat and the fluids they drink. Before the body can absorb and utilize nutrients, feeds must undergo a process called **digestion**. Digestion includes mechanical action, such as chewing and contractions of the intestinal tract; chemical action, such as the secretion of hydrochloric acid (HCl) in the stomach and bile in the small intestine; and the action of enzymes, such as maltase, lactase, and sucrase (which act on disaccharides), lipase (which acts on lipids), and peptidases (which act on proteins). Enzymes are produced either by the various parts of the digestive tract or by microorganisms. The role of digestion is to reduce feed particles to molecules so they can be absorbed into the blood and eventually support body functions.

CARNIVOROUS, OMNIVOROUS, AND HERBIVOROUS ANIMALS

Animals are classed as carnivores, omnivores, or herbivores according to the types of feed they normally eat. **Carnivores**, such as dogs and cats, normally consume animal tissues as their source of nutrients; **herbivores**, such as cattle, horses, sheep, and goats, primarily consume plant tissues. Humans and pigs are examples of **omnivores**, which eat both plant and animal products.

Carnivores and omnivores are monogastric (nonruminant) animals, meaning that their stomachs are simple in structure, having only one compartment. Some herbivores, such as horses and rabbits, are also considered monogastric. Other herbivores, such as cattle, sheep, and goats, are ruminant animals, meaning that their stomachs are complex in structure, containing four compartments.

The digestive tracts of pigs and humans are similar in anatomy and physiology; therefore, much of the information gained from studies on pig nutrition and digestive physiology can be applied to humans. Both the pig and the human are omnivores and both are monogastric animals. Neither can synthesize the B-complex vitamins or amino acids to a significant extent. Both pigs and humans tend to eat large quantities, which can result in obesity. Humans can control obesity by regulating food intake and exercising as a means of using, rather than storing, excess energy. Obesity in swine can be controlled by limiting the amount of feed available to them and through genetic selection of leaner animals. The latter has received the greater emphasis as pigs are typically fed **ad libitum** (free choice).

DIGESTIVE TRACT OF MONOGASTRIC ANIMALS

The anatomy of the digestive tract varies greatly from one species of animal to another. The basic parts of the digestive tract are the mouth, stomach, small intestine, and large intestine (or colon). The primary function of the parts preceding the intestines is to reduce the sizes of feed particles. The **small intestine** functions in splitting food molecules and in absorbing nutrients, whereas the **large intestine** absorbs water and forms indigestible wastes into **feces**. In a mammal having a simple stomach (such as the pig), the mouth has teeth and lips for grasping and holding feed that is masticated (chewed), and salivary glands that secrete saliva for moistening feed so it can be swallowed.

Feed passes from the mouth to the **stomach** through the **esophagus**. There is a sphincter (valve) at the junction of the stomach and esophagus, which can prevent feed from coming up the esophagus when stomach contractions occur. The stomach empties its contents into that portion of the small intestine known as the **duodenum**. The pyloric sphincter, located at the junction of the stomach and the duodenum, can be closed to prevent feed from moving into or out of the stomach. Feed goes from the duodenum to portions of the small intestine known as the jejunum and the ileum. It then passes from the small intestine to the large intestine, or colon. The ileocecal valve, located at the junction of the small intestine and the colon, prevents material in the large intestine from moving back into the small intestine.

The small intestine actually empties into the side of the colon near, but not at, the anterior end of the colon. The blind anterior end of the colon is the cecum, or, in some animals, the vermiform appendix. The large intestine empties into the rectum. The anus has a sphincter, which is under voluntary control so that the animal can prevent defecation until it actively engages in the process. The structures of the digestive system of the pig are shown in Figure 16.1.

Animals such as pigs, horses, and poultry are classed as monogastric animals, but they differ markedly. For example, the horse has a large structure called the **cecum**, where feed is fermented (Fig. 16.2). The cecum is posterior to the area where most feed is absorbed, thus horses do not obtain all of the nutrients made by microorganisms in the cecum. Due to the location of the cecum, horses are considered to be hindgut digesters. Digestive tract sizes and capacities of monogastric or nonruminant animals are contrasted in Table 16.1.

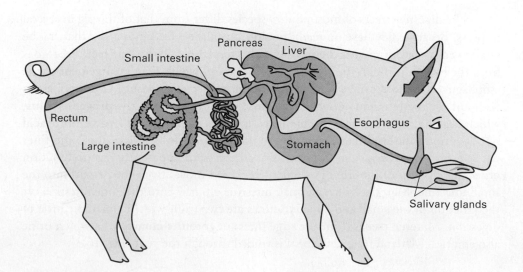

Figure 16.1

Digestive tract of the pig as an example of the digestive tract of a monogastric animal. Drawing by Sean Field. Adapted from Yen, 2001.

Figure 16.2
Digestive system of the horse. The posterior view shows the colon or large intestine proportionally larger than the rest of the digestive tract. Note particularly the location of the cecum at the anterior end of the colon.

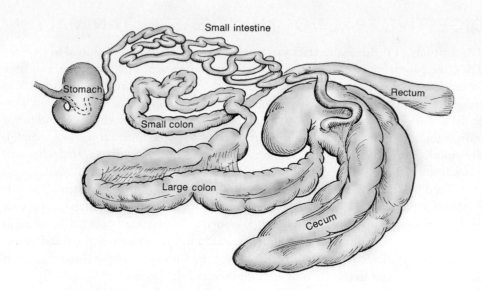

Table 16.1

DIGESTIVE TRACT SIZES AND CAPACITIES OF SELECTED MONOGASTRIC ANIMALS

Part of Digestive Tract	Species			
	Human	Pig	Horse	Chicken
Esophagus	—	—	4 ft	Total length of digestive tract in mature chickens is approximately 7 ft (beak to crop, 7 in.; beak to proventriculus, 14 in.; duodenum, 8 in.; ileum and jejunum, 48 in.; and cecum, 7 in.)
Stomach	1 qt	2 gal	4 gal	
Small intestine (capacity)	1 gal	2 gal	12 gal	
Small intestine (length)		60 ft	70 ft	
Large intestine (capacity)	1 qt	3 gal	11 gal	
Large intestine (length)		12 ft	20 ft	
Cecum (capacity)			8 gal	

The digestive tracts of most poultry species differ from that of the pig in several respects. Because they have no teeth, poultry break their feed into a size that can be swallowed by pecking with their beaks or by scratching with their feet. Feed goes from the mouth through the esophagus to the **crop**, which is an enlargement of the esophagus where feed can be stored. Some fermentation may occur in the crop, but it does not act as a fermentation vat. Feed passes from the crop to the **proventriculus**, which is a glandular stomach in that it secretes gastric juices and hydrochloric acid but does not grind feed. Feed then goes to the **gizzard**, where it is ground into finer particles by strong muscular contractions. The gizzard apparently has no function other than to reduce the size of feed particles. Feed moves from the gizzard into the small intestine. Material from the small intestine empties into the large intestine. At the junction of the small and large intestines are two ccca, which contribute little to digestion. Material passes from the large intestine into the **cloaca**, into which urine also empties. Material from the cloaca is voided through the vent (Fig. 16.3).

Figure 16.3
*Poultry digestive tract and
its components.* Drawing by
Sean Field.

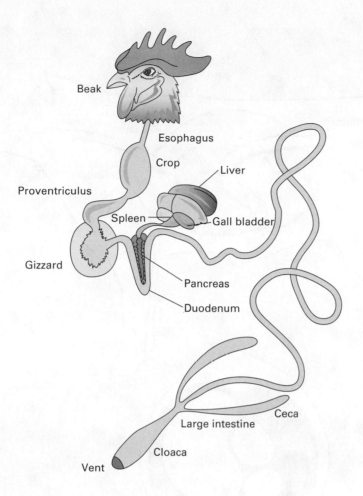

STOMACH COMPARTMENTS OF RUMINANT ANIMALS

In contrast to the single stomachs of monogastric animals, the stomachs of cattle, sheep, and goats have four compartments—**rumen**, **reticulum**, **omasum**, and **abomasum** (Fig. 16.4). The rumen is a large fermentation vat where bacteria and protozoa thrive and break down roughages to obtain nutrients for their use. It is lined with numerous papillae, which give it the appearance of being covered with a thick coat of short projections. The papillae increase the surface area of the rumen lining. The microorganisms in the rumen can digest cellulose and can synthesize amino acids from nonprotein nitrogen. The B-complex vitamins are also synthesized in the rumen. Later, these microorganisms are digested in the small intestine to provide these nutrients for the ruminant animal's use.

The reticulum has a lining with small compartments similar to a honeycomb; thus it is occasionally referred to as the **honeycomb**. Its function is to interact with the rumen in initiating the mixing activity of the rumen and providing an additional area for fermentation. The omasum has many folds, so it is often called the **manyplies**. The omasum may not have a major digestive function, although some believe that the folds produce a grinding action on the feed. The abomasum, or true stomach, corresponds to the stomach of monogastric animals and performs a similar digestive function.

The size and capacity of the ruminant stomach and intestinal tract are given in Table 16.2. The data in Table 16.2 are for mature ruminants, as the relative proportions of the stomach compartments are considerably different in the young lamb and calf. At birth, the abomasum comprises 60% of the total stomach capacity, whereas the rumen is only 25% of the total.

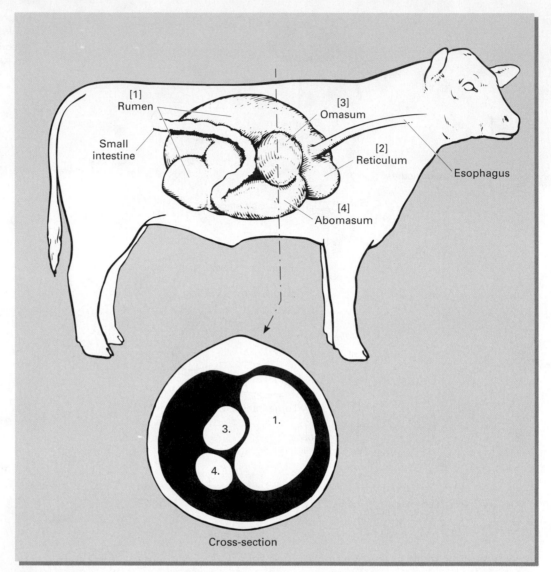

Figure 16.4
Beef cattle digestive tract.

Animals having the four-compartment stomach eat forage rapidly and later, while resting, regurgitate each bolus of feed (known as the **cud**). The regurgitated feed is chewed more thoroughly and swallowed, and then another bolus is regurgitated and chewed. This process continues until the feed is thoroughly masticated. The regurgitation and chewing of undigested feed is known as **rumination**. Animals that ruminate are known as **ruminants**. As microorganisms in the rumen ferment feeds, large amounts of gases (chiefly methane and carbon dioxide) are produced. The animal normally can eliminate the gases by controlled belching, also called **eructation**.

DIGESTION IN MONOGASTRIC (NONRUMINANT) ANIMALS

Feed that is ingested (taken into the mouth) stimulates the secretion of saliva. Chewing reduces the size of ingested particles and saliva moistens the feed. The enzyme amylase, present in the saliva of some species including pigs and humans, acts on starch. (Ruminants do not secrete salivary amylase.) However, little actual breakdown

Table 16.2
DIGESTIVE TRACT SIZES AND CAPACITIES OF MATURE RUMINANT ANIMALS

	Species	
Part of Digestive Tract	Cow	Ewe
Stomach		
Rumen	40 gal	5 gal
Reticulum	2 gal	2 qt
Omasum	4 gal	1 qt
Abomasum	4 gal	3 qt
Small intestine	15 gal (130 ft)	2 gal (80 ft)
Large intestine	10 gal	6 qt

Table 16.3
ENZYMES IMPORTANT IN THE DIGESTION OF FEED

Enzyme	Substrate	Substances Resulting from Enzyme Action
Amylase	Starch	Disaccharides, dextrin
Chymotrypsin	Peptides	Amino acids and peptides
Lactase	Lactose	Glucose and galactose
Lipase	Lipids	Fatty acids and glycerides
Maltase	Maltose	Glucose and glucose
Pepsin	Protein	Polypeptides
Peptidases	Peptides	Amino acids
Sucrase	Sucrose	Glucose and fructose
Trypsin	Protein	Polypeptides

of starch into simpler compounds occurs in the mouth, primarily because feed is there for only a short time.

An **enzyme** is an organic catalyst that speeds a chemical reaction without being altered by the reaction. Enzymes are rather specific; that is, each type of enzyme acts on only one or a few types of substances. Therefore, it is customary to name enzymes by giving the name of the substance on which it acts and adding the suffix -*ase*, which, by convention, indicates that the molecules so named are enzymes (Table 16.3). For example, lipase is an enzyme that acts on lipids (fats); maltase is an enzyme that acts on maltose to convert it into two molecules of glucose; lactase is an enzyme that acts on lactose to convert it into one molecule of glucose and one molecule of galactose; and sucrase is an enzyme that acts on sucrose to convert it into one molecule of glucose and one molecule of fructose. Some lipase is present in saliva, but little hydrolysis of lipids into fatty acids and glycerides occur in the mouth.

As soon as it is moistened by saliva and chewed, feed is swallowed and passes through the esophagus to the stomach. The stomach secretes HCl, mucus, and the digestive enzymes pepsin and gastrin. The strongly acidic environment in the stomach favors the action of pepsin that breaks proteins down into polypeptides. The HCl also assists in coagulation, or curdling, of milk. Little breakdown of proteins into amino acids occurs in the stomach. Mucous secretions help to protect the stomach lining from the action of strong acids.

In the stomach, feed is mixed well and some digestion occurs; the mixture that results is called **chyme**. The chyme passes next into the duodenum, where it is mixed with secretions from the pancreas, bile, and enzymes from the intestine.

Secretin, pancreozymin, and cholecystokinin, the three hormones that are released from the duodenal cells, stimulate secretion from the pancreas and discharge of bile from the gallbladder. The enzymes from the pancreas are lipase, which hydrolyzes fats into fatty acids and glycerides; trypsin, which acts on proteins and polypeptides to reduce them to small peptides; chymotrypsin, which acts on peptides to produce amino acids; and amylase, which breaks starch down to disaccharides, after which the disaccharides are broken down to monosaccharides. The liver produces bile that emulsifies fats; the bile is strongly alkaline and so helps to neutralize the acidic chyme coming from the stomach. Some minerals that are important in digestion also occur in bile.

By the time they reach the small intestine, amino acids, fatty acids, and monosaccharides (simple sugars or carbohydrates) are all available for absorption. Thus, the small intestine is the most important area for both digestion and absorption of feed. The small intestine is lined with fingerlike projections that protrude from the intestinal lining to increase the total surface area and thus absorptive capability. These structures are called **villi**.

Absorption of feed molecules may be either passive or active. Passive passage results from diffusion, which is the movement of molecules from a region of high concentration of those molecules to a region of low concentration. Active transport of molecules across the intestinal wall may be accomplished through a process in which cells of the intestinal lining or villi engulf the molecules and then actively transport these molecules to either the bloodstream or the lymph. Energy is expended in accomplishing the active transport of molecules across the gut wall.

When molecules of digested feed enter the capillaries of the blood system, they are carried directly to the liver. Small molecules of lipid may enter the lymphatic system, after which they go to various parts of the body, including the liver. The liver is an extremely important organ both for metabolizing useful substances and for detoxifying harmful ones.

In some monogastric animals, such as the horse, postgastric (cecal) fermentation of roughages occurs. In these animals, the feed that can be digested is digested and absorbed before the remainder reaches the cecum. These animals are perhaps more efficient than ruminants in their use of feeds such as concentrates. In the ruminant animals, the bacteria and protozoa use the concentrates given along with roughages. Because the microorganisms in ruminants use starches and sugars, little glucose is available to ruminants for absorption. The microorganisms do provide volatile fatty acids, which are absorbed by the ruminant and converted to glucose as an energy source. The postgastric fermentation that occurs in horses breaks down roughages, but this takes place posterior to the areas where nutrients are most actively absorbed; consequently, the animal does not obtain all nutrients resulting from postgastric fermentation.

DIGESTION IN RUMINANT ANIMALS

In mature ruminant animals (cattle, sheep, and goats), predigestive fermentation of feed occurs in the rumen and reticulum. The bacteria and protozoa in the rumen and reticulum use roughages consumed by the animal as feed for their growth and multiplication; consequently, billions of these microorganisms develop. The rumen environment is ideal for microorganisms because moisture, a warm temperature, and a constant supply of nutrients are present. Excess microorganisms are continuously removed from the rumen and reticulum along with small feed particles that escape microbial fermentation and pass through the omasum into the abomasum. When feed passes into the abomasum, strong acids destroy the bacteria and protozoa. The ruminant animal then digests the microorganisms in the small intestine and uses them as a source of nutrients. The digested microbial cells provide the animal with most of its amino acid needs and some energy. Thus, ruminant animals and microorganisms mutually benefit each other. All digestive processes in ruminants are the same as those in monogastric animals after the feed reaches the abomasum, which corresponds to the stomach of monogastric animals.

The rumen fermentation process also produces **volatile fatty acids** (acetic, propionic, and butyric acids), which are waste products of microbial fermentation of carbohydrates. The animal then uses these volatile fatty acids (VFAs) as its major source of energy. In the process of fermenting feeds, methane gas is also produced by the microorganisms. The animal releases the gas primarily through belching. Occasionally, the gas-releasing mechanism does not function properly and gas accumulates in the rumen, causing **bloat** to occur. Death will occur owing to suffocation if gas pressure builds to a high level and interferes with adequate respiration.

A young, nursing ruminant consumes little or no roughage. Consequently, at this early stage of life, its digestive tract functions similarly to that of a monogastric animal. Milk is directed immediately into the abomasum in young ruminants by the **esophageal groove**. The sides of the esophageal groove extend upward by a reflex action and form a tube through which milk passes directly from the esophagus to the abomasum. This allows milk to bypass fermentation in the rumen. Rumen fermentation is an inefficient use of energy and protein in a high-quality feed such as milk.

When roughage is consumed it is directed into the rumen, where bacteria and protozoa break it down into simple forms for their use. The rumen starts to develop functionally as soon as roughage enters it, but some time is required before it is completely functional. Complete development of the rumen, reticulum, and abomasum requires about 2 months in sheep and about 3–4 months in cattle. The type of feed given to the animal can influence development of the rumen. If only milk and concentrated feeds are given, the rumen shows little development. If very young ruminants are forced to consume only forage, the rumen develops much more rapidly.

Energy Pathways

Figure 16.5 shows the digestion and utilization of carbohydrates and fats contained in the ingested forages and grains. The primary energy end products of glucose and fatty acids supply energy in the body tissues and become milk fat and lactose in the lactating ruminant. Excess energy is stored as body fat in the body tissues.

The primary organs and tissues in energy metabolism are shown in Figure 16.5. These are the rumen, abomasum, small intestines, liver, blood vessels, mammary gland, and body tissues. Undigested carbohydrates (primarily complex carbohydrates such as lignin) are excreted through the large intestine and eliminated as feces. Other energy waste products, such as ketone bodies, are excreted through the kidneys in the urine.

Protein Pathways

The digestion, utilization, and excretion of dietary protein and **nonprotein nitrogen (NPN)** are shown in Figure 16.6. The end products of protein and NPN are amino acids, ammonia (NH_3), and synthesized amino acids. Excess NH_3 can be formed into urea in the liver, and then excreted through the urine, with some urea returning to the rumen as a component of saliva.

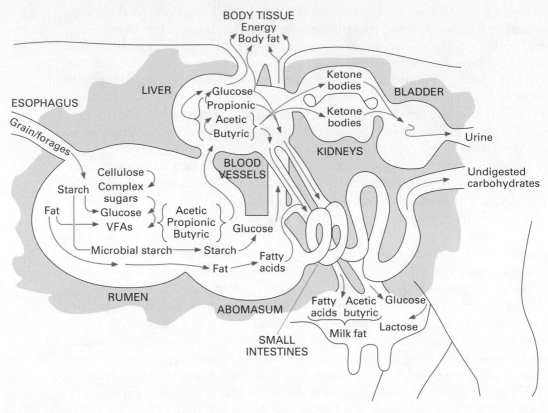

Figure 16.5

Energy pathways in the ruminant. Source: Figure by Patrick Hatfield from Montana State University. Used by Permission of Patrick Hatfield.

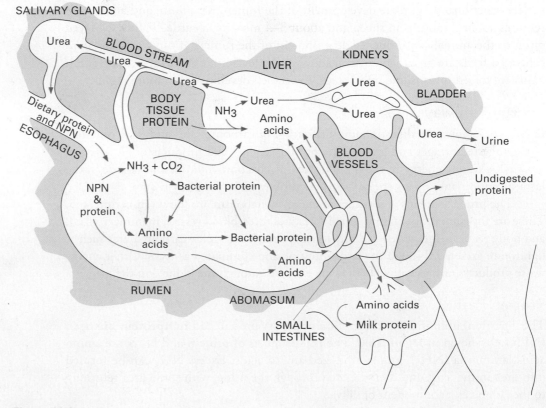

Figure 16.6

Protein pathways in the ruminant. Source: Figure by Patrick Hatfield from Montana State University. Used by Permission of Patrick Hatfield.

CHAPTER SUMMARY

- The digestive tract of a monogastric animal consists of the mouth, esophagus, stomach, small intestine, and large intestine (colon).

- The stomachs of ruminant animals have four compartments—rumen, reticulum, omasum, and abomasum. The latter functions like the stomach of a monogastric animal.

- Microorganisms in the rumen digest cellulose, synthesize amino acids, and synthesize B vitamins.

- Poultry differs from other monogastric animals in having a crop (for food storage), proventriculus (for chemical digestion), and gizzard (for physical grinding of the feed).

- The horse has a cecum (a large pouch between the small and large intestine) where feed is fermented as in the rumen of ruminant animals.

- Digestion breaks down the feed into smaller components that can be absorbed and utilized by the animal—carbohydrates to simple sugars, proteins to amino acids, and fat to fatty acids.

KEY WORDS

digestion
carnivore
herbivore
omnivore
ad libitum
small intestine
large intestine
feces
stomach
esophagus
duodenum
cecum
crop
proventriculus
gizzard
cloaca
rumen

reticulum
omasum
abomasum
honeycomb
manyplies
cud
rumination
ruminant
eructation
enzyme
chyme
villi
volatile fatty acids
bloat
esophageal groove
nonprotein nitrogen (NPN)

REVIEW QUESTIONS

1. Compare and contrast herbivores, carnivores, and omnivores.
2. List the primary organs of digestion and their respective roles in cattle, hogs, and chickens.
3. Compare and contrast digestion in the sheep and the horse.
4. Describe the process of digestion in ruminant animals.
5. List the important digestive enzymes and the respective substrates they affect.
6. Describe changes in the digestive physiology of newborn versus more mature ruminants.
7. Why is it important to understand pathways of energy and protein digestibility?

SELECTED REFERENCES

Cheeke, P. J. 2005. *Applied Animal Nutrition*. Upper Saddle River, NJ: Pearson Prentice Hall.

Ensminger, M. E., J. E. Oldfield, and W. W. Heinemann. 1990. *Feeds and Nutrition*. Clovis, CA: Ensminger Publishing Co.

Kellems, R. O. and D. C. Church. 2010. *Livestock Feeds and Feeding*. Upper Saddle River, NJ: Pearson Prentice Hall.

Yen, J. T. 2001. *Biology of the Domestic Pig*. Ithaca, NY: Cornell University Press.

17

Providing Nutrients for Body Functions

Proper nutritional management is of fundamental importance to the success of any livestock or poultry production enterprise because animals must be adequately fed to assure good health, productivity, and profitability. The basic task of the producer, then, is to supply animals with feed that satisfies their body functions for maintenance, growth, fattening, reproduction, lactation, egg laying, wool production, and work. Each of these functions has a unique set of nutrient requirements, which are additive when more than one function is occurring.

The task of the manager is to create diets that meet the animal's nutritional requirements at the least cost attainable. Livestock producers face high feed costs and increased volatility in price and availability, increasingly expensive costs to transport feed, and a host of other risk factors that makes least cost ration formulation difficult.

NUTRIENT REQUIREMENTS FOR BODY MAINTENANCE

Body maintenance requires that nutrients be supplied to keep the body functioning in a state of well-being. Theoretically in a state of maintenance there is no gain or loss of weight or production. It would be rare to find this condition in either wild or domestic animals. Maintenance functions that have a high priority for nutrients are (1) body tissue repair, (2) control of body temperature, (3) energy to keep all vital organs (respiratory, digestive, and so on) functioning, and (4) water balance maintenance.

Nutrients first meet maintenance needs before supplying any of the other body functions. Approximately half of all feed fed to livestock and poultry is used to meet the maintenance requirement. Feedlot animals on full feed may only use 30–40% of their nutrients for maintenance, while some mature breeding animals may need 90% of their feed for maintenance. Highly efficient dairy cows, producing over 100 lb of milk per day, have a daily feed consumption four to five times their maintenance requirement.

Body Size and Maintenance

Maintenance needs are related to body size. A large animal requires more feed than a small one, but maintenance requirements are not linearly related to body weight. Small animals require more feed per unit of body weight for maintenance than large ones. The approximate maintenance

Table 17.1
TDN NEEDED FOR MAINTENANCE OF CATTLE IN THE GROWING-FINISHING PERIOD

Weight of Cattle (lb)	TDN Needed Daily for Maintenance (lb)
400	5.7
600	7.7
800	9.7
1,000	11.4
1,100	12.3

requirement in relation to weight is expressed as $Wt^{0.75}$ rather than $Wt^{1.00}$. Thus, if a 500-lb animal requires 15 lb of feed per day for maintenance, a 1,000-lb animal of the same type does not require twice as much feed even though the latter animal weighs twice as much as the first. The quantity of $1,000^{0.75}$ can be determined and applied to show that the 1,000-lb animal requires approximately 1.7 times as much feed for maintenance as the 500-lb animal. The 1,000-lb animal therefore requires approximately 25.5 lb daily (15 lb $\times$ 1.7 = 25.5 lb). Table 17.1 shows how the total digestible nutrient intake (TDN) requirement changes with increasing weight. Where corn is 91% TDN, on an air-dry basis, it would take approximately 12.5 lb (12.5 lb $\times$ 0.91 = 11.4 lb TDN) of corn to fill the maintenance requirement of a 1,000-lb animal.

NUTRIENT REQUIREMENTS FOR GROWTH

Growth occurs when protein synthesis is in excess of protein breakdown. There are several important nutrient requirements for growth including energy, protein, minerals, vitamins, and energy. The dry matter of muscle and connective tissue is composed largely of protein; therefore, young, growing animals that need feed to sustain growth in addition to maintenance have greater protein requirements. A young, growing animal is like a muscle-building factory, and protein in the feed is the raw material for the manufacturing process. In the case where young animals are confronted with an energy restricted ration for a reasonable period of time such as may be the case with some winter feeding programs, the animal may experience compensatory growth once the feed supply is more adequate. **Compensatory growth** is a situation where above-average growth rates are realized when nutritional conditions improve. However, if provided with only a maintenance amount of feed for an extended period, a young animal may be permanently stunted.

The nutrient requirements for young growing beef calves are provided in Table 17.2. As weight and growth rate increase, TDN, protein intake, and net energy intake requirements also increase.

Monogastric animals not only need a certain quantity of protein, but they also must have certain amino acids for proper growth. The protein needs of hogs, for example, are usually supplied by feeding them soybean meal as a supplemental source of amino acids. Young ruminant animals cannot consume enough roughage to achieve maximum growth. If dams that produce adequate amounts of milk are nursing young ruminant animals, the young will do well on good pastures, good quality hay, or both together.

The mineral needs of a young, growing animal include calcium and phosphorus for proper bone growth, salt, or a normal sodium level in the body, and any mineral

Table 17.2
NUTRIENT REQUIREMENTS OF GROWING BEEF CALVES

Weight (lb)	Average Gain (lb/day)	TDN (lb)	Metabolizable Protein (lb)	Ne$_m$ (Mcal/day)	NE$_g$ (Mcal/day)
440	1.0	7.5	.34	4.1	1.3
	2.0	7.9	.66	4.1	2.7
	3.0	8.1	.97	4.1	4.2
550	1.0	8.0	.34	4.8	1.5
	2.0	8.8	.66	4.8	3.2
	3.0	9.9	.97	4.8	5.0
660	1.0	8.4	.35	5.6	1.7
	2.0	9.4	.67	5.6	3.7
	3.0	11.9	.97	5.6	5.7

Source: National Research Council, *Nutrient Requirements of Beef Cattle*, 2000.

that may be deficient in the area in which the animal lives. **Calcium** is usually plentiful in legume forages, and **phosphorus** is usually plentiful in grains, so a combination of hay and grain should provide all the calcium and phosphorus that young ruminant animals need. Animals fed on hay alone may need additional phosphorus, and those fed diets high in concentrates may need additional calcium. Some producers feed a mixture of steamed bone meal or dicalcium phosphate and salt at all times to ensure that their animals have the necessary calcium and phosphorus.

Iodine and **selenium** require special consideration. Some geographic areas may be deficient in one or both of these elements. An insufficient amount of iodine in the ration of pregnant females might cause an iodine deficiency in the fetus, which prevents thyroxine from being produced and thus causes a **goiter** in the newborn. Young animals with goiters die shortly after they are born. Iodized salt can be easily provided to the pregnant females to avoid the iodine deficiency.

A lack of selenium might cause the young to be born with **white muscle disease**. Giving the pregnant female an injection of selenium can prevent this deficiency. Injectable selenium is distributed commercially. Directions for proper dosages that are supplied by the distributor should be followed closely because an overdose in natural feeds or supplements can kill the animal. Nonetheless, selenium in normal quantities is an important antioxidant.

Young, growing animals need vitamins. Young ruminant animals are usually on pasture with their dams and thus are exposed to sunshine. The action of ultraviolet rays from the sun converts cholesterol in the skin into vitamin D, providing the animal with this vitamin. **Vitamin D** is needed for the proper use of calcium and phosphorus in bone growth. Because pigs, poultry, and rabbits are often raised inside, where sunshine is limited or lacking, they need a dietary source of vitamin D.

Most vitamins must be supplied to pigs and poultry through feeds. The only vitamin commonly fed to ruminant animals is **vitamin A**, and then only when they are on dry pasture or are being fed hay that is quite mature or that has been dried in the sun for several days due to being moistened in processing. Vitamin A is easily lost in sunlight and during extended dry storage. The activity of vitamin A in silage is often quite high because this vitamin is usually preserved by the acid fermentation that takes place when silage is made.

Young animals need sufficient energy to sustain their growth, high metabolic rate, and activities. They obtain some energy from their mother's milk, and additional

Table 17.3
WATER INTAKE OF ADULT LIVESTOCK UNDER TEMPERATE CONDITIONS

Species	Water (gal/day)
Cattle (beef)	7–20
Cattle (dairy)	10–29
Chickens	0.05–0.10
Horses	8–12
Sheep and goats	1–4
Swine	3–5
Turkeys	0.10–0.15

Source: Adapted from Cheeke, 2005.

energy is supplied as carbohydrates (starch and sugar) and fats from grazed forage or supplemental feeds. Feed grains are high in carbohydrates and also contain some fats. Young ruminants on good pasture typically obtain sufficient energy from pasture and from milk.

The energy needs of young pigs and poultry are generally supplied by feeding them grains such as corn, barley, or wheat. Young horses can usually obtain their energy needs when they are on pasture with their dams since most mares produce much milk; however, for adequate growth after weaning, they need some grain in addition to good pasture or good quality hay.

Water availability and intake are required for normal function. Intake will vary based on animal age, stage of production (lactating versus dry), level of feed intake, and a variety of environmental factors. The range of expected intake of water for adult farm animals is provided in Table 17.3.

NUTRIENT REQUIREMENTS FOR FINISHING

The process of **finishing** is complete once appropriate market weight and composition is achieved. At this stage optimal growth rates begin to decline and surplus feed energy as is stored as fat at levels that decrease the market desirability of the animal. Some degree of fattening is desirable to give meat part of its palatability characteristics and to provide energy reserves for productive activities such as reproductive performance.

Gain during the growth of lean tissue is usually less costly than gain from fattening. Production of a pound of fat requires 2.25 times as much energy as to produce a pound of protein tissue.

Finished weights are attained as the result of excess energy from carbohydrates, fats, or protein above that required for maintenance and growth. Usually finishing animals are full-fed high-energy rations during the last phase of the growing-finishing feeding program. However, if animals are fed beyond their weight and compositional ideal, profitability declines and market discounts may be encountered due to excess subcutaneous, intermuscular, and cavity fat deposition.

NUTRIENT REQUIREMENTS FOR REPRODUCTION

The requirements for reproduction fall into two categories—those for (1) gamete production and (2) fetal growth in the uterus. In general, healthy males and females are capable of producing gametes. The energy needs for germ cell production are no greater than those needed to keep animals in a normal, healthy condition. For example, ruminant animals grazing on pastures of mixed grass and legumes are generally neither deficient in phosphorus nor lacking in fertility. A lack of phosphorus may cause irregular estrous cycles and impaired breeding in females.

Animals that are losing weight rapidly because of poor feed conditions and animals that are overly fat may be low in fertility. To attain optimum fertility from female animals, they should be in moderately low to moderate body fat conditions as breeding season approaches, but should, ideally, be increasing in condition (i.e., gaining weight) for 2–4 weeks before and during the breeding season.

The nutrients required by the **growing fetus** are much greater in the last trimester of pregnancy, as little fetal growth occurs during the first two trimesters of pregnancy. Because the fetus is growing, its requirements are the same as those for growth of a young animal after it is born. Healthy females can withdraw nutrients from their bodies to support the growing fetus temporarily while the amount or quality of their feed is low, but reproductive performance will be lower if nutrition is inadequate for 2–3 months in cattle and a few weeks in swine and sheep.

NUTRIENT REQUIREMENTS FOR LACTATION

Among common farm animals, dairy cows and dairy goats produce the most milk; however, most mammalian females are expected to produce milk for their young as part of normal enterprise management schedules. Milk production requires considerable protein, minerals, vitamins, and energy. Lactating dairy cows attain peak lactation approximately 45 to 65 days after calving, which places considerable demands on the energy stores of the cow. This situation is compounded by the fact that her feed intake doesn't catch up with rising energy requirements for a period of 8 to 10 weeks or even longer for the high producing female. As a result most lactating dairy cows will lose body weight during the time of energy deficit.

The need for protein is greater because milk contains more than 3% protein. As an example, a cow that weighs 1,500 lb and produces 30 lb of milk per day needs at least 30 lb of feed per day, which contains 15% protein; this gives her 4.5 lb protein for her body and the milk she produces. If the protein she eats is 60% digestible, there is 2.7 lb of digestible protein, of which 1.5 lb is present in her milk. If a cow of this weight is to produce 100 lb of milk per day, she must now consume 50 lb of feed containing 15% protein to compensate for the 3 lb of protein in the 100 lb of milk that she produces. Generally, during peak milk production, feed consumption cannot compensate for nutrient output and the cow does mobilize some body protein.

Calcium and phosphorus are the two most important minerals needed for lactation. Milk is rich in these minerals; their absence or imbalance may result in decreased lactation or may even cause death. The dairy cow may develop milk fever shortly after calving if there is an exceptionally heavy drain of calcium from her system. Cows afflicted by milk fever might become comatose and die if not treated. However, an intravenous injection of calcium gluconate usually helps the cow recover in less than a day. Milk fever rarely occurs in species or animals that produce relatively small quantities of milk.

Dairy cows produce milk that contains considerable quantities of vitamin A and most B-complex vitamins. Because cows are ruminants, it is unnecessary to feed them B-complex vitamins, and they require vitamin D supplementation only if confined indoors.

In those parts of the world where sheep and dairy goats are the principal dairy animals, the energy needs of these animals are quite similar to those of dairy cows that produce much milk.

The requirements for milk production in sows are usually provided in several ways—by increasing the percentage of protein in the ration, by increasing the amount of feed allowed, and by providing a mineral mix (a combination of minerals that usually contains calcium, phosphorus, salt, and some trace minerals).

In the dairy cow, the roughage-to-concentrate ratio should be approximately 40 : 60, as a certain amount of roughage is needed to maintain the desired fat content in milk. Therefore, simply feeding more concentrates is not the only factor in increased milk production.

NUTRIENT REQUIREMENTS FOR EGG LAYING

The nutrient requirements of poultry are dependent on the specific purposes of production. For example, broilers need nutrients primarily for growth, with less emphasis on egg production; Leghorn-type hens (layers) need nutrients with primary emphasis on eggs and less on growth. Nutrient requirements for growth are discussed earlier in this chapter.

Leghorn-type chickens are smaller in body size than broilers, so their maintenance requirements are less. They are prolific in egg production, so they are usually fed **ad libitum** during the growing and laying period. Because layers eat ad libitum to satisfy their energy needs, rations need to have adequate concentrations of energy, protein (amino acids), vitamins, and minerals.

NUTRIENT REQUIREMENTS FOR WOOL PRODUCTION

Nutrient requirements for wool production are in addition to nutrients needed for maintenance, growth, and reproduction. Insufficient energy owing to amount or quality of feed is usually the most limiting nutritional factor affecting wool production. As wool fibers are primarily protein in composition, the ration should be adequate in protein content.

Shearing removes the sheep's natural insulation and may cause an increase in energy requirements owing to heat loss. This is especially true when periods of cold weather occur shortly after shearing.

NUTRIENT REQUIREMENTS FOR WORK

Animals used for work, either for pulling heavy loads or for being ridden, require large amounts of energy in addition to their needs for maintenance. Research indicates that the level of energy demand is proportional to the workload (level of physical exercise). Horses are the primary work animals in the United States, but elsewhere, donkeys, cattle, and water buffalo are used.

The primary requirement is energy above that needed for maintenance and growth. If energy in the ration is not sufficient to meet the work needs, then body fat stores will provide the additional energy needs. However, once fat stores are depleted body condition will continue to deteriorate unless nutrient intake and workload can be balanced.

Horses, mules, and donkeys will increase nitrogen excretion with increased muscular activity. However, this only minimally impacts dietary protein requirements. During prolonged perspiration Na and Cl are lost so additional supplementation of these minerals may be required. With prolonged workloads, draft animals should also have additional dietary P due to its importance in energy creating metabolic reactions.

RATION FORMULATION

Typically producers or their consultants balance rations with computer-assisted formulation software systems. However, the **Pearson square method** is an approach to balancing simple rations such as in the case of two-ingredient formulations.

Assume that a producer wants to determine the appropriate amounts of soybean meal and corn to utilize in combination to meet a goal of providing 14% crude protein in the diet. Utilizing the scenario in which a feed test reveals that the crude protein of the available corn and soybean meal is 11% and 45%, respectively, the ingredient values are listed on the left side of the box with the desired CP% at the center. Values are subtracted across the diagonal (convert all negative values to positive) to determine the ratio of feed ingredients required to meet the 14% CP goal.

A ration that achieves the desired level of protein requires 91% corn (31/34) and 9% soybean meal (3/32). The calculation can be verified as follows:

$$91 \text{ lb corn} \times 11\% \text{ CP} = 10.01 \text{ lb CP}$$
$$9 \text{ lb SBM} \times 45\% \text{ CP} = \underline{\ \ 4.05 \text{ lb CP}}$$
$$14.06 \text{ lb CP (just over 14\%)}$$

When using this method, it is important to remember that the number in the middle of the box must be intermediate in value to the numbers on the left side of the square and negative signs are ignored.

Chapters 15, 16, and 17 give a brief background on nutrition, which leads into ration formulation. The primary objective of ration formulation is economically matching the animal's nutrient requirements with the available feeds, taking into consideration the nutrient content of the feeds (Tables 17.4–17.11). Additional considerations are the palatability of the ration, the physical form of the feed, and other factors that affect feed consumption.

Additional material on feeding farm animals can be found in the following chapters on individual species: beef cattle (Chapter 26), dairy cattle (Chapter 28), swine (Chapter 30), sheep and goats (Chapter 32), horses (Chapter 34), and poultry (Chapter 35). For a more detailed explanation of ration formulation, the reader is encouraged to use *Livestock Feeds and Feeding* by Kellems and Church (2010). Software for formulating livestock diets can be accessed or purchased via:

> http://animalscience.ucdavis.edu/extension/software/—multiple species
>
> http://agmodelsystems.com—dairy and beef cattle
>
> http://www.iowabeefcenter.org/software_BRANDS.html - beef cattle
>
> http://www.caes.uga.edu/publications/pubDetail.cfm?pk_id=7886—swine and poultry
>
> www.pork.org/filelibrary/2012%20Swine%20In-Service/NSNG.pdf—swine

NUTRIENT REQUIREMENTS OF RUMINANTS

The following are some major comparisons that demonstrate changes in nutrient requirements for maintenance, growth, lactation, and reproduction:

1. For ewes (maintenance), note the increased requirement in dry matter, energy (TDN or ME), protein, calcium, phosphorus, vitamin A, and vitamin D as body weight changes from 110 to 176 lb (Table 17.4). More nutrients are needed to maintain a heavier body weight.

2. In Table 17.4, compare the requirements of ewes in the last 4 weeks of gestation with the maintenance requirements of ewes and the requirements of ewes that are nonlactating and in the first 15 weeks of gestation. During the latter part of

Table 17.4
DAILY NUTRIENT REQUIREMENTS OF SHEEP (DRY-MATTER BASIS)

Weight (lb)	Gain (lb)	Dry Matter (lb)	TDN (lb)	ME (Mcal)	Crude Protein (lb)	Ca (g)	P (g)	Vitamin A (IU)	Vitamin E (IU)
Ewes (maintenance)									
110	0.02	2.2	1.2	2.0	0.21	3.0	2.8	2,350	15
176	0.02	2.9	1.6	2.6	0.27	3.3	3.1	3,760	20
Ewes (nonlactating and first 15 weeks of gestation)									
110	0.07	2.6	1.5	2.4	0.25	3.0	2.8	2,350	18
176	0.07	3.3	1.8	3.0	0.31	3.3	3.1	3,760	22
Ewes (last 4 weeks of gestation with 200% lambing rate)									
110	0.5	3.7	2.4	4.0	0.43	4.1	3.9	4,250	26
176	0.5	4.4	2.9	4.7	0.49	4.8	4.5	6,800	30
Ewes (first 8 weeks of lactation, suckling singles)									
110	−0.06	4.6	3.0	4.9	0.67	10.9	7.8	4,250	32
176	−0.06	5.7	3.7	6.1	0.76	12.6	9.0	6,800	39
Ewes (first 8 weeks of lactation, suckling twins)									
110	−0.13	5.3	3.4	5.6	0.86	12.5	8.9	5,000	36
176	−0.13	6.6	4.3	7.0	0.96	14.4	10.2	8,000	45
Lambs (finishing)									
66	0.65	2.9	2.1	3.4	0.42	4.8	3.0	1,410	20
88	0.06	3.5	2.7	4.4	0.41	5.0	3.1	1,880	24
110	0.45	3.5	2.7	4.4	0.35	5.0	3.1	2,350	24

Source: National Research Council, *Nutrient Requirements of Sheep*, 2007.

Table 17.5
DAILY NUTRIENT REQUIREMENTS (NRC) FOR BREEDING HEIFERS AND COWS

Weight (lb)	Daily Gain (lb)	Dry-Matter Consumption (lb)	Total Crude Protein (lb)	TDN(lb)	ME(Mcal)	Ca (g)	P (g)
Pregnant Heifers—Last Third of Pregnancy							
1,000[a]	0.73	29.7	1.8	11.7	20.9	29	22
1,200[a]	0.88	23.7	2.0	13.3	24.2	33	24
1,400[a]	1.02	26.6	2.2	14.8	26.9	37	27
Cows Nursing Calves—Average Milking Ability[b] (First 3 months postpartum)							
900	0	23.0	1.9	12.5	20.9	24	17
1,100	0	26.0	2.1	13.9	23.4	27	19
1,400	0	29.0	2.3	15.4	25.7	30	21
Cows Nursing Calves—Superior Milking Ability[c] (First 3 months postpartum)							
900	0	25.4	2.6	14.9	24.9	35	24
1,100	0	28.4	2.8	16.4	27.3	37	25
1,400	0	31.3	3.0	17.8	29.7	40	27
Dry Pregnant Mature Cows—Month 3 of Pregnancy							
900	0	16.7	1.2	8.2	13.4	14	14
1,100	0	19.5	1.4	9.5	15.6	17	17
1,400	0	23.3	1.6	11.4	18.7	21	21
Dry Pregnant Cows—Month 8 of Pregnancy							
900	0.9	21.0	1.6	10.9	18.3	23	14
1,100	0.9	24.1	1.8	12.6	20.9	27	17
1,400	0.9	27.0	2.2	14.2	23.9	32	21

[a]Mature weight potential.
[b]Ten pounds of milk per day (equivalent of approximately 450 lb of calf at weaning if there is adequate forage).
[c]Twenty pounds of milk per day (equivalent of approximately 650 lb of calf at weaning if there is adequate forage).
Source: National Research Council, *Nutrient Requirements of Beef Cattle*, 2000.

Table 17.6
DAILY NUTRIENT REQUIREMENTS FOR GROWING-FINISHING HEIFERS AND STEERS

Live Weight (lb)	TDN (%DM)	Daily Gain (lb)	Dry-Matter Intake (lb)	Protein Intake (lb)	Crude Protein (%)	NEm (Mcal/lb)	NEg (Mcal/lb)	Ca (%)	P (%)
1,100 lb at Finishing (28% body fat)									
600	50	0.68	17.5	1.2	7.1	0.45	0.20	0.21	0.13
	70	2.86	18.0	2.2	12.3	0.76	0.48	0.45	0.23
	90	4.00	15.7	2.7	17.1	1.04	0.72	0.66	0.32
770	50	0.68	19.6	1.3	6.8	0.45	0.20	0.19	0.12
	70	2.86	20.2	2.2	10.9	0.76	0.48	0.39	0.20
	90	4.00	17.6	2.6	14.8	1.04	0.72	0.56	0.28
880	50	0.68	21.7	1.4	6.5	0.45	0.20	0.19	0.12
	70	2.86	22.4	2.2	9.8	0.76	0.48	0.34	0.18
	90	4.00	19.5	2.5	13.1	1.04	0.72	0.48	0.25
1,300 lb at Finishing (28% body fat)									
780	50	0.76	19.8	1.4	7.1	0.45	0.20	0.21	0.13
	70	3.21	20.4	2.5	12.1	0.76	0.48	0.45	0.23
	90	4.48	17.8	3.0	16.9	1.04	0.72	0.66	0.32
910	50	0.76	22.2	1.5	6.7	0.45	0.20	0.20	0.13
	70	3.21	22.9	2.4	10.7	0.76	0.48	0.39	0.20
	90	4.48	22.0	2.9	14.6	1.04	0.72	0.56	0.28
1,040	50	0.76	24.5	1.6	6.5	0.45	0.20	0.19	0.13
	70	3.21	25.3	2.4	9.6	0.76	0.48	0.34	0.19
	90	4.48	22.1	2.8	12.9	1.04	0.72	0.98	0.25

Source: National Research Council, *Nutrient Requirements of Beef Cattle*, 2000.

Table 17.7
DAILY NUTRIENT REQUIREMENTS OF DAIRY COWS

Body Weight (lb)	NE_l (Mcal)	TDN (lb)	Crude Protein (lb)	Ca (lb)	P (lb)	Vitamin A (IU)
Mature Lactating Cows (maintenance)						
800	7.16	6.9	0.70	0.029	0.024	30
1,100	8.46	8.1	0.80	0.044	0.031	38
1,300	9.70	9.3	0.89	0.053	0.037	46
1,550	10.89	10.5	0.99	0.062	0.044	53
Mature Dry Cows (last 2 months of gestation)						
800	9.30	9.1	1.96	0.057	0.035	30
1,100	11.00	10.8	2.32	0.073	0.044	38
1,300	12.61	12.4	2.66	0.086	0.053	46
1,550	14.15	13.9	2.98	0.101	0.062	53
Milk Production (meal or lb of nutrient per lb of milk for various fat percentages)						
Percentage fat						
3.0	0.291	0.282	0.077	0.0025	0.00170	
3.5	0.313	0.304	0.082	0.0026	0.00175	
4.0	0.336	0.326	0.087	0.0027	0.00180	
4.5	0.354	0.344	0.092	0.0028	0.00185	
5.0	0.377	0.365	0.098	0.0029	0.00190	

Source: National Research Council, *Nutrient Requirements of Dairy Cattle*, 2001.

Table 17.8

SUGGESTED DIETARY AMINO ACID AND PROTEIN ALLOWANCES FOR SWINE FED CORN-SOYBEAN MEAL DIETS

Amino Acids	Growing Pigs[a]			Barrows			Gilts		
(% of Diet)	20–45	45–80	80–120	120–170	170–220	220–280	120–170	170–220	220–280
Crude protein	20	18	16	14	13	12	16	14	13
Lysine	1.2	1.0	0.9	0.75	0.68	0.58	0.84	0.73	0.62
Methionine–cystine	0.72	0.62	0.56	0.48	0.43	0.37	0.53	0.47	0.39
Tryptophan	0.21	0.18	0.16	0.15	0.13	0.11	0.16	0.14	0.11
Threonine	0.78	0.67	0.60	0.52	0.48	0.40	0.58	0.51	0.43
Arginine	0.50	0.36	0.33	0.14	0.13	0.10	0.15	0.14	0.11
Histidine	0.39	0.32	0.29	0.24	0.21	0.19	0.27	0.23	0.20
Isoleucine	0.72	0.60	0.54	0.45	0.40	0.35	0.50	0.44	0.37
Valine	0.81	0.68	0.61	0.51	0.47	0.39	0.57	0.50	0.42
Leucine	1.20	1.00	0.90	0.75	0.68	0.58	0.84	0.73	0.62
Plenylalanine + Tyrosine	1.14	0.95	0.80	0.71	0.65	0.55	0.80	0.70	0.59

[a]Barrows and gilts.
Source: Adapted from Baker et al., 2002.

Table 17.9

PROTEIN, MINERAL, AND VITAMIN REQUIREMENTS FOR GESTATING AND LACTATING SOWS

	Gestation (% of diet)	Lactation (% of diet)
Crude protein	12	17
Calcium	0.75	0.75
Phosphorus, available	0.35	0.35
Salt	0.35	0.35
Iron (mg/lb)	36	36
Zinc (mg/lb)	23	23
Copper (mg/lb)	2.3	2.3
Manganese (mg/lb)	9.0	9.0
Iodine (mg/lb)	0.06	0.06
Selenium (mg/lb)	0.07	0.07
Vitamin A (IU/lb)	2,000	1,000
Vitamin D (IU/lb)	150	150
Vitamin E (IU/lb)	20	—
Vitamin K (IU/lb)	230	230
Riboflavin (mg/lb)	2	2
Niacin (mg/lb)	5	5
Pantothenic acid (mg/lb)	5	5
Vitamin B$_{12}$ (mg/lb)	7	7
Choline (mg/lb)	600	450

Source: Adapted from Baker et al., 2002.

Table 17.10
DAILY NUTRIENT REQUIREMENTS OF VARIOUS CLASSES OF HORSES

Class	Weight (lb)	DE (Mcal)	CP (g)	Ca (g)	P (g)	Vitamin A (IU)	Vitamin D (IU)	Vitamin E (IU)
Adult								
Maintenance	900	13.3	504	16	11.2	12,000	2,640	400
	1,100	16.7	630	20	14	15,00	3,300	500
	1,300	20.0	756	24	16.8	18,000	3,900	600
	2,000	30.0	1,134	36	25.2	27,000	5,900	900
Moderate work	900	18.6	614	28	16.8	18,000	2,640	720
	1,100	23.3	768	35	21	22,500	3,300	900
	1,300	28.0	921	42	25.2	27,000	3,900	1,080
	2,000	42.0	1,382	63	37.8	40,500	5,900	1,620
Heavy work	900	21.3	689	32	23.2	18,000	2,640	800
	1,100	26.6	862	40	29	22,500	3,300	1,000
	1,300	32.0	1,034	48	34.8	27,000	3,900	1,200
	2,000	48.0	1,551	72	52.2	40,500	5,900	1,800
Very heavy work	900	27.6	804	32	23.2	18,000	2,640	800
	1,100	34.5	1,004	40	29	22,500	3,300	1,000
	1,300	41.4	1,205	48	34.8	27,000	3,900	1,200
	2,000	62.1	1,808	72	52.2	40,500	5,900	1,800
Pregnant Mares[a]								
5 month	1,110	17.1	685	20	14	30,000	3,300	800
8 month	1,150	18.5	759	28	20	30,000	3,300	800
11 month	1,245	21.4	893	36	26.3	30,000	3,300	800
Lactating Mares								
1 month	1,100	31.7	1,535	59.1	38.3	30,000	3,300	800
3 month	1,100	30.6	1,468	55.9	36	30,000	3,300	800
6 month	1,100	27.2	1,265	37.4	23.2	30,000	3,300	800
Growing Foals[a]								
6 months	475	15.5	676	38.6	21.5	9,700	4,793	432
12 months	700	18.8	846	37.7	20.9	14,500	5,589	642

[a]Mature weight potential of 1,100 lb.
Source: Table from *Nutrient Requirements of Horses* by National Research Council (U.S.). Subcommittee on Horse Nutrition. Copyright © 1989. Used by permission of National Academies Press.

Table 17.11
SUGGESTED AMINO ACID RECOMMENDATIONS FOR BROILER DIETS[a]

	0–21 Days		22–42 Days		43–53 Days	
	Male	Female	Male	Female	Male	Female
Arginine	0.88	0.83	0.77	0.73	0.66	0.59
Lysine	0.81	0.76	0.70	0.67	0.53	0.50
Methionine + cystine	0.60	0.60	0.56	0.50	0.46	0.41
Tryptophan	0.16	0.15	0.12	0.11	0.11	0.10
Histidine	0.24	0.22	0.22	0.20	0.20	0.18
Leucine	0.84	0.64	0.81	0.76	0.76	0.72
Isoleucine	0.54	0.51	0.52	0.48	0.45	0.40
Phenylalanine + tyrosine	1.04	0.98	0.95	0.87	0.78	0.70
Threonine	0.53	0.49	0.43	0.41	0.42	0.38
Valine	0.66	0.63	0.62	0.57	0.51	0.46
Glycine + serine	1.00	0.94	0.80	0.74	0.70	0.63
Protein	15.25	15.25	13.75	13.75	12.25	12.25

[a]Expressed as a percent per metabolizable megacalorie per pound of feed. To calculate the percent of the nutrient required in the diet, multiply the tabular value by the metabolizable megacalorie per pound of feed.
Source: Adapted from Waldrop, 2002.

gestation, there are greater nutrient demands owing to rapid fetal growth. During the latter part of gestation compared with maintenance only, energy and protein requirements almost double, with mineral and vitamin requirements showing significant increases.

3. During lactation, nutrient requirements are even higher than those during gestation. Even with larger amounts of dry matter being supplied, the ewes lose weight. Compare the requirements of ewes nursing single lambs versus twins and note the even higher requirements for ewes nursing twins.

4. The nutrient requirements for lambs being finished for slaughter show that the gains are approximately the same, with increased nutrient requirements due to increased body weight. The increased nutrient requirements are primarily due to an increased maintenance requirement.

Diet Modification to Minimize Nitrogen and Phosphorus Excretion

Livestock and poultry managers have the challenge of meeting the animal's nutritional requirements while assuring a cost-effective ration as well as avoiding excess feeding of nitrogen (N) and phosphorus (P). Excess N and P will be excreted by the animal and may result in water and air quality issues.

Nitrogen is a structural component of the amino acids that form protein. Nitrogen is a by-product of protein digestion that has the potential to negatively affect surface water quality. Phosphorus is a mineral nutrient that also has the potential to be a pollutant when excreted in excess.

There are two environmental concerns from animal waste—the *volatilization* of N in the form of ammonia from manure that may enter the environment via rainfall, *dry precipitation*, or absorption directly from the waste product. As N content of manure increases the risk of ammonia loss rises. Therefore, control of N in the diet is a useful strategy to reduce contamination. The second issue is distribution of manure nutrients via the implementation of an integrated waste management plan. Manure is an excellent fertilizer and the environmental impact is negligible when manure nutrient application is matched to plant requirements. However, at excess application rates P can infiltrate the soil and contaminate surface water while N can pollute groundwater.

Specific strategies for minimizing the impact of N and P are discussed in Chapters 25, 27, 29 and 34. However, significant decreases in N and P excretion by livestock and poultry can be achieved via careful management of diet composition, better understanding of digestibility differences between feedstuffs, and assuring that appropriate diet changes are made in accordance with shifts in animal requirements for N and P. Historically, N and P have been overfed in the diets of many livestock. However, as pressure from environmental regulation increases, the need for more precise nutrient management will be assured.

CHAPTER SUMMARY

- Nutrients are used by animals for maintenance (maintaining body functions), growth, fattening, reproduction, egg laying, wool production, and work.

- Maintenance involves no change (gain or loss) in body weight. Maintenance requirements increase as body weight increases.

- Growth occurs when protein synthesis exceeds protein breakdown and loss.

- Fattening occurs when the energy in feeds (primarily from carbohydrates and fats) exceeds the body's need for maintenance and growth.

- Rations are formulated to match the animal's nutrient requirements with the nutrient content of available feeds.

KEY WORDS

body maintenance
growth
compensatory growth
calcium
phosphorus
iodine
selenium
goiter

white muscle disease
vitamin D
vitamin A
finishing
growing fetus
ad libitum
Shearing
Pearson square method

REVIEW QUESTIONS

1. Describe the highest priority maintenance functions.
2. Discuss the change in nutrient requirements as an animal grows.
3. Explain the concept of metabolic weight.
4. Compare water requirements of livestock and poultry species.
5. Explain the process of finishing and compare the nutrient requirements of the finishing phase to the growing phase.
6. Describe the change in nutritional requirements of the dam and fetus as pregnancy progresses.

7. Why do high milk production females typically lose weight as they attain peak lactation?
8. Discuss nutrient requirements for egg and wool production.
9. Demonstrate the use of the Pearson-square methodology.
10. Utilize the nutrient requirement tables to develop a feeding management plan for various livestock enterprises.
11. Describe the role of diet in modifying nitrogen and phosphorus excretion.

SELECTED REFERENCES

Baker, D. H., R. A. Easter, G. R. Hollis, M. Ellis, and R. Zijlstra. 2002. Dietary nutrient allowances for swine. *Feedstuffs Reference Issue*, 74:28.

Cheeke, P. J. 2005. *Animal Nutrition*. Upper Saddle River, NJ: Pearson Prentice Hall.

Council for Agricultural Science and Technology. 2002. *Animal diet modification to decrease the potential for nitrogen and phosphorus pollution*. Issue paper no. 21. Ames, IA.

Kellems, R. O. and D. C. Church, 2010. *Livestock Feeds and Feeding*. Englewood Cliffs, NJ: Pearson Prentice Hall.

Kline, K. H. 2002. Horse feeds and feeding. *Feedstuffs Reference Issue*, 74:28.

National Research Council. *Nutrient Requirements of Beef Cattle* (2000); *Dairy Cattle* (2001); *Goats* (2007); *Horses* (2007); *Poultry* (1994); *Sheep* (2007); and *Swine* (2012). Washington, DC: National Academy Press.

Pond, W. G. 1995. *Basic Animal Nutrition and Feeding*. New York: John Wiley & Sons.

18
Growth and Development

Profitable and efficient production of livestock and poultry involves understanding and effective management of growth and development. The use of integrated management protocols to affect genetic and environmental factors can alter growth patterns in farm animals.

Many aspects of growth and development are discussed in other chapters of this book. The material on reproduction, genetics, nutrition, and products should be integrated with this chapter for a deeper understanding of animal growth and development.

Generally, **growth** is an increase in body weight until mature size is reached. This growth is the result of an increase in cell size and cell numbers with protein deposition resulting. More specifically, growth is an increase in the mass of structural tissue (bone, muscle, and connective tissue) and organs accompanied by a change in body form and composition.

Development is defined as the directive coordination of all diverse processes until maturity is reached. It involves growth, cellular differentiation, and changes in body shape and form. In this chapter, growth and development are combined and discussed as one entity.

PRENATAL (LIVESTOCK)

The three phases of prenatal life—sex cells, the embryo, and the fetus—are briefly discussed in Chapter 10. Embryological development is a fascinating process; a spherical mass of cells differentiates into specific cell types and eventually into recognizable organs (Fig. 18.1). The endoderm differentiates into the digestive tract, lungs, and bladder; the mesoderm into the skeleton, skeletal muscle, and connective tissue; the ectoderm into the skin, hair, brain, and spinal cord. The growth, development, and differentiation processes, involving primarily protein synthesis, are directed by DNA chains of chromosomes and the organizers in the developing embryo (see Chapter 12). Thus the nucleus is a center of activity for different types of cells, directing the growth and development process.

The fetus undergoes marked changes in shape and form during prenatal growth and development. Early in the prenatal period, the head is much larger than the body. Later, the body and limbs grow more rapidly than other parts. The order of tissue growth follows a sequential trend determined by physiological importance, starting with the central nervous system and progressing to internal organs, bones, tendons, muscles, intermuscular fat, and subcutaneous fat.

During the first two-thirds of the prenatal period, most of the increase in muscle weight is due to **hyperplasia** (increase in number of fibers). During the last several months of pregnancy and postnatal

- Describe prenatal growth in farm animals and in poultry
- Compare and contrast the skeletons of mammals and birds
- Define the three muscle types
- Define power, strength, and endurance
- Compare and contrast red and white muscle fibers
- Describe the primary hormonal influences on growth
- Describe the growth curve
- Explain the importance of the growth curve in animal management
- Describe the effects of frame size, gender, muscling, and age on growth
- Calculate average daily gain and weight per day of age
- Discuss the evaluation of teeth to estimate animal age

Figure 18.1

The morphogenesis of (A) a single egg cell into (B) a morula, then to (C) a blastocyst. (D) The stage at which the two cavities have formed in the inner cell mass; an upper (amniotic) cavity and a lower cavity yolk sac. The embryonic disc containing the ectoderm and endoderm germ layers is located between cavities. (E) A cattle embryo showing the neural tube and somites. (F) The development of the 14-day cattle embryo.

Source: Kevin Pond from Colorado State University. Used by permission of Kevin Pond.

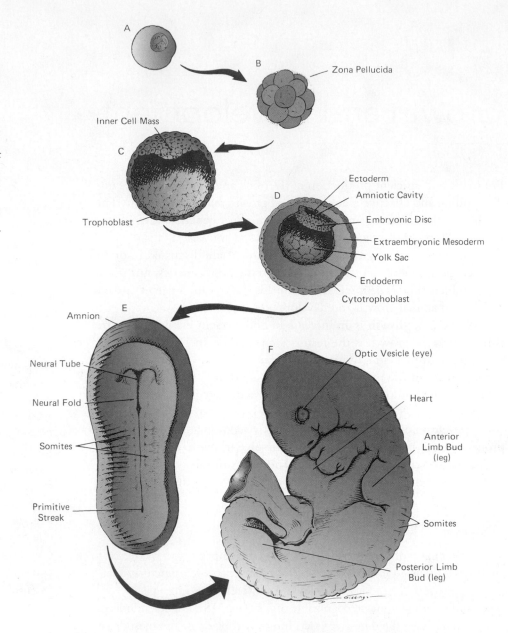

period, **hypertrophy** (increase in size of fibers) represents most of the muscle growth. Individual muscles vary in their rate of growth, with larger muscles (those of the legs and back) having the greatest rate of postnatal growth. Water content of fetal muscle declines with fetal age, and this decline continues through postnatal growth as well.

The relative size of the fetus changes during gestation, with the largest increase in weight occurring during the last trimester of pregnancy (Fig. 18.2).

BIRTH (LIVESTOCK)

Following birth, the number of muscle fibers does not appear to increase significantly; therefore, postnatal muscle growth occurs primarily via hypertrophy. In red meat animals, all muscle fibers appear to be red at birth, but shortly thereafter some differentiate into white and intermediate muscle types.

At birth, the various body parts have considerably different proportions when compared with mature body size and shape. The head is relatively large, the legs are

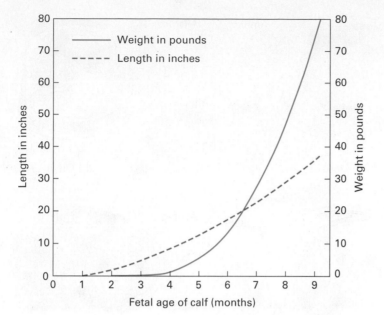

Figure 18.2
Growth of the fetal calf.
Source: Kevin Pond from Colorado State University. Used by permission of Kevin Pond.

long, and the body is small; in the mature animal, the head is relatively small, the legs are relatively short, and the body is relatively large. Birth weight represents approximately 5–7% of the mature weight, while leg length at birth is approximately 60%; height at withers is approximately 50% of those same measurements at maturity. Hip width and chest width at birth are approximately one-third of the same measurements at maturity. This shows that the distal parts (leg and shoulder height) are developed earlier than proximal parts (hips and chest).

POULTRY

Embryonic Development

The development of a chick differs from that of mammals because there is no physiological connection with its mother. The chick develops in the egg, entirely outside the hen's body. Embryonic development is much more rapid in chicks than in farm mammals.

Every egg, whether fertile or nonfertile, has a germ spot called the **blastoderm** (see Fig. 4.1). The blastoderm is where the chick embryo develops if the fertile egg is properly incubated.

A controlled environment must be maintained during incubation to produce live chicks from the fertile eggs. The major components of the controlled environment are (1) temperatures of 99.5–100°F; (2) 60–75% relative humidity; (3) turning of the egg every 1–8 hours; and (4) provision of adequate oxygen.

Three hours after fertilization, the blastoderm divides to form two cells. Cell division occurs until maturity except during the holding period before incubating the eggs.

There are four membranes that are essential to the growth of the chick embryo (Fig. 18.3). The **allantois** is the membrane that allows the embryo to breathe. It takes oxygen through the porous shell and oxygenates the blood of the embryo. The allantois removes the carbon dioxide, receives excretions from the kidneys, absorbs albumen used as food for the embryo, and absorbs calcium from the shell for use by the embryo. The **amnion** is a membrane filled with a colorless fluid that serves as a protection from mechanical shock. The **yolk sac** is a layer of tissue growth over the surface of the yolk. This tissue has special cells that digest and absorb the yolk material for the developing embryo. The **chorion** surrounds both the amnion and yolk sac.

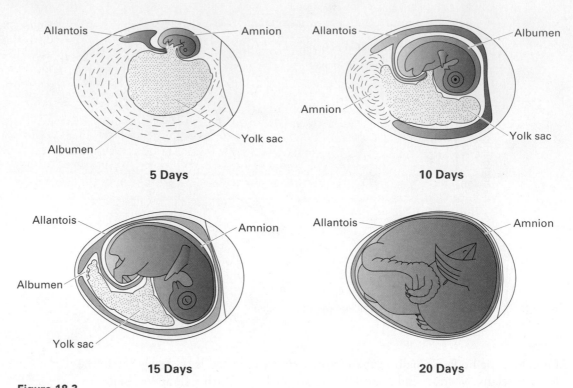

Figure 18.3
Changes in the development of the chick embryo with associated changes in membranes and other contents of the egg.

Table 18.1 identifies some of the primary changes in the growth and development of the chick embryo. Many of these phenomenal changes occur rapidly, sometimes in only hours.

BASIC ANATOMY AND PHYSIOLOGY

In the developmental process, cells become grouped in appearance and function. Specialized groups of cells that function together are called **tissues**. The primary types of tissues are (1) muscle, (2) nerves (3) connective, and (4) epithelial.

Organs are groups of tissues that perform specific functions. For example, the uterus is an organ that functions in the reproductive process. A group of organs that function in concert to accomplish a larger, general function is known as a **system**. The reproductive system is discussed in Chapter 10, the digestive system in Chapter 16, and the mammary system in Chapter 19.

It is not the intent of this chapter to discuss all the different systems even though they are important in growth and development. A few additional systems are briefly surveyed—those considered important in understanding farm animals and their productivity. Figures are provided, illustrating selected systems for one or two species. The systems are similar and generally comparable among the various species of farm animals, although some large differences between poultry and farm mammals exist.

Skeletal System

Figure 18.4 shows the skeletal system of the horse and Figure 18.5 portrays the chicken's skeletal system. Even though only bones and some joints are shown in these figures, teeth and cartilage are also considered part of the skeletal system.

Table 18.1
MAJOR CHANGES IN WEIGHT, FORM, AND FUNCTION OF THE CHICK EMBRYO (WHITE LEGHORN) DURING INCUBATION

Day	Weight (g)	Developmental Changes
1	00.0002	Head and backbone are formed; central nervous system begins
2	00.0030	Heart forms and starts beating; eyes begin formation
3	00.0200	Limb buds form
4	00.0500	Allantois starts functioning
5	00.1300	Formation and reproductive organs
6	00.2900	Main division of legs and wings; first movements noted
7	00.5700	
8	01.1500	Feather germs appear
9	01.5300	Beak begins to form; embryo begins to look birdlike
10	02.2600	Beak starts to harden; digits completely separated
11	03.6800	
12	05.0700	Toes fully formed
13	07.3700	Down appears on body; scales and nails appear
14	09.7400	Embryo turns its head toward blunt end of egg
15	12.0000	Small intestines taken into body
16	15.9800	Scales and nails on legs and feet are hard; albumen is nearly gone; yolk is main food
17	18.5900	Amniotic fluid decreases
18	21.8300	
19	25.6200	Yolk sac enters body through umbilicus
20	30.2100	Embryo becomes a chick; it breaks amnion, then breathes air in air cell
21	36.3000	Chick breaks shell and hatches

Source: Adapted from Rollins, 1984.

The skeleton protects other vital organs and gives a basic form and shape to the animal's body. Bones function as levers and in storing minerals, and the bone marrow is the site of blood cell formation.

Chicken bones are more pneumatic (bone cavities are filled with air spaces), harder, thinner, and more brittle than those of mammals, and have a different ossification process than occurs in mammals.

Muscle System

There are three types of muscle tissue—skeletal, smooth, and cardiac. Skeletal muscle is the largest component of meat animal products. Smooth muscle is located in the digestive, reproductive, and urinary organs. The heart is composed of cardiac muscle.

Figure 18.6 identifies some of the major muscles similar in name and location in the meat animal species and horses. Of special note is the longissimus dorsi, which is discussed in Chapter 8. The size of this muscle and the marbling it contains are important factors in determining yield grades and quality grades of meat animals.

The primary muscles of the turkey are shown in Figure 18.7. The muscles of poultry are referred to as **dark meat** (legs and thighs) and **white meat** (breast and wings). Dark meat originates from muscles that contain higher amounts of myoglobin that helps to transport oxygen to muscles. Dark meat muscles require more oxygen due to their higher work capacity (drumsticks compared to breast for example).

The role of muscle is to convert chemical energy to mechanical energy. Muscle performance depends on power, strength, and endurance. **Power** is determined by the amount of work that can be accomplished within a fixed time period. Thus, power is the result of the speed and strength of muscle contraction. **Strength** is the result

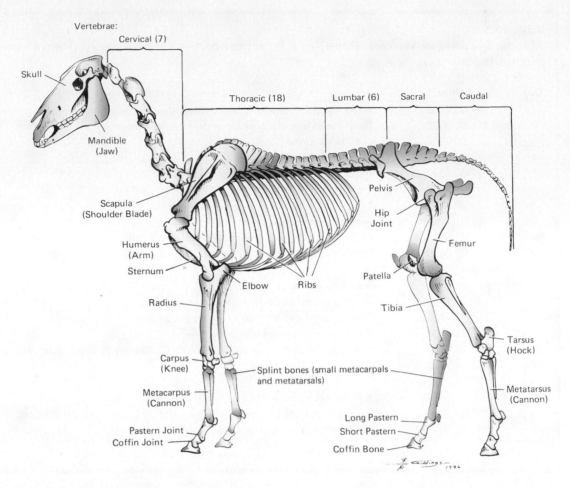

Figure 18.4
Skeletal system of the horse.

of muscle size with larger muscles capable of greater contractible force. **Endurance** is defined as the length of time in which muscle can work without becoming fatigued. Endurance is significantly affected by the nutrient storage capacity of a muscle. Effective training programs for the working horse are designed to improve these three components of muscle performance.

Skeletal muscle fibers can be generally classified as fast-twitch or slow-twitch. Fast-twitch fibers are typically predominant in sprinters and jumpers. Fast-twitch fibers are designed for power and are efficient at anaerobic activity for short periods of time. Because they have less capillary supply than slow-twitch fibers, fast-twitch fibers are often referred to as **white fibers**. Slow-twitch or **red fibers** have a more extensive capillary supply and are better suited to aerobic supply of nutrients and endurance activity. They are smaller in fiber diameter than fast-twitch fibers and contain more myoglobin as a means to enhance oxygen diffusion.

Circulatory System

Figure 18.8 shows the major components of the dairy cow's circulatory system. The circulatory systems of other farm animal species are similar to that of the dairy cow.

The heart, acting as a pump, and the accompanying vessels make up the circulatory system. **Arteries** are vessels that transport oxygenated blood away from the heart, while the vessels that carry oxygen depleted blood back to the heart are called **veins**. The lymph vessels transport lymph (intercellular fluid) from tissues to the heart.

Figure 18.5
Skeletal system of the chicken.

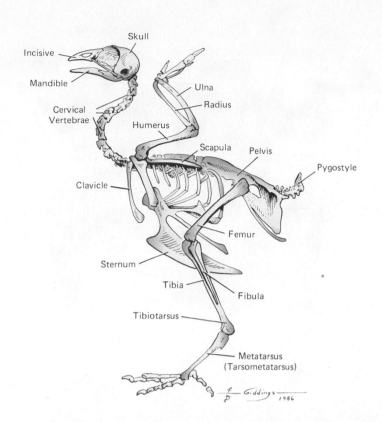

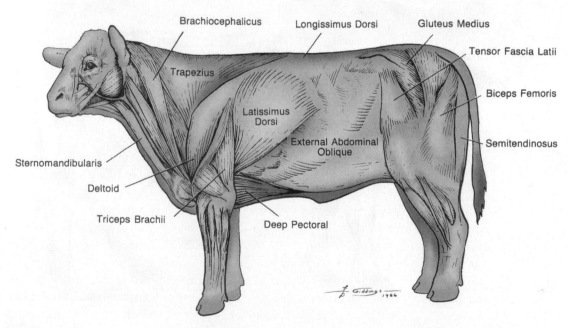

Figure 18.6
Primary muscles of the beef steer.

The circulatory system is important in growth and development; the blood transports oxygen, nutrients, cellular waste products, and hormones.

Milk production in all species is dependent on the nutrients in milk arriving through the circulatory system. Dairy cows producing 20,000–40,000 lb of milk per

Figure 18.7
Primary muscles of the turkey.

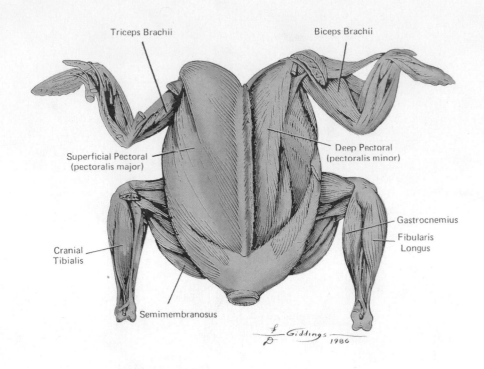

Triceps Brachii

Biceps Brachii

Superficial Pectoral
(pectoralis major)

Deep Pectoral
(pectoralis minor)

Gastrocnemius

Fibularis
Longus

Cranial
Tibialis

Semimembranosus

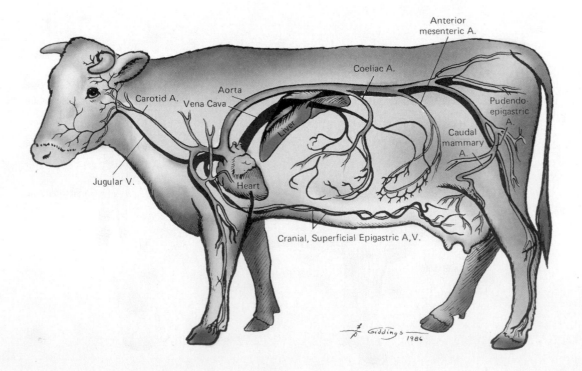

Anterior
mesenteric A.

Coeliac A.

Aorta

Carotid A.

Vena Cava

Liver

Pudendo-
epigastric
A.

Caudal
mammary
A.

Jugular V.

Heart

Cranial, Superficial Epigastric A,V.

Figure 18.8
Circulatory system of the dairy cow (A = artery, V = vein).

year (over 100 lb per day) must consume large amounts of feed, with these feed nutrients circulating in high concentration through the mammary blood supply. Approximately 400–500 lb of blood circulates through the dairy cow's mammary gland for each pound of milk produced.

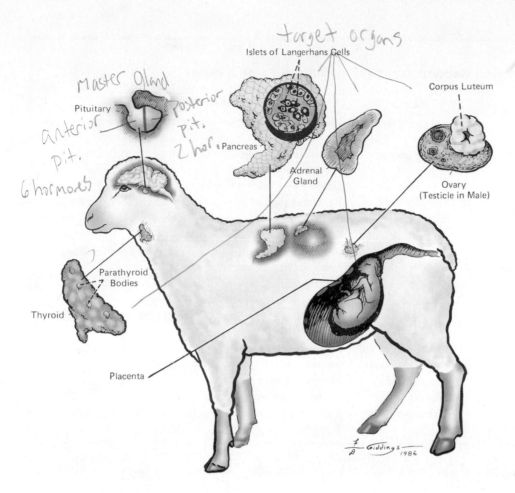

Figure 18.9
Primary endocrine organs of the sheep.

Endocrine System

Growth and development are highly dependent on the endocrine system. This system consists of several endocrine (ductless) glands that secrete **hormones** into the circulatory system. Hormones are chemical substances that affect a gland (or organ) or, in some cases, all body tissues.

The major endocrine glands are shown in Figure 18.9; the hormones produced and their major effects are identified in Table 18.2. Figure 19.3 (Chapter 19) shows the origin of most of the hormones noted in Table 18.2. The important roles of the pituitary and hypothalamus should also be observed in Figure 19.3.

GROWTH CURVES

Postweaning growth is a curved-line function regardless of how it is expressed mathematically. If growth is considered an increase in body weight, then most of the early growth follows a straight line or is linear in the age–weight relationship. As the animal increases in age and approaches puberty, rate of growth usually declines, and true growth ceases when the animal reaches maturity.

After an animal reaches maturity, it may initiate large fluctuations in body weight simply by increasing or decreasing the amount of fat or water that is stored. This increase in weight owing to fattening is not true growth because no net increase in body protein occurs. In fact, animals tend to lose body protein, as they grow older. The loss of protein is one of the phenomena in the aging process.

Table 18.2
MAJOR HORMONES AFFECTING GROWTH AND DEVELOPMENT IN FARM ANIMALS

Hormone	Source	Major Effect
Growth (Somatotrophic, STH)	Pituitary (anterior)	Body cell growth—especially muscle and bone cells
Adrenocorticotropic (ACTH)	Pituitary (anterior)	Stimulates adrenal cortex to produce adrenal cortical steroid hormones
Glucocorticoids	Adrenal (cortex)	Conversion of proteins to carbohydrates
Mineralocorticoids	Adrenal (cortex)	Regulates sodium and potassium balance; water balance
Thyroid-stimulating (TSH)	Pituitary (anterior)	Stimulates thyroid gland to produce thyroid hormones
Thyroid	Thyroid	Regulates metabolic rate
Testosterone	Testicles (interstitial cells)	Accessory sex gland development; male secondary sex characteristics and muscle growth
Estrogen	Ovary (follicle) Placenta	Female reproductive organ growth; female secondary sex characteristics; mammary gland duct growth
Progesterone	Corpus luteum; placenta	Uterine growth; alveoli growth in mammary gland growth
Antidiuretic (ADH) (Vasopressin)	Pituitary (posterior)	Controls water loss in kidney
Epinephrine	Adrenal (medulla)	Increases blood glucose concentration
Norepinephrine	Adrenal (medulla)	Maintains blood pressure
Insulin	Pancreas	Lowers blood sugar
Parathormone (PTH)	Parathyroid	Calcium and phosphorus metabolism
Glucagon	Pancreas	Raises blood sugar

Source: Compiled from multiple sources.

Figure 18.10 shows representative growth curves for most farm animals. The difference in curves for large and small breeds is primarily a function of differences in skeletal frame size. Maturity is reached at heavier weights in larger breeds, which have larger skeletal frame sizes than smaller-framed animals.

The relatively straight-lined growth shown for broilers does curve and level off as shown for the other species. This occurs when the birds reach approximately 11 weeks of age.

CARCASS COMPOSITION

Products from meat animals and poultry are composed primarily of fat, lean, and bone. A low proportion of bone, a high proportion of muscle, and an optimum amount of fat characterize a superior carcass. Understanding the growth and development of these animals is important in knowing when animals should be slaughtered to produce the desirable combinations of fat, lean, and bone. Figure 18.11 shows the variation in fat thickness, rib-eye area, and yield grade for two steers of comparable weight and age but of different sizes. The steer on the left was managed to a correct endpoint, while the steer on the right was fed too long and harvested at too heavy a weight.

Figure 18.12 shows the expected changes in fat, muscle, and bone as animals increase in live weight during the linear-growth phase. As the animal moves from the linear-growth phase into maturity, the growth curves shown in Figure 18.12 are extended similarly to those shown in Figure 18.10.

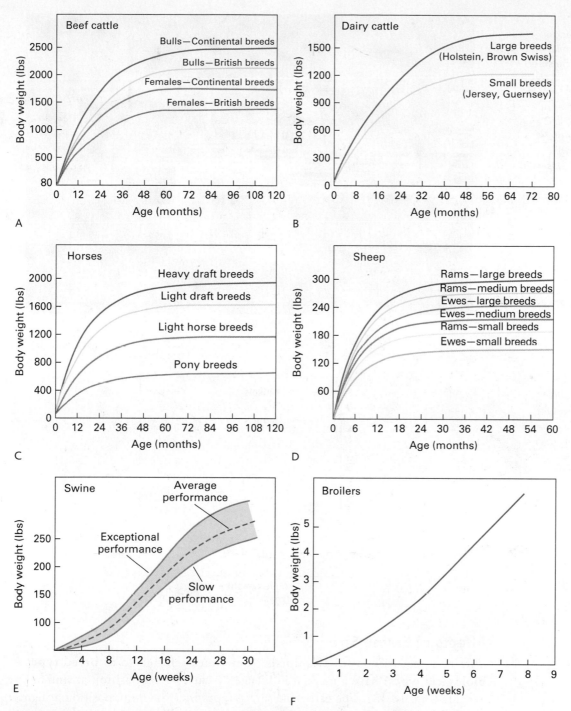

Figure 18.10

Representative growth curves for (A) beef cattle, (B) dairy cattle, (C) horses, (D) sheep, (E) swine, and (F) broilers. Source: *Handbook of Livestock Management Techniques* by Richard Battaglia, 4E, pp. 146, 149, 150. Copyright © 2007 by Pearson Education, Inc. Reprinted and Electronically reproduced by permission of Pearson Education Inc., Upper Saddle River, New Jersey.

At lightweights, fat deposition begins rather slowly, and then increases geometrically when muscle growth slows as the animal approaches physiological maturity. Muscle makes up the greatest proportion, and its growth is linear during the production of young slaughter animals. Bone has a smaller relative growth rate than either fat or muscle. Because of the relative growth rates, the ratio of muscle to bone increases as the animal increases in weight.

Figure 18.11
Two steers of varying composition at similar market weights. The steer on the left weighed 1,166 lb, was 52 in. tall, and had 0.30 in. of backfat, a 13.0-sq. in. ribeye, and a USDA yield grade of 2.4. The steer on the right weighed 1,160 lb, was 47 in. tall, and had 0.95 in. of backfat, a 9.9-sq. in. ribeye, and a USDA yield grade of 4.9. Source: Tom Field.

Figure 18.12
Tissue growth relative to increased live weight. Source: R. T. Berg and L. E. Walters in The Meat Animal: Changes and Challenges from The Journal of Animal Science. Copyright © 1983. Used by permission of American Society of Animal Science.

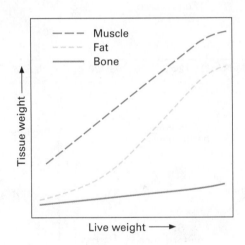

Effects of Frame Size

Different maturity types of animals (sometimes referred to as **breed types** or **biological types**) have a marked influence on carcass composition at similar live weights (Fig. 18.13). The earlier-maturing types increase fat deposition at lighter weights than either the average or later-maturing types.

Skeletal **frame size**, measured as hip height, is a more specific way of defining maturity type. Examples of frame sizes for bulls at various ages and heights are provided in Table 18.3. Figure 18.14 shows the difference in fat, lean, and bone composition for small- (frame 3), medium- (frame 5), and large-framed (frame 7) beef steers at different live weights. Note that the three different frame sizes have a similar composition at 1,100 lb (small frame), 1,300 lb (medium frame), and 1,500 lb (large frame).

Effect of Gender

The effect of sex is primarily on the fat component, although there are differences among species. Heifers deposit fat earlier than steers or bulls, with bulls being leaner at the

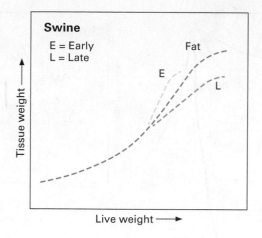

Figure 18.13
Effects of different maturity types on fat composition.
Source: R. T. Berg and L. E. Walters from The Meat Animal: Changes and Challenges, The Journal of Animal Science. Copyright © 1983. Used by permission of American Society of Animal Science.

Table 18.3
BULL HIP HEIGHT (IN.) AND ASSOCIATED FRAME SCORES

Age (mos.)	1	2	3	4	5	6	7	8
6	34.8	36.8	38.8	40.8	42.9	44.9	46.9	48.9
8	37.2	39.2	41.2	43.2	45.2	47.2	49.3	51.3
10	39.2	41.2	43.3	45.3	47.3	49.3	51.3	53.3
12	41.0	43.0	45.0	47.0	49.0	51.0	53.0	55.0
14	42.5	44.5	46.5	48.5	50.4	52.4	54.4	56.4
16	43.6	45.6	47.6	49.6	51.6	53.6	55.6	57.5
18	44.5	46.5	48.5	50.5	52.4	54.4	56.4	58.4

Source: Beef Improvement Federation.

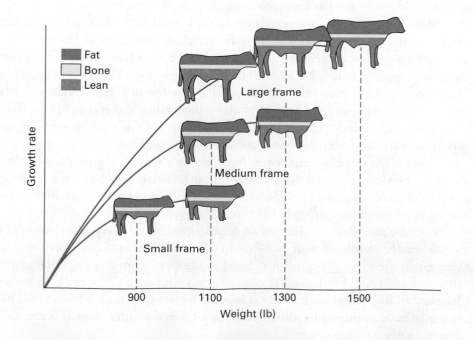

Figure 18.14
The relationship of frame size and weight to carcass composition in beef steers.
Source: Kevin Pond of Colorado State University. Used by permission of Kevin Pond.

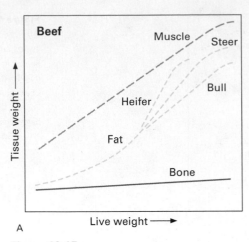

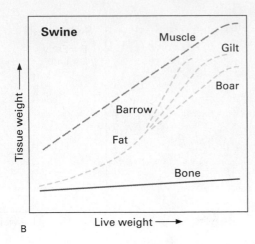

Figure 18.15

Influence of sex on carcass composition in (A) beef cattle and (B) swine. Source: R. T. Berg and L. E. Walters from The Meat Animal: Changes and Challenges, The Journal of Animal Science. Copyright © 1983. Used by permission of American Society of Animal Science.

same slaughter weight (Fig. 18.15). Heifers are typically slaughtered at lighter weights (100–200 lb less) than steers in order to ensure a similar fat-to-lean composition.

Swine are different from cattle, as barrows are fatter than gilts or boars at similar slaughter weights (Fig. 18.15) although the reason for this species difference is not known.

At typical slaughter weights, the sex differences in sheep do not have a marked effect on fat and lean composition. Apparently, sheep are harvested at an earlier stage of development (before puberty) than cattle and swine. Compositional differences in sheep are similar to those in cattle if sheep are fed beyond their usual slaughter weights.

Effect of Muscling

Relatively large differences exist in muscling among the various species of farm animals. The proportion of muscle in the carcass varies indirectly with fat, and a higher proportion of fat is associated with a lower proportion of muscle. Muscle has a much faster relative growth rate than bone. Muscle weight relative to live weight or muscle-to-bone ratio can be used as a valuable measurement of muscle yield.

Most of the current cattle population in the United States does not display large differences in muscling. There are, however, muscling differences in 10–15% of the cattle, which are best expressed by a muscle-to-bone ratio (pounds of muscle compared to pounds of bone). Muscle-to-bone ratios can range from 3 lb of muscle to 1 lb of bone in thinly muscled slaughter steers and heifers to 5 lb of muscle to 1 lb of bone in more thickly muscled cattle. Muscle-to-bone ratios higher than 5:1 are found in double-muscled cattle. These differences in muscle-to-bone ratio are economically important, particularly if more boneless cuts are merchandised.

Some of the muscling differences for cattle are shown in Figure 18.16: lighter-muscled, heavy-muscled, and double-muscled cattle are compared to cattle of average muscling. Double-muscled cattle do not have a duplicate set of muscles, but they do have an enlargement (hypertrophy) of the existing muscles.

Extreme muscling, as observed in double-muscled or heavy-muscled types of animals, can be associated with problems in total productivity. Some heavy-muscled swine might die when subjected to stressful conditions or produce undesirable meat that is pale, soft, and exudative (see porcine stress syndrome [PSS] and PSE pork in Chapter 29). Reproductive efficiency is lower in some heavy-muscled animals. There is some latitude in increasing muscle-to-bone ratio in slaughter animals without encountering negative side effects.

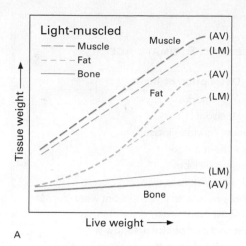

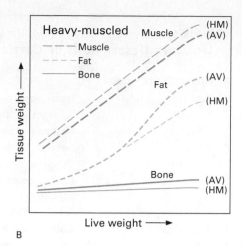

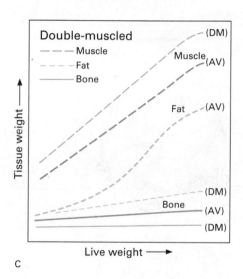

Figure 18.16

Effects of light A, heavy B, and double muscling C on carcass composition compared to average muscling (bold lines).

Source: R. T. Berg and L. E. Walters from The Meat Animal: Changes and Challenges, The Journal of Animal Science. Copyright © 1983. Used by permission of American Society of Animal Science.

Individual muscle dissection has shown a similarity of proportion of individual muscles to total body muscle weight. Small differences in muscle distribution have been shown, but the differences do not seem to be economically significant for the industry. Thus it appears that increasing muscle weight in areas of high-priced cuts at the expense of muscle weight in lower-priced cuts is not commercially feasible. Even though muscle distribution is not important, muscling as previously defined— as muscle-to-bone ratio—has high economic importance.

AGE AND TEETH RELATIONSHIP

For many years, producers have observed the teeth of farm animals to estimate their age or their ability to graze effectively with advancing age. Some animals are **mouthed** to classify them into appropriate age groups for livestock exhibition or show classification. The latter has not always proven exact because of variation in tooth condition and the ability of some exhibitors to manipulate the condition of the teeth.

Table 18.4 gives some of the guidelines used in estimating the age of cattle by observing their teeth. The **incisors** (front cutting teeth) are the teeth used. There are no upper incisors, and the molars (back teeth) are not commonly used to determine age.

Figure 18.17 shows the incisor teeth of cattle of different ages. If the grazing area is sandy and the grass is short, wearing of the teeth progresses faster than previously described. Also, some cows may be **broken-mouthed**, which means some of their permanent incisor teeth have been lost.

Table 18.4
UTILIZING DESCRIPTIONS OF CATTLE TEETH TO ESTIMATE AGE

Approximate Age	Description of Teeth
Birth	Usually only one pair of middle incisors
1 month	All eight temporary incisors
1.5–2 years	First pair of permanent (middle) incisors
2.5–3 years	Second pair of permanent incisors
3.25–4 years	Third pair of permanent incisors
4–4.5 years	Fourth (corner) pair of permanent incisors
5–6 years	Middle pair of incisors begins to level off from wear; corner teeth might also show some wear
7–8 years	Both middle and second pair of incisors show wear
8–9 years	Middle, second, and third pair of incisors show wear
10 years and over	All eight incisors show wear

Figure 18.17
Incisor teeth of cattle of different ages.

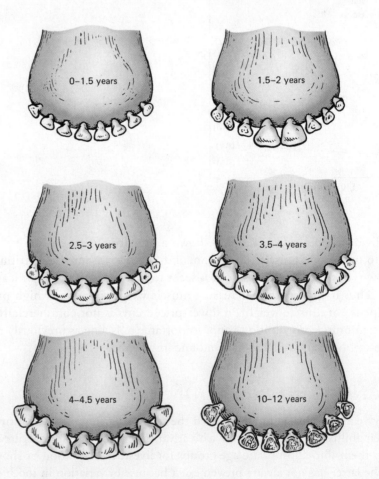

MEASUREMENTS OF GROWTH

The sale of livestock has been founded on value determination based on typically one of two standards—the individual animal or weight. Because of the importance of weight on the market value of livestock, measurement of increases in weight is of interest to managers.

A common calculation is to determine average daily gains (ADG) by the formula:

$$\frac{W_2 - W_1}{T_2 - T_1}$$

Where $W_2 - W_1$ = change in weight, while $T_2 - T_1$ = change in time.

Assume a calf weighs 500 lb at the beginning of a weigh period and 650 lb 50 days later. The average daily gain would be

$$\frac{650 - 500}{50} = 3.0 \text{ lb per day}$$

Another measure of an animal's growth is weight per day of age (WDA). WDA is calculated by dividing an animal's weight by its age in days. For example, a heifer weighing 400 lb at 200 days of age would have a WDA of:

$$\frac{400}{200} = 2.0 \text{ lb per day of age}$$

WDA is a snapshot evaluation of an animal's growth rate at a fixed point in time. It is typically measured at ages up to and just beyond weaning.

While weight and weight changes are important traits to be monitored by livestock managers, it is important to keep in mind that the composition, quality, and consistency of the final product determine true value.

CHAPTER SUMMARY

- Growth is an increase in body weight until mature size is reached. Development is the directive coordination of all physiological processes until maturity is obtained.

- Prenatal growth is a unique series of events starting from the union of two sex cells, which then grow and differentiate into a fetus with different organs and tissues.

- Embryonic development of the chick differs from that of mammals, as the chick develops more rapidly in the egg outside the hen's body.

- Muscle growth occurs through an increase in fiber size (hypertrophy) and an increase in fiber number (hyperplasia).

- The study of the anatomy and physiology of farm animals involves understanding the structure and function of cells, tissues, organs, and systems. The most common systems are skeletal, nervous, muscular, circulatory, and endocrine.

- Growth curves are influenced by nutrition, sex, frame size, and muscling. They are important in determining when animals should be harvested to produce desirable combinations of fat, lean, and bone.

KEY WORDS

growth
development
hyperplasia
hypertrophy
blastoderm
allantois
amnion
yolk sac
chorion

tissues
organs
system
dark meat
white meat
power
strength
endurance
white fibers

red fibers
arteries
veins
hormones
breed types (biological types)

frame size
mouthed
incisors
broken-mouthed

REVIEW QUESTIONS

1. Describe the progression of fetal growth and development.
2. Discuss the developmental progression of the chick embryo.
3. Describe the structure and function of muscle.
4. Compare and contrast fast versus slow-twitch muscle fibers.
5. List the primary hormones affecting growth and the major impacts of each.
6. Explain the growth curve and its value in livestock management.
7. Discuss the compositional changes that occur at various stages of the growth curve.
8. Define the effect of gender and muscling differences on growth.
9. Describe the changes to bovine teeth as the animal matures.
10. Demonstrate the calculation of average daily gain and weight per day of age.

SELECTED REFERENCES

Battaglia, R.A. 2007. *Handbook of Livestock Management*. 4th ed. Upper Saddle River, NJ: Prentice Hall.

Berg, R. T. and L. E. Walters. 1983. The meat animal. Changes and challenges. *J. Anim. Sci.* 57(Suppl. 2):133.

Berg, T. T. and R. M. Butterfield. 1976. *New Concepts of Cattle Growth*. Sydney, Australia: Sydney University Press.

Bone, J. F. 1988. *Animal Anatomy and Physiology*. Reston, VA: Reston Publishing Co.

Currie, W. B. 1988. *Structure and Function of Domestic Animals*. Boston, MA: Butterworth Publishers.

Frandson, R. D. 1986. *Anatomy and Physiology of Farm Animals*. Philadelphia, PA: Lea & Febiger.

Hammond, J., T. Robinson, and J. Bowman. 1983. *Hammond's Farm Animals*. Baltimore, MD: Edward Arnold University Park Press.

Prior, R. L. and D. B. Lasater. 1979. Development of the bovine fetus. *J. Anim. Sci.* 48:1546.

Rollins, F. D. 1984. Development of the embryo. *Arizona Coop. Ext. Serv. Publication* 8427.

Swatland, H. J. 1984. *Structure and Development of Meat Animals*. Englewood Cliffs, NJ: Prentice Hall.

Trenkle, A. H. and D. N. Marple. 1983. Growth and development of meat animals. *J. Anim. Sci.* 57(Suppl. 2):273.

19
Lactation

Lactation, the production of milk by the mammary gland, is a distinguishing characteristic of mammals, whose young at first feed solely on milk from their mothers (Fig. 19.1). Even after they start to consume other feeds, the young continue to nurse until they are weaned (separated from their mothers so they cannot nurse).

The mammary gland serves two functions: (1) it provides nutrition to animal offspring and (2) it is a source of passive immunity to the offspring. The importance of milk as a nutritional source to perpetuate each mammalian species has been known since the beginning of history. Only during the past few decades have the basic mechanisms of immunity through milk been determined.

Humans consume milk and recognize it as a palatable source of nutrients. Milk consumed in the United States comes primarily from dairy cows and, to a much lesser extent, from goats and sheep. In other countries, the milk supply also comes from water buffalo, yak, reindeer, camels, donkeys, and sows.

MAMMARY GLAND STRUCTURE

The mammary gland is an **exocrine gland** that produces the external secretion of milk transported through a series of ducts. The cow has four separate mammary glands that terminate into four teats; sheep and goats have two glands and two teats; and the mare has four mammary glands that terminate into two teats.

Sows have 6–20 mammary glands located in two rows along the abdomen, with each gland having a teat. Typically 10–14 of the sow's mammary glands are functional (Fig. 19.1). There is evidence that teat number in swine is not related to litter size or litter-weaning weight.

Figure 19.2 shows the basic structure of the cow's udder. The udder is supported horizontally and laterally by suspensory ligaments (Fig. 19.2A). The internal structure of the udder is similar for all farm animal species, except for the number of glands and teats. The volume of milk produced is significantly variable within species and certainly between species (Table 19.1).

The secretory tissue of the mammary gland is composed of millions of grapelike structures called **alveoli** (Fig. 19.2C). Each **alveolus** has its own separate blood supply from which milk constituents are obtained by epithelial cells lining the alveolus. Milk collects in the alveolus lumen and, during milk letdown, travels through ducts to a larger collection area called the *gland cistern*. During milk letdown and the milking process, milk is forced into the teat cistern and through the streak canal to the outside of the teat.

learning objectives
- Label the structures of the mammary gland
- Describe the development and function of the mammary gland
- Discuss the hormonal influences on milk production
- Describe nonhormonal influences on milk production
- Compare and contrast the milk composition of farm animals
- Discuss the role of colostrum

Figure 19.1
Mammary glands of the (A) cow, (B) sow, (C) ewe, and (D) mare. Source: 19.01a: Tom Field; 19.01b: David Noblecourt/ Fotolia; 19.01c: Tom Field; 19.01d: Onepony/Fotolia.

MAMMARY GLAND DEVELOPMENT AND FUNCTION

Development

Mammary gland growth and development occur rapidly as the female reaches puberty. The ovarian hormones (estrogen and progesterone) have a large effect on development. Estrogen is primarily responsible for duct and cistern growth, whereas progesterone stimulates growth of the alveoli.

Estrogen and progesterone are produced by the ovary under the stimulation of FSH and LH from the anterior pituitary (Fig. 19.3). The pituitary also has a direct influence on mammary growth through the production of growth hormone. Placental lactogen stimulates general cell growth of the mammary gland and may also influence fetal growth.

Figure 19.3 also identifies the additional indirect effect (other than FSH and LH) on mammary growth and development. Thyroid hormones are produced under the influence of TSH, and the adrenal gland produces the corticosteroids when stimulated by ACTH from the pituitary. All of these hormones work in concert to produce mammary growth and function.

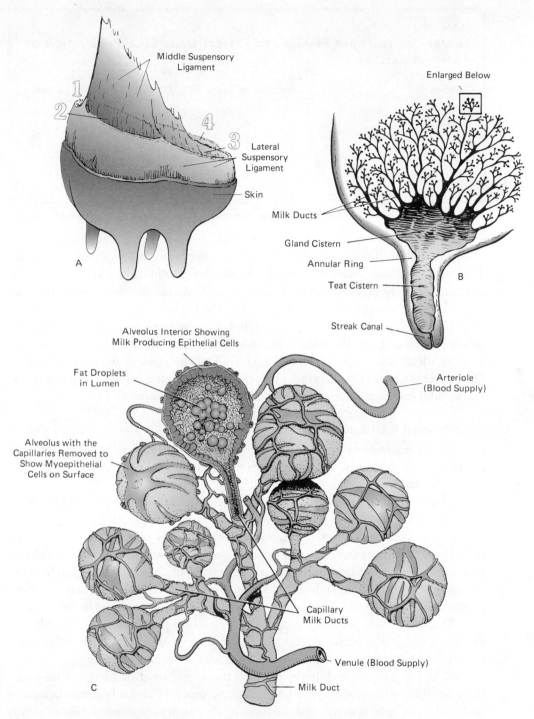

Figure 19.2

(A) Basic structure of the cow's udder, showing suspensory ligaments and the location of the four separate quarters. (B) A section through one of the quarters, showing secretory tissue, ducts, and milk-collecting cisterns. (C) An enlarged lobe with several alveoli and their accompanying blood supply.

Source: Kevin Pond of Colorado State University. Used by permission of Kevin Pond.

Milk Secretion

Growth hormone, adrenal corticoids, and prolactin are primarily responsible for the initiation of lactation. These hormones become effective as parturition nears and when estrogen and progesterone hormone levels decrease.

Table 19.1
NUMBER OF TEATS PER FEMALE AND TYPICAL DAILY MILK PRODUCTION OF FARM ANIMALS

Species	Number of Teats	Daily Milk Production (lbs)
Dairy cow	4	55
Beef cow	4	15
Doe (goat)	2	7
Ewe	2	3
Mare	2	27
Sow	14	13

Through milking or nursing, the milk in the gland cistern is soon removed. There remains a large amount of milk in the alveoli that is forced into the ducts by the contractions of the myoepithelial cells (Fig. 19.2). These cells contract under the influence of the hormone oxytocin, which is secreted by the posterior pituitary (Fig. 19.3).

Oxytocin, released by the suckling reflex, can also be released by other stimuli; a suckling calf will nudge the udder with its head to initiate the milk letdown response. The milk letdown can be associated with feeding the cows or washing the udder. (The latter is a typical management practice before milking cows.) Oxytocin release can be inhibited by pain, loud noises, and other stressful stimuli.

Maintenance of Lactation

Lactation is maintained primarily through hormonal influence. Prolactin, thyroid hormones, adrenal hormones, and growth hormone are all important in the maintenance of lactation.

Daily milk production typically increases during the first few weeks of lactation, peaks at approximately 4–6 weeks, and then decreases over the next several weeks of the lactation period. Persistency of lactation measures how milk production is maintained over time. For example, in dairy cattle, persistency is determined by calculating milk production of the current month as a percentage of the last month's production.

Figure 19.4 shows lactation curves for the various species of farm animals. The curve for dairy cows (Fig. 19.4A) represents approximately 14,000 lb produced in 305 days.

Decreased milk production during the lactation period is due primarily to a decreased number of active alveoli and less secretory tissue (epithelial cells) in the alveoli. These and other changes are associated with hormonal changes.

When milking or suckling is stopped, the alveoli are distended and the capillaries are filled with blood. After a few days the secretory tissue becomes involuted (reduced in size and activity) and the lobes of the mammary gland consist primarily of ducts and connective tissue. The female then becomes **dry**, as milk secretion is not occurring. After the dry period (approximately 2 months in the dairy cow) and when parturition approaches, hormones and other influences prepare the mammary gland to resume its secretion and production of milk.

FACTORS AFFECTING MILK PRODUCTION

Inheritance determines the potential for milk production, but feed and management determine whether or not this potential is attained. The best feed and care will not make a high-producing female out of one that is genetically a low producer. Likewise, a female with high genetic potential will not produce at a high level unless she receives

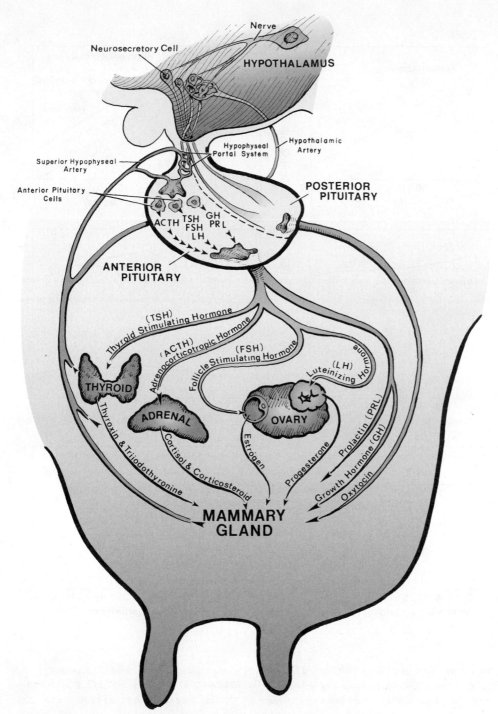

Figure 19.3

Major hormones (and their sources) influencing the anatomy and physiology of the mammary gland. Refer to Figure 18.11 for the location of the endocrine glands throughout the body.

Source: Kevin Pond of Colorado State University. Used by permission of Kevin Pond.

proper feed and care. Production is also influenced by the health of the animal. For example, **mastitis** (an inflammation of the udder) in dairy cows can reduce production by 30% or more. Proper management of dairy cows, particularly adherence to routine milking and feeding schedules, contributes to a high level of production.

In lactating farm animals, the level of milk production is important because milk from these females provides much of the required nutrients for optimal growth in the

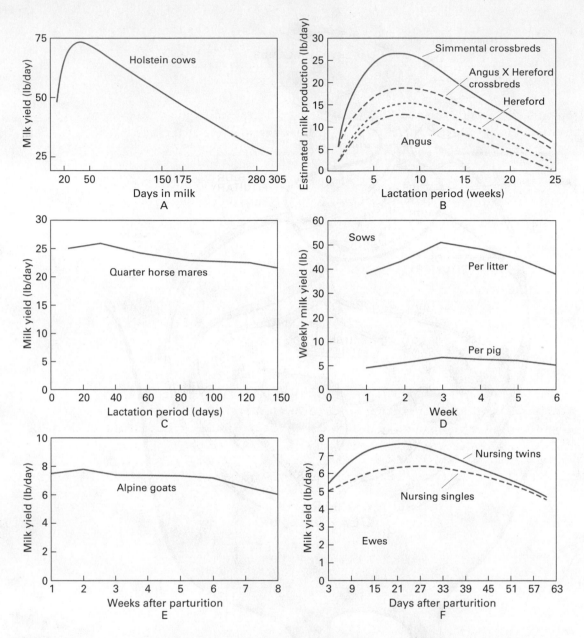

Figure 19.4

Lactation curves for several species of farm animals. (A) Dairy cows. Source: Colorado State University (CO, ID, and UT DHIA records). *(B) Beef cow.* Source: Colorado State University, Jenkins and Ferrell, 1984 Beef Cow Efficiency Forum. *(C) Quarter horse mares.* Source: *J. Anim. Sci.* 54:496. *(D) Sows.* Source: *Missouri Agric. Research Bull.* 712. *(E) Dairy goat does. (F) Ewes.* Source: Michigan State Agric. Expt. Sta. Research Report 491.

young. If beef cows, for example, are inherently heavy milkers, their milk-producing levels will be established by the ability of the young to consume milk. Normally, if the cow is not milked, she will adjust her production to the consumption level of the calf. However, if the cow is milked early in lactation, her milk flow may be far greater than the calf can consume. This situation makes it necessary either to continue milking the cow until the calf becomes large enough to consume the milk or to put another calf onto the cow.

Anything that causes a female to reduce her production of milk is likely to cause some regression of the mammary gland, thereby preventing resumption of full production. If the young animal lacks vigor, it does not consume all the milk that its mother can supply, and her level of production is lowered accordingly. This is one reason inbred animals grow more slowly and crossbred animals more rapidly during the nursing period. Crossbreds are usually larger and more vigorous at birth and can stimulate a high level of milk production by their dams, whereas inbred animals are smaller and less vigorous and cause their dams to produce milk at a comparatively low rate.

Females with male offspring produce more milk than females with female offspring. Male offspring are usually heavier at birth and have a faster growth rate than females. This puts a greater demand on the milk-producing ability of the dam.

Females with multiple births usually produce more total milk than females with single births. Age of the female also affects milk yield, with younger and older females producing less milk compared to females that have had several lactations. For example, sows are usually at peak milk yields during third or fourth lactation, beef cows produce the most milk at 5–9 years of age, and dairy cows have the highest milk yields between 5–8 years of age.

Large amounts of nutrients are needed to supply the requirements of lactating females producing high levels of milk, because much energy is required for milk secretion and because milk contains large quantities of nutrients. Adequate nutrition is much greater for lactation than for gestation; for example, a cow producing 100 lb of milk daily could yield 4.0 lb of butterfat, 3.3 lb of protein, and almost 5.0 lb of lactose in that milk. If nutrition is inadequate in quality or quantity, a cow that has the inherent capacity to produce 100 lb of milk daily will draw nutrients from her own body as the body attempts to sustain a high level of milk production. Withdrawal of nutrients reduces the body stores. This withdrawal usually occurs to a limited extent in high producers even when they are well fed. Cows reduce their milk production in response to inadequate amounts or quality of feed, but they usually do so only after the loss of nutrients is sufficiently severe to cause a loss in body weight. In gestation, a cow becomes relatively efficient in digesting her feed, so feed restrictions in gestation are less harmful than they might be at other times.

MILK COMPOSITION

Species Differences

Milk composition is markedly different among the mammalian species (Table 19.2). Milk from reindeer is exceptionally high in **total solids**, whereas the mare's milk is

Table 19.2
AVERAGE MILK COMPOSITION OF SEVERAL DOMESTICATED LIVESTOCK SPECIES (HUMAN DATA PROVIDED FOR COMPARISON)

Species	Total Solids (%)	Fat (%)	Protein (%)	Lactose (%)	Minerals (%)
Human	13.3	4.5	1.6	7.0	0.2
Cow (*Bos taurus*)	12.7	3.9	3.3	4.8	0.7
Cow (*Bos indicus*)	13.5	4.7	3.4	4.7	0.7
Goat	12.4	3.7	3.3	4.7	0.8
Water buffalo	19.0	7.4	6.0	4.8	0.8
Ewe	18.4	6.5	6.3	4.8	0.9
Sow	19.0	6.8	6.3	5.0	0.9
Mare	10.5	1.2	2.3	5.9	0.4
Reindeer	33.7	18.7	11.1	2.7	1.2

Adapted from multiple sources.

low in this regard. Large differences in milk fat percentages exist between mammals; mares are low (1.4%) and reindeer are high (18.7%). Milk from aquatic mammals is exceptionally high in solids and percent fat. For example, milk from fur seals is 53.3% fat.

Milk from cows or goats is much higher in protein than milk from humans. However, human milk is comparatively quite high in percent lactose. Substitution of cow or goat milk would be most common in many cultures.

In dairy cows, the amount or percentage of fat is easily changed through changing the diet fed. Carbohydrates (**lactose**) remain relatively constant even with dietary fluctuations. Varying the protein content of the ration has little change on the protein content of the milk.

COLOSTRUM

The fetus develops in a sterile environment. The microorganisms existing in the external environment have not yet challenged its immune system. Before or shortly after birth, the fetal immune system must be made functional or death is imminent.

Immunoglobulins (Ig) are involved in the passive immunity transfer from the mother to the offspring. In some animals immunoglobulins are transferred in utero through the bloodstream, whereas in other animals immunoglobulins are transferred through **colostrum**. Certain animals utilize both methods of Ig transfer; however, the colostrum method is most common for larger, domestic animals.

These Ig antibodies give the newborn protection from harmful microorganisms that invade the body and cause illness. The intestinal wall of the newborn is quite porous, permitting colostrum antibodies to be absorbed and enter the bloodstream. Within a few hours (no more than 24), the gut wall becomes less porous, allowing little absorption of the antibodies to occur. Thus, passive immunity of the newborn is dependent on an adequate supply of antibodies in the colostrum and on consuming the colostrum within a few hours after parturition.

CHAPTER SUMMARY

- Lactation is the production of milk from the mammary gland.
- Milk from mammals provides nutrients and passive immunity to the young and is a nutrient source for humans.
- Colostrum, the first milk produced after the young are born, is the source of immunoglobulins that provide early immunity to the offspring.
- The mammary gland is an exocrine gland that produces an external secretion (milk) from structures called alveoli.

- Several hormones—namely, estrogen, progesterone, thyroid hormone, growth hormone, adrenal hormone, and prolactin—affect mammary gland development, milk secretion, and the maintenance of lactation.
- Lactation curves for most farm animals peak a few weeks after lactation is initiated, then decrease throughout the lactation period.
- Milk is composed primarily of water with lesser amounts of protein, fat, lactose, and minerals.

KEY WORDS

exocrine gland
alveoli
alveolus
dry

mastitis
total solids
lactose
colostrum

REVIEW QUESTIONS

1. What are the anatomical structures of the mammary system and the role of each?
2. Compare average daily milk production of the various livestock species.
3. Describe the hormones that affect lactation and the role of each.
4. Describe the lactation curve.
5. Discuss differences in milk composition between species.
6. Explain the role of colostrum in providing immunity to newborn mammals.

SELECTED REFERENCES

Ackers, R. M. 1994. Lactation. *Encyclopedia of Agricultural Science*. San Diego, CA: Academic Press,.

Allen, A. D., J. F. Lasley, and L. F. Tribble. Milk production and related factors in sows. *Missouri Agric. Research Bull.* 712.

Gibbs, P. G., G. D. Potter, R. W. Blake, and W. C. McMullan. 1982. Milk production of quarter horse mares during 150 days of lactation. *J. Anim. Sci.* 54:496.

Henry, M. S., and M. E. Benson. 1990. Milk production and efficiency of lactating ewe lambs. *Michigan State Agric. Expt. Sta. Report* 491.

Keown, J. F., R. W. Everett, N. B. Empet, and L. H. Wadell. 1986. Lactation curves. *J. Dairy Sci.* 69:769.

Larson, B. L. (ed.). 1985. *Lactation*. Ames, IA: Iowa State University Press.

Offedal, O. T., H. F. Hintz, and H. Schryver. 1983. Lactation in the horse: Milk composition and intake by foals. *J. Nutr.* 113:2196.

Sakul, H. and W. J. Boylan. 1992. Lactation curves for several U.S. sheep breeds. *Animal Production* 54:229.

20
Adaptation to the Environment

Livestock throughout the world are expected to produce under an extremely wide range of environmental conditions. Variations in temperature, humidity, wind, light, altitude, and exposure to parasites and disease organisms are some of the environmental conditions to which livestock are exposed (Fig. 20.1).

Under **intensive management**, such as integrated poultry enterprises, confinement swine operations, and large dairies, many environmental conditions are highly controlled. Feed and water are plentiful, rations are carefully balanced, excellent health programs are carefully monitored, and, in many cases, temperature, humidity, and other weather influences are controlled via provision of thoughtfully engineered housing systems.

Extensive management involves less producer control over the environmental conditions in which animals are expected to produce (Fig. 20.2). Many ruminant animals are extensively managed while also exposed to numerous climatic conditions as they graze forage, which may be seasonally limited. Livestock in these less than perfect conditions are capable of functioning because they are **genetically adapted** to the existing environment or because they have undergone a process known as **acclimatization**. Genetic adaptation occurs over extended periods of time where populations develop distinct characteristics that help them better cope with climatic conditions. For example, British breeds of cattle such as Hereford, Shorthorn, and Angus are capable of growing thick hair coats that help them to function in cold, winter-like conditions and to then shed to better handle the warmer temperatures of late spring, summer, and early fall. Brahman cattle however have been selected through time to handle environments where heat, humidity, and parasite exposure are the norm. As a result, Brahman cattle have slick hair, increased skin folds to increase surface area, dark skin pigmentation, and sweat more than cold-weather adapted breeds.

Acclimatization is a series of biological adaptations that reduce physiologic stress due to climatic conditions. This process requires several weeks to months to attain. During the period of transition, livestock moved from one set of climatic stressors to a new environment require careful observation and management to assure success.

It is important to understand the impacts of the climatic environment on animal productivity in order to make sound management decisions. These decisions may involve selecting animals that are more genetically adapted to the existing climate or modifying the environment. For example, a feedyard manager would evaluate whether it was

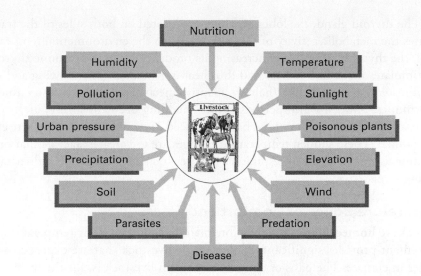

Figure 20.1
Environmental influences on livestock that must be managed by humans to assure profitability.

Figure 20.2
The ability of cattle to adapt to changing environmental conditions allows them to be managed under extensive conditions.
Source: Tom Field.

more cost-effective to construct shades and/or sprinkler systems to enhance performance of *Bos taurus* breeds of cattle exposed to heat or to utilize *Bos indicus* cattle that are genetically more heat tolerant. In some cases, managers would choose a combination of environmental modifications and utilization of genetic selection to obtain optimal results.

RELATIONSHIP BETWEEN ANIMALS AND THE ENVIRONMENT

Adjusting to Environmental Changes

As the seasons shift, two major kinds of changes occur in the environment that affects livestock performance: changes in temperature and changes in length of daylight. In the Western Hemisphere as summer approaches, temperature and length of daylight increase, thus increasing the amount of heat available to the animal. As autumn approaches, length of daylight and temperature decrease. Hormonal changes help the animal respond physiologically to these seasonal changes.

The thyroid gland, the lobes of which are located on both sides of the trachea, regulates the metabolic activity of the animal. When the environmental temperature is cool, the thyroid responds by increasing its production of the hormone **thyroxine** that stimulates metabolic activity and thus heat is generated. The growing and shedding of insulating coats of fiber/hair are adaptive mechanisms that are also influenced by hormones. Sudden warm periods before shedding occur in spring, and sudden cool periods before a warm coat has been produced in autumn, can have severe effects on an animal. These sudden temperature changes are the greatest factors that predispose animals for the onset of respiratory diseases such as pneumonia, influenza, and shipping fever.

Temperature Zones of Comfort and Stress

Livestock are **homeotherms** and therefore maintain constant body temperature. This requirement provides significant challenges for livestock that are exposed to wide changes in climate. The basis of climatic impact on livestock is, for the most part, a change in rate of energy flow. Farm animals must maintain constant body temperature in spite of the fact that the rate of energy gain and loss changes when animals are exposed to varying conditions.

Livestock gain or lose heat through four routes conduction, convection, evaporation, and radiation. **Conduction** occurs when heat flows through a medium where warm molecules transfer energy to colder molecules due to close association. For example, a dairy cow lying on cold bedding will lose heat via conduction whereas a calf resting on warm, dry bedding will gain heat from conduction. **Convection** is the process of heat transfer arising from air or water movement. For example, a show steer being washed on a warm day is losing heat due to the flow of water across the animal's body. Similarly, a brood mare standing in a pasture when the air temperature is 50°F but exposed to a 35 mph wind flow is losing heat due to convective forces.

Evaporation is an essential physiological cooling mechanism that releases heat either through sweating or panting. **Sweating** is an effective cooling process in horses (similar to humans) and marginally effective in cattle, sheep, goats, and swine. Poultry do not have sweat glands and thus are particularly susceptible to heat stress. Chickens experience disproportionately high death losses during prolonged heat stress. They also respond to heat stress by lowering and terminating egg production, which explains why egg production is higher in the milder seasons of spring and fall. **Panting** is utilized by livestock and poultry as a form of evaporative cooling and is a behavioral signal of heat stress.

Radiation is the process of heat transfer between two objects that are not touching. Heat moves from warmer to cooler objects so animals can both gain and lose heat due to radiation. An animal standing in the sun on a cold, windless winter day gains heat from radiation. However, that same animal standing in the shade of a cool concrete wall will lose heat to the wall due to reflective radiation.

One difficulty when discussing effect of climate on livestock is an accurate description of the climate. This is most conveniently done by use of an equivalent temperature referred to as **effective ambient temperature**. The term *effective ambient temperature (EAT)* is a theoretical index of the heating or cooling power of the environment in terms of dry-bulb temperature and includes any environmental factor that alters environmental heat demand such as solar radiation, wind, humidity, or precipitation. Specific formulas for combinations of such variables as wind and temperature (windchill index) and temperature and humidity (heat humidity index) are available, although a comprehensive assessment that accounts for all climatic variables is not available. An example of effective temperature is the windchill index for cattle

Table 20.1
WINDCHILL FACTORS FOR CATTLE WITH WINTER COAT

Wind Speed (mph)	Temperature (°F)												
	−10	−5	0	5	10	15	20	25	30	35	40	45	50
Calm	−10	−5	0	5	10	15	20	25	30	35	40	45	50
5	−16	−11	−6	−1	3	8	13	18	23	28	33	38	43
10	−21	−16	−11	−6	−1	3	8	13	18	23	28	33	38
15	−25	−20	−15	−10	−5	0	4	9	14	19	24	29	34
20	−30	−25	−20	−15	−10	−5	0	4	9	14	19	24	29
25	−37	−32	−27	−22	−17	−12	−7	−2	2	7	12	17	22
30	−40	−41	−36	−31	−26	−21	−16	−11	−6	−1	3	8	13
35	−60	−55	−50	−45	−40	−35	−30	−25	−20	−15	−10	−5	0
40	−78	−73	−68	−63	−58	−53	−48	−43	−38	−33	−28	−23	−18

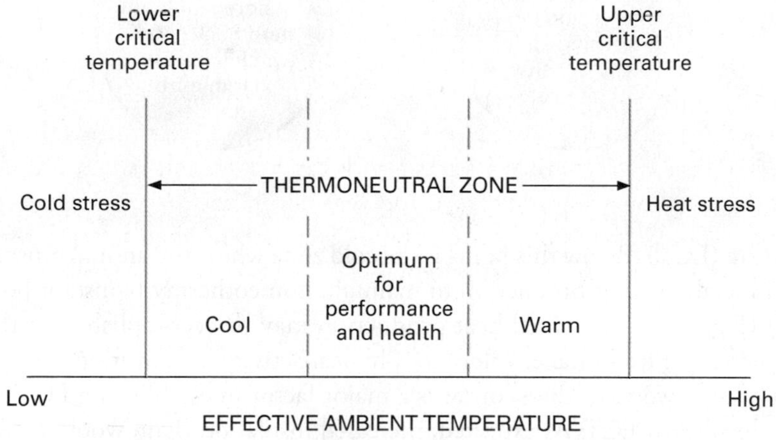

Figure 20.3
Range of stress related to changes in effective ambient temperature.

that is shown in Table 20.1. For example, a temperature of 20°F combined with a wind speed of 30 mph results in an effective ambient temperature of −16°F.

Evaluation of the relationship between animals and their thermal environment begins with the **thermal neutral zone (TNZ)** (Fig. 20.3). Thermal neutral zone is defined as the range in effective ambient temperature where rate and efficiency of performance is maximized and health is optimal. It must be emphasized that thermal comfort is a human term and that comfort for the stockman may be different from the TNZ of the animal; therefore, selection or assessment of animal environments must not be based on human comfort.

At temperatures immediately below optimum, but still within the TNZ, there is a cool zone (Fig. 20.3) where animals invoke mechanisms to conserve body heat (Fig. 20.4). These are mainly behavioral responses such as postural adjustments and seeking protection from the wind but also may include reduced blood flow to the skin by a process termed vasoconstriction. Specific examples of behavioral responses include chickens fluffing their feathers to create a large space of dead air about themselves to provide added insulation and cattle congregating in an area that provides protection from the wind as they huddle together for protection.

The effectiveness of vasoconstriction and behavioral responses to cool conditions is maximal at the lower limit of the TNZ, a point called the **lower critical**

Figure 20.4
Mechanisms to reduce heat loss or increase heat production.

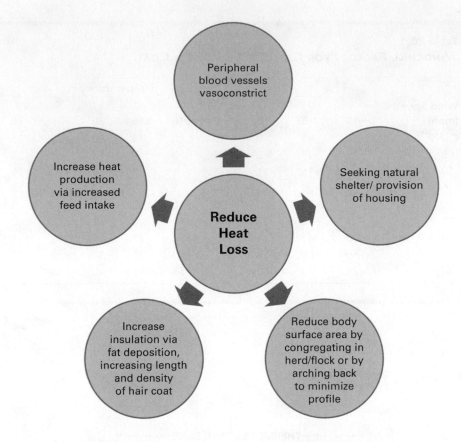

temperature (LCT). Below this point is the cold zone where the animal must increase its rate of metabolic heat production to maintain homeothermy (constant body temperature) (Fig. 20.3). Increased heat production may be accomplished in the short term by increasing feed intake, voluntary physical activity, or shivering. Insulation in the form of hair, wool, feathers, or fat is a major factor in establishing LCT and rate of energy loss below LCT. As expected, increased insulation (hair, wool, feathers, fat) will raise the TNZ and LCT to a higher effective temperature. Clearly, thermal zones and critical temperatures change for a given animal depending on factors such as insulation, level of feed intake, and level of activity.

As effective temperature rises above optimum, the animal is in the warm zone (Fig. 20.3). Increasing blood flow to the skin by vasodilation, reducing feed intake, and changing posture are typical mechanisms used to facilitate rate of heat loss and minimize heat production in the warm zone (Fig. 20.5). When effective temperature exceeds the **upper critical temperature (UCT)**, animals are heat stressed and must employ evaporative heat loss mechanisms such as sweating and panting to maintain homeothermy. Higher producing animals have greater metabolic heat production and therefore tend to be more susceptible to hot climates. This is different from cold conditions where high-producing animals with their higher metabolic heat production are in a more advantageous position than poor-producing animals. To avoid excessive heat loss, animals maintain the temperature of their extremities at a level below that of the rest of the body through a unique mechanism called **countercurrent blood flow action**. Blood in arteries coming from the core of the body is relatively warm and blood in veins in the extremities is relatively cool. The blood in veins in the extremities cools the warmer arterial blood so that the extremities are kept at a cool temperature. With the extremities thus kept at a relatively cool temperature, loss of heat to the environment is reduced.

Figure 20.5
Mechanisms to increase heat loss or decrease heat production.

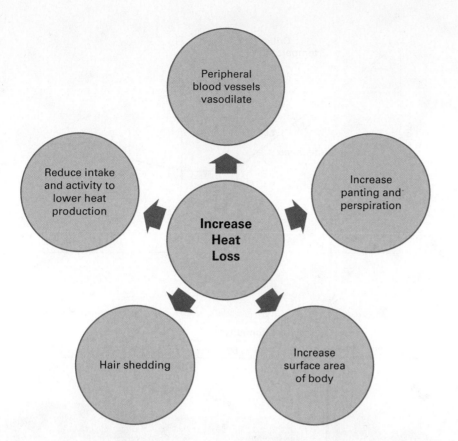

Relationship of Intake and Maintenance during Stress

Optimal performance is reached in the TNZ or range of effective temperature where rate and efficiency of performance is high (Fig. 20.6).

Animal productivity is a function of feed intake and feed energy required for maintenance. Nutrient requirements and predicted gains of farm animals are published from research studies in which animals are protected from environmental extremes. The most common environmental factor that alters both performance and nutrient requirements is temperature; thus, livestock producers should be aware of critical temperatures that affect the performance of their animals and should consider making changes in their feeding and management programs if economics so dictate.

Feed intake and the feed energy required for maintenance are affected by temperature change. During periods of cold stress an animal increases feed intake but at a slower rate than the rise in maintenance requirement, this offset results in an energy imbalance. Thus, as cold increases, managers must assure that sufficient additional feed is provided to avoid excessive weight loss.

In **heat stress** conditions, animals restrict their feed intake and experience rising maintenance costs. In both cases, managers must carefully evaluate cost of gain and other measures of productivity to determine whether or not environmental management strategies must be set in place. Prolonged exposure to heat stress reduces animal performance and in the most extreme situations results in increased death losses. Animals exhibiting heat stress tend to seek shade, increase panting rates, display excessive salivation, open-mouth breathing, as well as exhibiting incoordination and trembling. **Heat stress indices** have been established for cattle, swine, laying hens, and turkeys based on temperature and relative humidity and have been used for a number of decades. Figure 20.7 is an example of the heat index for swine. Generally, when the heat stress index is in the alert category, producers should prepare to initiate cooling

Figure 20.6
*Effect of temperature
on rate of feed intake,
maintenance energy
requirement, and gain.*
Source: D. R. Ames from Livestock
Nutrition in Cold Weather. Copyight
© 1980. Used by permission of the
American Society of Animal Science.

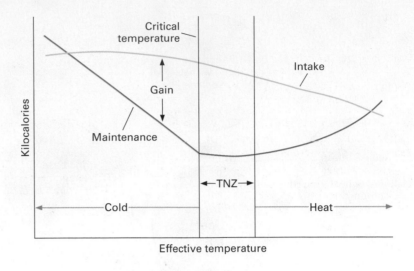

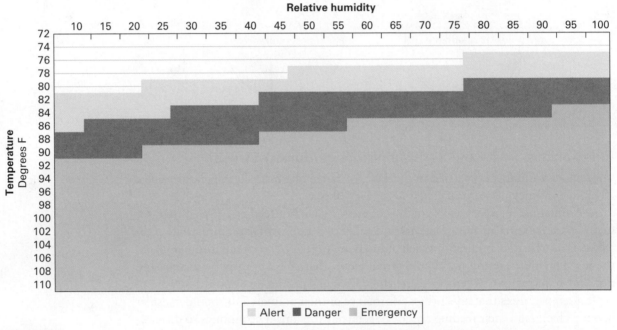

Figure 20.7
Heat stress index for swine. Adapted from Agricultural and Biosystems Engineering, Iowa State University.

strategies for livestock and poultry, ventilation rates should be increased, cooling fans should be turned on, and careful observation for signs of heat stress should be established. Under *danger* conditions, increased air movement is critically important, spraying and misting techniques should be incorporated to cool animals, close monitoring of livestock and poultry should be undertaken, and the use of cooling pads and ventilation should be incorporated into buildings. During *emergency* conditions, transport should be minimized, feed should be withdrawn during the hottest part of the day, and light-controlled houses should receive reduced light levels to minimize activity. The critical temperatures and optimal temperature ranges for some animals are shown in Figure 20.8. Effective ambient temperatures have not been calculated for all farm animals.

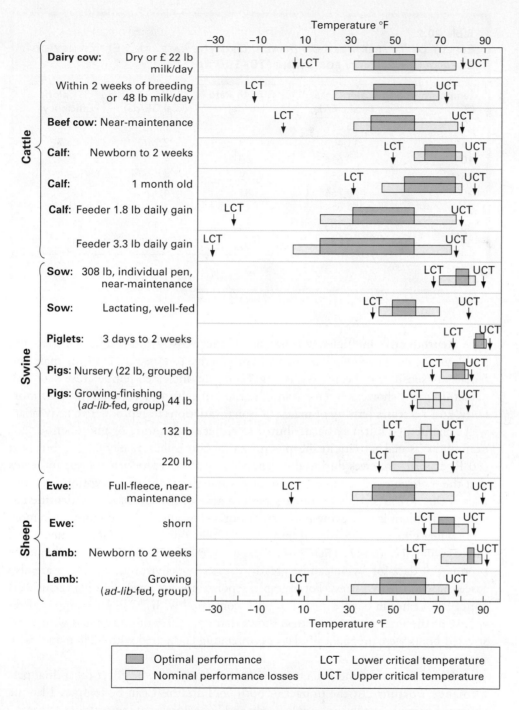

Figure 20.8
Critical ambient temperatures and temperature zones for optimal performance in cattle (Bos taurus), swine, and sheep. Adapted from Yousef, 1985.

An example of temperature effect on rate of performance and efficiency of food animals is illustrated by data shown in Table 20.2 collected from swine grown in temperatures ranging from cold stress (32°F) to heat stress (86°F). Efficiency was reduced during both cold and heat stress, while being highest in the TNZ.

Most dairy cows will experience a reduction in milk production when the environmental temperature exceeds 80°F. This occurs primarily because appetite is depressed and feed consumption declines. Milk yields will improve when thermal stress is alleviated by providing management such as shade, fans, sprinklers, or refrigerated air. While the temperature and efficiency values may differ for animals with different insulation, diet, etc., or for different species and products, the same general pattern of reduced efficiency is consistent among animals exposed to temperature stress.

Table 20.2
**EFFECT OF TEMPERATURE ON INTAKE, GROWTH RATE, AND EFFICIENCY OF
ENERGY CONVERSION FOR SWINE (70–100 KG)**

Temperature (°C)	Caloric Intake (kcal DE/da)	Growth Rate (kg/da)	Product (kcal GE/da)	% Caloric Efficiency
0°	15,377	.54	2,991	19.4
5°	11,404	.53	2,936	25.7
10°	10,616	.80	4,432	41.7
15°	9,554	.79	4,376	45.8
20°	9,766	.85	4,709	48.2
25°	7,976	.72	3,988	50.1
30°	6,703	.45	2,493	37.1
35°	4,579	.31	1,717	37.4

Source: Ames, 1980.

Reproductive inefficiency is one of the most costly and production-limiting problems facing the livestock industry. Reproductive processes in both the male and female are sensitive to heat stress. As a general rule, increased temperature decreases ovulation rates, shortens intensity and duration of estrus, increases embryonic mortality, and decreases fertility of males. Seasonal variations in reproductive activity are well documented but may be attributed to either temperature or photoperiod. For example, seasonal activity in sheep is primarily controlled by day length, but heat (90°F) exposure of ewes during different phases of the reproductive cycle indicates that the embryo is most susceptible to heat stress shortly after conception. The impact of thermal stress on reproductive performance of the sow has been identified as a major problem in the swine industry. Reports indicate increased embryonic mortality when gilts are heat stressed immediately following conception and substantial increases in stillborn pigs when gilts are exposed to heat in late gestation. Obviously, production data for swine can be greatly affected by environment. Cattle are also susceptible to thermal stress. For example, reports indicate lowered (48% versus 0%) conception rates in cows exposed to 90°F compared with 70°F. An example of response of the male to thermal stress shows that only 59% of gilts mated with heat-stressed boars were pregnant 30 days postbreeding compared with 82% mated with control boars.

In summary, the thermal environment can have a drastic effect on animal performance. Fortunately, the impact of both cold and heat can be tempered by the ability of animals to adjust physiologically and behaviorally to temperature extremes. The cost of producing food and fiber of animal origin is increased by exposure to the climatic extremes.

MANAGING THE THERMAL ENVIRONMENT

Livestock producers are usually willing to incorporate management systems to improve energetic efficiency when it is economically advantageous. Initially, one thinks of modifying the existing environment to reduce the impact of thermal stress and improve energetic efficiency. Among the many possibilities for improving livestock environment to reduce cold stress are windbreaks (natural and human-made), sheds, confinement buildings without supplemental heat, and buildings environmentally controlled with supplemental heat. Modifications that provide optimum environments from an efficiency viewpoint require the most input (buildings with supplemental heat). Basically, the decision on the degree to modify animal environments depends on the cost of providing improved environment compared with the value of improved performance (cost-to-benefit ratio). Of course, such factors as cold tolerance of the animals, effect of diseases, and other determinants must be considered as well.

Mitigating Heat Stress

During periods of heat stress, animals lower their intake of feed, become less active to reduce the amount of heat they generate, and seek shade to reduce their exposure to the sun.

Table 20.3 illustrates a partial list of adjustment factors for air temperature to account for the influence of floor type, air speed, and evaporative cooling systems for determining effective temperature in swine. Obviously, many factors influence the thermal environment of livestock.

Mitigating heat stress via housing/shelter can be accomplished through three primary avenues: shade, ventilation, and evaporative cooling systems. **Shade** is the most cost-effective strategy available to abate the impact of heat. Shade structures that are 10–12 feet high to assure optimal air flow can lower the solar heat load on livestock by 30–50%. Animals with access to shade will have lower rectal temperatures, reduced respiration rates, and higher productivity. **Ventilation** should be designed to increase air flow and when combined with **evaporative cooling systems** such as sprinklers or foggers/misters. Proper air flow and wetting systems can dramatically improve rate of gain, feed efficiency, milk production, and laying rates.

Adjusting Rations for Weather Changes

In addition to housing considerations, a second management approach for dealing with both cold and heat is to consider the impacts of temperature extremes on performance and consider making changes in their feeding programs if economics so dictate.

Table 20.3

ADJUSTMENT FACTORS TO AIR TEMPERATURE TO DETERMINE EFFECTIVE TEMPERATURE FOR SWINE

Factor	Temperature Adjustment (°F)
Type of flooring	
Mat	+5
Dry concrete	–9
Wet concrete	–18
Plastic-coated wire	–7
Uncoated wire	–9
Air speed	
10 yd/minute	–7
30 yd/minute	–13
100 yd/minute	–18
Evaporative cooling	
Fogger/mist system	–6
Drip coolers	–10
Sprinkler system	–10

Source : Adapted from Bell, 1996.

Table 20.4
ESTIMATED LOWER CRITICAL TEMPERATURES FOR BEEF CATTLE

Coat Description	Critical Temperature (°F)
Summer coat or wet	59
Fall coat	45
Winter coat	32
Heavy winter coat	18

Coldness of a specific environment is the value that must be considered when adjusting rations for cows. **Coldness** is simply the difference between effective temperature (**windchill**) and lower critical temperature. Using this definition for coldness, instead of using temperature on an ordinary thermometer, helps explain why wet, windy days in March might be colder for a cow than extremely cold but dry, calm days in January (Table 20.4).

The major effect of cold on the nutrient requirements of cows is an increased need for energy, which usually means increasing the total amount of daily feed. Feeding tables recommend that a 1,200-lb cow receive 16.5 lb of good mixed hay to supply energy needs during the last one-third of pregnancy. How much feed should the cow receive if she is dry and has a winter hair coat but the temperature is 20°F with a 15-mph wind? The coldness is calculated by subtracting the windchill or effective temperature (4°F) from the cow's lower critical temperature (32°F). Thus the magnitude of coldness is 28°F. A rule of thumb (more detailed tables are available) is to increase the amount of feed 1% for each degree of coldness. A 28% increase in the original requirement of 16.5 lb would mean that 21.1 lb of feed must be fed to compensate for the coldness. However, if the cow is wet, the same increase in feed would be required at 31°F windchill, which is also 28°F of coldness. This illustrates the importance of effective temperature when assessing animal environment.

Animals typically increase their consumption of water as temperatures increase. If the water consumed is cooler than the temperature of the animal, it can help considerably to cool the body. Cattle alter their water intake throughout the year in accordance with seasonal changes in temperature, doubling intake in the hottest months as compared to the winter months. Hens may consume up to five times more water than feed when temperatures approach 90°F. Providing adequate water in both intensive and extensive management systems is vitally important.

Inability of Animals to Cope with Climatic Stress

Some animals are unable to adjust to environmental stressors. When the temperature becomes excessively high, some animals (notably pigs) might lose control of their senses and do things that aggravate the situation. If nothing is done to prevent them from doing so, pigs that get too hot will often run up and down a fence line, squealing, until they collapse and die. Pigs seen doing this should be cooled with water and encouraged to lie down on damp soil.

Most dairy cows will experience a reduction in milk production when the environmental temperature exceeds 80°F. This occurs primarily because appetite is depressed and feed consumption declines. Milk yields will improve when providing shade, fans, or refrigerated air to alleviate thermal stress.

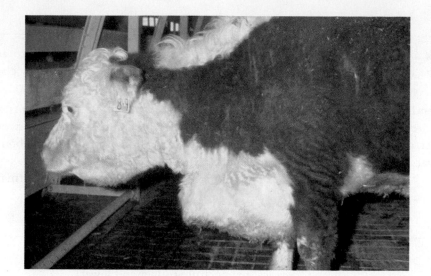

Figure 20.9
A classic case of high altitude or brisket disease. Note the swelling in the dewlap and brisket.
Source: Tom Field.

Equines may experience a variety of ailments due to winter conditions but effective management shifts in accordance with climatic changes can mitigate these challenges. Common problems of horses in winter include colic (intestinal impaction), muscle injury, and respiratory ailments. Horse owners can minimize these problems by assuring water availability and appropriate intake, avoiding overfeeding of grain, maintaining a daily exercise program with thorough warm-up and cool-down periods, matching the ration to the activity level, and keeping barns and stables well ventilated, clean, and dry.

An example of poor adaptation to the environment occurs when cattle are grazed at high elevations (6,000 ft and above), 1–5% will develop high mountain (brisket) disease (Fig. 20.9). It is characterized as right-heart failure (deterioration of the right side of the heart). As the animal becomes more affected, marked subcutaneous edema (filling of tissue with fluid) develops in the brisket area. It was considered that this disease was caused by chronic **hypoxia** (lack of oxygen), which leads to pulmonary **hypertension** (high blood pressure).

Studies have shown that some breeds are more prone to this disease than others and that there is sufficient heritability of the disease that selection can be utilized to reduce the incidence. Selection strategies are based on the measurement of **pulmonary arterial pressure (PAP)**. Data collection should occur at elevations greater than 5,000 ft. Yearling cattle with PAP scores of <41 mmHg are considered to be low risk, 41–49 mmHg moderate risk, and >49 mmHg to be at high risk of experiencing high mountain disease.

However, there are nongenetic factors that may contribute to the expression of brisket disease. Cattle that experience calfhood pneumonia or other respiratory stress are more likely to experience brisket disease. Furthermore, some recent studies at the U.S. Poisonous Plant Center at Logan, Utah, have shown that certain poisonous plants that grow only at high elevations can cause brisket disease when fed to cattle maintained at low elevations.

CHAPTER SUMMARY

- Extensively managed livestock are expected to be productive under a wide range of environmental conditions.
- Variations in environmental conditions are primarily due to changes in temperature, humidity, wind, daylight, altitude, feed, and water, and to exposure to parasites and disease-causing organisms.
- Producers may change animals genetically to adapt to different environmental conditions or may change the environment when economically feasible.
- The thermal neutral zone (TNZ) or comfort zone identifies a range of temperatures where heat production and heat loss from the animal's body are about the same. Temperatures above or below the TNZ significantly affect feed intake and the animal's performance.

KEY WORDS

intensive management
extensive management
genetically adapted
acclimatization
thyroxine
homeotherm
conduction
convection
evaporation
sweating
panting
radiation
effective ambient temperature

thermal neutral zone (TNZ)
critical temperature (upper and lower)
countercurrent blood flow action
heat stress
heat stress indices
shade
ventilation
evaporative cooling systems
coldness
windchill
hypoxia
hypertension
pulmonary arterial pressure (PAP)

REVIEW QUESTIONS

1. Compare and contrast intensive and extensive management.
2. Provide examples of genetic adaptation and of acclimatization.
3. Describe the impacts of changes in day length on livestock performance.
4. Describe the role of conduction, convection, evaporation, and radiation on homeotherms.
5. Contrast sweating and panting as cooling mechanisms.
6. How can managers use knowledge of effective ambient temperature to make better decisions?
7. How do wind speed and precipitation affect the TNZ?
8. Describe countercurrent blood flow.
9. Compare the impacts of heat stress and cold stress on livestock.
10. What is the importance of the heat stress index to livestock managers?
11. Describe management strategies to mitigate heat stress.
12. Describe ration adjustments in response to weather changes.
13. Describe dysfunctions that occur when livestock fail to adapt to environmental conditions.

SELECTED REFERENCES

Ames, D. R. 1980. Livestock nutrition in cold weather. *Anim. Nutr. Health*, October.

Bell, A. 1996. Air temperature may be deceiving. *Pork 96*, October.

Bonsma, J. 1983. *Livestock Production*. Cody, WY: Ag Books.

Gaughan, J. B., T. L. Mader, S. M. Holt and A. Lisle. 2008. A new heat load index for feedlot cattle. *J. Anim. Sci.* 86:226–234.

Hahn, G. L. 1985. Weather and climate impacts on beef cattle. *Beef Research Progress Report* no. 2, MARC, USDA, ARS-42.

McDowell, L. R. (ed.). 1985. *Nutrition of Grazing Ruminants in Warm Climates*. New York: Academic Press.

National Research Council. 1981. *Effect of Environment on Nutrient Requirements of Domestic Animals*. Washington, DC: National Academy Press.

Panda, A. K. 2011. Alleviate poultry heat stress through antioxidant vitamin supplementation. Innovations in Poultry Nutrition. WATTAgNet.com.

West, J. W. 2002. Effects of heat stress on production in dairy cattle. *J. Dairy Sci.* 86:2131–2144.

Yousef, M. K. 1985. *Stress Physiology in Livestock. Vol. II: Ungulates*. Boca Raton, FL: CRC Press.

21
Animal Health

Productive animals are typically in excellent health. Death loss (**mortality rate**) is the most dramatic sign of health problems; however, lower production levels and higher costs of production due to sickness (**morbidity**) are economically more serious. For example, feedlot cattle that are healthy may return as much as $100 per head profit as compared to cattle that are sick. The $100 profit is realized from reduced death loss, improved cost of gain, decreased medicine costs, and better carcass grading resulting from better health.

Disease is any deviation from normal health in which there are marked physiological, anatomical, or chemical changes in the animal's body. There are two major disease types: noninfectious and infectious. **Noninfectious diseases** result from injury, genetic abnormalities, ingestion of toxic materials, and poor nutrition. Microorganisms are not involved in noninfectious diseases. Examples of noninfectious diseases are plant poisoning, bloat, and mineral deficiencies.

Microorganisms such as bacteria, viruses, and protozoa cause **infectious diseases**. A **contagious disease** is an infectious disease that spreads rapidly from one animal to another. Trichomoniasis, ringworm, and transmissible gastroenteritis (TGE) are examples of infectious diseases.

IMMUNE FUNCTION

Understanding and managing the immune system are critical in assuring the health of livestock and poultry. **Immunity** is the process by which particles foreign to the body are identified and subsequently destroyed or metabolized. A highly functional immune system is required to resist infections or toxins. However, immunity is not a fixed state but varies based on age the animal, nutritional status, degree of exposure to organisms capable of initiating disease (**pathogens**), and a host of other stressors.

Immunity takes two forms: The first is **natural** or **native immunity**, which is present at birth under normal circumstances. For example, skin, secretions that coat the respiratory and intestinal tracts and the acidic environment of the stomach are protections against disease native to the animal. The second form is **acquired resistance** provided by the actions of specialized white blood cells called lymphocytes. These cells are produced by the spleen, lymph nodes, intestine, mammary glands, respiratory and reproductive tracts. Acquire immunity is activated when the body encounters foreign substances or **antigens**.

Lymphocytes take two forms: B cells and T cells. **B cells** secrete **antibodies** in response to specific antigens that are transferred via body fluids to provide **humoral immunity** such that free pathogens are recognized by the antibodies and neutralized. **T cells** provide intracellular protection by stimulating production of substances that directly attack an infected cell. **Cell-mediated immunity** is especially effective relative to virus infected cells, intracellular bacteria, and cancer. An example of a cell-mediated response is the production of **macrophages** that engulf foreign bodies (bacteria and viruses) as well as dead or damaged cells and then destroying them by enzymatic action.

Immune function develops in two phases: passive and active. Because newborns are not immediately capable of protecting themselves with their own antibody production, they receive **passive immunity** in the form of antibody-rich colostrum also known as first milk. Antibodies can only be absorbed intact from the dam's milk by the newborn's digestive system for a period of 12 to 36 hours post-birth. **Active immunity** is attained when the individual can initiate its own antibody production against specific invasive antigens.

Prevention

The old adage, "An ounce of prevention is worth a pound of cure," is an important part of herd health. Preventative herd health programs can eliminate or reduce most of the health-related livestock losses. Most major animal disease problems are associated with health management.

The components of a herd health-management program include (1) veterinarian-assisted planning, (2) sanitation, (3) proper nutrition, (4) record analysis, (5) physical facilities, (6) source of livestock, (7) proper use of biologics and pharmaceuticals, (8) minimization of stress, and (9) personnel training. All of these components require cooperative efforts between the producer and veterinarian.

The Role of the Veterinarian

Veterinarians are professionally educated and trained in animal health management. Practicing veterinarians are familiar with the disease and health problems common in their particular region. Producers who have the most successful herd health programs use veterinarians for planning and implementation of animal health-management systems. Important parts of the plan include regularly scheduled visits by the veterinarian throughout the year to assess preventative programs as well as to provide animal treatments. The veterinarian can also train farm and ranch personnel in simple herd health-management practices and serve as a reference for new products.

In most livestock and poultry operations, it is more economically feasible to have veterinarians assist in planning and implementing preventative health programs than to use their services in crisis situations only.

Producers should keep cost-effective health records as recommended by the veterinarian. These records are needed to assess problem areas and to develop a more effective herd health plan. The records can include such information as what was done and when in the preventative and treatment programs, descriptions of the conditions during health problems, and deaths. Possible causes of death also should be recorded, along with **necropsy** records when economically feasible.

Sanitation

The severity of some diseases is dependent on the number and virulence of microorganisms entering the animal's body. Many microorganisms live and multiply outside the animal, so implementing sanitation practices can reduce exposure to pathogens. These practices, in turn, reduce the incidence of disease outbreaks.

Manure and other organic waste materials are ideal environments for the proliferation of microorganisms. A good sanitation program includes cleaning of organic materials from buildings, pens, and lots. This allows the effective destruction of microorganisms via high temperatures and drying. Buildings, pens, and pastures should be well drained, preventing prolonged wet areas or mud holes. These sanitation practices help both in disease prevention and in controlling parasites.

Antiseptics and **disinfectants** are carefully selected and effectively utilized in a good sanitation program. Antiseptics are substances, usually applied to animal tissue, that kill or prevent the growth of microorganisms. Disinfectants are products that destroy pathogenic microorganisms. They are usually agents used on inanimate objects. In the absence of disinfectants, sanitizing with clean water may be helpful.

Other important sanitation measures include the prompt and proper disposal of dead animals, either to rendering plants or by burial.

Sound Nutritional Management

Well-nourished animals receive an adequate daily supply of essential nutrients. Undernourished animals usually have weak immune systems, thus making them more vulnerable to invading microorganisms.

Nutrition can be particularly critical during times of unusual stress, such as at weaning. Helping weaned calves, lambs, or pigs transition to the next production stage is of critical importance. Nutritional management of newly weaned animals can be improved with strategies such as supplementing with vitamins B and E, as well as trace minerals such as zinc.

Record Keeping

Proper records permit identification of health problems and determination of the cause, thereby allowing alternative methods of prevention and treatment to be assessed. Records provide an analytical approach to health management and provide both historical context but also a means to utilize trend analysis to better assist livestock managers and their veterinarians in making decisions. Recording health and disease information requires a commitment to record all disease incidences, their duration and treatment; to develop consistent protocols and forms as the basis for valid benchmarks; and to conduct data analysis to yield meaningful information.

Examples of records that should be maintained include but are not limited to the following: mortality rates, morbidity rates, vaccination records, treatment records, animal health product inventory, and employee training. Whenever possible, maintaining individual animal treatment records should be maintained while preventative measures such as vaccination records can be effectively documented in groups of animals especially when supported by an effective animal identification system. Treatment records should include animal id, date of treatment, product administered, withdrawal dates (assure compound clearance prior to marketing animals or products from them), dose, route of administration, and site of administration.

Facilities

Physical facilities contribute to animal health problems by causing physical injury or stress, or by allowing dissemination of pathogens through a group of animals. They can contribute to the spread of disease by not preventing its transmission (e.g., venereal disease transmission owing to poor or inadequate fences). Even proper facilities can enhance transmission of disease if they are not managed properly. For example, well designed housing systems that are poorly maintained, not cleaned regularly, or are overcrowded with subject animals to stressful conditions.

Source of Livestock

Producers can reduce the spread of infectious diseases into their herds and flocks by (1) purchasing animals (entering their herds) from other producers who have effective herd health-management programs; (2) controlling exposure of their animals to people and vehicles; (3) providing clothing, boots, and disinfectant to people who must be exposed to the animals and facilities; (4) isolating animals to be introduced into their herds so that disease symptoms can be observed for several weeks; (5) controlling insects, birds, rodents, and other animals that can carry disease organisms; and (6) keeping their animals out of drainage areas that run through their farm from other farms.

Biosecurity

The development of a detailed **biosecurity** plan is particularly critical where animals are intensively managed or where an enterprise is dependent on a marketing contract that specifies freedom from pathogens. The swine and poultry industries tend to have more sophisticated biosecurity measures than other livestock industries.

Biosecurity is of increasing importance as world trade in livestock and meat rises in both frequency and value. Furthermore, the threat of bioterrorism against the food supply must be considered a possibility. As a result, management systems designed to stop the spread of infectious diseases by minimizing the movement of biological organisms within and between groups of animals should be developed.

The components of an effective biosecurity system include verification of sources of animals in regards to herd health management, serological testing, treatment and vaccination protocols, monitoring incidence of disease, and record keeping. If outside sources of animals are utilized, a biosecurity plan also specifies design of isolation facilities, duration of isolation, protocols for employees working in the isolation units to prevent contamination of other farm sites, serological testing, and additional elements of a critical control point management system.

The biosecurity plan will also detail assuring integrity of farm perimeters, managing farm employees and visitors, pest management, quality assurance of feed and water supplies, managing outside trucks used to transport animals to and from the farm, disposal of dead and chronically sick animals, and manure disposal. A more detailed list of biosecurity strategies is outlined in Table 21.1.

More specific details can be obtained from the farm veterinarian, the extension service, and state and national livestock associations.

Use of Biologicals and Pharmaceuticals

Drugs are classified as biologicals or pharmaceuticals. **Biologicals** are used primarily to prevent diseases, whereas **pharmaceuticals** are used mainly to treat diseases. Both are needed in a successful herd health-management program.

Most biologicals are used to stimulate immunity against specific diseases. Vaccines are biological agents that stimulate active immunity in the animal.

Biologicals stimulate the body's immune system to produce antibodies that fight diseases. This is similar to the immunity attained through natural infection. An animal is immunized when sufficient antibodies are produced to prevent the disease from developing. Specific antibodies must be produced for specific diseases; thus vaccination for several different diseases may be necessary. Also, periodic revaccination is often needed to maintain circulating antibodies at an adequate level.

Immunity via vaccination occurs only if the immune system responds properly. In some situations vaccination is not effective. In most animal species, for example, the immune systems of newborn young are not yet developed enough to respond. This is why their immunity is dependent on the maternal antibodies ingested from their mother's milk (colostrum). Also, undernourished or stressed animals, or those animals already exposed to a disease, rarely give a positive response to vaccination.

Vaccines can be used to stimulate both passive and active immunity. The dam can be vaccinated to assure provision of specific antibodies to her offspring via colostrum or individual animals can be vaccinated to stimulate immunity against specific

Table 21.1
COMPONENTS OF AN EFFECTIVE BIOSECURITY PLAN

Avoid Introduction of Diseased or Infected Animals
Select replacement stock from within the herd or flock
Introduce only young, virgin males
If replacements are purchased, buy only from known sources
Test for specific diseases of concern (recognize that tests may be limited in effectiveness)
Quarantine new animals for 30 days or more

Increase Specific Disease Resistance
Implement an appropriate and effective vaccination program
Store and use the vaccine according to label guidelines
Follow booster schedules precisely

Increase Overall Disease Resistance
Minimize dystocia
Provide balanced diet in adequate volume to meet animal requirements
Assure appropriate trace mineral supplementation
Control internal and external parasites
Minimize animal stress during handling and transport

Minimize Exposure to Disease Agents
Isolate sick animals
Properly dispose of dead animals
Change needles appropriate to risk
Maintain clean facilities
Assure sanitation and proper waste management
Properly control rodents, birds, and insects
Assure freshwater supply
Store feed properly
Test and monitor feedstuffs
Maintain effective record-keeping systems
Avoid outside vehicle traffic through livestock areas
Provide protective clothing and foot covering for visitors

Source: Adapted from Smith, 2002.

diseases. Vaccines take one of three forms: killed, modified live (MLV), or chemically attenuated (altered). In the case of killed vaccines, dead organisms or their parts are used to stimulate humoral immunity and are especially effective in stimulating antibody production by the dam. The advantages of killed vaccines is that they rarely cause abortion, they cannot replicate and thus cannot induce disease, and have a stable shelflife. Disadvantages of killed vaccines are cost (compared to MLV), slower and shorter duration immune response, and a requirement of a second vaccine administration to boost response.

Modified live vaccines contain an altered form of the antigen that allows replication without fully initiating the disease. MLV stimulate both cell-mediated and humoral immunity and achieve a more rapid and longer lasting immune response. Additional advantages include cost and a wider spectrum of response. Disadvantages are that MLV must be mixed to rehydrate them, once mixed the vaccine is sensitive to temperature changes, exposure to sunlight, and disinfectants, abortion may result, and disease expression may occur in immunosuppressed animals and it may mutate to the virulent form.

Attenuated or chemically altered vaccines are only effective when administered directly to the mucosal membranes. However, they have the advantage of combining the safety of killed vaccines with the enhance immune response of MLV.

The amount of vaccine, frequency, route of administration, and duration of immunity vary with the specific vaccine. The manufacturer's directions should be followed carefully.

Vaccines (biologicals) and drugs (pharmaceuticals) can be administered in the following manner:

• Topical: applied to the skin (Fig. 21.1).
• Oral: administered through the mouth by feeding, drenching, or using a balling gun. The latter are used to deliver pills such as capsules or boluses.
• Injection into body tissue via a needle and syringe
• Intranasal: administration via inspired air.

Types of injections include (1) subcutaneous (under the skin but not into the muscle) (Fig. 21.2), (2) intramuscular (directly into the muscle) (Fig. 21.3), (3) intravenous (into a vein), (4) intramammary (through a cannula in the teat canal and into the milk cistern), (5) intraperitoneal (into the peritoneal cavity), and (6) intrauterine (using an infusion pipette or a tube through the cervix).

Figure 21.1
Topical administration of a parasite treatment.
Source: Tom Field.

Figure 21.2
Subcutaneous injection in the area in front of the shoulder using the tented technique as approved for beef cattle. Source: Tom Field.

Figure 21.3
Selection of injection site is an important management decision in preventing blemishes to valuable cuts of meat. Source: National Pork Board.

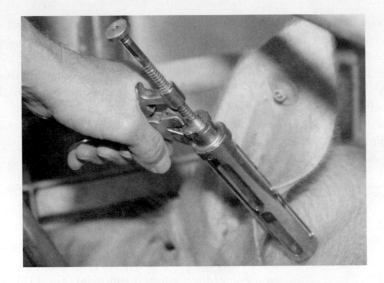

It is important to recognize that intramuscular injections in food animals should be minimized whenever possible as this route of administration may lead to injection-site blemishes. These blemishes can affect the tenderness and visual appeal of meat not only at the injection site but throughout the surrounding tissue. Significant educational efforts by organizations such as the National Cattlemen's Beef Association and National Pork Producers' Council have limited the use of intramuscular injections in the livestock industry.

Pharmaceuticals are used to kill or reduce the growth of microorganisms in the treatment of diseases and infections. Pharmaceuticals come in a variety of forms: liquids, powders, boluses, drenches, and feed additives. They should be administered only after a specific diagnosis has been made. Examples of pharmaceuticals are antibiotics, steroids, sulfonamide compounds, and hormones.

Other pharmaceuticals are used to control external and internal parasites. Losses owing to external and internal parasites are usually in the form of reduced weight gains, poor feed conversion, lower milk and egg production, reduced hide value, and excessive carcass trim, rather than high death losses.

External parasites of concern include flies, lice, mites, and ticks, which live off the flesh and blood of animals. Parasites such as ticks use biting action while mites burrow through the skin or hide of the animal. These parasites can transmit infectious diseases from one animal to another. Insecticides are used to control external parasites. They can be applied as systemics or by spraying, fogging, dipping, or using back rubbers, ear tags, or dust bags.

Internal parasites are classed as helminths. Roundworms and tapeworms are the most common internal parasites, with flukes causing problems in certain environments. Internal parasites are controlled by interrupting their life cycle by (1) the presence of unfavorable climatic conditions (wet, warm weather favors the proliferation of many internal parasites); (2) destroying intermediate hosts; (3) managing animals so they will not ingest the parasites or their eggs; and (4) giving therapeutic chemical treatment. **Anthelmintics** are drugs that are given to kill the internal parasites.

Stress

Stress is any environmental factor that can cause a significant change in the animal's physiological processes. Physical sources of stress are temperature, wind velocity, nutritional deficiency or oversupply, mud, snow, dust, fatigue, weaning, ammonia buildup, transportation, castration, dehorning, and abusive handling. Social or behavior-related stress can result from aggression or overcrowding.

Prolonged stress can impair the body's immune system, causing a reduced resistance to disease. Stressful conditions in animals should be minimized within economic constraints.

Personnel Training

Maintaining good animal health requires continuous training of personnel. This is one of the most difficult areas to accomplish, particularly in a livestock operation that employs many people. While the owner understands what needs to be done, this may differ from what the employees actually accomplish. Cooperative training programs and informative sessions with the veterinarian can assist in the implementation of effective herd health-management programs. Training is the basis for the many quality assurance programs being implemented in the livestock and poultry industries.

DETECTING UNHEALTHY ANIMALS

Visual Observation

Detecting sick animals and separating them from the healthy animals is an important key to a successful treatment program. Animals treated in the early stages of sickness usually respond more favorably to treatment than do animals whose illness has progressed to advanced stages.

Early signs of sickness are not easily detected; however, the following are some of the observable signs, several of which are shown in Figure 21.4:

1. Loss of appetite is observed.
2. Animal appears listless and depressed.
3. Ears may be droopy or not held in an alert position.
4. Animal has a hump in its back and holds its head in a lower position.
5. Animal stays separated from the rest of the herd or flock.
6. Coughing, wheezing, or labored breathing occurs.
7. Movement appears stiff and labored.

Vital Signs

Body temperature, respiration rate, and heart rate are commonly observed **vital signs** of the animal. Additional vital signs include gut sounds, ease of respiration, capillary refill rates, and coloration of sclera (eye). Health problems are usually evident when

Figure 21.4
This calf is sick with scours. Calves suffering from this condition typically appear listless and dehydrated.
Source: Tom Field.

Table 21.2
VITAL FUNCTIONS OF SELECTED LIVESTOCK AND POULTRY SPECIES

Animal	Rectal Temperature (± 1°F)	Respiration Rate[a] (per min)	Heart Rate[b] (per min)
Cattle	101.5	30[a]	60–70
Swine	102.0	16	60–85
Sheep	103.0	19	60–120
Goat	102.0	15	70–135
Horse	100.5	12	25–70
Poultry	107.1	12–36[b]	250–300

[a]Average.
[b]Normal range.

one or more of the vital signs deviate from the normal range. The average and normal ranges for the vital functions are given in Table 21.2.

Body temperature is taken rectally. However, thermographic scanning technologies are being evaluated for possible use in the livestock industry. An elevated body temperature occurs in overheated animals or with most infectious diseases. A subnormal temperature indicates chilling or a critical condition of the animal.

Evaluating the animals' vital signs and visually observing their appearance can identify health problems identified in their early stages. This allows for early treatment, which usually prevents serious losses.

MAJOR DISEASES OF FARM ANIMALS

Table 21.3 identifies some of the major diseases and health problems of beef cattle, dairy cattle, swine, horses, poultry, and sheep.

Table 21.3
SELECTED MAJOR DISEASES AND HEALTH PROBLEMS OF BEEF CATTLE, DAIRY CATTLE, SWINE, HORSES, POULTRY, AND SHEEP

Species/Disease/Cause	Signs	Prevention	Treatment
Beef Cattle			
Bovine viral diarrhea (BVD) Virus. Laboratory diagnosis imperative for accuracy	Feeder cattle: ulcerations throughout digestive tract; diarrhea (often containing mucus or blood)	Vaccination prior to exposure; avoid contact with infected animals	Symptomatic treatment; antibiotics; sulfonamides; force feed
	Breeding cattle: abortions, repeat breeding	Vaccination annually 30 days prior to breeding	None; symptomatic treatment
Bovine respiratory syncytial virus (BRSV)	Labored breathing; pneumonia	Vaccination	Antihistamines, corticosteroids

Table 21.3 *continued*

Species/Disease/Cause	Signs	Prevention	Treatment
Brucellosis Bacteria	Abortions	Calfhood vaccination (in some states) between the ages of 4 and 12 months	Test and slaughter; report reactors to state veterinarian
Campylobacteriosis (vibriosis) Bacteria	Repeat breeding; abortions (1–2%)	Vaccination of females and bulls prior to breeding; use of artificial insemination; virgin bulls on virgin heifers; avoid sexual contact with infected animals; cull open cows in infected herds	None (consult herd veterinarian)
Infectious bovine rhinotracheitis (IBR) Virus. Laboratory diagnosis imperative for accuracy	Pneumonia; fever, vaginitis, abortion and infertility in females; preputial infections in males	Vaccinate cows 40 days prior to breeding; vaccinate feeder cattle prior to exposure; semen from reputable bulls	Oxytetracycline, penicillin to minimize bacterial infections
Leptospirosis *Leptospira* spp. Several bacteria	Breeding cattle: fever, off feed, abortions, icterus, discolored urine	Vaccination at least annually (in high-risk areas, more frequently); control of rodents, proper water management; avoid contact with wildlife and other infected animals	Penicillin; dihydrostreptomycin
Scours (diarrhea), calf colibacillosis "septicemia" *Escherichia coli K99* (pathogenic bacteria); several viruses plus stress factors	Diarrhea, weakness, and dehydration; rough hair and coat	Dry, clean calving areas; adequate colostrum	Fluid therapy; antibiotics to prevent secondary infections
Scours—calf diarrhea (viral) Virus (reovirus and corona virus)	Acute diarrhea, high mortality; affects calves shortly after birth (see Fig. 21.5)	Precalving dam vaccination with specific viral vaccines	Fluid therapy
Trichomoniasis Protozoa	Infertility and abortion (2–4 mos.); pyometra	Use of artificial insemination; virgin bulls on virgin heifers; maintain closed herd	Cull carrier animals
Dairy Cattle			
Mastitis Primarily infectious bacteria (*Streptococci*, *Staphylococci*, or *E. coli*)	Inflammation of the udder; decreased milk production	Use California mastitis test (CMT) or somatic cell counts (SCC)—checks for white blood cells in milk; avoid injury to udder; use proper milking techniques	Determine specific organism causing the problem, then select most effective antibiotic; frequent milking; anti-inflammatories
Milk fever (parturient paresis) Metabolic disorder (low blood calcium level associated with a deficiency of vitamin D and phosphorus)	Occurs at onset of lactation; muscular weakness; drowsiness; cow lies down in curled position	Proper nutrition and management of cows during dry period	Intravenous injection of calcium borogluconate
Respiratory diseases (See *BVD, BRSV, and IBR* under **Beef Cattle** heading)			

Table 21.3 *continued*

Species/Disease/Cause	Signs	Prevention	Treatment
Uterine infections Usually results from bacterial contamination at the time of calving	Infections classified according to tissues involved; animal may be systemically or only locally affected; endometritis; inflammation of lining of uterus; metritis: infection of all tissues of uterus; pyometra: pus in the uterus	Sanitation; cleanliness when giving calving assistance	Early detection; antibiotics, hormone therapy, prostaglandins
Venereal diseases (See *BVD, IBR, leptospirosis, vibriosis and trichomoniasis,* under **Beef Cattle**)			
Swine Atrophic rhinitis *Bordetella bronchiseptica* (bacteria); *Pasteurella* (secondary invader)	Sneezing (most common); sniffling; snorting; coughing; twisting of snout; nasal infection (inflammation of membranes in nose). Diagnosis confirmed by observing turbinate bone atrophy (nasal) during postmortem examination	Monitor performance; monitor contact with animals from outside the herd; correct environmental deficiencies (sanitation, temperature, humidity, ventilation, dust, drafts, excessive ammonia, and overcrowding)	Vaccinate against *Bordetella* and *Pasteurella* organisms; improve environment and facilities; medicate sow feed with sulfamethazine or oxytetracycline
Colibacillosis *E. coli* Bacteria	Diarrhea (pale, watery feces); weakness; depression; most serious in pigs under 7 days of age	Good sanitation; good management practices (nutrition, pigs suckle soon after birth, prevent chilling); vaccinate sows to increase protective value of colostrum	Effective treatment is limited; identify strain of *E. coli*; consult with herd veterinarian
Mycoplasmal infections (Pneumonia and arthritis) Mycoplasma bacteria	Pneumonia: death loss low; dry cough; reduced growth rate; lesions in lungs Arthritis: inflammation in lining of chest and abdominal cavity; lameness; swollen joints; sudden death	Pneumonia: reduce animal contact; good nutrition, warm and dust-free environment, and parasite (ascarid and lung worm) control minimizes effects of the disease	Pneumonia: adequate treatment not available; sulfas and antibiotics prevent secondary pneumonia infections Arthritis: no satisfactory treatment; can depopulate and restock with disease-free animals
Porcine Epidemic Diarrhea Virus (PEDV) Coronavirus	Baby pigs (less than 3 weeks old): up to 80% mortality; fever, vomiting, diarrhea, dehydration in sows, nursery and finishing pigs with much lower morbidity	Sanitation is a critical control point as the disease is spread by pigs ingesting contaminated feces. Cross contamination due to factors such as boots or overalls must be avoided	No treatment or effective vaccine available, emphasis on prevention through superior management and biosecurity and quarantine of infected pigs
Transmissible gastroenteritis (TGE) Virus of coronavirus group	Vomiting; diarrhea; weakness; dehydration; high death rate in pigs under 32 weeks of age	Sanitation is most cost effective; prevent transfer of virus from infected animals through exposure to other pigs, birds, equipment, and people	No drugs are effective; freshwater and a draft-free environment will reduce losses; antibiotics may prevent some secondary infections

Table 21.3 *continued*

Species/Disease/Cause	Signs	Prevention	Treatment
Horses			
Colic Noninfectious: a general term indicating abdominal pain	Looking at flank; kicking at abdomen; pawing; getting up and down; rolling; sweating	Parasite control, avoiding moldy or spoiled feeds; prevent from overeating or drinking too much water when hot; have sharp points of teeth filed (floated) annually to assure proper chewing of feed	Quiet walking of horse; avoid undue stress; bran mash and aspirin; milk of magnesia; mild, soapy, warm water enema for impaction; always consult veterinarian as soon as possible
Lameness Most common are stone bruises, puncture wounds, sprains, and navicular disease	Departure from normal stance or gait; limping; head bobbing; dropping of the hip; alternate resting on front feet; pointing (extends one front foot in front of the other); stiffness; navicular disease: soft tissue bursitis in the foot	Bruise: avoid running horse on graveled roads or rocky terrain; puncture wounds: avoid areas where nails and other sharp objects may be present; sprains: avoid stress and strains on feet and legs; navicular disease: exact cause not known; avoid continual concussion on hard surfaces; provide proper shoeing and trimming	Varies with cause of lameness. Bruises: soaking foot in bucket of ice water or standing horse in mud; aspirin; puncture wound: tetanus protection; antibiotic; aspirin; soaking foot in hot Epsom salts; navicular disease: no known cure; drug therapy and corrective shoes for temporary relief; surgery
Respiratory Three key viral diseases: viral rhinopneumonitis; viral arteritis; influenza; one key bacterial disease: strangles	Nasal discharge; coughing; lung congestion; fever; rhinopneumonitis may cause abortion	Isolate infected animals; avoid undue stress; draft-free shelter; parasite control; avoid chilling; rhinopneumonitis, influenza, and strangles vaccines are available	Follow prevention guidelines; strangles: drain abscess
Parasites 150 different internal parasites with strongyles (bloodworms), ascarids (roundworms), bots, pinworms, and stomach worms most common; external parasites (most common are lice, ticks, and flies)	Unthrifty appearance ("pot-bellied," rough hair coat); weakness; poor growth; lice, ticks, and flies can be visually observed on the horse	Good sanitation practices (clean feeding and watering facilities; regular removal of manure); periodic rest-periods for pastures; avoid over-crowding of horses; avoid spreading fresh manure on pastures	Internal parasites: deworm horses twice a year (usually spring and fall); bot eggs can be shaved from the hairs of the legs; external parasites: application of insecticides
Poultry			
Avian influenza Virus	Drop in egg production; sneezing; coughing; in the severe systemic form, deep drowsiness and high mortality are common	Vaccine (but has short immunity); select eggs and poults from clean flocks	No effective drug available
Coccidiosis Coccidia (depends on type of coccidiosis—there are nine or more types. Some signs, prevention, and treatment of three more common types are identified)	Weight loss; unthriftiness; palor; blood in droppings; lesions in intestinal wall; highly transmissible to other species	Use coccidiostat (kills coccidia organism)	Sulfa drugs in drinking water

Table 21.3   *continued*

Species/Disease/Cause	Signs	Prevention	Treatment
Lymphoid leukosis Virus	Combs and wattles may be shriveled, pale, and scaly; enlarged, infected liver; lesions common in liver and kidneys	Sanitation; development of resistant strains through breeding methods	None
Marek's disease Herpesvirus	Can cause high mortality in pullet flocks; paralysis; death might occur without any clinical signs	Vaccination	None
Mycoplasma infections Mycoplasma organisms (there are several diseases caused by mycoplasma organisms; most important are chronic respiratory disease (CRD) and infectious synovitis)	CRD—difficult breathing; nasal discharge; rattling in windpipe; death loss may be high in turkey poults; swelling of face in turkeys; synovitis—swollen joints, tendons, tendon sheaths and footpads; ruffled feathers	For both CRD and synovitis: use chicks or poults from disease-free parent stock; sanitation (cleaning and disinfecting the premises)	Antibiotic in feed or drinking water
Newcastle disease	Gasping, coughing, hoarse chirping; paralysis; mortality may be high	Vaccination	None
Sheep			
Pseudotuberculosis (*Caseous Lymphadenitis*) Bacteria	Enlarged lymph nodes	Sanitation; reduce opportunity for wound infection at docking, castration, and shearing	Antibiotics
Enterotoxemia **(overeating disease)** Bacteria	Often sudden death occurs without warning; sick lambs might show nervous symptoms; e.g., head drawn back, convulsions	Vaccination (most effective in young lambs under 6 weeks of age nursing heavy-milking ewes and in weaned lambs on lush pasture or in feedlots)	None; antitoxins can be used but they are expensive and immunity is temporary (2–3 weeks)
Epididymitis Bacteria	Most common in western United States; swelling of epididymis; poor semen quality; low conception rates	Rigid culling and vaccination	Rigid culling
Footrot Bacteria	Lameness; interdigital skin is usually red and swollen	Remove from wet pastures or stubble pastures; vaccines in some cases	Disinfectants such as 5% formalin or 10% copper sulfate; antibiotics
Pneumonia Many types; caused by viruses, stress, bacteria, and parasites	Sudden death; nasal discharge; depression; high temperature	Reduce stress factors	Antibiotics; sulfonamides
Pregnancy toxemia Undernutrition in late pregnancy; stress associated with poor body condition	Listlessness; loss of appetite; unusual postures; progressive loss of reflexes; hypoglycemia; coma; death	Prevent obesity in early pregnancy; provide good nutrition during last 6 weeks of pregnancy; reduce environmental stresses	In early stages the administration of some glucogenic materials may reduce mortality; in advanced stages no treatment improves survival

QUALITY ASSURANCE

Quality assurance programs were originally designed by a variety of professional and agricultural organizations to help livestock producers improve management practices, record keeping, and personnel training by creating a series of best practices and approved protocols. These efforts were undertaken as a means to assure consumers that foods produced were wholesome and safe. Over time quality assurance efforts have added significant improvements via the inclusion of protocols related to animal care, employee certification and safety, and environmental stewardship (Fig. 21.5).

Effective quality assurance programs are designed to focus management on the following requirements:

1. A valid client–patient relationship with a licensed veterinarian.
2. Design and implantation of an effective herd health plan focused on disease prevention.
3. Responsible use of antimicrobials including antibiotics.
4. Correct storage and administration of animal health care products.
5. Proper feed storage, processing, and delivery.

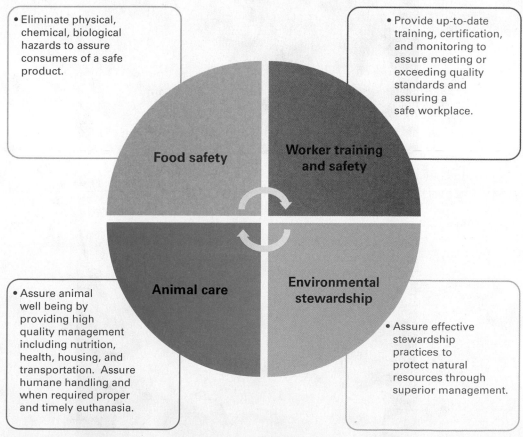

Figure 21.5

Quality assurance systems are founded on the four pillars of food safety, animal care, environmental stewardship, and employee safety.

6. Integrated records system founded on effective animal identification (Fig. 21.6) plus capture and analysis of essential data related to animal health and well-being.
7. Provide up-to-date employee training, certification, and accountability.
8. Provide humane animal handling and appropriate animal care.
9. Management strategies to improve environmental stewardship.
10. Create a culture of continuous improvement.

Improved record keeping is central to implementing a successful quality assurance program (Fig. 21.7). Records should be kept in regards to health product inventories, individual and group treatments, and processing maps (Fig. 21.8) that indicate product and route of administration, pesticide use, and other appropriate documentation. Specific quality assurance protocols can be obtained by accessing the websites provided at the end of the chapter.

Figure 21.6
Quality assurance systems require effective animal identification. Source: National Pork Board.

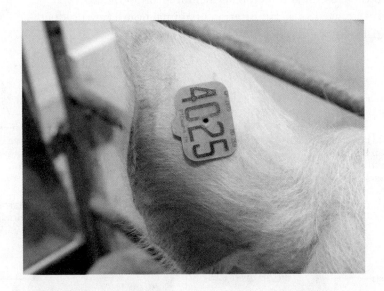

Figure 21.7
Individual animal treatment records are a critical component of a quality assurance system. Source: National Pork Board.

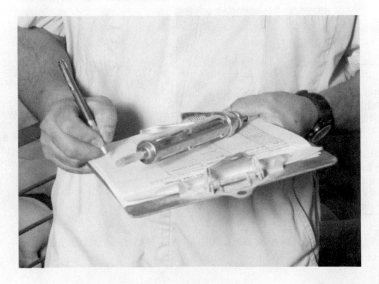

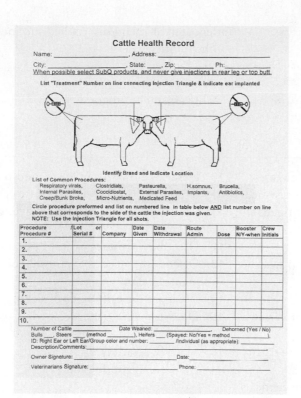

Figure 21.8
Processing records facilitate maintenance of useful information about the health care of groups of animals.
Source: Tom Field.

CHAPTER SUMMARY

- Disease is any deviation from normal health in which there are marked physiological, anatomical, or chemical changes in the animal's body.

- Death loss (mortality rate) and sickness (morbidity) rates are minimized in an excellent herd health-management program that includes (1) planning with a veterinarian, (2) sanitation, (3) proper nutrition, (4) record keeping and analysis, (5) physical facilities, (6) choosing source of livestock, (7) proper use of biologics and pharmaceuticals, (8) minimizing stress, and (9) training personnel who are working with the animals.

- Health problems are usually evident when one or more of the vital signs (temperature, respiration rate, and heart rate) deviate from the normal range.

- The detection, prevention, and treatment protocols are given for the major diseases and health problems for cattle, swine, horses, poultry, and sheep.

- Implementation of an effective quality assurance program is a key to increased profitability and improved consumer confidence in products from animal agriculture.

KEY WORDS

mortality rate
morbidity
disease
noninfectious diseases
infectious diseases
contagious disease
macrophages
immunity (natural, native, humoral, cell-mediated, passive, active)

pathogens
acquired resistance
antigens
lymphocytes
B cells
antibodies
T cells
necropsy
antiseptics

disinfectants
biosecurity
biologicals
pharmaceuticals
modified live vaccines

attenuated or chemically altered vaccines
anthelmintics
stress
vital signs
quality assurance programs

REVIEW QUESTIONS

1. Compare infectious versus noninfectious diseases.
2. Compare native immunity with acquired resistance.
3. Describe the role of lymphocytes and antigens in immune function.
4. Compare passive and active immunity.
5. What are the nine components of a herd health-management system.
6. Contrast disinfectants and antiseptics.
7. Why is record keeping important to herd health management?
8. Describe the components of an effective biosecurity program.
9. What are the roles of biologics and pharmaceuticals in animal health?
10. Describe how vaccines are used to affect immune response.
11. Compare killed, modified live, and attenuated vaccines.
12. Describe the routes of administration.
13. Discuss the need for parasite control in livestock management.
14. What are the early signs of sickness?
15. Describe the vital signs for various species of livestock.
16. Identify a variety of diseases that impact livestock and poultry production.

SELECTED REFERENCES

American Sheep Industry Association, USDA, and Colorado State University. 1995. *Producing High Quality Products from Sheep.* Denver, CO.

Battaglia, R. A. and V. B. Mayrose. 2002. *Handbook of Livestock Management Techniques.* Upper Saddle River, NJ: Prentice Hall. (esp. Chap. 10, "Animal Health Management").

Gaafar, S. M., W. E. Howard, and R. E. Marsh (eds.). 1985. *Parasites, Pests and Predators.* New York: Elsevier Science.

Galyean, M. L., G. C. Duff, and L. J. Perino. 1999. Interaction of cattle health/immunity and nutrition. *J. Anim. Sci.* 77:1120.

Griffin, D., S. Ensley, D. Smith, and G. Dewall. 2002. *Understanding Vaccines.* NebGuide G1455. Univ. of Nebraska Coop. Extension.

Jackson, N. S., W. J. Greer, and K. Baker. 2000. *Animal Health.* Danville, IL: Interstate.

Keeler, R. F., K. R. VanKampen, and L. F. James (eds.). 1978. *Effects of Poisonous Plants on Livestock.* New York: Academic Press.

Kirkbride, C. A. 1986. *Control of Livestock Diseases.* Springfield, IL: Charles C. Thomas.

Merck Veterinary Manual. 8th ed. 2002. Rahway, NJ: Merck & Co.

National Cattlemen's Beef Association. 2012. Beef Quality Assurance Program. Centennial, CO.

National Pork Producers' Council. 2013. *Pork Quality Assurance Program.* Des Moines, IA.

Naviaux, J. L. 1985. *Horses in Health and Disease.* Philadelphia, PA: Lea & Febiger.

Radostits, O. M. and D. C. Blood. 1985. *Herd Health.* Philadelphia, PA: W. B. Saunders.

Roeber, D. L., K. E. Belk, S. B. LeValley, J. S. Scanga, J. N. Sofos, and G. C. Smith. 2001. Producing consumer products from sheep—the sheep safety and quality assurance program. American Sheep Industry Association.

Smith, D. 2002. Biosecurity principles for livestock producers. Publication G1442. University of Nebraska-Lincoln, NC.

Stalheim, O. H. V. 1994. Animal diseases. *Encyclopedia of Agricultural Science.* San Diego, CA. Academic Press.

22

Animal Behavior

Knowledge of animal behavior is essential to understanding the whole animal and its ability to adapt to various management systems utilized by livestock and poultry producers. The value and performance of farm animals can be increased when managers apply their knowledge of animal behavior.

STOCKMANSHIP

The thoughtful and intentional management of livestock requires the heart, head, and eye of a stockman. Over the ages, those who have chosen to devote their lives to livestock production have developed and passed on a collective knowledge about the best ways to provide care, handle, restrain, transport, and harvest animals. The successful livestock manager understands animal behavior, can work in harmony with them, and is able not only to provide economic benefits as the fruit of their labor but also derive a deep satisfaction from developing the eye of the master, the hand of craftsman, and the heart of the shepherd as related to livestock care and handling.

In contemporary times, the most noted animal handling and facility design expert is Temple Grandin. Her life story is the subject of an acclaimed HBO movie, she has been featured in numerous national and international media stores, and her humane livestock handling systems have been implemented across the globe. She is also an articulate and effective advocate for autistic people. Temple's autism provided her a sensory perspective that allows her to think in pictures and thus to recognize patterns of behavior that lead her to design significantly more effective systems for handling animals during handling, restraint, transportation, and harvest. Dr. Grandin and many others such as Bud Williams, Curt Pate, Ron Gill, Tom Noffsinger, and Lynn Locatelli have forged a path that assures better management and aligns livestock managers with the deep values arising from a commitment to stockmanship and animal husbandry.

ANIMAL BEHAVIOR

Animal behavior is a complex process involving the interaction of inherited abilities and learned experiences to which the animal is subjected. Behavioral changes enable animals to adjust to changing conditions, improve their chances of survival, and serve humans. Producers who understand patterns of behavior can manage and handle animals more effectively and humanely.

learning objectives

- Describe the fields of animal behavior
- Discuss how an understanding of animal behavior is useful in managing:
  Reproduction
  Nutrition
  Handling and restraint
  Facility design and administration
- Describe the following behaviors:
  Caregiving
  Agonistic
  Eliminative
  Investigative
  Allelomimetic
  Maladaptive

Basically, there are two major fields of animal behavior: one is psychology and the other is **ethology**. Historically, psychology has been directed toward studying learning in humans and applying insights gained from nonhuman animal studies to understanding human behavior. Ethology, the study of animal behavior, however, originated with naturalists as early as Aristotle who originally emphasized instinctive behavior, but who also studied learned behavior in animals.

Instinct (reflexes and behavioral patterns) is inherently present at birth. All mammals, at birth, have the instinct to nurse even though they must first learn the location of the teat. Shortly after hatching, chicks begin pecking to obtain feed.

Habituation is lack of response to a repeated stimulus such as a low-flying aircraft. When animals first see and hear the airplane, the novelty of it may frighten them. However, after they are repeatedly exposed to this experience, they become habituated and are no longer frightened.

Conditioning is the process whereby an animal makes an association between a previously neutral stimulus (e.g., a bell) or behavioral response (e.g., lifting its foot) and a previously significant stimulus, such as a shock or food. There are two types of conditioning:

1. **Classical conditioning** —e.g., Pavlov's noted study showed an association formed between an unconditioned stimulus (the sight of food, which caused salivation) and a neutral stimulus (the sound of a bell). The animal initially salivated at the sight of food; later, the mere sound of the bell produced salivation because of the previous association between the two stimuli.
2. **Operant conditioning** is learning to respond in a particular way to a stimulus as a result of **reinforcement** when the proper response is made. Reinforcement is a punishment or reward for making the proper response. Animals avoiding an electric fence and cattle coming to the feed bunk when they see a feed truck are examples of operant conditioning. In the first example, the animal is negatively reinforced by the shock. In the second example, cattle are positively reinforced with feed from the feed truck when they arrive at the feed bunk.

Trial and error is trying different responses to a stimulus until the correct response is performed, at which time the animal receives a reward. For example, newborn mammals soon become hungry and want to nurse. They search for some place to nurse on any part of the mother's body until they find the teat. This is trial and error until the teat is located; then, when the young mammals nurse, they receive milk as a reward. Soon they learn where the teat is located and find it without having to go through trial and error. Thus, the young become conditioned in nursing behavior through reinforcement.

Reasoning is the ability to respond correctly to a stimulus the first time a new situation is presented. **Intelligence** is the ability to learn to adjust successfully to certain situations. Both short-term and long-term memory is part of intelligence.

Imprinting covers those processes where the helpless young bond to their caretaker—usually their dam. The way imprinting occurs varies between species. Odors and the dam licking the fluids from the newborn lead to bonding and rapid recognition in cattle and sheep. Creating a nest helps to assure that a sow bonds with her young pigs.

SYSTEMS OF ANIMAL BEHAVIOR

Farm animals exhibit several major systems or patterns of behavior: (1) sexual, (2) caregiving, (3) care-soliciting, (4) agonistic, (5) ingestive, (6) eliminative, (7) shelter-seeking, (8) investigative, and (9) allelomimetic. Some of these behavior systems are

Figure 22.1
A cow or heifer in estrus will allow either bulls or females to mount. Source: Tom Field.

interrelated, though they are discussed separately in this chapter. It is not the intent here to describe in detail the different behavior patterns for all farm animals. The major focus is on identifying the behavioral activities that most significantly affect animal well-being, productivity, and profitability. By understanding animal behavior, producers can plan and implement more effective management systems for their animals.

SEXUAL BEHAVIOR

Observations on sexual behavior of female farm animals are useful in implementing breeding programs. Cows that are in heat, for example, allow themselves to be mounted by others. Producers observe this condition of "standing heat" or estrus to identify those cows to be **hand-mated** or bred artificially. Other ewes do not mount ewes in heat, but vasectomized rams can identify them.

Males and females of certain species produce **pheromones**, chemical substances that attract the opposite sex. Cows, ewes, and mares may have pheromones present in vaginal secretions and in their urine when they are in heat. Bulls, rams, and stallions will smell the vagina and the urine using a nasal organ that can detect pheromones. A common behavioral response in this process is called **flehmen**, during which the male animal lifts its head and curls its upper lip.

It appears that in a sexually active group of cows, the bull is attracted to a cow in heat most often by visual means (observing cow-to-cow mounting) rather than by olfactory clues. The bull follows a cow that is coming into heat, smells and licks her external genitalia, and puts his chin on her rump. When the cow is in standing heat, she stands still when the bull chins her rump. When she reaches full heat, she allows the bull or other females to mount (Fig. 22.1).

When females are sexually receptive, they usually seek out a male if mating has not previously occurred. Females are receptive for varying lengths of time; cows are usually in heat for approximately 12 hours, whereas mares show heat for 5–7 days, with ovulation occurring during the last 24 hours of estrus.

Vigorous bulls breed females several times a day. If more than one cow is in heat at the same time, bulls tend to mate with one cow once or a few times and then go to others. Other bulls may become attached to one female and ignore the others that are also in heat.

The ram chases a ewe that is coming into heat. The ram champs and licks, puts his head on the side of the ewe, and strikes with his foot. When the ewe reaches standing heat, she stands when approached by the ram.

The buck goat snorts when he detects a doe in heat. The doe shows unrest and may be fought by other does. Mating in goats is similar to that in sheep.

The boar does not seem to detect a sow that is in heat by smelling or seeing. If introduced into a group of sows, a boar chases any sow in the group. The sow that is in heat seeks out the boar for mating, and when the boar is located she stands still and flicks her ears. Boars produce pheromones in the saliva and preputial pouch, which attracts sows and gilts in estrus to the boars. Ejaculation requires several minutes for boars, in contrast to the instantaneous ejaculation of rams and bulls.

The sequence of events in estrus detection in horses appears to be similar to that in swine. The stallion approaches a mare from the front and a mare not in heat runs and kicks at the stallion. When the mare is in standing heat she stands, squats somewhat, and urinates as he approaches. Her vulva exhibits **winking** (opens and closes) when she is in heat.

In chickens and turkeys, a courtship sequence between the male and female usually takes place. If either individual does not respond to the other's previous signal, the courtship does not proceed further. After the courtship has developed properly, some females run from the rooster, which chases them until they stop and squat for mating. The male chicken or turkey stands on a squatting female and ejaculates semen as his rear descends toward the female's cloaca. Semen is ejaculated at the cloaca and the female draws it into her reproductive tract while the male is mounting her.

Male chickens and turkeys show a preference for certain females and may even refuse to mate with other females. Likewise, female chickens and turkeys may refuse to mate with certain males. This is a serious problem when pen matings of one male and 10–15 females are practiced, as the eggs of some females may be infertile. This preferential mating is a greater problem in chickens, as AI is the common breeding practice in turkeys.

Little relationship appears to exist between sex drive and fertility in male farm animals. In fact, some males that show extreme sex drive have reduced fertility because of frequent ejaculations that result in semen with reduced sperm numbers.

Research studies show that many individual bulls have sufficient sex drive and mating ability to fertilize more females than are commonly allotted to them. An excessive number of males, however, are used in multiple-sire herds to offset the few that are poor breeders and to cover for the social dominance that exists among several bulls running with the same herd of cows. The bull may guard a female that he has determined is approaching estrus. His success in guarding the female or actually mating with her is dependent on his rank of dominance in a multiple-sire herd. If low fertility exists in the dominant bull or bulls, then calf crop percentages will be seriously affected even in multiple-sire herds.

Tests have been developed to measure differences in libido and mating ability in young bulls. While behavioral differences are evident between different bulls, these tests, based on pregnancy rates, have not proven accurate for use by the beef industry. Performance may be improved in young bulls by exposing them to a female in heat prior to being placed in the breeding pasture.

Bulls being raised with other bulls commonly mount one another, have a penile erection, and occasionally ejaculate. Individual bulls can be observed arching their backs, thrusting their penises toward their front legs, and ejaculating.

Bulls can be easily trained to mount objects that provide the stimulus for them to experience ejaculation. AI studs commonly use restrained steers for collection of semen. Bulls soon respond to the artificial vagina when mounting steers, which provides them with a sensual reward.

Mating behavior has an apparent genetic base, as there is evidence of more frequent mountings in hybrid or crossbred animals.

Some profound behavior patterns are associated with sex of the animal and changes resulting from castration. This verifies the importance of hormonal-directed expression of behavior. Intact males display more aggressive behavior, whereas castrates are more docile after losing their source of male hormone.

CAREGIVING BEHAVIOR

Caregiving behavior can originate from the sire or dam; however, most caregiving is maternally oriented.

When the young of cattle, sheep, goats, and horses are born, the mothers clean the young by licking them. This stimulates blood circulation and encourages the young to stand and to nurse. Sows do not clean their newborns but encourage them to nurse by lying down and moving their feet as the young approach the udder region. They thus help the young to the teats.

Most animal mothers tend to fight intruders, especially if the young squeal or bawl. Often cows, sows, and mares become aggressive in protecting their young shortly after parturition. Serious injury can occur to producers who do not use caution with these animals.

Strong attachments exist between mother and newborn young, particularly between ewe and lamb and cow and calf. Beef cows diminish their output of milk about 100–120 days after birth of young, and ewes do the same after 60–75 days. This reduction in milk encourages the young to search for forage, the consumption of which stimulates rumen development. It is at this time that caregiving by the mother declines.

If young pigs have a high-energy feed available at all times, they nurse less frequently. Without a strong stimulus of nursing, sows reduce their output of milk. Some sows may wean their pigs early and show little concern for them a few days after they are weaned. Producers usually wean pigs at 21–35 days of age.

There is evidence that more cows calve during periods of darkness than during daylight hours. The calving pattern, however, can be changed by when cows are fed. Cows that are fed during late evening have a higher percentage of their calves during daylight hours.

CARE-SOLICITING BEHAVIOR

Care-soliciting behavior is manifested when young animals cry for help when disturbed, distressed, or hungry. Lambs bleat, calves bawl, pigs squeal, and chicks chirp. Even adult animals call for help when under stress. The female and her offspring may recognize each other's vocal sounds; however, it appears that the most effective way the dam recognizes her offspring is by smell. The offspring usually nurses with its rear end toward the female's head. This allows a dam to smell her offspring and decide to accept or reject it. The dam will bunt a rejected young animal with her head and kick it with her rear legs when it attempts to nurse. The young animals are less discriminate in their nursing behavior than are their dams.

AGONISTIC BEHAVIOR

Agonistic behavior includes behavior activities of fight and flight and those of aggressive and passive behavior when an animal is in contact (physically) with another animal or with livestock and poultry producers.

Interaction with Other Animals

Unless castrated when young, the males of all farm animals fight when they meet other unfamiliar males of the same species. This behavior has great practical implications for management of farm animals. Male farm animals are often run singly with a group of females in the breeding season, but it is often necessary to keep males together in a group at times other than the breeding season. The typical producer simply cannot afford to provide a separate lot for each male.

Bulls and other males may engage in prolonged physical activity when fighting, even to the point of exhaustion. Therefore, bulls and other potential fighting males should be put together either early in the morning or late in the evening when the environmental temperature is lower than at midday. If possible mix strange animals in a new pen to reduce the intensity of fighting. A resident animal may fight more intensely to defend its territory. Fighting to establish social dominance essentially moves through four stages: offense, defense, escape, and passivity. Mixing unfamiliar males often results in fighting behavior that is concluded once the "defeated" animal has escaped and assumed a passive posture. It is important, therefore, to assure that there is sufficient room to allow less dominant animals to escape from the more dominant individuals.

Cows, sows, and mares usually develop a pecking order, but fight less intensely than males. Sows that are strangers to each other sometimes fight. Ewes seldom, if ever, fight, so ewes that are strangers can be grouped together without harm. Young animals should be raised in a social group with their own species so they can learn proper social behavior with their own kind. Young intact males that are reared in isolation away from their own kind are more likely to attack people or be overly aggressive toward other animals.

Some cows withdraw from the group to find a secluded spot just before calving. Almost all animals withdraw from the group if they are sick.

Early and continuous association of calves is associated with greater social tolerance, delayed onset of aggressive behavior, and relatively slow formation of social hierarchies.

Status and social rank typically exist in a herd of cows, with certain individuals dominating the more submissive ones. The presence or absence of horns is important in determining social rank, especially when strange cows are mixed together. Also, horned cows usually outrank polled or dehorned cows where close contact is encountered, such as at feed bunks or on the feed ground.

Large differences in age, size, strength, genetic background, and previous experience have powerful effects in determining social rank. Once the rank is established in a herd, it tends to be consistent from one year to the next. There is evidence that genetic differences exist for social rank.

Animals fed together consume more feed than if they are fed individually. A competitive environment evidently is a stimulus for greater feed consumption. Dairy calves separated from their dams at birth appear to gain equally well whether fed milk in a group or kept separate. There is, however, evidence that they learn to eat grain earlier when group-fed compared with being individually fed. Cattle individually fed in metabolism stalls consume only 50–60% of the amount of feed they will eat if the animals are group-fed.

Table 22.1

INTERPRETATION OF EQUINE MOOD VIA ASSESSMENT OF POSTURE

Mood	Head-Neck	Ears	Eyes	Nostrils-Muzzle	Feet-Legs	Tail
			Physical Signs			
Anger	Neck outstretched	Pinned back	Narrowed	Lips pursed, nostrils flared	Stomping, stroking	Swishing
Challenge	Head and neck stretched upward and outward	Active movement	Clearly focused	Nostrils flared	Active pacing	Held high
Curiosity	Head and neck extended toward subject of interest	Pricked forward	Focused on subject	Sniffing	Firmly planted	Held up
Fear	Neck pulled in with head turned toward source	Fixed on source	Wide open	Nostrils flared	Rigid or fleeing	Clamped down
Relaxation	Head down	Drooped	Nearly closed	Lips drooped	Inactive	Low and relaxed
Sexual arousal	Arched and flexed	Forward and flicking	Dilated	Active	Pawing	Held high
Submission	Head low and averted	Held low	Averted	Low, lip smacking in young	—	—

Source: Adapted from multiple sources.

When fed in a group of older cows, 2-year-old heifers have difficulty getting their share of supplemental feed. Two-year-old heifers and 3-year-old cows can be fed together without any significant age effect in competition for supplemental feed. These behavior differences no doubt explain some of the age-related differences in nutrition, weight gains, and postpartum intervals that exist when cows of all ages compete for the same supplemental feed.

Dominant cows raised in a confinement operation usually consume more feed and wean heavier calves. More submissive cows wean lighter calves (25%), and fewer of them are pregnant compared with more aggressive cows.

Interactions with Humans

Producers rank the disposition or temperament of animals from docile to wild or "high-strung." Evaluating animal posture can help handlers understand the mood and intent of an animal (Table 22.1). This evaluation is usually made when the animals are being handled through various types of corrals, pens, chutes, and other working facilities. The typical behavior exhibited by animals with poor dispositions is one of fear or of aggressive fighting or kicking.

There is evidence that farm animals develop good or poor dispositions from the way they have been treated or handled (Fig. 22.2), though there is also evidence that disposition has an inherited basis as well. A few heritability estimates for disposition are in the medium-to-high category, indicating that the trait would respond to selection. Some producers cull or eliminate animals with poor dispositions from their herds and flocks because of the potential for personal injury and economic loss (broken fences and facilities), as well as to reduce the excitability of other animals. In fact, at least one purebred cattle breed organization records temperament score as part of its national cattle evaluation program.

Figure 22.2
Animal behavior is heavily influenced by their early interaction with humans.
Source: Tom Field.

Recent research has shown that cattle with a nervous, excitable temperament (e.g., being highly agitated when restrained) experience lower weight gains in the feedlot. When scored for exit speed, cattle that ran out of a squeeze chute also had lower weight gains than those who exited more calmly. Additionally research has demonstrated that cattle that have been acclimated to quiet handling and thus experience less stress are also healthier.

Behavior during Handling and Restraint

Most animals are handled and restrained several times during their lifetimes. Ease of handling depends largely on the animal's temperament, size, and previous handling experience, and on the design of the handling facilities. Animals remember positive and negative experiences. Livestock with previous experience of calm, quiet handling will be less stressed and easier to handle in the future than animals that have had previous experiences with rough handling. Animals of all species will remain calmer and be easier to handle if they become acclimated to people quietly moving among them.

Understanding animal behavior can assist in preventing injury, undue stress, and physical exertion for both animals and producers. Understanding the natural responses of animals to the movement of a handler is critical to minimizing stress during movement of livestock. Most animals have a flight zone (Fig. 22.3). When a person is outside this zone, the animal usually exhibits an inquisitive behavior. When a person moves inside the flight zone, the animal usually moves away. The size of the flight zone depends on the tameness or wildness of the animal.

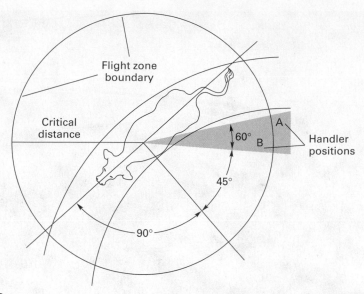

Figure 22.3

Handler positions for moving cattle. Positions A and B are the best places for the handler to stand. The flight zone is penetrated to cause the animal to move forward. Retreating outside the flight zone causes the animal to stop moving. Handlers should avoid standing directly behind the animal because they will be in the animal's blind spot. If the handler gets in front of the line extending from the animal's shoulder, the animal will back up. This is the point of balance. The solid curved lines indicate the location of the curved single-file chute. When the handler is close to the animal in a chute the point of balance is at the shoulder. When the handler is further away the point of balance may move forward to just behind the eye.

Source: Temple Grandin. Copyright © . Used by permission of Temple Grandin.

Figure 22.4

Shadows that fall across a chute can disrupt the flow of animals through the facilities. The lead animal often balks and refuses to cross the shadows. Source: Tom Field.

Blood odor appears to be offensive to some animals; therefore, the reduction or elimination of such odors may encourage animals to move through handling facilities with greater ease. Animals are easily disturbed by loud or unusual noises such as motors, pumps, and compressed air.

With their 310–360° vision, cattle are sensitive to shadows and unusual movements observed at the end of a chute or outside a chute (Fig. 22.4). For these reasons, cattle will move with more ease through curved chutes with solid sides (Fig. 22.5).

Figure 22.5
Animals move more easily through curved chutes with solid sides because the solid sides prevent them from seeing people and other distractions through the barrier. Source: Tom Field.

Round pens, having an absence of square corners, handle cattle that are more excitable with lower rates of injury.

Quiet, stress-free handling of livestock is critical to assuring animal well-being and health status, improving productivity, minimizing injury to humans, and maximizing the longevity of facilities.

INGESTIVE BEHAVIOR

Farm animals exhibit **ingestive behavior** when they eat and drink. Rather than initially chewing their feed thoroughly, ruminants swallow it as soon as it is well lubricated with saliva. After the animals have consumed a certain amount, they ruminate (regurgitate the feed for chewing). Cattle graze for 4–9 hours/day and sheep and goats graze for 9–11 hours/day. Grazing is usually done in periods, followed by rest and **rumination**. Rumination involves a repetitive process where the cud (a fibrous feed mass) is regurgitated, chewed again, mixed with additional saliva, and then reswallowed. Sheep rest and ruminate more frequently when grazing than do cattle—cattle ruminate 4–9 hours/day; sheep 7–10 hours/ day. Cows may regurgitate and chew between 300 and 400 **boluses** of feed per day; sheep between 400 and 600 boluses per day.

Under range conditions, cattle usually do not go more than 3 miles away from water, whereas sheep may travel as much as 8 miles a day. When cattle and sheep are on a large range, they tend to overgraze near the water area and to avoid grazing in areas far removed from water. Development of water areas, fencing, placing of salt away from water, and herding the animals are management practices intended to assure a more uniform utilization of the range forage.

Cattle, horses, and sheep have palatability preferences for certain plants and many have difficulty changing from one type of plants to other types. Most animals prefer to graze lower areas, especially if they are near water. These grazing behaviors tend to cause overgrazing in certain areas of the pasture and to reduce weight gains.

The behavior of cows grazing native range during winter is affected by cow age and weather conditions. At the Range Research Station in Miles City, Montana, cows grazed less as temperatures dropped below 20°F, and at a −30°F, 3-year-olds grazed approximately 2 hours less than 6-year-olds. Also, with colder temperatures, cows waited longer before starting to graze in the morning. At 30°F, cows started grazing between 6:30 and 7:00 A.M., but at −30°F, they waited until about 10 A.M. to begin grazing.

ELIMINATIVE BEHAVIOR

Eliminative behavior involves fecal and urine deposition. Cattle, sheep, goats, and chickens void their feces and urine indiscriminately. Hogs, by contrast, defecate in definite areas of the pasture or pen. Knowing defecation patterns of the pigs can help to plan the ease of cleaning swine pens. Horses tend to void their feces on scent piles of other horses and genders.

Cattle, sheep, goats, and swine usually defecate while standing or walking. All these animals urinate while standing, but not usually when walking. Cattle defecate 12–18 times a day; horses, 5–12 times. Cattle and horses urinate 7–11 times per day. Animals on lush pasture drink less water than when they consume dry feeds; therefore, the amount of urine voided may not differ greatly under these two types of feed conditions.

All farm animals urinate and defecate more frequently and void more excreta than normal when stressed or excited. They often lose a minimum of 3% of their live weight when transported to and from marketing points. Much of the **shrink** in transit occurs in the first hour, so considerable weight loss occurs even when animals are transported only short distances. Weight loss can be reduced by handling animals carefully and quietly, and by avoiding any excessive stress or excitement of the animals.

SHELTER-SEEKING BEHAVIOR

Animal species vary greatly in the degree to which they demonstrate **shelter-seeking behavior**. Cattle and sheep seek a shady area for rest and rumination if the weather is hot, and pigs try to find a wet area. When the weather is cold, pigs crowd against one another when they are lying down to keep each other warm. In snow and cold winds, animals often crowd together. In extreme situations, they pile up to the extent that some of them smother. Unless the weather is cold and windy, cattle and horses often seek the shelter of trees when it is raining. This may be hazardous where strong electrical storms occur because animals under a tree are more likely to be killed by lightning than those in the open.

INVESTIGATIVE BEHAVIOR

Pigs, horses, and dairy goats are highly curious, investigating any nonthreatening strange object. They usually approach carefully and slowly, sniffing and looking as they approach. Cattle also do a certain amount of **investigating behavior** (Fig. 22.6). Sheep are less curious and more timid than some other farm animals. They may notice a strange object, become excited, and run away from it. An object such as a paper cup on the ground can be either threatening or attractive to the animals. A novel object may attract the animals when they are on pasture. However, this same object may cause bolting and balking if the handler attempts to force the animals to walk over it. In one situation, the object triggers a fear reaction and in the other situation an investigative response.

Figure 22.6
Calf expressing investigative behavior as the result of interest in a novel stimuli, in this case a person. Source: Tom Field.

ALLELOMIMETIC BEHAVIOR

Animals of a species tend to do the same thing at the same time. Cattle and sheep tend to graze at the same time and rest and ruminate at the same time. Range cattle gather at the watering place at about the same time each day because one follows another. This behavior is of practical importance because the producer can observe the herd or flock with little difficulty, notice anything that is wrong with a particular animal, and have that animal brought in for treatment. When artificially inseminating beef cattle, the best time to locate range cows in heat is when they gather at the watering place. **Allelomimetic behavior** is useful in driving groups of animals from one place to another.

ADDITIONAL BEHAVIORS

Communication and maladaptive behavior are two other behaviors that are common to the nine behavior systems previously presented. Some highlights of these two behaviors are given in the following text.

Communication

Communication exists when some type of information is exchanged between individual animals. This may occur with the transfer of information through any of the senses.

Females more easily adopt the young of others through transfer of the odor of one young animal to another. Cows may foster several calves if their own calves are removed at birth and the foster calves are smeared with amniotic fluid previously collected from the second "water bag." This is an example of imprinting.

Many farm animals learn to respond to the vocal calls or whistles of the producer who wants the animals to come to feed. The animals soon learn that the stimulus of the sound is related to being fed. This is an example of operant conditioning.

The bull vocally communicates his aggressive behavior to other bulls and intruders into his area through a deep bellow. This bellow and aggressive behavior is under the influence of testosterone, as the castrated male seldom exhibits similar behavior. This communication behavior is part of the agonistic behavior system.

The bull also issues calls to cows and heifers, especially when he is separated from but still within sight of them. This type of communication could be included in the system of sexual behavior.

Horses have at least four vocal and three nonvocal sounds: (1) *squeals* (made during threats and encounters between individual animals—high pitched); (2) *nickers* (made by the stallion during mating and between the mare and foal prior to feeding—low pitched); (3) *whinnies* (begin as a squeal and end in a nicker—occurs between horses that are distressed or seek social contact); (4) *groans* (occur during discomfort or anguish); (5) *blows* (nonvocal sounds as air passes through the nostrils during prefeeding or when in alarm); (6) *snorts* (produced during nasal irritation, conflict, or relief); and (7) *snores* (produced by inhaling air—this has little part in communication).

Cattle are especially perceptive in their sight, as they have 310–360° vision. This affects their behavior in many ways—for example, when they are approached from different angles and when they are handled through various types of facilities.

Maladaptive or Abnormal Behavior

Animals that cannot adapt to their environment may exhibit inappropriate or unusual behavior. Some animals under intensive management systems, such as poultry and swine, are often kept in continuous housing to assure biosecurity and to maintain consistent temperature and humidity to reduce animal stress. Both chickens and swine may demonstrate cannibalistic behavior, which may lead to significant losses if preventive measures are not taken. Some swine producers remove the tails of baby pigs to prevent tail chewing. Tail chewing can cause bleeding, and whenever bleeding occurs, the pigs are likely to become cannibalistic. Chickens will peck at each other and may cause substantial injury. These cannibalistic behaviors may be expressed under a variety of housing conditions but are exaggerated when flock density is too high, birds are subjected to excessive stress or if long periods of boredom are common.

Some uncastrated male animals raised with other males masturbate and demonstrate homosexuality. In the latter situation, males mount other males, attempting to breed them. Some of the more submissive males may have to be physically separated from the more aggressive males to prevent injury or death.

The **buller-steer syndrome** is exhibited in steers that have been castrated before puberty. This demonstrates a masculine behavior of other than testosterone origin. Certain steers (bullers) are more sexually attractive for other steers to mount. As one steer mounts a buller, other steers are attracted to do the same. Thus, the activity associated with the buller-steer syndrome can cause physical injury, a reduction in feedlot gains, and additional labor and equipment expense as bullers are usually sorted into separate feedlot pens. On some feedlots 1–3% of the steers are bullers.

Growth implants and reduced pen space have been cited as reasons for an increased incidence of buller-steers. Some behaviorists cite evidence of similar homosexual behavior of males in free-ranging natural environments.

CHAPTER SUMMARY

- Animal behavior is a response to instincts and various stimuli to which the animals are exposed. It is a complex process involving the interaction of inherited abilities and learned experiences.

- Behavioral changes enable animals to adjust to changing conditions, improve their chance of survival, and serve humans by adapting to various management systems.

- Farm animals exhibit several major systems or patterns of behavior: (1) sexual, (2) caregiving, (3) care-soliciting, (4) agonistic, (5) ingestive, (6) eliminative, (7) shelter-seeking, (8) investigative, and (9) allelomimetic.

KEY WORDS

ethology
instinct
habituation
conditioning (classical and operant)
reinforcement
trial and error
reasoning
intelligence
imprinting
hand-mated
pheromones
flehmen
winking

caregiving behavior
care-soliciting behavior
agonistic behavior
ingestive behavior
rumination
boluses
eliminative behavior
shrink
shelter-seeking behavior
investigative behavior
allelomimetic behavior
buller-steer syndrome

REVIEW QUESTIONS

1. Discuss the role of stockmanship in livestock and poultry production.
2. Provide examples of classical and operant conditioning.
3. Contrast trial and error behavior with imprinting.
4. Describe the nine patterns of behavior and how this knowledge can improve management.
5. Describe the physical signs of various moods expressed by equines.
6. What is the importance of understanding animal behavior when designing handling facilities?
7. What are the vocal and nonvocal sounds emitted by horses?
8. Describe maladaptive behavior and how management can be helpful in mitigation.

SELECTED REFERENCES

Fraser, A. F. and D. M. Broom. 1990. *Farm Animal Behavior and Welfare*. London: Bailliere Tindall.

Grandin, T. 1993. Teaching principles of behavior and equipment design for handling livestock. *J. Anim. Sci.* 71:1065.

Grandin, T. (ed.). 2000. *Livestock Handling and Transport*. 2nd ed. Wallingford, Oxon, UK: CAB International.

Grandin, T. 2008. *Humane Livestock Handling*. North Adams, MA: Storey Publishing.

Houpt, K. A. 1991. *Domestic Animal Behavior for Veterinarians and Animal Scientists*. Ames, IA: Iowa State University Press.

Hurnik, J. F., A. B. Webster, and P. B. Siegel. 1995. *Dictionary of Farm Animal Behavior*. Guelph, Canada: Office of Educational Practice.

Lynch, J. J., G. N. Hinch, and D. B. Adams. 1992. *The Behavior of Sheep*. Sydney: CSIRO Publications.

McGlone, J. J. 1994. Animal behavior (ethology). *Encyclopedia of Agricultural Science*. San Diego, CA: Academic Press.

Monahan, P. and D. Wood-Gush. 1990. *Managing the Behavior of Animals*. New York: Chapman and Hall.

23

Issues in Animal Agriculture

From atop the ridge line near the middle of our ranch, the panoramic view of the Gunnison valley and its unique landscape is indeed spectacular. But even more amazing than the proverbial 10,000 foot view is the intricacy of the systems at play on our ranch and on those of our neighbors—family, plant and animal communities, soil, geography, and culture. It is on that ridge where I find solace and from where the hard questions of life can be considered and pondered. It is a good place to think about the future, to make peace with the past, and to consider alternatives for the present. These days as I gaze on the landscape that has sustained my family for nearly eighty years, I wonder what the future holds - will we be able to remain viable as a profitable enterprise? How will future generations judge the legacy of intended stewardship we leave behind? What will become of our community and our culture?

Viability, stewardship, continuity, and sustainability have been concepts that mattered to each generation of our family on the land in the high country of Colorado—certainly each time period with its own context, tools, knowledge, and pressures but nonetheless these ideas— these words—tie us to the past and to the future. Yet, words can be used as battering rams and failure to reach consensus as to the meaning and context of a word or idea can generate friction between parties engaged in commerce, government, and even personal relationships.

Our legacy depends on our ability to blend the concepts of viability, stewardship, continuity and sustainability into the fabric of our decision making. Perhaps Wyoming rancher Bob Budd said it best - "we have to learn to act in our own lifetimes while thinking at the pace of rocks and mountains." If we are truly committed to creation of sustained wealth then we must embrace several key principles:

- We must recognize the uniqueness of landscapes, communities, and enterprises. Thus, there is not a "one size fits all" recipe.
- Solutions must be found by blending the deep knowledge of landowners, local communities, and science. Much can be learned from studying and interacting within and external to our industry.

- Multi-dimensional wealth creation is a journey not a destination. There are no quick fixes nor simple answers, rather we must seek solutions and continuous improvement by increasing our knowledge, being deeply thoughtful in making trade-offs, and preventing the noise of politics and activism from weakening our resolve to leave a lasting legacy of value.

Any discussion of issues and challenges should be preceded by a conversation about core values and the culture of enterprises. Responsible animal agriculture is dependent on the concepts of stockmanship and

learning objectives

- Discuss the role of stockmanship and stewardship in livestock production
- Discuss the balance of feeding 9 billion people while assuring a sustainable future
- Describe the triple bottom line approach to sustainability
- Discuss the importance of reducing food waste
- Explain the role of risk assessment and risk management in decision making
- Describe the history of environmental issues in the United States
- Describe the primary environmental issues confronting the livestock industry with specific focus on land use, waste management, air quality, and water quality
- Describe the range of philosophies relative to animal well-being
- Discuss various approaches that can be taken to better assure animal care and well-being
- Discuss the diet-health issue and how misconceptions and junk science have been involved
- Discuss the history of the cholesterol issue
- Discuss the relationship between animal product consumption and human health

- What are the keys to maintaining appropriate diets and adherence to dietary guidelines
- Describe the role of TQM in assuring food safety
- Discuss the primary pathogens of concern relative to food safety
- Compare conventional and organic production and the impact on food safety
- Describe effective quality assurance systems
- Discuss the role of production technologies in livestock and poultry production

stewardship. **Stockmanship** is the application of skill, creativity, and knowledge to sound principles of livestock management founded on the model of the good shepherd. This concept integrates the results of sound science and practical experience. **Stewardship** is defined as the careful and responsible management of something entrusted to one's care. In the realm of animal agriculture this involves the thoughtful and wise management of livestock, people, natural resources, finances, and technology to create benefit for humanity. Decision making resulting in high integrity outcomes is founded on four core considerations—values and ethics, community and people, economics and profit, and resource health (Fig. 23.1).

Animal agriculture is a complex, multidimensional human activity influenced by cultural, environmental, political, economic, and social forces. Fundamentally, managers in the livestock and poultry industries have the opportunity to wrestle with systems questions and must make integrated systems decisions to be successful. Livestock management requires the ability to deal with the intertwined nature of humans, natural resources, animals, markets, communities, supply chains, and public policies. Such an environment is typified by the need to weigh competing interests, make thoughtful trade-offs, and balance the polarities of short- versus long-term needs as well as a host of cost–benefit relationships.

Given its complexity, it is not surprising that the practices, policies, and direction of the various industries that comprise animal agriculture are subject to social commentary and the diverse wants and needs of a diverse marketplace. Far too often the discussion around issues related to animal agriculture is driven by emotion and rigid doctrine. The advice professed by Renaissance philosopher Sir Francis Bacon when addressing complex problems was to "read not to believe or disbelieve but rather to weigh and compare." This approach provides a sound and reasonable model to deal with the major issues confronting animal agriculture. These issues can be

Figure 23.1
Effective decision making in the livestock industry depends on awareness of a variety of factors.

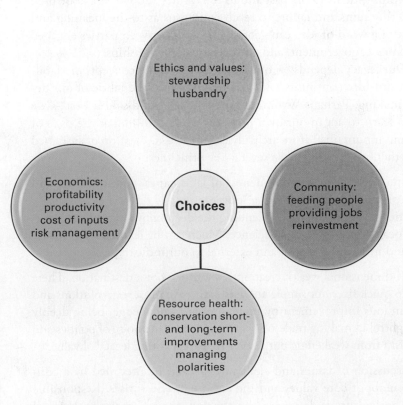

categorized into five buckets: (1) sustainability and natural resource management, (2) food safety, (3) animal well-being, (4) diet-health issues, and (5) production technologies. This chapter will focus generally on these five categories and it is not designed as a discussion of all possible issues related to livestock and poultry production. Additional focus on these issues is provided in Chapters 3, 4, 5, 21, and 22 as well as the chapters dealing with feeding and managing the various species subsequent to this chapter.

FEEDING 9 BILLION PEOPLE BY 2050

At the beginning of 2011 there were 7 billion people on earth. By 2050 that number will climb to 9 billion, essentially adding the equivalent of a population equivalent to one and a half times that of China's population to the world total. Creating the food production, processing, and infrastructure capacity to feed 9 billion people over the next four decades is daunting. However, there are substantial reasons that suggest just such a feat can be accomplished. First of all, population growth is slowing. The global population growth rate over the next four decades is expected to be approximately 30% compared to a nearly 80% rate of expansion from 1970 to 2010. Second, agricultural yields have allowed a level of productivity sufficient to hold a global hunger crisis at bay. While concerns exist as to the ability to sustain rising yield rates, experts point to opportunities to increase food production by narrowing the gap between the low yield producer and those with high production rates, intensification of production with the application of precision irrigation systems, enhanced housing systems for livestock and poultry coupled with improved nutritional management, genetic selection, and health management, and by harnessing the potential of new plant breeding technologies and production systems. Given that somewhere between 33.5% of the calories produced by the agricultural system never reach a human stomach due to losses from pests, spoilage, and waste; if half of this loss could be eliminated, it would equate to increasing food availability by 15–25%. Recapturing these losses requires enhanced production management, improved storage, handling and transportation of both raw and finished product, and changes in packaging, consumer behavior, and regulations that prevent leftover and minimal out-of-date food from reaching hungry people.

Nonetheless reaching the goal of feeding 9 billion people by 2050 will not be without controversy. Too often, there is a tendency by individuals, organizations, and policymakers to seek popular and seemingly simple approaches to complex problems by focusing on one-size-fits-all approaches. History has shown that such an approach is not effective. Decision making almost always involves choices that require trade-offs. Careful analysis of agricultural issues clearly shows that there are very few black and white choices available. Good decision making requires acknowledgement of the assumptions that define both the problem at hand and the alternate solutions, careful consideration of both intended and unintended consequences, and rational assessment of cost–benefit relationships. In the end, the best decisions are those that are site- and situation-specific, based on knowledge and good information, and aligned with core values of the people involved.

SUSTAINABILITY

Many problems are framed by **polarities** or competing interests. Managers of livestock systems must deal with a host of polarities. For example, profitability attainment involves having an enterprise that is sufficiently large to capture economies of size and scale. However, as more animals are added to the enterprise, the waste production of the business increases. If the waste management issue isn't sufficiently addressed then animal health declines as does the productive capacity of the natural resource base. Thus long-term profits depend not only on productive capacity but also on the careful management of impacts on natural resources.

Sustainability has been defined as a method of harvesting or using a resource so that the resource is not depleted or permanently damaged. However, this definition is perhaps too narrow as it fails to consider the dynamics of other polarities such as economy and culture. In addition to meeting ecosystem performance metrics, managers must also consider other factors as well. Assessing the **viability** of alternatives, defined as whether something is capable of growing, or something practical and able to be done; is foundational to good decision making. Livestock and poultry enterprises tend to be continuous in nature. Thus **continuity**—an uninterrupted succession or flow; a coherent whole—cannot be stopped and restarted in the same way as enterprises that generate storable products.

As people have struggled with the complexity of making effective natural resource decisions, a useful model has emerged called the **triple bottom line** approach to sustainability. The triple bottom line of sustainability is based on three columns of importance: ecosystems, economies, and communities. Sustainability is not simple, it is not a recipe, it is not a classification system by which industries, companies, or processes can be conveniently labeled as "sustainable" or "not sustainable," and it is not a task to be completed and then checked off the list. Rather sustainability is a commitment to an ongoing journey that involves balancing decisions within the scope of the triple bottom line concept. The choice is to become more sustainable. Attempts to create black and white categories such as sustainable or unsustainable are misguided.

One opportunity to improve the overall sustainability of food production available to all participants in the global food supply chain, including consumers, is to reduce food waste. Experts estimate that 25% to 40% of the calories produced by agriculture never reach a consumer's stomach. The source of waste is diverse; in developing countries about 40% of the waste is due to poor harvest, storage, transportation, and processing infrastructure. However, in developed nations, 40% of food waste occurs at food retail, food service, or in the home.

While some waste is to be expected, opportunities to reduce food losses are available through improvements in infrastructure, changes in consumer behavior to minimize the amount of food purchased but subsequently thrown away as plate waste, creating better diversions of inedible waste into composting or recycling as animal feed, and removing regulatory and transportation barriers to food donations. Beginning in 2011, the Food Waste Reduction Alliance comprised of a broad coalition of food supply chain companies, industry associations, consumer groups, and nongovernmental organizations (NGOs) began in earnest to reduce food waste in the United States.

VALID COMPARISONS AND ASSESSING RISK

Good decision making depends on access to data and the ability to make valid comparisons when evaluating alternatives. Essentially, valid comparisons help decision makers avoid the trap as described by the adage "not all that glitters is gold." For example, a French study evaluated the assumption that a diet centered on fruits and vegetables would have a lower environmental impact than a diet that included animal products as center of the plate items. When the total greenhouse gas emissions for each product were calculated on a per 100 grams of food production, fruits, vegetables, and starch-based diets had an advantage over diets containing dairy, fat, pork, poultry, fish, and beef. However, when the comparison was made on a 100 kcal of energy intake basis, the advantage of the plant-based diet was eroded and any potential environmental advantage was negated.

Another example involves comparing transportation costs in the food supply chain. Some advocate that consuming only locally produced food is the most socially responsible decision based on a concept known as "food miles." While the purchase of locally produced foods is perfectly acceptable, doing so on the assumption of reducing environmental footprint arising from reduced transportation proves to be far more complicated to justify. Capper (2011) evaluated the case where eggs were delivered in volume (23,400 dozen eggs per load and 2,400-mile round-trip) to a retail outlet as opposed to a local farmer delivering a pickup truck load of eggs to the same outlet (1,740 dozen eggs and 300-mile round-trip). In this scenario, the smaller load traveling less distance produced nearly four times greater fuel consumption per dozen eggs and an equivalent amount of carbon emission per dozen eggs.

Assessing risk becomes important when evaluating most of the issues outlined in this chapter. Technology, shifts in societal attitudes, consumer perceptions, and the consequences of excessive regulations require the development of better systems of determining short- and long-term risk associated with applications of technology or the distribution of resources. Improvements in the precision of analytical measurement place additional pressure on the need for thoughtful risk analysis.

As residential and urban uses of land increase into traditional farming areas, the potential for conflict tends to increase. A survey of rural residents in Pennsylvania determined that approximately one-third of respondents had a complaint about a neighboring farm. Of those registering a complaint, 57% were concerned about odor, while an additional 18% complained about fly problems. No other topic elicited more than a 5% response rate from complainants.

The perceptions of farm neighbors in regards to various issues were quantified by livestock species and are presented in Table 23.1. These results point to the need for a proactive approach by the livestock industry to deal effectively with a breadth of issues and concerns.

An entire field of study has emerged to deal with risk management. Risk assessment is a process the involves identifying potential hazards, determining the impact of those hazards by determining relative rates of risk, determining appropriate risk reduction

---

Table 23.1

MEAN PERCEPTION OF RURAL RESIDENTS TO LIVESTOCK ENTERPRISES (SCALE OF 1 TO 8, WITH 1 = STRONG AGREEMENT)

Perception	Dairy	Swine	Poultry	Beef	Veal
Use too many pesticides	5.1	5.7	5.6	5.6	6.4
Dispose of animal waste properly	4.4	4.8	4.8	4.7	5.9
Harm surface or ground water	4.7	4.8	5.0	4.9	6.0
Provide pleasant scenery	2.9	4.9	5.1	3.3	5.4
Handle and care for animals humanely	3.5	4.3	4.7	3.9	5.7
Animals are generally healthy	3.3	4.3	4.7	3.8	5.9
Give too many antibiotics	6.0	6.1	6.0	5.9	6.6
Give too many growth promotants	6.0	6.1	5.9	5.9	6.6
Make good neighbors	2.8	4.4	4.3	3.3	5.4
Contribute significantly to local economy	2.7	3.8	3.3	3.3	5.4

Source: Adapted from Jones et al., 2000.

alternatives, choosing and implementing one or more interventions, recording outcomes, and using performance metrics to continuously measure and adjust accordingly.

Discussions of risk must also include an awareness of the methodology by which risk is calculated. For example, absolute risk is the probability of an event occurring (i.e., shark attack) and because the calculation occurs without accounting for differences in risk factors (surfers vs. those never having swam in the ocean) there isn't an informing context by which a person can alter risk factors to change the likelihood of occurrence. For example, a resident of the United States has a 1 in 200 chance of being a victim of violent crime but only a 1 in 700,000 chance of being attacked by a shark. Consumers have a 1 in 375,000 chance of acquiring a foodborne disease from eating fruits or vegetables but only a 1 in 900,000 chance of acquiring a foodborne disease from eating beef. These examples of absolute risk are interesting but not particularly informative.

Relative risk on the other hand calculates difference in risk between two different sets of factors associated with risk. For example, the absolute risk for a person incurring lung cancer is 1% while people who smoke have 10 times greater likelihood of suffering the disease than their nonsmoking contemporaries. However, relative risk does not measure actual risk; it only quantifies the differences in likelihood of occurrence between risk factor categories.

ENVIRONMENTAL MANAGEMENT ISSUES

Societal approaches to natural resource use have been transformed over time. In the early history of European settlement of North America, the prevailing attitude assumed that resources were limitless and environmental management focused almost exclusively on using resources from a short-term perspective with little regard for the future. Westward expansion was driven by the notion of "a land of milk and honey." Horace Greeley's exhortation to "Go west, young man" typified U.S. history prior to the late 1800s.

Gifford Pinchot, father of the U.S. Forest Service, advocated wise use with concerned planning for the future. Pinchot and other early federal lands managers believed that technically proficient professionals should manage public resources and that the goal of management should be to create the "greatest good for the greatest number." Henry David Thoreau and other late-19th-century writers advanced the belief that wilderness has unique value and that certain resource or geographic areas ought to be protected from human influence. Contemporary environmental philosophy spans a wide spectrum of beliefs and attitudes ranging from those who pollute and damage the ecosystem without conscience to ecoterrorism. Polluters are those whose narrow and singular short-term focus creates damage to natural resources while ecoterrorism is an extremist view that encourages the use of guerrilla warfare tactics against any entity or individual deemed to have negatively affected the ecosystem. Fortunately, these extremes have few advocates. Multiple-use conservation is the most typically embraced approach to resource management where the focus is to advocate for sound resource stewardship with both short- and long-term objectives related to the ecosystem, communities, and economies.

Conflicts typically arise from the efforts of a special interest group(s) to bring an agenda into the public arena while an opposing group(s) works to neutralize the impact of the other or to advocate for an alternative viewpoint or solution. Federal and state agencies contribute to political conflict either by direct participation or by providing the forum in which conflict can occur. Regulatory agency participation is heightened by the increasing independence from legislative control and the tendency that each agency operates within a relatively narrow focus of societal need.

Policy is created by the ability of interest groups to define an issue, call attention to the group's position, and then pressure government to take a desired action. Resource allocation decisions are difficult to make in an arena where conflict is the norm and unfortunately politicization of issues tends to polarize opinions and make progress difficult. This is particularly challenging when special interest groups compete to gain the attention of policy makers in a zero-sum game where extremism on both sides of an issue dominates the conversation. Too often in these situations, decision makers fail to recognize the long-term impact of their choices and operate without careful evaluation of the trade-offs and cost–benefit relationships inherent in any complex system.

LAND USE

Thirty percent of the earth's land surface is used for agricultural purposes. Of the available land in the United States, 19.5% is classified as cropland with nearly 26% in grasslands and pasture (mostly located in the intermountain West and the southwestern states of Oklahoma and Texas). Comparatively, 29% of the United States is forested with an additional 13% in preserved/protected areas, national parks, and roads. Forested acres increased in the United States by 17 million acres in the period 1987 to 2002. However, 1.2 million acres of grasslands were converted to cropland from 2007 to 2012 as the result of higher prices for corn resulting from the federal ethanol blending mandate.

Maintaining the health, stability, and productivity of land is an important management goal. Reducing erosion is important on a global basis and agriculturalists in the United States have initiated a number of erosion control measures. The amount of cropland erosion occurring at rates about soil loss tolerable rates fell by 27.5 million acres in the United States in the 25-year period beginning in 1987. Soil scientists and geographers note that the United States has the lowest levels of soil erosion among global agricultural regions and that the primary challenge lies in those regions defined by poor economic conditions and high-density populations.

The use of fertilizers, herbicides, and pesticides has been important to increasing land productivity. Fertilizer use by U.S. agriculture peaked in the late 1970s. However, with improved plant genetics and precision agricultural techniques, U.S. corn production doubled in the 30-year period between 1980 and 2010 but with increased efficiency related to fertilizer application rates. In 1980, 3.2 pounds of fertilizer were required to produce a bushel of corn but in 2010 a bushel of corn required only 1.6 pounds of fertilizer inputs. In the same time period, aggregate pesticide/herbicide application rates declined 0.6% per annum.

Conversion of Agricultural Land

The conversion of agricultural land to nonagricultural uses is becoming an increasingly significant challenge to preserving wildlife habitat and assuring the security of food supplies for future generations (Fig. 23.2). Since 1978, Colorado has lost an average of 90,000 acres of agricultural land per year. This loss equates to an area 140 miles long and 1 mile wide. On a global basis, rural populations are declining while the growth of urban centers is accelerating. These trends are expected to continue well into the future. With most of the urban growth expected to occur in developing economies, a significant infrastructure investment will be required to transport, store, and distribute food to these megacities.

Urban dwellers desire small-acreage homesites, cities, and communities grow, and growing recreational demand breaks up existing blocks of land. As large areas of land that once were managed as an ecosystem are divided into smaller parcels and then crisscrossed with infrastructure improvements such as roads and power lines, both wildlife populations and agricultural production capacity are negatively affected.

Figure 23.2
The loss of agricultural lands to urbanization is one of the most critical issues facing public policy makers. Continued loss of agricultural lands to other uses threatens to weaken the food security of the United States. Source: Tom Field.

As the popularity of 5- to 35-acre homesite tracts grows, more pressure will be placed on traditional agricultural enterprises as they interface with urbanization. As large tracts of land are removed from production, not only agriculture is affected. Wildlife lose migrating routes, are exposed to increased predation from cats and dogs that accompany suburbanization, and lose vital habitat.

Federal Lands

The Louisiana Purchase of 1803 initiated nearly 200 years of **federal land** acquisitions, redistributions, and management policies. Various Homestead Acts, cash sales, and land grants to support schools and agricultural colleges disposed of some federal lands before the turn of the century. The first National Parks were created in 1872; the Federal Reserve Act of 1891 and the Weeks Act of 1911 allowed expansion of the lands managed as National Forests. Table 23.2 outlines the percentage of the 11 contiguous western states made up of federal lands.

Table 23.2
FEDERAL LANDS IN THE WEST AS A PERCENTAGE OF STATE LAND MASS

State	%
Nevada	86
Utah	64
Idaho	63
Oregon	52
Wyoming	48
California	45
Arizona	43
Colorado	36
New Mexico	33
Montana	29
Washington	29

Source: Public Lands Statistics, U.S. Bureau of Land Management.

The establishment of the open range livestock industry in the western states and territories followed the conclusion of the Civil War. Stimulated by the expansion of the railroad system complete with refrigerated cars, increasing demand for beef both domestically and abroad, and the ready availability of cattle in Texas to be driven northward, the open range livestock industry boomed in the 1870s and into the 1880s. The legends of the cattle barons were created by foreign investment, primarily from Scotland and England, which allowed for the development of large ranches such as the Swan Cattle Company, which ran over 120,000 cattle at its zenith. Large tracts of land were controlled in the West by homesteading on or near water sources. A common thread in the history of the western states is that the control of water allowed for the domination of other resources as well.

Unfortunately, the cattle boom associated with the open range livestock industry was speculative and thus many of the management decisions of the time were shortsighted and resulted in damage to the rangelands. A devastating drought in 1886 and the blizzards of the winter of 1887 combined to cause cattle death losses as high as 60% across the Great Plains region. Just as quickly as it had boomed, the open range livestock industry went bust. It gave way to the forces of inclement weather, failed speculative ventures, and the continuous process of westward expansion and settlement. But significant damage to the western ecosystems had already been done. Restoration of the range resource yielded the first practitioners of range ecosystem management.

The 1905 grazing regulations were the first in a long line of policies that focused on improving and protecting the western rangelands. The Taylor Grazing Act helped establish the framework for the development of the Bureau of Land Management. Significant legislation that affects livestock grazing on federal lands is summarized in Table 23.3. Each of these legislative actions has yielded a degree of conflict and confrontation between groups who place differing values on the federal lands. The federal lands are host to a multitude of uses including livestock and wildlife grazing, timber production and harvest, mining, a variety of recreation activities ranging from backpacking to skiing, and preservation of watersheds.

Multiple uses of federal lands will continue, as will disagreements relative to the management of the resources (Fig. 23.3). However, consensus-building models of decision making coupled with judicious management of these lands will continue to assure their productivity and sustainability.

Table 23.3
HISTORY OF LEGISLATIVE ACTIONS AFFECTING PUBLIC LAND GRAZING

Homestead Acts of 1862, 1909, 1916

Weeks Act of 1911

Taylor Grazing Act of 1934

Multiple Use-Sustained Yield Act of 1960

Wilderness Act of 1964

Classification and Multiple Use Act of 1964

National Environmental Policy Act of 1969

Forest and Rangeland Renewable Resources Planning Act of 1974

National Forest Management Act of 1976

Federal Land Management and Policy Act of 1976

Public Rangeland Improvement Act of 1978

Rangeland Reform Proposal (Department of Interior) 1993–94

Figure 23.3
Multiple uses of federally owned/managed lands continue to be important. However, conflicting interests between user groups must be successfully managed to assure the viability of the resource.

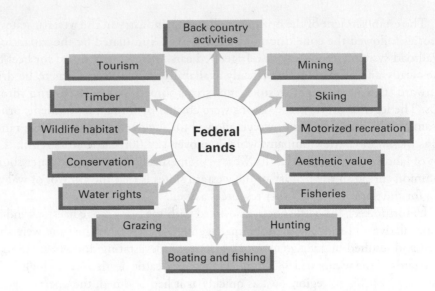

Figure 23.4
Management of animal waste is critical to sustainability. Manure management systems separate solid and liquid components that will be eventually used as fertilizer and irrigation water. Photo by Justin Field.

Waste Management

Waste management issues are of particular concern to intensive agricultural systems such as farrow-to-finish confinement facilities, dairies, and feedlots. Specific issues often involve **manure management** and odor control. Industry has shown an increasing willingness to overcome these challenges. For example, technologies have been incorporated that allow waste to be processed so that solids can be separated for use as fertilizer and so that water can move into a settling pond before being reused as a source of irrigation water (Fig. 23.4). The efforts of U.S. pork producers to improve environmental practices and management are outlined in Table 23.4.

Manure disposal is often accomplished by applying wastewater and solids to land resources as a means of fertilization. Animal waste is rich in organic materials, nitrogen, phosphorus, and potassium. The use of animal waste as fertilizer is a classic example of creating both economic and environmental benefits by linking the animal and crop/forage system.

Odor from manure can be a particular problem for livestock producers who manage animals in intensive systems (e.g., large feedlots, dairies, poultry, and swine

Table 23.4
ENVIRONMENTAL CONTROL PRACTICES ON U.S. SWINE FARMS

Waste Storage System	Small (<2,000 hd) %	Medium (2,000–9,999 hd) %	Large (>10,000 hd) %	All Farms
Above ground slurry storage	4	9	1	5
Deep pit ground slurry storage	55	67	49	57
Waste solids separated from liquids	16	8	20	15
Other systems	25	3	1	21
Anaerobic lagoon	17	45	65	23
Formal manure management plan	19	67	91	28
Manure applied to land	95	95	95	95
Diet manipulation for odor control[a]	47	62	84	50
Manure management to reduce odor[b]	25	43	54	29
Air quality management to reduce odor[c]	29	24	34	28

[a]Low crude protein, low phytase corn, pelleting, vegetable oil to control dust, etc.
[b]Chemical additives, composting, separate solids from liquids, etc.
[c]Biofiltering exhaust, windbreaks to reduce dust, etc.
Source: Adapted from NAHMS, 2000.

Table 23.5
METHODOLOGIES TO EFFECTIVELY MANAGE MANURE

Method	Characteristics
Ambient digester	Methane gas from anaerobic digestion is captured beneath a plastic tarp for eventual use as a fuel.
Aerobic/anaerobic digester with microbial treatment	Solids are removed and effluent moves through both aerobic and anaerobic steps prior to recycled use to flush manure or wash floors, etc. Solid residue often used as fertilizer.
Dewatering, drying, and desalinization	Water is removed, solids are dried, and mineral components are removed for use in related value-added products.
Thermophilic anaerobic digester	Solids are anaerobically digested at high temperatures to generate methane gas that can then be used to fuel boilers to produce the hot water used in the process.

operations). Odor and noise regulations are currently handled by state and local regulations. As urban sprawl continues to move into traditional agricultural areas, odor, and air-quality issues must be carefully monitored by animal agriculturalists. A variety of new technologies are being developed to better manage manure and animal waste (Table 23.5).

Air Quality

Air-quality management in animal agriculture is focused on two general categories: (1) particulate matter (dust) generated by animals and manure and (2) odor and gaseous compounds. Air emissions are influenced by a variety of factors including composition of the diet, housing type, flooring type for housed systems and pen surface for outdoor systems, manure management, feed storage systems, animal densities, and weather conditions. Odor issues tend to be fairly localized to an area less than several miles from the animal handling facility. Other gaseous compounds and dust are also often more localized in their impact but can have regional impacts when they are redeposited on waterways or react chemically to form secondary molecules that can be transported very long distances. For example, a high percentage of the air pollutants detected along the west coast of the United States originated in China or India.

Compounds of note that may originate from livestock and poultry operations currently subject to federal regulatory action include ozone, particulate matter (dust), carbon monoxide, sulfur dioxide, nitrogen dioxide, and lead. As of 2009, facilities that emit 25,000 metric tons of CO_2 equivalent (CO_2e) or more per year of greenhouse gases (GHG) from its manure management system to report GHG emission levels annually to the EPA. While not yet regulated, it is likely that emissions of nitrous oxide, methane, ammonia, and hydrogen sulfide will become subject to federal regulatory authority.

Options to mitigate air emissions include dietary manipulation to minimize excretion of nitrogen and phosphorus, manure treatments to alter volatilization, use of wet scrubbers or biofilters to treat exhaust from housed animal buildings, and the use of vegetative buffers. The livestock industry has aggressively pursued management strategies and mitigation protocols to reduce impacts on air quality. Swine production systems create generalized odor, ammonia, hydrogen sulfide, volatile organic compounds, bioaerosols, and greenhouse gases. However, diet manipulation and improved manure management are capable of significantly reducing production of these compounds. The poultry industry has been focused on improving housing and ventilation systems to more effectively manage ammonia. The dairy industry is focused on improvement management to reduce ammonia, odor and greenhouses gases. The beef feedlot industry, which manages mostly open pens, is focused on managing dust and particulates as well as ammonia particularly when weather creates extended periods of wet conditions.

Water Utilization and Quality

The livestock industry uses water in three categories: (a) production and management of livestock; (b) processing of meat, milk, and eggs; and (c) irrigation to produce feedstuffs for livestock and poultry rations. Water usage by livestock production and processing activities are less than 1% of U.S. freshwater withdrawals. Freshwater usage for irrigation of crops fed to poultry and livestock in the United States account for approximately 9% of total freshwater withdrawals. Experts estimate that water use for livestock production purposes in the Texas High Plains, a region characterized by dependence on irrigation, dairy production, feed yards, and integrated swine production, to be less than 2% of total water use in 2000 with increased use to 2.45% by 2030.

There is significant variation when comparing regions. For example, irrigated crop production east of the 100th meridian in the United States is quite low while the use of irrigation is relatively high in the western region with California and Idaho having the largest freshwater usage for agricultural irrigation. In many states in the intermountain west and extending to the west coast, water storage projects have been critical to the develop of crops, horticulture, and animal agriculture. Continued efforts to create efficient use strategies, watershed management, and water storage projects are critical to both rural and urban regions.

Water-quality issues are often at the forefront of environmental debates. Specific concerns include nonpoint source pollution plus direct ground or surface water contamination.

Eight specific water-quality best management practice categories are essential to improved agricultural water resource stewardship: conservation tillage, crop nutrient management, pest management, conservation buffers, irrigation water management, grazing management, animal feeding operations management, and erosion and sediment control.

Conservation tillage practices are designed to maintain crop residue on the soil surface and can be categorized into three basic systems: no-till/strip-till, mulch-till, and ridge-till farming. These systems reduce erosion and surface runoff by avoiding soil surface disturbance and enhancement of soil quality increasing infiltration and soil organic material.

Crop nutrient management strategies are important to minimizes undesirable environmental impacts caused by excessive nutrients that runoff agricultural lands into waterways. These strategies recognize that nutrients are cycled through ecosystems, are transformed into various chemical forms, that "leaks" occur in the cycle when excess nutrients (beyond the capacity of plants or soil microbes to effectively utilize them) are lost into the environment via leaching, erosion, volatilization, or runoff processes, and that management systems can reduce "leaks." Nitrogen, phosphorus, and potassium are typically the nutrients of greatest concern.

Pesticide management is important to avoid impacting nontargeted species and non–point source contamination of ground and surface water. Integrated pest management strategies that focus on prevention of infestation, the use of nonchemical measures, and a precise, targeted utilization of chemical pesticides provides an effective approach that balances profitable yields, input costs, and environmental impact.

Conservation buffers are areas of permanent vegetation placed to help control pollutant by intercepting sediment, nutrients, and farm chemicals from runoff. The use of grasses, shrubs, and trees can be incorporated into the design of vegetative barriers suited to the weather, soil type, and geography of a particular site.

The science of irrigation system design can be leveraged to create water transport and distribution methods that are well suited to topography, soil type, and crop alternatives. Irrigation systems should be designed to be cost-effective, meet the needs of crops, and minimize resource wastage of water, energy, and time. Furthermore, losses from erosion and transfer of pollutants to ground or surface water systems should be minimized. These goals can be obtained via correct irrigation application rates, efficient scheduling of irrigation timing, efficient transport of irrigation water to minimize losses and contamination, capturing and reusing runoff and tailwater, and intentional management of drainage water.

Well-designed grazing systems are also important to enhancing the health of watersheds and waterways. More detail can be found in Chapters 25 and 31. Correct management of more intensive animal feeding operations is also essential to assuring water quality. Proper planning of animal facilities and drainage systems is needed to prevent animal wastes from contaminating water sources. In fact, specialized engineering firms have been created that specialize in facility design, environmental monitoring, and regulatory compliance to assist livestock and poultry producers create enhanced capacity to effectively manage water quality. Chapters 25, 27, 29, 31, and 34 provide additional insight into species specific best practices. Finally, many of the previous management strategies are designed to mitigate erosion and prevent sediment loss, contamination of water sources, and to minimize turbidity.

ANIMAL WELL-BEING

Livestock and poultry owners and managers have an obligation to the well-being of the animals and birds under their care. The legacy of animal husbandry is founded on the concept that by caring for their flocks and herds, the stockman assures that his or her own interests will be served. It is this symbiotic relationship that provides much of the foundation for those who make their living and their way of life through animal agriculture.

Table 23.6
PRODUCER CODE OF CATTLE CARE

- Provide appropriate food, water, and care to protect the health and well-being of animals.
- Provide disease prevention to protect herd health, including access to veterinary care.
- Provide facilities that allow safe, humane, and efficient movement and/or restraint of cattle.
- Use appropriate methods to humanely euthanize terminally sick or injured livestock and dispose of them properly.
- Provide personnel with training/experience to properly handle and care for cattle.
- Make timely observations of cattle to ensure basic needs are being met.
- Minimize stress when transporting cattle.
- Keep updated on advancements and changes in the industry to make decisions based upon sound
- Production practices and consideration for animal well-being.
- Persons who willfully mistreat animals will not be tolerated.

Source: Adapted from Beef Quality Assurance Program.

Figure 23.5
(A) Producers strive to assure the welfare of the animals under their care. (B) Youth projects provide an educational opportunity to teach stockmanship and stewardship skills to the next generation. Source: 23.5a and b: Tom Field.

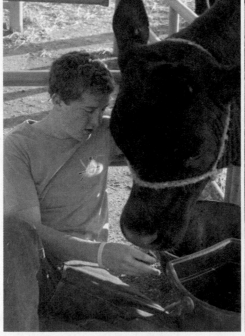

A   B

The debate about the role of human beings in animal well-being has generated discussion both external to and within U.S. animal agriculture sectors. This discussion has yielded a variety of viewpoints. For example, the Colorado Cattlemen's Association has adopted an animal welfare **code of ethics** that articulates the values of its membership (Table 23.6). This ethic might well represent the traditional perspective of animal husbandry (Fig. 23.5).

However, as is often the case with issues of a social nature there are numerous and often conflicting viewpoints. The perspective continuum ranges from animal exploitation to animal liberation (Table 23.7).

Animal exploitation would be characterized as events or uses that are conducted without concern for the animals involved. Examples might include cockfighting or dogfighting. These activities typically occur in violation of existing laws. Animal use groups

Table 23.7
ANIMAL WELFARE/ANIMAL RIGHTS GROUPS: A SUMMARY OF THE VIEWS

Category/ Group	Viewpoint
Animal exploitation	These groups argue that animals exist for human use or even abuse; animals are human property. These groups advocate or conduct activities that are illegal in most states (e.g., dogfighting, cockfighting, live pigeon shooting). Most of their activities involve pain or death of the animals primarily for entertainment.
Animal use	These groups believe that animals exist primarily for human use (e.g., livestock production, hunting, fishing, trapping, rodeos, zoos). These organizations have guidelines for the responsible care of the animals. They believe that harvesting of animals for food should be as painless as possible.
Animal control	These organizations enforce the laws, ordinances, and regulations affecting animals. Animals may be supplied for research; surplus animals are destroyed. Many advocate spaying or neutering animals.
Animal welfare	These are national groups, humane societies, and welfare agencies that support humane treatment of animals. They work within existing laws to accomplish goals. They publicize and document animal abuses to get laws changed. They do not provide animals for research. They require spaying and neutering, and are willing to euthanize surplus pets rather than let them suffer.
Animal rights	These national and local groups believe animals have intrinsic rights that should be guaranteed like human rights. These rights include not being killed, eaten, used for sport or research, or abused in any way. Some hold that pets have the right to breed. Most require spaying or neutering.
Animal liberation	These groups believe that animals should not be forced to work or produce for human benefit. They call for animal liberation.

Source: Adapted from K. B. Morgan, *An Overview of Animal-Related Organizations with Some Guidelines for Recognizing Patterns* (Kansas City, MO: Community Animal Control, 1989).

typically function under the premise that animals are for human use but that people have a responsibility to provide good animal care and to utilize management practices that minimize animal suffering. Examples of such groups might include livestock producers, breed associations, zoos, and equine sport enthusiasts. Animal control groups include regulatory agencies or organizations at all levels of government that deal with domestic or wild animal control. These groups typically focus on enforcement of existing laws.

Animal welfare groups, such as the American Humane Society, go a step further to extend care for animals that are suffering or homeless due to neglect. Historically, these groups have focused on pets or companion animals but have also endeavored to create standards for production agriculture. **Animal rights** groups advocate that animals have rights that include not being consumed, used in sport or research, or in some cases even used as pets or companions. Examples of animal rights groups include People for the Ethical Treatment of Animals (PETA) and the Humane Society of the United States (HSUS). These groups almost universally promote vegetarianism, are opposed to the taking of animal life, and may even believe that spaying/neutering are violations of an animal's right to breed. **Animal liberationists** are even more extreme groups that extend the animal rights agenda in violent ways and believe that their cause is so justified that any means used to advance that cause is acceptable.

Philosopher Bernard Rollin (1993) articulates a reasoned perspective that "the ethic which has emerged in mainstream society does not say we should not use animals or animal products. It does say that the animals we use should live happy lives where they can meet the fundamental set of needs dictated by their natures and where they do not suffer at our hands."

A common assumption is that the animal issue can be conveniently categorized into two notions—animal welfare and animal rights. However, as Rollin (1995) points out, this assumption is flawed in that it (1) attempts to place the discussion

Table 23.8
ANIMAL WELFARE CONCERNS RELATIVE TO ANIMAL AGRICULTURE

Beef	Swine	Dairy	Poultry	Horses
Handling techniques	Gestation crates	Lameness	Housing/cages	Training techniques
Castration	Housing	Calf housing	Debeaking	Housing
Dehorning	Transport	Tail docking	Forced molting	Performance injuries
Transport	Tail docking	Production technologies	Transport	Transport
Harvest	Harvest	Transport	Harvest	Harvest

into an us-versus-them context and (2) fails to account for the truth that animals have rights that dictate the need for husbandry practices that assure their welfare. Welfare concerns relative to each species are listed in Table 23.8.

Livestock producer organizations, individual companies, and supply chain alliances have developed animal care guidelines designed to serve as best practices in regards to transportation and harvest protocols, housing, and rearing. Examples of the recommendations include minimizing beak trimming in poultry, phasing out feed-withdrawal molting, increasing space allocations per bird, and eliminating tail docking in dairy cattle. Furthermore, these standards of care have been underpinned by widespread training and educational programs.

Much of the criticism leveled about inhumane treatment of livestock has been related to the management of nonambulatory animals. Even with the best available facilities, handling, management protocols and intentions, individual cases of animals that are injured, affected by terminal illness, or otherwise rendering immobile will occur. In these cases, having a scientifically based and implementable strategy to humanely euthanize animals is critical. At the same time for those animals that are severely lame but still ambulatory, protocols must be in place to assure that these individuals are not subjected to the additional stress of transport and associated movement through the marketing process. Furthermore, it is imperative that disabled, nonamublatory or diseased animals are not delivered to a processing plant. Transport and harvest guidelines have been developed to assure that animal well-being is at the forefront of transportation and marketing decisions. These guidelines include the following components:

1. Animals should be loaded, transported, and unloaded in a manner that prevents injury and distress.
2. Nonambulatory animals should never be loaded onto a transport vehicle destined for market or harvest facilities.
3. Nonambulatory animals as a result of transport should be unloaded appropriately.
4. Animals must be handled humanely and in accordance with applicable federal, state, and local laws.
5. Euthanasia protocols must be in place and followed to provide the most humane death possible when other viable options are not available.

A practical approach for producers making transport and marketing decisions is to focus on the STOP principle. For example, cattle that are sick (fever greater than 103°F for cattle have been treated but have not yet cleared the withdrawal time to avoid tissue residues), thin (exhibit of body condition score of less 3 on an 8-point scale), have active ocular lesions or are blind in both eyes, or are experiencing pain due to lameness, peritonitis, bone fractures, or other injuries should not be transported or marketed.

Additional scientific discipline related information about animal well-being can be found in Chapters 20 (environmental adaptation), 21 (animal health), and 22 (behavior) while more in depth species specific information can be found in Chapters 25, 27, 29, 31, 33, and 34.

The livestock industry will be best served by assuring that producers maintain a strong record of sound husbandry practices via creation of quality assurance initiatives, producer and employee training, and enhancing communication throughout the supply chain. Furthermore, an investment of research funding to improve facilities, management techniques, and animal handling will pay dividends.

PUBLIC HEALTH

Diet and Health

Most nutritionists agree that a healthy diet should contain all the required nutrients and enough calories to balance energy expenditure. Food choices, or lack of choices, determine the nutritional status of most people. Obesity caused by caloric excesses has become a leading nutritional problem in the United States with approximately one-third of the population classified as obese. Conversely, nutritional deficiencies are a daily reality for millions of consumers across the globe. In developed economies, consumers have the opportunity to make food choices based on multiple considerations: taste, price, quality, lifestyle, impacts of diet on health, and food safety.

The relationship between diet and disease is a complex topic. Before examining the root causes of human disease, it is worthwhile to revisit the concept of risk analysis. American citizens enjoy an ever-increasing life span, better health, and a higher standard of living than their ancestors could have imagined. Despite this evidence of well-being, concerns about **diet–health relationships** persist. Consumer concerns are perpetrated when reports fail to account for the following:

- Wholesale extrapolation of results obtained from lab animal models to humans
- The fact that natural compounds may contribute significantly more risk than human-made compounds
- Effect of dose rate on disease incidence

In an environment where sensationalistic headlines are all too common, urban myths are perpetuated via social media, and where scientific knowledge is low, consumers are advised to be wary of "**junk science**." Warning signs that results of a study are being incorrectly represented are as follows:

1. The results recommend changes that offer quick-fix promises.
2. Foods are described as "good" versus "bad."
3. Simplistic conclusions are offered from a complex study.
4. The study was not peer reviewed.
5. Recommendations ignore differences between individuals or among groups.
6. Results are interpreted to offer significant negative consequences from a specific food item or diet selection.

Sound, data-based research work often appears to be lost in an emotionally charged issue. Many organizations and individuals have occasionally based judgments and decisions on emotion rather than on the best accumulated research facts. Scientific principles should be identified so that decisions are based on true relationships. It is also important to recognize that epidemiological studies that attempt to characterize influences on disease occurrence often times are only able to determine associated risk factors as opposed to causative factors. Risk factors should not be assumed to cause disease but rather should be evaluated from the opportunity to alter dietary and lifestyle habits to improve overall health and well-being.

The major known risk factors associated with **coronary heart disease (CHD)** are genetics (a family history of CHD), high blood cholesterol, smoking, hypertension, physical inactivity, and obesity. Obesity caused by excess caloric intake is a major nutritional problem in the United States. Of the leading 10 causes of death in the United States, obesity is considered a risk factor in five (CHD, stroke, hypertension, type II diabetes, and some forms of cancer). More than 50% of U.S. adults exceed their recommended weights. Interestingly enough, survey results suggest that consumers are less concerned about caloric and fat intake than they were in 1990. In a 10-year span, the percentage of consumers who reported that they were always conscious of caloric intake fell from 40% to 25%. Consumers reporting that they were always cautious of caloric intake fell from 51% to 33% over the same time period. These trends motivated significant expenditures by U.S. Department of Agriculture (USDA) and other federal government agencies to increase consumer awareness but to date there is not clear evidence of success by these campaigns.

It is generally accepted that consumption of animal fat by humans causes an increase in the level of cholesterol in the blood, while consumption of vegetable oils (polyunsaturated fats) causes a decrease in blood cholesterol concentration. These relationships led to the theory that there is a relationship between consumption of animal fats and the incidence of **atherosclerosis** ("plugging" of the arteries with fatty tissue), which in turn results in an increased likelihood of CHD.

Evidence supporting the proposed blood cholesterol—heart disease relationship is still in question. Studies in which dietary fat intake has been modified, either in kind or amount, did not demonstrate significantly reduced mortality rates in test populations. Changing diets from animal fats to vegetable fats has not improved the heart disease record. There are data that suggest that poor health conditions can also result from people eating diets high in polyunsaturated fats.

Most consumers do not know the difference between saturated and polyunsaturated fats. Through many margarine and vegetable oil commercials, however, they have been informed that saturated is "bad" and **unsaturated** is "good." In a review of the diet and heart disease relationship, some medical doctors argue that the dietary–heart hypothesis became popular because a combination of the urgent pressure of special interest groups or health agencies, oil-food companies, and ambitious scientists had transformed that hypothesis into absolute fact.

Senator George McGovern's committee on Nutrition and Human Needs in the late 1960s and early 1970s shifted its emphasis from developing policy aimed at eliminating malnutrition to dealing with the issue of consuming too many calories. Largely influenced by a self-proclaimed diet expert, McGovern's committee released a document titled "Dietary Goals for the United States" based on two days of testimony. Written by a journalist with absolutely no background in science, nutrition, or health, the report would become the basis for a national policy focused on dietary fat. Twenty-five years later, there is still not compelling and clear evidence as to the effect of dietary cholesterol and fat intake on human longevity.

The relationship between cholesterol intake and death from CHD is inconclusive. All **saturated fatty acids** are not equal in terms of their effect on serum cholesterol. For example, it is important to recognize that less than half of the fatty acids found in beef fat are saturated. The 18-carbon length fatty acid (stearic acid) has either a neutral effect on serum cholesterol or may actually lower serum cholesterol concentrations when substituted for other saturated fatty acids. Stearic acid comprises almost one-third of the saturated fatty acids in beef.

For example, a select grade steak will have about 50% of its total fat in monounsaturated form of which nearly 90% is oleic acid (the beneficial fat found in olive oil). The remaining one-half of the total fat is saturated but one-third of that is stearic acid, which is potentially beneficial, but at the very worst neutral in its effect. In total, more than one-half, and as much as three-quarters, of the fat in the steak will lower cholesterol levels (Taubes, 2001). Trans fatty acids are unsaturated fatty acids structurally characterized by a trans arrangement of alkyl chains. Trans fatty acids are formed during the hydrogenation of vegetable oils and have been linked to an increase in blood cholesterol. Foods that may contain trans fatty acids include margarine, vegetable cooking oils, many bakery goods and prepackaged mixes, french fries, and packaged popcorn. Food manufacturers voluntarily lowered the amounts of trans fats in their food products by more than 73% since 2005, mostly by reformulating products. The FDA reported that the average daily intake of trans fats by Americans fell from 4.6 grams daily in 2003 to 1 gram in 2012. Yet, in 2013, FDA proposed an outright ban of trans fatty acids in food processing and preparation. While some welcomed the proposed regulation many view this as significant government overreach with limited impacts on public health in light of the aforementioned trends.

Cholesterol is a naturally occurring substance in the human body. Every cell manufactures cholesterol on a daily basis. The average human turns over (uses and replenishes) 2,000 mg of cholesterol daily. Average dietary consumption of cholesterol is approximately 600 mg daily. Therefore, the body makes 1,400 mg each day to meet its needs.

The body cannot use cholesterol unless it is joined with a water-soluble protein, creating complexes known as lipoproteins. There are several different types of lipoproteins. Research workers have identified two of these lipoproteins: **HDL (high-density lipoprotein)** and **LDL (low-density lipoprotein)**. High blood levels of the LDLs have been generally associated with increased cardiovascular problems, while some research data show that higher blood levels of HDLs may reduce heart attacks by 20%. Additional research with laboratory animals has shown that those fed beef had HDL levels 33% higher than animals fed soybean diets.

There is evidence of genetic differences in the proportion of HDLs and LDLs in individuals. People who are overweight, nonexercisers, and cigarette smokers have higher proportions of LDLs than those who are lean, exercisers, and nonsmokers.

Although the exact roles of dietary cholesterol and blood levels in the development of CHD are not known, it would appear logical from the existing information to use prudence in implementing drastic changes in dietary habits. Certainly, those individuals with high health risks primarily due to genetic background should take the greatest precautions.

On an average per-capita basis, red meat and poultry supply less than 16 grams of fat per day. This level of fat (15.6 g) is less than 24% of the 67 grams of fat per day recommended by the American Heart Association for a 2,000-calorie-per-day diet. The 15.6 grams of fat represent 30% of the calories from fat where each gram of fat contains 9 calories. In the past 25 years, the source of fat in the diet has shifted as meat has become leaner and as consumers have added more processed snacks, desserts, and other foods that contain relatively high levels of fats and oils of which the vast majority are of plant origin (Fig. 23.6).

Some individuals argue that consumers eat too much red meat and that excessive red meat consumption causes cancer. Evidence to support this statement is questionable as valid estimates of consumption range from 2.8 to 3.4 ounces per day which is well within dietary recommendations. For example, the American Heart Association recommends 3.5 oz of cooked meat per person on a daily basis. Based on this recommendation, the average U.S. per-capita consumption of meat is not

Figure 23.6

Dietary shifts in meat, sugar and fat consumption between 1985 and 2009.

Source: USDA.

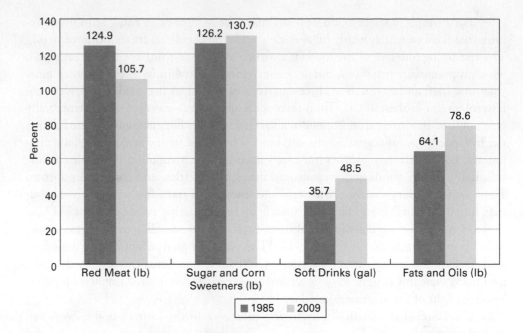

Figure 23.7

(A) Percent contribution to protein intake from various food groups. (B) Percent contribution to total red meat consumption from various sources.

Source: Adapted from USDA.

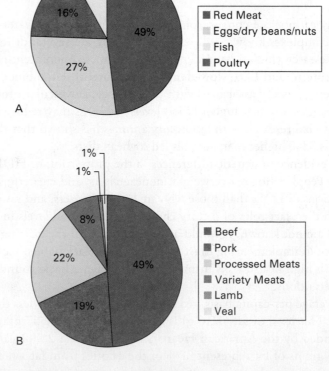

excessive. Figure 23.7A illustrates the sources of protein intake in the United States as well as the makeup of red meat consumption by source in Figure 23.7B.

Based on a comprehensive review of the scientific literature investigating the role of meat consumption and cancer, Alexander et al. (2010) concluded that red meat consumption has not been shown to be a causative factor for cancer. Their conclusion was based on the facts that red/processed meat intake and cancer incidence were weak in magnitude and not statistically significant, results of studies were highly inconsistent across studies and within different samples of the population, and results are confounded by lifestyle and other dietary choices.

Austin et al. (1997) concluded that if an association is shown between red meat and cancer, one of three explanations is possible: (1) the association is due to chance or some bias in the study method; (2) the association is due to confounding evidence (i.e., meat and cancer are associated only because they both are related to some common underlying condition such as a low intake of cancer-protective fruits/vegetables); (3) the association is real. Whether or not it can be concluded that red meat is a risk factor for cancer depends on several criteria: consistency of the association, strength of the association (relative risk), specificity of the association, and congruence with existing knowledge (e.g., is there an explanation or biological mechanism).

The criteria of "consistency" require that the association be repeated under different circumstances, such as among various population groups and among individuals within a population. When a positive association between red meat and a specific cancer is demonstrated, it generally is weak. Moreover, the association between consumption of red meat of different types (beef, pork, lamb, and processed meats) and specific cancers is inconsistent. Although it has been suggested that meat components such as fat, protein, or iron, or chemicals formed during the cooking of meat might be carcinogenic, these hypotheses remain unproven. On the contrary, meat contains some components such as conjugated linoleic acid, a fatty acid, which may protect against cancer.

Inferring a relationship from epidemiological studies of diet and chronic disease is particularly difficult due to several characteristics of both diet and chronic disease. First, there are several problems with accurately quantifying dietary intake. This was a major problem in most of the epidemiological studies reviewed by Alexander et al. (2010) and Austin et al. (1997). Second, diseases such as cancer are caused by a variety of genetic and environmental factors. Diet (e.g., red meat intake) is only one of many lifestyle factors that may influence risk of developing a disease. Also, chronic diseases such as cancer tend to have a long latency period during which time changes in many factors may occur. For these reasons, it is difficult to determine if an association, such as one between red meat and cancer is real.

Epidemiological investigations can identify risk factors, not a cause-and-effect relationship. Only when epidemiological findings are supported by information from other types of scientific studies, such as experimental animal studies and human clinical trials, can a decision regarding a causal relationship be made on firmer ground.

Dietary Guidelines

The U.S. government has released several reports that outline dietary guidelines and goals for American citizens. Some argue that it is the government's responsibility to provide people with information about diet; others support scientifically based guidelines, but believe Americans should have freedom of choice.

The *Dietary Guidelines* published by the USDA and Department of Health and Human Services (HHS) include broad recommendations and are a reasonable attempt to initiate sound nutritional practices among individuals. The revised 1990 *Dietary Guidelines*, drew heavily on two diet and health reports by the National Academy of Sciences and the U.S. Surgeon General, that recommended not more than 30% of daily calories come from fat and that less than 10% come from saturated fat. In 1992, the USDA began using the Food Guide Pyramid in nutritional education programs. The pyramid replaced the wheel graphic used to display the four basic food groups that has been used in nutritional education programs since the 1950s. In 2011, USDA replaced the food pyramid with a concept known as "My Plate" (Fig. 23.8). Table 23.9 outlines the age-specific intake recommendations from USDA.

Obesity has been identified as a significant challenge to the overall health of human beings. For the past quarter of a century, per capita red meat consumption has

Figure 23.8
Federal dietary recommendations are framed around the My Plate concept.
Source: USDA.

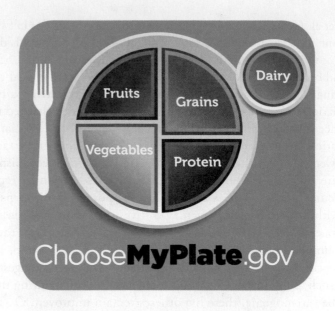

Table 23.9
AGE GROUP SPECIFIC DIETARY RECOMMENDATIONS

	Fruit[a]	Vegetables[a]	Grain[b]	Protein[b]	Dairy[a]	Oils[c]
Children 2–3 yrs	1	1	3	2	2	3
Children 4–8 yrs	1–1.5	1–1.5	5	4	2.5	4
Girls 9–13 yrs	1.5	2	5	5	3	5
Girls 14–18 yrs	1.5	2.5	6	5	3	5
Boys 9–13 yrs	1.5	2.5	6	5	3	5
Boys 14–18 yrs	2	3	8	6	3	6
Women 19–30 yrs	2	2.5	6	5	3	6
Women 31–50 yrs	1.5	2.5	6	5	3	5
Women 51+ yrs	1.5	2	5	5	3	5
Men 19–30 yrs	2	3	8	6	3	7
Men 31–50 yrs	2	3	7	6	3	6
Men 51+ yrs	2	2.5	6	5	3	6

[a]Number of cups.
[b]Ounce equivalent (half to be whole grain).
[c]Teaspoons.
Source: USDA.

declined 16%, consumption of other fats and oils has grown 23%, sugar consumption grew by 3%, and soft drink consumption rose by 32%. These data suggest that the overconsumption of fats and oils from processed foods, sugary foods, and soft drinks must be addressed to address the obesity issue.

Adherence to the suggested *Dietary Guidelines* can be easier by following these suggestions of the Dietary Guidelines Alliance:

1. Be realistic—make small, incremental changes in eating and exercise habits.
2. Be adventurous—expand your tastes to enjoy a variety of foods.
3. Be flexible—balance food consumption and exercise regime over several days instead of focusing on one meal or one day.
4. Be sensible—enjoy all foods—just don't overdo it. Choose sensible portion sizes.
5. Be active—exercise is a key to maintaining appropriate weight.

Vegetarian and vegan diets have been advocated as healthier than diets containing animal products. However, diets devoid of meat are relatively low in protein, saturated fat, long-chain n–3-fatty acids, retinol, vitamin B_{12}, and zinc while vegan diets are also low in calcium. A study by Key and coworkers in 2006 reported that there were no significant differences between meat-eaters and non-meat-eaters for incidence of hypertension, colorectal cancer, breast cancer, prostate cancer, and total mortality rate.

In the final analyses, balance and moderation are keys to appropriate and healthy dietary choices.

Food Safety

Although the food industry does not have a perfect record in regard to food safety, U.S. consumers enjoy a plentiful food supply that can arguably be called the safest in the world. Again, relative risk must be carefully evaluated when considering food safety issues. Assuring **food safety** is a joint responsibility shared by producers, processors, and consumers.

Food safety assurance requires the attention of consumers, producers, processors, retailers, and food-service outlets. Historically federal and state food inspection systems have been expected to assure food safety. Supply chain interventions can be implemented at various stages that significantly reduce the risk of food borne illness. These efforts on founded on **total quality management** systems based on the premise quality cannot be inspected into a product. Instead, intentional, science-based systems must be developed to assure quality and safety at every key step in the process.

One of the most critical control points is management of food temperature. The danger zone—40 to 140°F—offers conditions under which bacterial pathogens can multiply. Therefore, food-service establishments and consumers should strive to maintain raw and cooked foods outside the danger zone.

Additional steps that consumers should take include cooking and reheating food thoroughly, storing food at correct temperatures, avoiding cross-contamination between foods during storage and preparation, maintaining a high degree of personal sanitation, and keeping food preparation surfaces meticulously clean. These steps are most critical for establishments serving large numbers of consumers as most food-borne illness outbreaks are traced to foods served at dining establishments, catering facilities, or institutional food service (schools, hospitals, jails, etc).

Food scientists rank food hazards as: (1) microbial contamination, (2) naturally occurring toxins, (3) environmental contaminants (e.g., heavy metals), (4) pesticide residues, and (5) food additives.

Examples of food safety issues for the livestock industry include infection by **E. coli 0157:H7** and other foodborne pathogens (Table 23.10). Concerns about *E. coli* 0157:H7 bacterial infection outbreaks peaked in 1993, when 1,000 cases were reported in the United States. Over three-quarters of the 1993 incidents were attributed to improperly prepared and cooked ground beef, most of which was traced to a single food-service supplier.

Data reported by the Centers for Disease Control comparing foodborne infections occurring in 2012 to the baseline years of 1996–98 documented a 22% decline in foodborne illnesses due to *Campylobacter, Listeria, STEC 0157 (E. coli 0157:H7), Vibrio,* and *Yersina.* Illness rates caused by **Salmonella** over the same time period were unchanged. While the goal must be to eliminate foodborne illness outbreaks, these data show that large industries can make progress in food safety as both the number and rate of foodborne illness outbreaks has declined over the past two decades.

Table 23.10
AN OVERVIEW OF FOODBORNE, ILLNESS-CAUSING ORGANISMS

Name of Organism	Description	Infective Dose[a]	Symptoms	Annual Occurrence in the United States	Associated Foods
Salmonella	Rod-shaped, motile bacterium; widespread in animals, especially poultry and swine; gram-negative.	As few as 15–20 cells	Nausea, vomiting, abdominal cramps, fever.	2–4 million cases	Major: raw poultry, meats, eggs, dairy foods. Minor: coconuts, sauces, salad dressings, cake mixes, peanut butter, chocolate.
Clostridium botulism	Anaerobic, spore-forming rod; produces a pathogenic neurotoxin; 7 types of botulism are recognized; gram-positive.	A few ng of toxin	Weakness, vertigo; difficulty in breathing, swallowing; muscle weakness, constipation.	100 confirmed	Most frequently associated with poorly processed, home-canned foods.
Staphylococcus aureus	Spherical bacteria; some strains produce entero toxins; gram-positive.	Less than 1 mg of toxin in contaminated food will yield symptoms	Nausea, vomiting, retching, abdominal cramping.	Relatively unknown due to misdiagnosis	Foods that require a high degree of handling during preparation and that are kept at slightly elevated temperatures are typically involved.
Campylobacter jejuni	Slender, curvical, motile rod; gram-negative, relatively fragile.	400–500 bacteria	Diarrhea (watery or sticky), may contain blood and white cells; fever, abdominal pain, headache, muscle pain.	Leading cause of bacterial diarrhea	Major: raw chicken. Minor: unpasteurized milk, nonchlorinated water, seafood, hamburger, cheese, pork, eggs.
Listeria monocytogenes	Bacterium, non-spore forming, very hardy; gram-positive.	<1,000 total organisms	Flu-like symptoms may be followed bysepticemia, meningitis, encephalitis, or cervical infections in pregnant women.	350 with 48 fatalities	Major: soft cheese and ground meat. Minor: poultry, dairy products, hot dogs, seafood, vegetables.
Escherichia coli 0157:H7 (STEC)	*E. coli* is a normal inhabitant of the gut of all animals. There are 4 classes of virulant *E. coli*. Majority of *E. coli* serve to suppress harmful bacteria and to synthesize vitamins. 0157:H7 produces a potent toxin(s).	10 organisms	Severe cramping, diarrhea (may become bloody); typically self-limiting with 8-day duration; young children and the elderly may develop hemolytic uremic syndrome and renal failure.	8,000 confirmed with 20 fatalities	Major: undercooked ground beef, human- to-human contact, environmental exposure to organisms. Minor: poultry, apple cider, raw milk, vegetables, hot dogs, mayonnaise, salad bar items.

[a]Dose varies depending on immune state of the individual.
Source: USDA.

In fact, microbial contamination has resulted in a variety of new processing technologies designed to improve food safety. Some of the innovations include steam vacuuming, hot water washing, weak acid rinses, and steam pasteurization. Most processors are working to implement a multiple set of hurdles or blockades to microbial growth. Such systems have tremendous potential to enhance food safety.

Another food safety technology is **radiation pasteurization** or irradiation. Unlike the previously mentioned technologies, the use of radiation has been surrounded by a degree of controversy. The use of low doses of gamma rays, X-rays, and electron beams can safely control pathogenic microbes. The use of this technology doesn't make foods radioactive and essentially has little effect on food because its cells are no longer active. Small changes to vitamins, such as B_1 and C, do occur but are not dangerous to human health. As consumers learn more about the technology, their acceptance levels have increased. Irradiation of wheat and flour was approved in 1963 and since that time pork, poultry, red meat, and eggs have also been approved. Society's desire for an assured level of food safety will continue to create demand for new and improved technologies.

Microorganisms

Microorganisms, such as molds, viruses, bacteria, yeasts, and parasites, are universally present. These microorganisms play potentially beneficial, harmful, or neutral roles in regard to food. Molds and yeasts are important in baking and cheese making, while some bacteria act as inhibitors to other harmful bacteria. Harmful microorganisms either cause spoilage or disease. Disease-causing microorganisms are referred to as **pathogens**.

Food spoilage microbes are usually visually apparent, while foodborne pathogens may be difficult to detect. The Centers for Disease Control reports that 77% of foodborne illness outbreaks originate due to improper handling and cooking in food-service establishments, 20% due to improper food handling and cooking in the home, and 3% due to food manufacturing defects or failure.

Sanitation and temperature control are critical elements in controlling potentially harmful bacteria. Processors, retailers, and food-service operators can minimize contamination via proper equipment sanitation, disciplined personnel hygiene practices, proper handling and storage, cooking to appropriate temperatures, and systematically assessing food safety efforts.

Consumers should never purchase outdated goods or those whose containers have been broken or dented. Furthermore, refrigerator temperatures should be maintained at or below 40°F and freezer temperatures at or below 0°F. Consumers are advised to follow label directions for preparation and serving, to avoid cross-contamination of foods, and to wash, rinse, and sanitize equipment, countertops, and cutting surfaces. Foods should never be thawed at room temperature, food should be cooked to appropriate temperatures (typically 140°F or greater, 155°F for ground meats, and 165°F for leftovers), and during storage or preservation foods should be kept outside the range of 40–190°F.

One consequence of a food safety failure is often the initiation of a national product recall. These product recalls are expensive from both the loss of product and the loss of consumer confidence. The five largest meat product recalls amounted to 19 to 35 million pounds of product, resulting in millions of dollars in lost revenue. Prevention and early detection systems are being developed to better provide wholesome products to consumers.

Residues

Much of the U.S. public feels that chemical **residues** from pesticides, hormones, and food additives are the greatest food safety threats. What is critical to assess is the level of the residue in the food, the amount consumed, and the risks associated with that level of consumption. Understanding risk and accounting for the major innovations in measuring technologies are fundamental to accurately interpreting residue data.

Consumers may perceive that giving farm animals anabolic steroids/hormones will produce meat with residues that are harmful to health. Examples would be using a cattle ear implant containing growth promoting hormones such as estrogen to increase growth, and repartitioning feed energy to produce more muscle and less fat. As a result of this process, the average amount of estrogen in a 3-oz serving of beef is increased from 1.2 to 1.9 ng (a **nanogram** is one-billionth of a gram). Similar-size servings of potatoes, ice cream, and soybean oil contain more nanograms of estrogen (200, 500, and 1 million times more, respectively), compared to beef from implanted cattle. Considering that the average nonpregnant human female produces 480,000 nanograms of estrogen each day by normal physiological processes, the increased estrogen consumed in beef is of no physiological or medical consequence; that is, there are no problems associated with ingestion of meat from an animal.

There is a perception that so-called natural or **organic** animal products contain far fewer residues of antibiotics, hormones, drugs, and pesticides. Scientific tests of organic versus conventional beef have shown no evidence to confirm this perception. Monitoring by USDA shows that there are almost zero (< .0001) violative residues of antibiotics in beef from fed steers and heifers. For nonfed cows and bulls the incidence rate of violative residues was higher (< .003) but still extremely low. Nonetheless, a segment of consumers desire the opportunity to purchase natural or organic foods.

Subtherapeutic **antibiotic** use in the livestock industry is predicated on the principle of facilitating the growth of resistant bacteria while restraining the growth of competing bacteria. Criticisms of the practice arise from concerns about the possibility of the resistant bacteria flourishing in farm environments and eventually infecting humans.

About 45% of U.S. antibiotic usage is by the animal industry (including pets) and nearly 90% of those go toward the treatment or prevention of specific diseases. Thus, only about 6% of all antibiotics are used at the subtherapeutic level to promote growth and efficiency. The leading compounds found to be in violation by Food Safety and Inspection Service testing of market dairy cows were penicillin (245 violations), flunixin (167 violations), and sulfadimethoxine (142 violations).

Overprescribing and misuse of antibiotics in human medicine is also of concern. In fact, on a weight basis, humans use nearly 10 times the tonnage of antibiotics as compared to livestock. The leading predictor of a patient's likelihood of developing a drug-resistant infection is whether the patient was ingesting antibiotics in the previous 3 months. Furthermore, experts point to the fact that many people fail to complete the full regimen of prescribed antibiotic therapy, thus creating opportunities for development of microbial resistance.

Furthermore, human levels of antibiotic resistance are rising while resistance in livestock has remained historically stable. Proponents of banning subtherapeutic antibiotics fail to study the lessons in Europe where these products were banned in the early to mid-1990s. For instance, in Denmark the use of growth-promoting antibiotics was banned in 1997. As a result, illnesses requiring antibiotic therapy rose dramatically in swine herds with the end result being a dramatic rise in antibiotic use by the industry. Furthermore, pig productivity dropped significantly, mortality increased, and feed efficiency declined. Researchers at Iowa State predict that such a ban in the United States would increase the cost of production by $4.00 per pig.

The most logical approach to this issue is to assure responsible use of antibiotics, avoidance of the use of antibiotics of critical value to human medicine in livestock, continual antibiotic-resistance monitoring, and conducting research that evaluates the specific and direct cause-and-effect relationships.

Quality Assurance

Animal production began a new era in the 1990s as production practices were evaluated in light of their impact on the process of assuring that consumers received wholesome products of highly perceived value. Quality assurance programs have been instituted in the food and fiber industries as a means to coordinate animal health care practices and management protocols.

Quality assurance programs in the food animal industries have focused on (1) superior record keeping, (2) safe use of pharmaceutical products, with a special emphasis on product administration in accordance with label restrictions, (3) minimizing intramuscular injections, (4) assuring quality of animal feeds, and (5) avoiding residues.

Most quality assurance efforts are designed in accordance with **Hazard Analysis Critical Control Point (HACCP)** protocols. HACCP programs utilize a seven-step process to ensure that product quality is attained throughout the production process (Table 23.11).

Quality assurance efforts require an effective process of communicating critical information throughout the production and processing chain. For example, livestock producers can avoid drug residues in animal products by establishing effective systems of management that include the following principles:

- Emphasize disease prevention in animal health management.
- Select and utilize antibiotics thoughtfully by developing treatment protocols in consultation with the herd veterinarian.
- Treat the fewest number of animals possible, follow the treatment protocol precisely, and follow withdrawal periods to the letter.
- Record treatment, outcome, product, and withdrawal times while providing employee training.

Quality assurance efforts must be applied at every level of production, processing, and preparation to assure that food is safe and wholesome. For example, the

---

Table 23.11
SEVEN STEPS FOR UTILIZING HAZARD ANALYSIS CRITICAL CONTROL POINTS

Number	Steps
1	Conduct a hazard analysis.
2	Identify critical control points in the process.
3	Establish preventative measure limits for each control point.
4	Establish monitoring process for critical control points.
5	Take corrective action.
6	Institute effective record-keeping system.
7	Monitor system on continuous basis and assure it works.

following critical control points are important to the ability of food-service establishments to assure food safety:

Food receiving
• Provide visual, packaging, and temperature inspection

Storage
• Hold refrigerated foods at 41°F or lower
• Store frozen foods at 0°F or lower
• Maintain dry storage temperatures between 50 and 70°F
• Implement first in–first out inventory control

Preparation
• Avoid cross-contamination
• Meats prepared separately from produce
• Poultry cooked to 165°F for 15 seconds
• Ground beef cooked to 155°F for 15 seconds
• Pork cooked to 145°F for 15 seconds
• Beef roasts heated to 145°F for 3 minutes
• Steaks heated to 145°F for 15 seconds
• Minimize human traffic in food prep areas

Cooking surfaces
• Scheduled cleaning throughout the business day
• Deep-cleaning at the close of business each day

Employee break rooms and wash areas
• Scheduled cleaning and sanitation
• Thorough and frequent handwashing
• Separate restrooms for employees and customers when possible
• Thorough cleaning at least daily

Dining areas
• Tables wiped and sanitized between customers
• Use separate cleaning tools for dining area, kitchen area, and restrooms to avoid cross-contamination

PRODUCTION TECHNOLOGIES

Agriculture has been successful in feeding the world's population in large part due to innovations in management, scientific breakthroughs that enhance production capability and efficiency, and transfer of technology in a free market system. Many of these approaches are referred to as biotechnologies. **Biotechnology** has been defined as the application of physical, chemical, and engineering principles to biological systems. Therefore, biotechnology ranges from the use of vaccines to the manipulation of the genome. The application of biotechnology to the field of agriculture offers significant opportunities for the improvement of human health, food production, animal health, and modification of the impact of agricultural production on the environment. Biotechnical applications will help reduce dependence on agri-chemicals by producing pest/disease-resistant plants and animals. Biotechnology will improve the efficacy of vaccines or even produce milk that boosts human immunity against bacterial infection. The use of biotechnology offers excellent benefits to human medicine via mass production of proteins such as blood factor VIII, tissue plasminogen activator, and antithrombin III, which serve to control bleeding common to hemophilia, to break up blood clots during

Table 23.12

BENEFITS FROM THE APPLICATION OF BIOTECHNOLOGY TO ANIMAL AGRICULTURE

Human Health

Decrease morbidity and mortality from disease

Biopharmaceuticals

Xenotransplantation

Food Production

Feeding livestock

- Improve efficiency of feed utilization
- Enhance nutritional value of feed grains and forages

Metabolic modifiers

- Enhance production efficiency
- Improve lean: fat ratio
- Increase milk yields
- Decrease production of animal waste

Microbial genomic application to enhance food safety

- Development of diagnostic tools for rapid identification of pathogens
- Creation of vaccines against foodborne pathogens

Animal Health

Vaccine development

Selection of animals with enhanced disease resistance capability

Modification of Environmental Impact

Utilization of genetically modified crops on manure composition and quantity

- Decrease phosphorus excretion
- Decrease nitrogen excretion

Source: Adapted from Council for Agricultural Science and Technology, 2003.

cardiac arrest, and as a blood thinner, respectively. Table 23.12 provides a summary of the opportunities for biotechnology to add value to animal agriculture.

In the case of **genetic engineering**, the perception of people ranges from "progress" to "mutant/monster." Even management practices such as crossbreeding that have been in effect for years in both plant and animal agriculture receive some degree of disfavor among consumers.

Clearly consumers have contradictory notions about genetic engineering, as these techniques are viewed more favorably when applied to plants than animals, and are believed to have the potential to improve life but require strict regulations and labeling restrictions. Furthermore, consumers are very unsure as to whom to trust as a reliable source of information about genetic engineering. Several studies have found that scientists are not always seen as reliable sources of information about biotechnology. However, the 2000 Philip Morris survey of consumer opinions on agriculture reported that 57% of consumers supported biotechnology if it improved the taste of food, 69% if it increased food production, 65% if nutritional values were enhanced, and 73% if pesticide use was reduced as a result.

It is important to recognize that biotechnology does not always offer simple, cost-effective solutions. Furthermore, the volume of genetic information carried in one cell limits the speed of breakthroughs arising from genetic engineering—it is estimated that there are nearly 3 billion pieces of genetic information in bovine cells. Additionally, evidence is accumulating that genetic markers may affect different traits in different breeds. Reproductive success may also be less than desirable. The oocyte maturation rate resulting from nuclear transfer is 70% and 80%, respectively, for cattle and hogs. Yet only 25% and 15% of embryos from these species develop to the blastocyst stage. Numerous problems with clones have become apparent including survival rate, accelerated aging, low reproductive efficiency, and increased birth weights. The cost of bovine cloning is a minimum of $15,000, which serves as a significant barrier to adoption.

Perhaps genetic engineering of microorganisms to enhance food processing will provide the most value to society. The first federal approval of a genetically engineered food product was the enzyme chymosin used in cheese making. Produced by genetically modified bacteria, chymosin is used to replace calf rennet is not used in more than half of all cheese manufacturing and its benefits are improved reliability of supply, purity, and cost-effectiveness.

Biotechnology brings with it a new set of moral and ethical dilemmas that will have to be resolved by society. Patenting of genetically engineered animals, regulation of the release of biotechnically altered species into the environment, and the use of biotechnology in human medicine are only a few of the issues that will face consumers and producers alike.

Organic and Natural Products

Consumer demand for food that has been grown under organic conditions has grown at a slow but steady rate. As such, USDA developed the National Organic Program guidelines to provide uniform standards for the voluntary labeling of products grown under organic management.

Some of the guidelines include the following: Livestock must be maintained under continuous organic management from the last third of gestation or hatching with three exceptions—(1) edible poultry products must be from birds that have been managed organically beginning no later than the second day of life; (2) milk products must be from females managed continuously under organic conditions no later than one year prior to lactation; and (3) purchased breeding stock from nonorganic sources must be introduced into the herd prior to the last third of gestation. Further provisions are prescribed in regards to feedstuffs, health products, record keeping, and other management protocols.

Producers should carefully evaluate the cost and returns associated with aligning their enterprise with an organic supply chain. A careful evaluation of the requirements and restrictions along with a detailed market analysis is recommended. The benefits to consumers must also be considered. Recent studies have clearly shown that there are no significant nutritional differences between organically grown versus conventionally grown foods.

While some have advocated that despite the economic costs in higher food prices and reduced productivity, the United States should shift all agricultural production to organic standards. Unfortunately, this viewpoint fails to account for the dramatic impacts of such an approach on the ability of the agricultural system to adequately meet human demand for food as well as a series of negative environmental impacts that would result. As Norman Borlaug, the father of the green revolution in food production, stated in 2000, "even if you could use all the organic material that you have—the animal manures, human waste, the plant residues—and get them back on the soil, you couldn't feed more than 4 billion people. In addition, if all agriculture were organic, you would have to increase cropland area dramatically, spreading out into marginal areas and cutting down millions of acres of forests."

Jason Clay of the World Wildlife Fund points out that "Backyard vegetables are fine. So are organics … But solutions to the global food crisis will come from big business, genetically engineered crops and large-scale farms." Research shows that total greenhouse gas emissions per pound of beef production are almost double for organic grass-fed beef as compared to conventional systems of production. The answers are not conveniently black and white but will require critical analysis to determine the optimization of technology in food production.

ISSUES AND OPPORTUNITIES

The issues confronting animal agriculture involve a variety of perspectives and the debate is frequently bastardized by political motivation and the desire for power. Unfortunately, these issues are often discussed in the public arena by those who prefer to focus on the politics of fear, the use of incomplete or erroneous data to support their own position, and by a media who depends on sound bites instead of depth and critical analysis. Too often, politicians and regulators seek simple solutions to complex problems with resulting unintended consequences that put food security, agricultural sustainability, and rural infrastructure at risk. It is vital that fact-based decisions are made with a full awareness of the cost and benefits that accompany each alternative. It is equally important that those who make their living in the livestock industry continue to wrestle with the issues presented in this chapter while always seeking to find solutions that best meet the needs of consumers, producers, and the animals that provide so much benefit to society.

With challenging issues come opportunities. Those individuals who face the problems and build effective teams and innovative solutions to meet the challenges will provide immeasurable value to humanity. The challenge is to separate the myths, perceptions, and emotions from the scientific truths and realities. U.S. consumers can be provided highly palatable animal products that are nutritious, wholesome, sustainable, and affordable. With the creativity of people, resources and opportunities can be sustained for generations to provide the economic, physical, and emotional well-being of the U.S. and world population—even with significant population increases.

CHAPTER SUMMARY

- Issues facing the livestock industry can be categorized as animal welfare, biotechnology, environment, diet and health, food safety, and marketing.

- Management of these issues in terms of public perception is important to the stability of the livestock industry.

- Application of the principles of sound stewardship and husbandry practices is critical to the future of livestock enterprises.

KEY WORDS

stockmanship
stewardship
polarities
sustainability
viability
continuity
triple bottom line
assessing risk
federal lands

manure managementcode of ethics
animal welfare
animal rights
animal liberationists
diet–health relationships
junk science
coronary heart disease (CHD)
atherosclerosis
unsaturated

saturated fatty acids
cholesterol
high-density lipoprotein (HDL)
low-density lipoprotein (LDL)
Dietary Guidelines
food safety
total quality management
E. coli 0157:H7
Salmonella
radiation pasteurization

pathogens
residues
nanogram
organic
antibiotic
Hazard Analysis Critical
   Control Point (HACCP)
biotechnology
genetic engineering

REVIEW QUESTIONS

1. What are the challenges and opportunities associated with the goal of feeding 9 billion people by 2050?
2. Discuss the concepts of trade-offs and polarity as they relate to making decisions at the micro and macro levels.
3. Define sustainability and the various factors that should be considered in making decisions.
4. Explain the concepts of continuity and viability.
5. How does food waste impact the sustainability of food production.
6. Discuss the importance of valid comparisons and accurate risk assessment to effective problem solving.
7. What are several approaches to determining risk potential?
8. Discuss the history of environmentalism.
9. Describe the shifts in environmental policy and their impacts on agriculture.
10. Discuss land use categories.
11. Describe issues and policies related to the federal lands.
12. Discuss waste management protocols and effective manure management strategies.
13. Describe management protocols that can be implemented to affect air quality.

14. Discuss water use by animal agriculture and methodologies to improve water quality.
15. Discuss the spectrum of perspectives relative to animal well-being.
16. What are the major animal well-being issues relative to each species?
17. What are the signs of junk science?
18. Discuss the history of the diet-health issue.
19. Describe the dietary guidelines and approaches for consumers to align with the recommendations.
20. Discuss the role of diet in obesity.
21. Discuss the role of temperature control in assuring food safety.
22. Rank the five food hazards.
23. Discuss trends in the occurrence of foodborne illness.
24. Discuss the primary pathogens of concern relative to foodborne illness.
25. Discuss the role of quality assurance efforts and HACCP in mitigating foodborne illness.
26. Discuss the pros and cons of production technologies.

SELECTED REFERENCES

Animal Industry Foundation. 1988. *Animal Agriculture: Myths and Facts*. Arlington, VA: Animal Industry Foundation.

Assessment of Marketing Strategies to Enhance Returns to Lamb Producers. 1991. College Station, TX: Texas Agricultural Market Research Center.

Alexander, D D., et al. 2010. Red meat and processed meat consumption and cancer—A technical summary of the epidipiologic evidence. Health Science Practice, Exponent, Inc.

Austin, H. and L. D. McBean. 1997. Red meat and cancer. A review of current epidemiological findings. Chicago, IL: National Cattlemen's Beef Association.

Bauman, D. E. 1992. Bovine somatotropin: Review of an emerging animal technology. *J. Dairy Sci.* 75:3432–3451.

Bjerklie, S. 1990. Poultry's decade of issues. *Meat Poul.* 36:24.

Capper, J. 2011. Replacing rose-tinted spectacles with a high-powered microscope: The historical versus modern carbon footprint of animal agriculture. *Anim. Frontiers* 26–32.

Cheeke, P. R. 1993. *Impacts of Livestock Production on Society. Diet/Health and the Environment.* Danville, IL: Interstate Publishers, Inc.

Council for Agricultural Science and Technology. 2013. *Animal Feed vs. Human Food: Challenges and Opportunities in Sustaining Animal Agriculture toward 2050.* Ames, IA.

Council for Agricultural Science and Technology. 2012. *Water and Land Issues Associated with Animal Agriculture: A U.S. Perspective.* Ames, IA.

Council for Agricultural Science and Technology. 2011. *Air Issues Associate with Animal Agriculture: A North American Perspective.* Ames, IA.

Council for Agricultural Science and Technology. 2003. *Biotechnology in Animal Agriculture: An Overview.* Ames, IA.

Davidson, M. H., D. Hunningshake, K. C. Maki, P. O. Kwiterovich, and S. Kafonek. 1999. Comparison of the effects of lean red meat vs. lean white meat on serum lipid levels among free-living persons with hypercholesterolemia. *Arch. Intern. Med.* 159:1331.

Food Marketing Institute. Animal Welfare Program. 2002. Washington, DC.

Hafs, H. D. (ed.). 1993. Genetically modified livestock: Progress, prospects and issues. *J. Anim. Sci.* 71(Suppl. 3).

Hallman, W. K. and J. Metcalfe. 1993. *Public Perceptions of Agricultural Biotechnology: A Survey of New Jersey Residents.* Ecosystem Policy Research Center, Rutgers University.

Jones, K., T. W. Kelsey, P. A. Nordstrom, L. L. Wilson, A. N. Maretski, and C. W. Pitt. 2000. Neighbors perceptions of animal agriculture. *Prof. An. Scientist* 16:105–110.

Key, T. J., P. N. Appleby, and M. S. Rosell. 2006. Health effects of vegetarian and vegan diets. *Proc. Nutrit. Soc.* 65:35–41.

Kliebenstein, J. and S. Hurley. 2001. Manure management: Consumers' perception levels. *PORK,* Sept., 32–34.

Meat industry: A sampling of the issues and controversies affecting meat and poultry industries all over the planet. 1989. *Meat and Poultry* 35:12–14, 41, 43.

National Animal Health Monitoring System. 1996. *Reference of 1996 Dairy Management Practices.* USDA: Veterinary Services.

National Animal Health Monitoring System. 2002. *Reference of 2000 Swine Health and Environmental Management.* USDA: Veterinary Services.

National Cattlemen's Foundation. 1990. *Special Interest Group Profiles.* Washington, DC: Hill and Knowlton.

National Pork Board and National Pork Producers Council. 2012. Responsible Farming: Pork industry progress report.

Parker, J. 2011. The 9 billion-people question: A special report on feeding the world. The Economist (Feb. 26).

Rollin, B. E. 1990. Animal welfare, animal rights, and agriculture. *J. Anim. Sci.* 68:3456.

Rollin, B. E. 1993. Animal Production and the New Social Ethic for Animals. Proc. Food Animal Well-Being: USDA and Purdue University.

Rollin, B. E. 1995. *Farm Animal Welfare: Social, Bioethical and Research Issues.* Ames, IA: Iowa State University Press.

Simmons, J. 2010. *Technologies role in the 21st Century; Food Economics and Consumer Choice. Proceedings of the 2010 International Livestock Congress.* Denver, CO.

Smidt, D. 1983. *Indicators Relevant to Farm Animal Welfare.* Boston: Martinus Nijhoff Publishers.

Smith, G. C. 1993. Anti-meat propaganda: Combating myths with scientific facts. Presented to the International Meat Industry Convention. Chicago, IL.

Susskind, L. and P. Field. 1996. *Dealing with an Angry Public—The Mutual-Gains Approach to Dissolving Disputes.* New York: The Free Press.

Taubes, G. 2001. The soft science of dietary fat. *Science* 291:2536–2545.

Taylor, R. E. and T. G. Field. 2007. *Beef Production and Management Decisions.* Upper Saddle River, NJ: Pearson-Prentice Hall.

Vieux, F., L. Soler, D. Touazi, and N. Darmon. 2013. High nutritional quality is not associated with low greenhouse gas emissions in self-selected diets of French adults. *Am. J. Clinical Nutrition* 112:035105.

24

Beef Cattle Breeds and Breeding

learning objectives

- Explain the importance of breed type in the cattle industry
- Discuss the reasons for broadening the genetic base in the cattle business
- Outline the various approaches to improving cattle performance in the major economic traits
- Discuss the process of bull and replacement female selection
- Explain the role of sire summaries in genetic improvement
- Describe the various approaches to crossbreeding

A **breed** of cattle is defined as a race or variety, the members of which are related by descent and similar in certain distinguishable characteristics. More than 250 breeds of cattle are recognized worldwide, and several hundred other varieties and types have not been identified with a breed name. The Texas Longhorn breed characterized the early history of the western cattle industry in the United States. The Longhorn was hardy, range-adapted, and disease-resistant but fell out of favor as the industry progressed with a focus on feedlot performance and superior carcass characteristics.

Some of the oldest recognized breeds in the United States were officially recognized as breeds during the middle to late 1800s. Most of these breeds originated from crossing and combining existing strains of cattle. When a breeder or group of breeders decided to establish a breed, distinguishing that breed from other breeds was of paramount importance; thus, major emphasis was placed on readily distinguishable visual characteristics, such as color, color pattern, polled or horned condition, and rather extreme differences in form and shape.

New cattle breeds, such as Brangus, Santa Gertrudis, Barzona, and Beefmaster, have come into existence in the United States during the past 50–75 years. Attempting to combine the desirable characteristics of several existing breeds has helped to develop these breeds. Currently there are new "breeds" of cattle being developed. They are sometimes called **composite breeds** or **synthetic breeds**. Examples include Balancers (Gelbvieh × Angus), LimFlex (Limousin × Angus), and SimmAngus (Simmental × Angus).

It was not long after some of the first breeds were developed that the word **purebred** was attached to them. Herd books and registry associations were established to assure the "purity" of each breed and to promote and improve each breed. *Purebred* refers to purity of ancestry, established by pedigree, which shows that only animals recorded in that particular breed have been mated to produce the animal in question. Purebreds, therefore, are cattle within the various breeds that have pedigrees recorded in their respective breed registry associations.

When viewing a herd of purebred Angus or Herefords, uniformity of color or color pattern can be noted. Because of this uniformity of one or two characteristics, the word *purebred* has come to imply genetic uniformity (homozygosity) of all characteristics. Cattle within the same breed are not highly homozygous because high levels of homozygosity occur only after many generations of close inbreeding. Close inbreeding has not been a breeding strategy favored in the beef industry. If breeds were uniform genetically, they could not be improved even if changes were desired.

GENETIC VARIATION

The genetic basis of cattle breeds and their comparison is not well understood by most livestock producers. Often the statement is made, "There is more variation within a breed than there is between breeds." The validity of this claim needs to be examined carefully. Considerable variation does exist within a breed for most of the economically important traits. This variation is depicted in Figure 24.1, which shows the number of calves of a particular breed that fall within certain weight-range categories at 205 days of age. Connecting the high points of each bar forms a bell-shaped curve. The solid line that separates a bell curve distribution into equal halves represents the breed average. Most of the calves are near the breed average; however, at the outer edge of the bell curve are high- and low-weaning weight calves. Note that there are progressively fewer calves as the distribution moves away from the average in both directions.

Figure 24.2 compares three hypothetical breeds of cattle in terms of weaning weight. The breed averages are different; however, the variation within each breed is comparable among all three. The statement "There is more variation within a breed than between breed averages," is more correct than the statement, "There is more variation within a breed than there is between breeds."

Figures 24.1 and 24.2 are hypothetical examples; however, they are based on realistic samples of data obtained from the various breeds of cattle. Figure 24.3 shows how

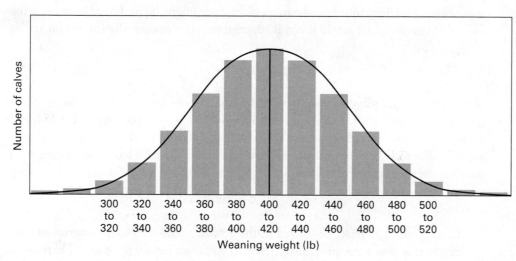

Figure 24.1
Variation or differences in weaning weight in beef cattle. The variation shown by the bell-shaped curve could be representative of a breed or a large herd. The vertical line in center is the average or the mean, which is 410 lb. Note that the number of calves is greater around the average and is less at the extremely light and extremely heavyweights. Source: Kevin Pond of Colorado State University. Used by permission of Kevin Pond.

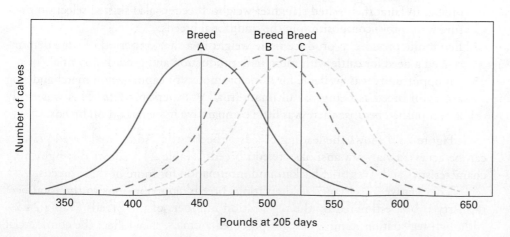

Figure 24.2
Comparison of breed averages and the variation within each breed for weaning weight in beef cattle. The vertical lines are the breed averages. Note that some individual animals in breeds B and C can be lower in weaning weight than the average of breed A. Source: Kevin Pond of Colorado State University. Used by permission of Kevin Pond.

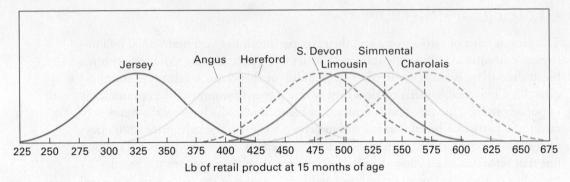

Figure 24.3
Breed differences in retail product when breeds are slaughtered at the same age. Source: USDA: ARS: U.S. Meat Animal Research Center.

breeds can differ greatly when only one trait is considered. Note how distinctly different the Charolais-sired calves are from the Jersey or Angus or Hereford calves. The statement, "There is more variation within the Charolais breed than between the Simmental and Jersey breeds" is not valid in this case. Rather, "There is more variation within the Charolais breed than between breed averages" is a more correct statement.

The information in Figure 24.3 supports Charolais as being the most superior breed based on breed averages in regards to retail product yield. However, this comparison is only for one trait. While maximizing retail yield might be appropriate for some cattle breeders, more moderate performance might be preferred in other situations. Furthermore, no single breed is superior for all economically important traits.

MAJOR U.S. BEEF BREEDS

Shorthorn, Hereford, and Angus were the major beef breeds in the United States during the early 1900s. During the 1960s and 1970s, the number of cattle breeds remained relatively stable at between 15 and 20. Today, more than 60 breeds of cattle are available to U.S. beef producers. However, fewer than 20 have a widespread impact on the national herd. Why the large importation of the different breeds from several different countries? Following are several reasons that stimulated the expansion of breed importations:

1. Grain feeding, a practice started in the 1940s, resulted in many overfat cattle—cattle that had been previously selected to fatten on forage diets. Therefore, a need was established for grain-fed cattle that could produce a higher percentage of lean to fat at the desired slaughter weight. Breeders also shifted selection pressure to improve composition in the traditional breeds.
2. Economic pressure to produce more weight in a shorter period of time demonstrated a need for cattle with more milk production and growth potential.
3. An opportunity was available for some promoters to capitalize on merchandising a certain breed as being the ultimate in all production traits. This was easily accomplished because there was little comparative information on breeds.

Figure 24.4 shows the leading breeds of beef cattle. Additional breed pictures can be accessed via www.ansi.okstate.edu/breeds/. Table 24.1 shows distinguishing characteristics and gives brief background information for more of these breeds.

The relative importance of the various breeds' contributions to the total beef industry is best estimated by the registration numbers of the breeds (Table 24.2). Although registration numbers are for purebred animals, they reflect the commercial

Figure 24.4

Some of the breeds important to the United States. Source: American Angus Association, Beefmaster Breeders International, Coleman Locke, International Brangus Breeders Association, American International Charolais Association, American Hereford Association, American Gelbvieh Association, North American Limousin Foundation, Red Angus Association of America, Santa Gertrudis Breeders International, American Shorthorn Association, and Altenburg Superbaldies.

Red Angus

Santa Gertrudis

Shorthorn

Simmental

Figure 24.4 (continued)

Table 24.1

BACKGROUND AND DISTINGUISHING CHARACTERISTICS OF MAJOR BEEF BREEDS IN THE UNITED STATES

Breed	Distinguishing Characteristics	Brief Background
Angus	Black color; polled	Originated in Aberdeen Scotland. Imported into the United States in 1873.
Beefmaster	Various colors; horned	Developed in the United States from Brahman, Hereford, and Shorthorn breeds. Selected for ability to reproduce, produce milk, and grow under range conditions.
Brahman	Various colors, with gray predominant; they are one of the Zebu breeds, which have the hump over the top of the shoulder; most Zebu breeds also have large, drooping ears and excess skin in the throat and dewlap	Major importations to the United States from India and Brazil. Largest introductions in early 1900s. These cattle are heat tolerant and well adapted to the harsh conditions of the Gulf Coast region.
Brangus	Black and polled predominate, although there are Red Brangus which is a separate breed	U.S. breed developed around 1912—3/8 Brahman, 5/8 Angus.
Charolais	White color with heavy muscling; horned or polled	One of the oldest breeds in France. Brought into the United States soon after World War I, but its most rapid expansion occurred in the 1960s.
Chianina	White or black color with black eyes and nose	An old breed originating in Italy. Acknowledged to be the largest breed, with mature bulls weighing more than 3,000 lb.
Gelbvieh	Golden colored or black; horned or polled	Originated in Austria and Germany. Developed as a dual-purpose breed used for draft, milk, and meat.
Hereford	Red body with white face; horned	Introduced in the United States in 1817 by Henry Clay. Followed the Longhorn in becoming the traditionally known range cattle.

Table 24.1 *continued*

Breed	Distinguishing Characteristics	Brief Background
Limousin	Golden color or black with marked expression of muscling; polled or horned	Introduced into the United States in 1969, primarily from France.
Longhorn	Multicolored with characteristically long horns	Came to West Indies with Columbus. Brought to the United States through Mexico by the Spanish explorers. Longhorns were the noted trail-drive cattle from Texas into the Plains states.
Polled Hereford	Red body with white face; polled	Bred in 1901 in Iowa by Warren Gammon, who accumulated several naturally polled cattle from horned Hereford herds.
Red Angus	Red color; polled	Founded as a performance breed in 1954 by sorting the genetic recessives from Black Angus herds.
Salers	Uniform mahogany red or black with medium to long hair; horned or polled	Raised in the mountainous area of France, where they were selected for milk, meat, and draft.
Santa Gertrudis	Red color; horned	First U.S. breed of cattle that was developed on the King Ranch in Texas. Combination of 5/8 Shorthorn and 3/8 Brahman.
Shorthorn	Red, white, or roan in color; horned and polled	Introduced into the United States in 1783 under the name *Durham*. Most prominent in the United States around 1920.
Simmental	Yellow, red-and-white, red, or black color pattern; polled and horned	A prominent breed in parts of Switzerland and France. First bull arrived in Canada in 1967. Originally selected as a dual-purpose for milk and meat.

Table 24.2
ANNUAL REGISTRATION NUMBERS FOR MAJOR U.S. BEEF BREEDS

Breed	Annual Registrations (in thousands)						Year U.S. Association Formed
	2013	2008	2005	1995	1985	1975	
Angus	315	348	299	225	175	306	1883
Hereford	70	69	69	120	180	419	1881
Simmental	49	52	44	71	73	75	1969
Red Angus	46	47	43	30	12	10	1954
Gelbvieh	36	36	27	34	16	NA	1968
Charolais	32	35	26	56	23	45	1957
Limousin	22	38	40	79	44	32	1971
Brangus	22	25	23	31	30	13	1949
Beefmaster	14	18	21	47	32	12	1961
Shorthorn	14	20	18	20	17	29	1872
Brahman	8	8	8	15	30	27	1924

Source: Beef breed associations.

cow-calf producers' demand for the different breeds. The numbers of cattle belonging to various breeds in this country have changed over the past years. No doubt some breeds will become more numerous in future years and others will decrease significantly in numbers. These changes will be influenced by economic conditions and how well the breeds meet the needs of the commercial beef industry.

IMPROVING BEEF CATTLE THROUGH BREEDING METHODS

Genetic improvement in beef cattle can be achieved by selection and by using a particular mating system. Significant genetic improvement by selection results when the selected animals are superior to the herd average and when the heritability of the traits is relatively high. It is important that the traits included in a breeding program be of economic importance and that they be measured objectively.

TRAITS AND THEIR MEASUREMENT

The most economically important traits of beef cattle are classified as follows: (1) reproductive performance, (2) weaning weight, (3) postweaning growth, (4) feed efficiency, (5) carcass merit, (6) longevity (functional traits), (7) conformation, and (8) freedom from genetic defects.

Reproductive Performance

Reproductive performance has the highest economic importance of all the traits. Most cow-calf producers have a goal for percent calf crop weaned (number of calves weaned compared to the number of cows in the breeding herd) of 85% or higher. Beef producers also desire each cow to calve every 365 days or less and to have a calving season for the entire herd of less than 90 days.

The heritability of fertility in beef cattle is low (less than 20%), so minimal genetic progress can be made through direct selection for that specific trait. Heritability for birth weight and **scrotal circumference** are high; therefore, selection for optimum levels of these traits will improve the percent calf crop weaned. The most effective way to improve certain reproductive traits (such as rebreeding after calving) is to improve the environment, for example through adequate nutrition and good herd health practices. Selecting bulls based on a reproductive soundness examination, which includes semen testing, will also improve reproductive performance in the herd.

Reproductive performance can be improved through breeding methods by crossbreeding to obtain heterosis for percent calf crop weaned; by using bulls with relatively light birth weights (heritability of birth weight is 40%), which decreases calving difficulty; and by selecting bulls that have a relatively large scrotal circumference. Scrotal circumference has a high heritability (40%); bulls with a larger scrotal size (over 32 cm for yearling bulls) produce a larger volume of semen and have half-sister heifers that reach puberty at earlier ages than heifers related to bulls with a smaller scrotal size.

Weaning Weight

Weaning weight, as measured objectively by weighing calves on scales, reflects the milking and mothering ability of the cow and the preweaning growth rate of the calf. Weaning weight is commonly expressed as the adjusted 205-day weight, where the weaning weight is adjusted for the age of the calf and age of the dam. This adjustment puts all weaning-weight records on a comparable basis since older calves weigh more than younger calves and mature cows (5–9 years of age) milk more heavily than younger cows (2–4 years of age) and older cows (10 years and older).

Weaning weights of calves are usually compared as a ratio expressed by dividing the calf's adjusted weight by the average weight of the other calves in the contemporary group. For example, a calf with a weaning weight of 440 lb, where the contemporary group average is 400 lb, has a ratio of 110 – (440/400) × 100. This calf's weaning-weight ratio is 10% above the herd average. Ratios can be used primarily for selecting cattle within the same herd in which they have had similar environmental circumstances (weather, feed availability, etc.). Comparing ratios between herds is misleading from a genetic standpoint because most differences between herds are caused by differences in the environment. Weaning weight will respond to selection because it has a heritability of 30%. Producers should avoid placing too much emphasis on increasing weaning weight due to the possibility of sacrificing economic performance via expenditure of excessive input costs to increase weaning weight.

Producers may opt to utilize a creep feeding system that allows calves access to a concentrate feed source but restricts access by the dams. While short-term creep feeding may be an effective means to prepare calves for feeding following weaning, it is typically expensive. Furthermore, creep feeding may cloud the ability to select accurately for increases in weaning weight that can be sustained on a forage-based diet.

Postweaning Growth

Postweaning growth measures the growth from weaning to a finished weight. Postweaning growth might take place on a pasture or in a feedlot. Usually, animals with relatively high postweaning gains make efficient gains at a relatively low cost to the producer.

Postweaning gain in cattle is usually measured in pounds gained per day after a calf has been on a feed test for 100–160 days. Weaning weight and postweaning gain are usually combined into a single trait; namely, adjusted 365-day weight (or yearling weight). It is computed as follows:

Adjusted 365-day weight = (160 × average daily gain) + adjusted 205-day weight

Average daily gain and adjusted 365-day weight both have high heritabilities (40%); therefore, genetic improvement can be quite rapid when selection is practiced on postweaning growth or yearling weight.

Feed Efficiency

The pounds of feed required per pound of live weight gain measure **feed efficiency**. Only feeding each animal individually and keeping records on the amount of feed consumed can obtain specific records for feed efficiency. Feeding system technologies such as the GrowSafe System or Callen Gate System allow for more widespread collection of efficiency data than was possible previous to these technologies.

Interpretation of feed-efficiency records can be somewhat difficult, depending on the endpoint to which the animals are fed. Feeding endpoint can be a certain number of days on feed (e.g., 140 days) to a specified slaughter weight (e.g., 1,200 lb) or to a carcass compositional endpoint (e.g., low choice quality grade or 0.4 in. of fat). Most differences in feed efficiency shown by individual animals are related to the pounds of body weight maintained through feeding periods and the daily rate of gain or feed intake of each animal. Cattle fed from a similar initial feedlot weight (e.g., 600 lb) to a similar slaughter weight (e.g., 1,100 lb) will demonstrate a high relationship between rate of gain and efficiency of gain. In this situation, cattle that gain faster will require fewer pounds of feed per pound of gain. Thus, a breeder can select for rate of gain and thereby make genetic improvement in feed efficiency. However, when cattle are fed to the same compositional endpoint (approximately the same carcass fat), there are small differences in the amount of feed required per pound of gain. This is true for different sizes and shapes of cattle, and even for various breeds that vary greatly in skeletal size and weight.

The heritability of feed efficiency at similar beginning and final weights is high (45%), so selection for more efficient cattle can be effective.

Carcass Merit

Carcass merit is presently measured by quality grades and yield grades, which are discussed in detail in Chapter 8. Many cattle-breeding programs have goals to produce cattle that will grade choice and have yield grade 2 carcasses. Objective measurements using ultrasound backfat thickness on the live animal and hip height measurement can assist in predicting the yield grade at certain slaughter weights. Visual appraisal, which is more subjective, can be used to predict amount of fat or predisposition to fat and skeletal size. These visual estimates can be relatively accurate in identifying actual yield grades if cattle differ by as much as one yield grade.

Because quality grade cannot be evaluated accurately in the live animal, it is necessary to evaluate the carcass. However, researchers are attempting to identify gene markers associated with marbling in the hope that evaluations of live animals may be possible in the future. Steer and heifer progeny of different bulls are slaughtered to identify the genetic superiority or inferiority of bulls for both quality grade and yield grade. The heritability of most beef carcass traits is high (over 30%), so selection can result in marked genetic improvement for these traits.

Longevity

Longevity measures the length of productive life. It is an important trait, particularly for cows, when replacement heifer costs are high or when a producer reaches an optimum level of herd performance and desires to stabilize it. Bulls are usually kept in a herd for 3–5 years, or inbreeding might occur. Some highly productive cows remain in the herd at age 15 years or older, whereas other highly productive cows have been culled before reaching 4 years of age. These cows may have been culled because of such problems as skeletal unsoundness, poor udders, eye problems (i.e., cancer eye), and lost or worn teeth. Little selection opportunity for longevity exists because few cows remain highly productive past the age of 10 years. Most cows that leave the herd early have poor reproduction performance.

Some producers need to improve their average herd performance as rapidly as possible rather than improve longevity. In this situation, a relatively rapid turnover of cows is needed. Some selection for longevity occurs because producers have the opportunity to keep more replacement heifers born to cows that are highly productive in the herd for a longer period of time than heifers born to cows that stay in the herd for a short time. Some beef producers attempt to identify bulls that have highly productive dams with a long history of lifetime productivity. Stayability expected progeny differences are calculated for some breeds to assist breeders in selecting for longevity. Certain conformation traits, such as skeletal soundness and udder soundness, may be evaluated to extend the longevity of production.

Conformation

Conformation is the form, shape, and visual appearance of an animal. How much emphasis to put on conformation in a beef cattle selection program has been, and continues to be, controversial. Some producers believe that putting a productive animal into an attractive package contributes to additional economic returns. It is more logical, however, to place more selection emphasis on traits that will produce additional numbers and pounds of lean growth for a given number of cows. Placing some emphasis on conformation traits such as skeletal, udder, eye, and teeth soundness is justified. Conformation differences such as fat accumulation or predisposition to fat can be used effectively to make meaningful genetic improvement in carcass composition.

Most conformation traits are medium to high (30–60%) in heritability, so selection for these traits will result in genetic improvement.

Genetic Defects

Genetic defects, other than those previously identified under longevity and conformation, need to be considered in breeding productive beef cattle. Cattle have numerous known hereditary defects; most, however, occur infrequently. Some defects increase in their frequency, and selection needs to be directed against them. A single pair of genes that are usually recessive determines most of these defects. When one of these hereditary defects occurs, it is a logical practice to cull both the cow and the bull.

Some commonly occurring genetic defects in cattle today are double muscling, syndactyly (mule foot), arthrogryposis (palate-pastern syndrome), osteopetrosis (marble bone disease), hydrocephalus, and dwarfism. An enlargement of the muscles evidences **double muscling** with large grooves between the muscle groups of the hind leg. Double-muscled cattle usually grow slowly and their fat deposition in and on the carcass is much less than that of the normal beef animal. **Syndactyly** is a condition in which one or more of the hooves are solid in structure rather than cloven. Mortality rate is high in calves with syndactyly. **Arthrogryposis** is a defect in which the pastern tendons are contracted, and the upper part of the mouth has not properly fused together. **Osteopetrosis** is characterized by the marrow cavity of the long bones being filled with bone tissue. All calves having osteopetrosis have short lower jaws, protruding tongue, and impacted molar teeth. A bulging forehead where fluid has accumulated in the brain area is typical of the defect of **hydrocephalus**. Calves with arthrogryposis, hydrocephalus, or osteopetrosis usually die shortly after birth. The most common type of **dwarfism** is snorter dwarfism, in which the skeleton is quite small and the forehead has a slight bulge. Some snorter dwarfs exhibit a heavy, labored breathing sound. This defect was most common in the 1950s, and it has decreased significantly since that time.

BULL SELECTION

Bull selection must receive the greatest emphasis if optimum genetic improvement of a herd is to be achieved. Sire selection accounts for 80–90% of the genetic improvement in a herd over a period of several years. This does not diminish the importance of good beef females because genetically superior bulls have superior dams. However, most of the genetic superiority or inferiority of the cows will depend on the bulls previously used in the herd. Also, the accuracy of records is much higher for bulls than for cows and heifers because bulls produce more offspring and are used in more herds especially when artificial insemination is incorporated. While artificial insemination (Chapter 11) offers significant opportunities for increasing the rate of genetic change, the technology is mostly utilized by seedstock producers or commercial producers who are developing their own replacement heifers and desire to assure that young cows are bred to low-birth-weight sires. Less than 10% of all beef cows are artificially inseminated. This low rate of adoption is likely due to the need for more intensive management.

Purebred breeders should provide accurate performance data on their bulls, and commercial producers should request the information. Excellent performance records can be obtained and made available on the farm or ranch. The trait ratios are useful and comparative if the bulls have been fed and managed in similar environments.

Breeding Values

Phenotype is the appearance or performance of an animal determined by the *genotype* (genetic makeup) and the environment in which the animal was raised.

Figure 24.5

Breeding value as depicted by the "brick wall concept."

Source: Gibb, Boggess, and Wagner, 1992.

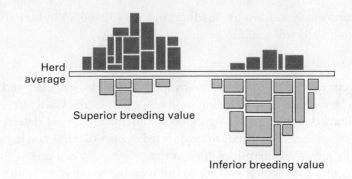

Herd average

Superior breeding value

Inferior breeding value

Genotype is determined by two factors: (1) breeding value (what genes are present) and (2) nonadditive value (how genes are combined). Environmental factors influence the phenotype through known and unknown effects. Age of dam, resulting in different levels of milk production, is a known environmental effect adjusted for in an adjusted 205-day weight (phenotype). Unknown effects are things like injury or health problems for which it is difficult to adjust the phenotypic record.

An example depicting breeding values is shown in Figure 24.5. In the "brick wall concept," each brick represents a gene. Some genes have positive effects, whereas other genes have negative effects. The different-sized bricks represent the magnitude of the gene's effect. In this example, the 10 pairs of genes (20 bricks) are used to show a superior breeding value and a negative breeding value. The superior breeding value (higher brick wall) is where the sum of the additive gene effects (bricks) is above herd average.

Sire Summaries

The development of sire summaries, published by breed associations on an annual or biannual basis, provides the tools to substantially improve sire selection.

Expected progeny difference (EPD) and **accuracy (ACC)** are the important terms used in understanding sire summaries. An EPD combines into one figure a measurement of genetic potential based on the individual's performance and the performance of related animals such as the sire, dam, and other relatives. EPD is expressed as a plus or minus value, reflecting the genetic transmitting ability of a sire. The most common EPDs reported for bulls, heifers, and cows are birth weight EPD, milk EPD, weaning-growth EPD, maternal EPD (includes milk and preweaning growth), and yearling-weight EPD. Some additional traits are included as well (Table 24.3).

Table 24.3
SIRE SUMMARY DATA (EPDS OR RATIOS) FOR SELECTED BREEDS

Traits	Angus[a]	Brahman[a]	Charolais[a]	Gelbvieh[a]	Hereford[a,f]	Limousin[a]	Red Angus[a]	Simmental[a]
Calving ease[b]								
First calf[c]	+	o	+	+	+	+	+	+
Maternal[d]	+	o	+	+	+	+	+	+
Scrotal circumference	+	o	+	+	+	+	+	o
Birth weight	+	+	+	+	+	+	+	+
Gestation length	o	o	o	+	o	+	o	o
Weaning weight	+	+	+	+	+	+	+	+
Yearling weight	+	+	+	+	+	+	+	+
Milk[e]	+	+	+	+		+	+	+

Table 24.3 *continued*

Traits	Angus[a]	Brahman[a]	Charolais[a]	Gelbvieh[a]	Hereford[a,f]	Limousin[a]	Red Angus[a]	Simmental[a]
Maternal (milk + 1/2 weaning weight)[e]	+	o	+	+	+	+	+	+
Disposition (docility)	+	o	o	o	o	+	o	o
Stayability	o	o	o	+	o	+	+	+
Heifer pregnancy	+	o	o	o	o	o	+	o
Mature cow energy requirements	o	o	o	o	o	o	+	o
Yearling hip height	+	o	o	o	o	o	o	o
Mature cow weight	+	o	o	o	o	o	o	+
Mature daughter height	+	o	o	o	o	o	o	+
Carcass weight	+	+	+	+	o	+	o	l
Marbling	+	+	+	+	+	+	+	+
Ribeye area	+	+	+	+	+	+	+	+
Fat thickness	+	+	+	+	+	+	+	+
Percent retail product	+	+	o	o	o	o	o	+
Ultrasound rump fat	+	o	o	o	o	o	o	o
Carcass/Grid value	+	o	o	+	+	+	o	o
Feedlot value	+	o	o	+	o	o	o	o
Days to Finish/ Terminal Index	o	o	+	+	o	o	o	+
Warner-Bratzler Shear Force	o	+	o	o	o	o	o	+

[a]+ = trait in sire summary; o = trait not in sire summary.
[b]Computed from birth weights.
[c]Expressed as a ratio, not an EPD. Measures calving ease of bulls' own calves.
[d]Expressed as a ratio, not an EPD. Measures calving ease of a bull's daughters.
[e]Computed from weaning weights.
[f]Indices for Baldy Maternal, Brahman Influenced Maternal, and Calving Ease are also calculated.
Source: Breed sire summaries.

Accuracy (ACC) is a measure of expected change in the EPD as additional progeny data become available. EPDs with ACC values of 0.90 and higher would be expected to change very little, whereas EPDs with ACC values below 0.70 might change dramatically with additional progeny data.

Table 24.4 shows sire summary data for several beef breed associations.

If bull B and bull C were used in the same herd (each on an equal group of cows), the expected average performance of their calves would be:

Bull B's calves would be 16 lb heavier at birth.

Bull B's calves would weigh 59 lb more at weaning.

Bull B's calves would weigh 78 lb more as yearlings.

Bull B's daughters would wean calves weighing 7 lb less.

Table 24.4
SELECTED DATA FROM AN ANGUS SIRE EVALUATION REPORT

| | Weaning Weight | | | | Weaning Weight Maternal | | | | | |
| | Birth Weight | | Direct | | | Milk[a] | | Total Maternal[b] | Yearling Weight | |
Sire	EPD	ACC	EPD	ACC	EPD	ACC	DTS[c]	EPD	EPD	ACC
A	−1	0.95	+20	0.95	+1	0.93	331	+11	+45	0.94
B	+9	0.95	+47	0.95	−17	0.89	88	6	+74	0.90
C	−7	0.83	−12	0.85	−10	0.79	29	−16	−4	0.85
D	+1	0.95	+22	0.95	+3	0.92	230	+14	+54	0.94
E	1	0.73	+15	0.70	+12	0.42	0	+20	+45	0.66

[a]Milk EPD is measured by comparing calf weaning weights from daughters of bulls.
[b]Calculated by taking ½ of weaning weight (direct) + milk EPD.
[c]DTS is daughters.

Bull A and bull D have an optimum combination of calving ease (birth weight) and growth traits (weaning weight and yearling weight). Bull E is a promising young sire with an excellent combination of EPDs. However, the ACC is relatively low and could significantly change with more progeny and as daughters start producing.

It should also be noted that while EPDs are very effective selection tools for within-breed comparisons, comparing expected progeny differences across breeds is more problematic. Conversion tables are available to assist breeders who desire to make across-breed EPD comparisons.

Bull selection becomes more complex in that desired genetic improvement is typically viewed as creating a combination of several traits. "Stacking pedigrees" for only one trait may result in some problems in other traits. A good example is selecting for yearling weight alone, which results in increased birth weight because the two traits are genetically correlated. Birth weight is associated with calving difficulty; so increased birth weight might be a problem. Large increases in birth EPD can occur and calving difficulty can be serious.

The most difficult challenge in bull selection is selecting bulls that will improve maternal traits. Frequently, too much emphasis is placed on yearling growth, frame size, and large mature size. These traits can be antagonistic to maternal traits such as birth weight, early puberty, and maintaining a cow size consistent with an economical feed supply. Figure 24.6 shows an example of stacking a pedigree for maternal traits for yearling bull selection where the bull is used naturally on heifers. If an older bull can be used through AI, then emphasis would be on accuracies higher than those available on young bulls.

Table 24.5 identifies bull-selection criteria that would apply to a large number of commercial producers. First, the breeding program goals and the maternal traits are identified. Second, the bull-selection criteria are identified, giving emphasis to the maternal traits. These traits would receive emphasis if a commercial producer were using only one breed or a rotational crossbreeding program or in producing the cows used in a terminal crossbreeding program. Finally, Table 24.5 identifies the traits emphasized in selecting terminal cross bulls. In selecting terminal cross sires, little emphasis is placed on maternal traits as no replacement heifers from this cross are kept in the herd.

SELECTING REPLACEMENT HEIFERS

Heifers, as replacement breeding females, can be selected for several traits at different stages of their reproductive lives. The objective is to identify heifers that will conceive early in the breeding season, calve easily, give a flow of milk consistent with the feed

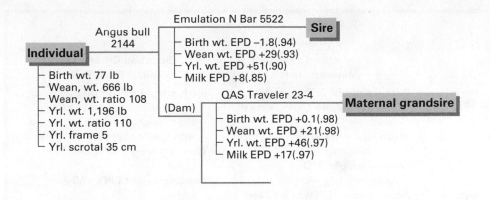

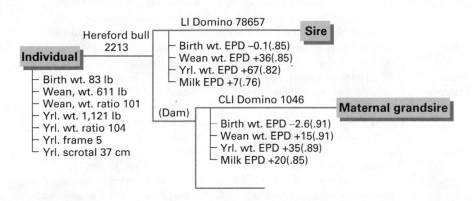

Figure 24.6
Yearling bull selection emphasizing maternal traits. Bulls used naturally on replacement heifers.
Source: Colorado State University.

Table 24.5
BULL SELECTION CRITERIA FOR COMMERCIAL BEEF PRODUCERS

Goals and Maternal Traits	Bull-Selection Criteria for Maternal Traits[a]	Selection Criteria for Terminal Cross Bulls[b]
Breeding Program Goals 1. Select for an optimum combination of maternal traits to maximize profitability (avoiding genetic antagonisms and environmental conflicts that come with maximum production or single-trait selection) 2. Provide genetic input so cows can be matched with their environment—cows that wean more lifetime pounds of calf without overtaxing the forage, labor, or financial resources 3. Stack pedigrees for maternal traits	**Yearling frame size:** 4.0–6.0 (smaller frame size will adapt better to harsher environments—e.g., less feed, more severe weather, less care) **Mature weight:** Under 2,000 lb in average condition (preference for future). Now we likely will have to consider bulls under 2,500 lb **Milk EPD:** 0 to +20 lb (prefer +10) **Maternal EPD:** +20 to +40 (prefer +25) **Body condition:** Backfat of 0.20 to 0.35 in. at yearling weights of 1,100–1,250 lb. Monitor reproduction of daughters	**Yearling frame size:** 6.0–8.0 (should be evaluated with frame size of cows so slaughter progeny will average 5.0 and 6.0) **Mature weight:** No upper limit as long as birth weight and frame size are kept in desired range **Milk EPD:** Not considered **Maternal EPD:** Not considered **Body condition:** Evaluate with body condition of cows so that progeny will be between 0.25 and 0.45 in at 1,100–1,300-lb slaughter weights
Maternal Traits **Mature weight:** Medium-sized cows (1,000–1,250 lb under average feed supply)	**Scrotal circumference:** 34–40 cm at 365 days of age. Passed breeding soundness examination	**Scrotal circumference:** 32–38 cm at 365 days of age. Passed breeding soundness exam

Table 24.5   continued

Goals and Maternal Traits	Bull-Selection Criteria for Maternal Traits[a]	Selection Criteria for Terminal Cross Bulls[b]
Milk production: Moderate (wean 500–550-lb calves under average feed)	**Birth weight EPD:** Preferably under +1.0 lb. Evaluate calving ease of daughters	**Birth weight EPD:** Preferably no higher than +5.0; want calves birth weights in 85–95-lb range with few, if any, over 100 lb
Body condition ("fleshing ability"): "5" condition score at calving without high-cost feeding	**Weaning weight EPD:** Approximately +20 lb	**Weaning weight EPD:** +40 lb or higher
Early puberty/high conception: Calve by 24 months of age	**Yearling weight EPD:** Approximately +40 lb	**Yearling weight EPD:** +60 lb or higher
Calving ease: Moderate birth weights (65–75-lb heifers; 75–90-lb cows). Calf shape (head, shoulders, hips), which relates to unassisted births	**Accuracies:** All EPDs of 0.90 and higher (older, progeny-tested sires). Young bulls with accuracies below 0.90—evaluate trait ratios and pedigree EPDs to predict mature bulls with above-average EPDs and accuracies. Select sons of bulls that meet EPDs listed above	**Accuracies:** All EPDs of 0.90 and higher (older, progeny-tested sires). Young bulls with accuracies below 0.90—evaluate trait ratios and pedigree EPDs to predict mature bulls with above average EPDs and accuracies. Select sons of bulls that meet EPDs listed above
Early growth and composition: Rapid gains—relatively heavy weaning and yearling weights within medium (4–6) frame size. Yield grade 2 (steers slaughtered at 1,150 lb)	**Functional traits:** Visual evaluation of bull and his daughters	**Visual:** Functional traits: disposition and structural soundness
Functional traits (longevity): Udders (shape, teats, pigment); eye pigment; disposition; structural soundness	**Visual:** Sufficiently attractive to sire calves that would not be economically discriminated against in the marketplace Preference for "adequate middle" as medium-frame-size cattle need middle for feed capacity	Sufficiently attractive to sire calves that would not be economically discriminated against in the marketplace

[a]EPDs are for British breeds.
[b]EPDs are for Continental breeds.

supply, wean heavy calves, and make a desirable genetic contribution to the calves' postweaning growth and carcass merits.

Beef producers have found it challenging to determine which young heifers will make the most productive cows. Table 24.6 shows the selection process that producers use to select the most productive replacement heifers. This selection process assumes that more heifers will be selected at each stage of production than the actual numbers of cows to be replaced in the herd. The number of replacement heifers that producers keep is based primarily on their cost of production and their market value at various stages of production. More heifers than the number needed should be kept through pregnancy-check time. Heifers are selected on the basis that they become pregnant early in life primarily for economic reasons rather than expected genetic improvement from selection.

COW SELECTION

Cows should be culled from the herd based on the productivity of their calves and additional evidence that they can be productive the following year, such as early pregnancy and soundness of udders, eyes, skeleton, and teeth. Productivity of cows is measured by pregnancy test, weaning and yearling weights (ratios) of their calves,

Table 24.6

REPLACEMENT HEIFER SELECTION GUIDELINES AT THE DIFFERENT PRODUCTIVE STAGES

Stage of Heifer's Productive Life	Emphasis on Productive Trait	
	Primary	Secondary
Weaning (7–10 months of age)	Cull only the heifers whose actual weight is too light to prevent them from showing estrus by 15 months of age. Also consider the economics of the weight gains needed to have puberty expressed. Cull heifers that are too large in frame and birth weight.	Weaning weight ratio Weaning EPD Milk EPD Predisposition to fatness Moderate skeletal frame Skeletal soundness Docility
Yearling (12–15 months of age)	Cull heifers that have not reached the desired target breeding weight (e.g., minimum of 650–750 lb for small- to medium-sized breeds or cross, minimum of 750–850 lb for large-sized breeds and crosses). Cull heifers that are extreme in size and weight.	Milk and weaning weight EPDs Yearling weight ratio Yearling EPD Predisposition to fatness Moderate skeletal frame Skeletal soundness Docility
After breeding (19–21 months of age)	Cull heifers that are not pregnant and those that will calve in the latter one-third of the calving season.	Milk and weaning weight EPDs Yearling weight ratio Yearling EPD Predisposition to fatness Moderate skeletal frame Skeletal soundness Docility
After weaning first calf (31–34 months of age)	Cull to the number of first-calf heifers actually needed in the cow herd based on the weaning weight performance of the first calf. Preferably all the calves from these heifers have been sired by the same bull. Select for early pregnancy.	

Table 24.7

PERFORMANCE DATA ON HIGH- AND LOW-PRODUCING COWS IN THE SAME HERD

Cow Number	Number of Calves	Weaning Weight (lb)	Weaning Weight Ratio	Yearling Weight Ratio
1	9	572	105	102
2	9[a]	464	89	94

[a]One calf died before weaning (not computed in averages).

and the EPDs of the cows. Table 24.7 shows the weaning and yearling values of the high- and low-producing cows in a herd. Cow 1 is a moderately high-producing cow, whereas cow 2 is the low-producing cow. Cow 2 should be culled and replaced with a heifer of higher breeding potential.

CROSSBREEDING PROGRAMS FOR COMMERCIAL PRODUCERS

Most commercial beef producers use **crossbreeding** programs to take advantage of **heterosis** in addition to the genetic improvement from selection. A crossbreeding system should be determined according to which breeds are available and then adapted to the commercial producers' feed supply, market demands, and other environmental conditions. A good example of adaptability is the Brahman breed, which is more heat and insect resistant than most other breeds. Because of this higher resistance, the level of productivity (in the southern and Gulf regions of the United States) is much higher for Brahmans and Brahman crosses.

Most commercial producers travel 150 or fewer miles to purchase bulls used in natural mating. Therefore, a producer should assess the breeders with excellent breeding programs in a 150-mile radius, as well as available breeds. This assessment, in most cases, should be determined before planning which breeds to use and in which combination to use them.

Breeds should be chosen for a crossbreeding system based on how well the breeds complement each other. Table 24.8 gives some comparative rankings of the major beef

Table 24.8
BREED CROSSES GROUPED IN BIOLOGICAL TYPES ON BASIS OF FOUR MAJOR CRITERIA

Breed Group[b]	Traits Used to Identify Biological Types[a]			
	Growth Rate and Mature Size	Lean-to-Fat Ratio	Age at Puberty	Milk Production
Jersey	X	X	X	XXXXX
Longhorn	X	XXX	XXX	XX
Hereford-Angus	XXX	XX	XXX	XX
Red Poll	XX	XX	XX	XXX
Devon	XX	XX	XXX	XX
Shorthorn	XXX	XX	XXX	XXX
Galloway	XX	XXX	XXX	XX
South Devon	XXX	XXX	XX	XXX
Tarentaise	XXX	XXX	XX	XXX
Pinzgauer	XXX	XXX	XX	XXX
Brangus	XXX	XX	XXXX	XX
Santa Gertrudis	XXX	XX	XXXX	XX
Beefmaster	XXX	XX	XXXX	XX
Sahiwal	XX	XXX	XXXXX	XXX
Brahman	XXXX	XXX	XXXXX	XXX
Nellore	XXXX	XXX	XXXXX	XXX
Brown Swiss	XXXX	XXXX	XX	XXXX
Gelbvieh	XXXX	XXXX	XX	XXXX
Holstein	XXXX	XXXX	XX	XXXXXX
Simmental	XXXXX	XXXX	XXX	XXXX
Maine Anjou	XXXXX	XXXX	XXX	XXX
Salers	XXXXX	XXXX	XXX	XXX
Piedmontese	XXX	XXXXXX	XX	XX
Limousin	XXX	XXXXX	XXXX	X
Charolais	XXXXX	XXXXX	XXXX	X
Chianina	XXXXX	XXXXX	XXXX	X

[a]Breed crosses grouped into several biological types based on relative differences (X = lowest, XXXXXX = highest).
[b]Breed group based on these breeds of sires mated to Hereford and Angus cows.

breeds for productive characteristics. Although the information in Table 24.8 is useful, it should not be considered the final answer for decisions on breeds to use. First, a producer needs to recognize that this information reflects breed averages; therefore, there are individual animals and herds of the same breed that are much higher or lower than the ranking given (Figs. 24.1–24.3). Producers need to use some of the previously described methods to identify superior animals within each breed. Second, it should be recognized that these average breed rankings can change with time, depending on the improvement programs used by the leading breeders within the same breed. Obviously, those traits that have high heritability estimates would be expected to change most rapidly, assuming the same selection pressure for each trait. A careful analysis of the information in Table 24.8 shows that no single breed is superior for all important productive characteristics. This gives an advantage to commercial producers using a crossbreeding program if they select breeds whose superior traits complement each other. An excellent example of breed complementarity is shown by the Angus and Charolais breeds, which complement each other in both quality grade and yield grade.

Most of the heterosis achieved in cattle as a result of crossbreeding is expressed by weaning time. The cumulative effect of heterosis on pounds of calf weaned per cow exposed is shown in Figure 24.7, in which maximum heterosis is obtained when crossbred calves are born from crossbred cows. The traits that express an approximate heterosis of a 20% increase in pounds of calf weaned per cow exposed to breeding are early puberty of crossbred heifers, high conception rates in crossbred females, high survival rate of calves, increased milk production of crossbred cows, and a higher preweaning growth rate of crossbred calves.

Consistently high levels of heterosis (10–20%) can be maintained generation after generation if crossbreeding systems such as those shown in Figures 24.8–24.10 are used. The crossbreeding system shown in Figure 24.10 combines a two-breed rotation with a terminal cross. In this system, the two-breed rotation is used primarily

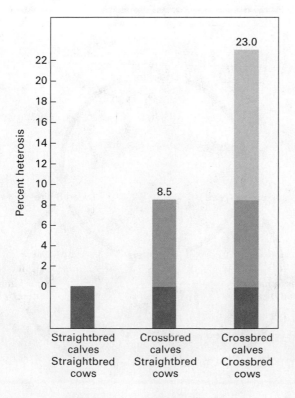

Figure 24.7

Heterosis resulting from crossbreeding for pounds of calf weaned per cow exposed to breeding.

Source: USDA.

Figure 24.8
Two-breed rotation cross. Females sired by breed A are mated to breed B bulls, and heifers sired by breed B are mated to breed A bulls. This will increase the pounds of calf weaned per cow bred by approximately 15%.

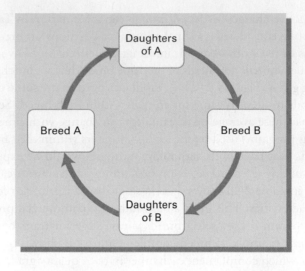

Figure 24.9
Three-breed rotation cross. Females sired by a specific breed are bred to the breed of the next bull in rotation. This will increase the pounds of calf weaned per cow by approximately 20%.

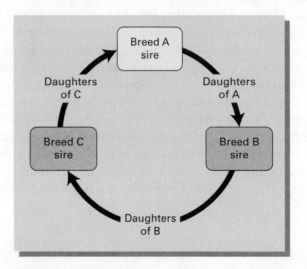

Figure 24.10
Two-breed rotation and terminal site crossbreeding system. Sires are used in the two-breed rotation primarily to produce replacement heifers. Terminal cross sires are mated to the less productive females. This system will increase the pounds of calf weaned per cow bred by more than 20%.

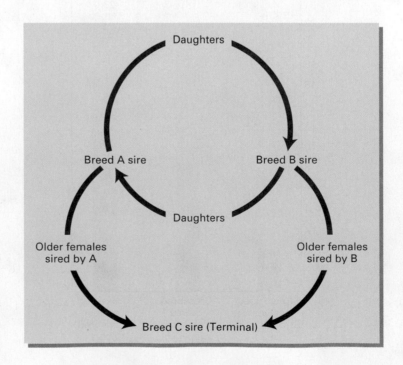

to produce replacement females for the entire cowherd. In most cowherds, approximately 50% of the cows are bred to sires to produce replacement females, with the remaining 50% being bred to terminal cross sires. All terminal cross calves are sold. This crossbreeding system maintains heterosis levels as high as those in the three-breed rotation system.

In the rotational crossing, breeds with maternal-trait superiority (high conception, calving ease, and milking ability) would be selected. The terminal cross sire could come from a larger breed where growth rate and carcass cutability are emphasized. A primary advantage of the rotational-terminal cross system is that smaller- or medium-sized breeds can be used in rotational crossing and a larger breed could be used in terminal crossing. Terminal crossbreeding systems will maximize heterosis at approximately 25–30%. Some of the systems that will fit a one-breeding pasture program include (1) using a composite (synthetic) breed, (2) rotating two or three breeds of bulls (each breed every 3–5 years), and (3) putting multiple-sire breeds of bulls with crossbred cows. Each of these crossbreeding systems will maintain heterosis at approximately 10–15%. These systems may be more cost-effective and profitable than some of the more complex crossbreeding systems discussed earlier.

Table 24.9 shows the advantage a commercial producer has over a purebred breeder in being able to use more of the breeding methods for genetic improvement. Commercial producers can use crossbreeding, whereas purebred breeders cannot use crossbreeding if they wish to maintain breed purity.

Traits with a low **heritability** respond little to genetic selection, but they show a marked improvement in a sound crossbreeding program. The commercial producer needs to select sires carefully to improve the traits with a high heritability.

Table 24.9
HERITABILITY AND HETEROSIS FOR THE MAJOR BEEF CATTLE TRAITS

Traits	Heritability	Heterosis
Reproduction	Low[a]	High
Growth	Medium	Medium
Carcass	High	Low

[a]Exceptions are age at puberty, scrotal circumference, and birth weight (dystocia). They are highly heritable.
Source: Adapted from *MARC Research Progress Reports* (*USDA-ARS*).

CHAPTER SUMMARY

- Phenotypic differences in breeds of beef cattle are primarily expressed as color (color pattern), polled or horned head, mature size, and muscling.

- There are more than 250 breeds of cattle in the world, with more than 60 breeds in the United States. However, fewer than 20 of these breeds make a significant impact on the national herd.

- Based on annual registration numbers, the Angus breed is the most numerous breed in the United States, followed by Hereford/Polled Hereford, Limousin, Beefmaster, and Charolais.

- Genetic improvement occurs by selecting for superiority (optimum combination) for the following traits: (1) reproductive performance, (2) weaning weight, (3) postweaning growth, (4) feed efficiency, (5) carcass merit (weight, yield grade, and quality grade), and (6) longevity.

- Effective bull selection, primarily through EPDs, accounts for 80–90% of the genetic improvement in a herd over a period of several years.
- It is economically important to select the biological type (mature weight, milk production, age at puberty, and carcass composition) that matches cows to a low-cost forage environment and their progeny to the marketplace.
- Optimum levels of heterosis are obtained by using crossbreeding or composite breeds.

KEY WORDS

breed
composite breeds
synthetic breeds
purebred
reproductive performance
scrotal circumference
weaning weight
postweaning growth
feed efficiency
carcass merit
longevity
conformation

genetic defects
double muscling
syndactyly
arthrogryposis
osteopetrosis
hydrocephalus
dwarfism
expected progeny difference (EPD)
accuracy (ACC)
crossbreeding
heterosis
heritability

REVIEW QUESTIONS

1. Discuss the reasons breeds were created and their value to the current industry.
2. Define the value of genetic variation.
3. What factors lead to the increased importation of new breeds to the United States?
4. Compare and contrast the major U.S. beef cattle breeds and identify their strengths and weaknesses.
5. Describe the trends in breed registration numbers for the major breeds.
6. Discuss the traits of critical economic value.
7. Describe several genetic abnormalities.
8. Describe the genetic selection tools available to cattle breeders.
9. Define how bull and female selection criteria might be developed for a specific herd.
10. Discuss the role of crossbreeding and the various protocols to implement a crossing system.
11. Discuss the concept of biological type.
12. Compare and contrast the use of heterosis and heritability.

SELECTED REFERENCES

Boggs, D. 1992. *Understanding and Using Sire Summaries*. Beef Improvement Federation, BIF-FS3.

Bull and Heifer Replacement Workshops (proceedings). 1990. Fort Collins, CO: Colorado State University.

Field, T. G. 2007. *Beef Production and Management Decisions*. Upper Saddle River, NJ: Prentice Hall.

Gibb, J., M. V. Boggess, and W. Wagner. 1992. *Understanding Performance Pedigrees*. Beef Improvement Federation, BIF-FS2.

Gregory, K. E. and L. V. Cundiff. 1980. Crossbreeding in beef cattle: Evaluation of systems. *J. Anim. Sci.* 51:1224.

Hickman, C. G. (ed.). 1991. *Cattle Genetic Resources*. Amsterdam: Elsevier Science Publishers.

Jarrige, R. and C. Berauger (eds.). 1992. *Beef Cattle Production*. Amsterdam: Elsevier Science Publishers.

Kress, D. D. 1989. Practical breeding programs for managing heterosis. *The Range Beef Cow Symposium XI*.

Legates, J. E. 1990. *Genetics of Livestock Improvement*. Englewood Cliffs, NJ: Prentice Hall.

Middleton, B. K. and J. B. Gibb. 1991. An overview of beef cattle improvement programs in the United States. *J. Anim. Sci.* 69:3861.

Silcox, R. and R. McGraw. 1992. *Commercial Beef Sire Selection*. Beef Improvement Federation, BIF-FS9.

25

Feeding and Managing Beef Cattle

An overview of the beef industry and beef production is presented in Chapter 2. This chapter will identify the primary factors affecting beef cattle productivity and profitability.

COW–CALF MANAGEMENT

Fundamentally, the beef industry is founded on the capacity of cattle to convert a variety of feedstuffs including those classified as high roughage and low quality into a nutrient-dense food for consumers. Cattle can utilize forage (grasses, forbs, etc.), by-product feeds (distiller by-products, food manufacturing waste, etc), and thrive in low-intensity management systems focused on utilization of lands not suitable for cultivation. The effective and profitable management of a cow–calf enterprise is founded on the ability of a manager to allocate time and resources to critical activities. A survey of successful cow–calf managers and their consultants determined that the seven most critical areas of management were:

1. Herd nutrition
2. Pasture and range management
3. Herd health
4. Financial management
5. Marketing
6. Production management (reproduction)
7. Genetics

Cow–calf managers must effectively integrate these seven factors into a management plan to assure long-term profitability and viability of the enterprise. However, each enterprise in the cattle industry has its own unique strengths and limitations, thus making "one-size-fits-all" recommendations impossible. Nonetheless, to provide context as to the critical areas of month to month focus a generalized management calendar for a cow–calf business is outlined in Table 25.1.

Cow–calf producers are interested in managing their operations to generate profits while building a business that is capable of providing opportunities to others. One approach to assessing the profitability of a commercial cow–calf operation is by analyzing the following criteria: (1) **calf crop percentage** weaned (e.g., number of calves produced per 100 cows in the breeding herd), (2) average weight of calves at weaning (7–9 months of age), and (3) annual cow cost (dollars required to keep a cow each year).

An example of the economic assessment of a commercial cow–calf enterprise that has an 85% calf crop weaned, 500-lb weaning weights, and a $600 annual cost would be as follows:

learning objectives

- Discuss the cyclic nature of profitability in the beef industry
- Describe the primary management approaches to optimizing percent calf crop and weaning weight
- List the primary factors affecting cost of production
- Explain the use of break-even price discovery
- Compare and contrast commercial and farmer-feeders
- Discuss profitability in the feed yard business
- Discuss the role of environmental management in cattle feeding

Calf crop % (0.85) × weaning weight (500 lb) = 425 lb of calf weaned per cow in the breeding herd

Annual cow cost ($600) ÷ lb of calf weaned (425 lb) = $ 141.18 per hundredweight break-even price

Table 25.1
MANAGEMENT CALENDAR FOR A SELECTED COW–CALF OPERATION WITH SPRING AND FALL CALVING HERDS

Major Activities	Jan	Feb	Mar	Apr	May	June	July	Aug	Sept	Oct	Nov	Dec
Planning, assessing goals, record analysis	x	x	x	x	x	x	x	x	x	x	x	x
Calving			s	s	s			f	f	f		
Breeding preparation (semen, bulls, equipment				s			f					
Vaccinate cows (Vibriosis, Leptospirosis)	s						f					
Vaccinate heifers (Clostridium c, IBR, Lepto, BVD)				s						f		
Synchronize heifers			s	s						f		
Synchronize cows				s						f		
Breed heifers				s	s	s					f	f
Breed cows				s	s	s	s				f	f
Process calves (vaccinations, castration, dehorning)				s	s					f	f	
Pregnancy test		f								s	s	
Fertilize pastures				x								
Control pinkeye						x	x	x				
Fly control tags						x						
Wean calves			f						s	s	s	
Vaccinate (IBR, PI3, Lepto, etc.)			f						s	s	s	
Retag cows			f							s	s	
Vaccinate heifers for Brucellosis				f						s		
Culling decisions made					f						s	s
Pre-calving vaccinations	s						f					
Increase pre-calving nutrition of bred females		s	s									
Maintain facilities and equipment	x	x	x	x	x	x	x	x	x	x	x	x
Cash-flow update and evaluation	x	x	x	x	x	x	x	x	x	x	x	x
Income tax preparation/filing	x											
Income tax evaluation/ adjustment										x		
Provide salt and mineral supplements	x	x	x	x	x	x	x	x	x	x	x	x
Evaluate management plan	x	x	x	x	x	x	x	x	x	x	x	x
Assure adaquate water quantity and quality	x	x	x	x	x	x	x	x	x	x	x	x
Continuing education	x	x	x	x	x	x	x	x	x	x	x	x
Budget preparation											x	x

s = spring calving herd activity; f = fall calving herd activity; x = applies to both fall and spring herds.
Note: this is a generic calendar and not intended for a specific enterprise

Table 25.2

BREAK-EVEN PRICE (PER 100 LB) FOR COMMERCIAL COW–CALF OPERATIONS WITH VARYING CALF CROP PERCENTAGES, ANNUAL COW COSTS, AND WEANING WEIGHTS

Calf Crop Percent Weaned	Annual Cow Cost	Break-Even Prices for Calves of Different Weaning Weights		
		400 lb	500 lb	600 lb
95	$700	$184.21	$147.37	$122.81
95	600	157.89	132.96	105.26
95	500	131.58	110.80	87.72
85	700	205.88	164.71	137.25
85	600	176.47	141.18	117.65
85	500	147.06	117.65	98.04
75	700	233.33	186.67	155.56
75	600	200.00	160.00	133.33
75	500	166.67	133.33	111.11

Figure 25.1
Percent calf crop, as measured by a live calf born and raised per cow, is the most economically important trait for the cow-calf producer. The bull, cow–calf, and producer each make a meaningful contribution to the level of productivity for this trait.
Source: Coleman Locke.

A break-even price of $141.18 per hundredweight means that the producer would have to receive more than $141.18 per 100 lb of calf sold, to cover the yearly cost of maintaining each cow in the breeding herd. Table 25.2 illustrates the break-even price for several levels of calf crop percentage, weaning weight, and annual cow cost. Profitability of the commercial cow–calf operation is determined by comparing the market price of the calves at the time they are sold with the break-even price.

Cow–calf operations are managed best by operators who know the factors that affect calf crop percentage (Fig. 25.1), weaning weight, and annual cow cost. The primary management objective should be to generate profits via improvement in pounds of calf weaned per cow and optimization or control of the annual cow costs.

Costs and Returns

Figure 25.2 shows the dollars returned over cash costs for an average U.S. cow–calf operation. Note the deviation in net returns over the 30-year period, which was the most profitable three decades in history for cow–calf producers Also notice the general inverse relationship between cattle inventory numbers and returns over costs. During 1986 to 1993, higher prices for calves resulted in higher levels of profitability; however, returns for 1994 to 1998 were negative or just at break-even during a period of herd expansion. As supplies came back into line with demand, cow–calf profits increased beginning in 1999. As cow inventory reduction continued for the next 14 years, profitability was sustained for the cow–calf sector.

Costs and returns are computed from an **enterprise budget** (sometimes called an **enterprise analysis**). The enterprise budget is one of the most useful financial forms upon which to make management decisions. The component of receipts and expenses for an average cow–calf enterprise are shown in Table 25.3. Notice that as weaning percent increases, the amount of returns needed to cover costs (break-even price) is reduced.

Figure 25.2

Annual cow–calf returns and cattle inventory.

Source: USDA: AMS and NASS compiled by Livestock Marketing Information Center.

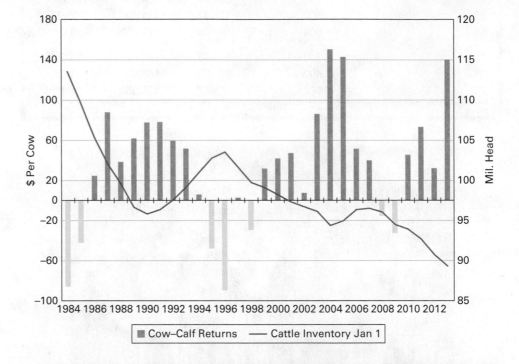

Table 25.3
RELATIVE CONTRIBUTION OF VARIOUS FACTORS TO A TYPICAL COW–CALF BUDGET

	%
Revenue:	
Steers	48
Heifers	30
Breeding Stock	22
Expenses:	
Feed cost (summer grazing and crop aftermath)	26
Purchased feed cost (hay, mineral, supplement)	34
Labor	15
Veterinary	3
Marketing	2
Fuel, Oil, Utilities	5
Interest	5
Other	10

MANAGEMENT FOR OPTIMUM CALF CROP PERCENTAGES

The primary management factors affecting calf crop percentages are as follows:

1. Heifers need to be fed adequate levels of a balanced ration to reach puberty at 15 months of age if they are to calve at the desired age of 2 years. Medium-frame-sized heifers of English breeds and crosses (e.g., Angus and Hereford) should weigh 650–750 lb at 15 months of age. Cattle of larger-frame-sized terminal breeds or crosses should weigh 100–150 lb more to ensure a high percentage of heifers cycling at breeding.

2. Heifers should be bred to calve early in the calving season. Heifers calving early are more likely to be pregnant as 2- and 3-year-olds, whereas heifers calving late will likely not conceive during the next breeding season. Some producers save more heifers as potential replacements at weaning and as yearlings so that selection for early pregnancy can be made at pregnancy test time.

3. Heifers typically have a longer postpartum interval than cows. This interval becomes shorter if first-calf heifers (heifers with their first calves) are separated from mature cows 60 days prior to calving and after calving. This separation allows the heifers to obtain their share of the feed essential for a rapid return to estrus.

4. Feeding programs are designed to have cows and heifers in a moderate body condition (visually estimated by the fat over the back and ribs) at calving time. Table 25.4 illustrates a **body condition scoring (BCS)** system. Cows that are thin at calving usually have longer postpartum intervals (Fig. 25.3). Cows that are too fat reflect higher feed inputs and thus higher costs than is necessary for profitable production.

5. Depending on the circumstances, mature cows should be observed twice daily during calving season. If possible, first-calf heifers should be observed every 4 hours as some females will have difficulty calving (**dystocia**) and will need assistance in delivery of the calf. Calving difficulties should be kept to a minimum to prevent potential death of calves and cows. Calving difficulty is also undesirable because cows given assistance will usually have longer postpartum intervals. Breeding young cows to easy calving sires is critical in reducing dystocia.

6. Calving difficulty should be minimized but usually cannot be eliminated. A balance should be maintained between the number of calves born alive and the weight of the calves at weaning. Calves that are heavier at birth usually have heavier weaning weights. However, the death loss of calves increases if birth weights are too large. Conversely, calves that are too light at birth often have increased death rates as well. Birth weight is the primary cause of calving difficulty; therefore, management decisions should be made to keep birth weights moderate. Bulls of the larger breeds or larger frame sizes should not be bred to heifers, and large, extremely high growth bulls, even in the breeds known for calving ease, should not be bred to heifers. Birth weight within a herd is influenced by genetics. Genetic differences are more important than certain environmental differences such as amount of feed during gestation. Bulls to be used artificially should have extensive progeny test records for birth weight and calving ease in addition to an individual birth-weight record.

7. The bull's role in affecting pregnancy rate has a marked influence on calf crop percentage. Before breeding, bulls should be evaluated for breeding soundness by addressing physical conformation and skeletal soundness, palpating the genital organs, measuring scrotal circumference, and testing the semen for

Table 25.4
SYSTEM OF BODY CONDITION SCORING (BCS) FOR BEEF CATTLE

Group	BCS	Description
	1	*Emaciated*—Cow is extremely emaciated, with no fat detectable over spinous processes, transverse processes, hip bones, or ribs. Tail-head and ribs project quite prominently.
Thin condition	2	*Poor*—Cow still appears somewhat emaciated but tail-head and ribs are less prominent.
	3	*Thin*—Ribs are still individually identifiable but there is obvious fat along spine and over tail-head, with some tissue cover over dorsal portion of ribs.
Borderline condition	4	*Borderline*—Individual ribs are no longer visually obvious. There is some fat cover over ribs, transverse processes, and hip bones.
Optimum moderate condition	5	*Moderate*—Cow has generally good overall appearance with fat cover apparent over ribs, down spine, and along the tail-head.
	6	*High moderate*—A higher degree of fat is detected over ribs and around tail-head.
	7	*Good*—Cow appears fleshy and obviously carries considerable fat. In fact, "rounds" or "pones" are beginning to be obvious on either side of the tail-head. There is some fat around vulva and in twist.
Fat condition	8	*Fat*—Cow is very fleshy and over conditioned. Cow has large fat deposits over ribs and around tail-head, and below vulva. "Rounds" or "pones" are obvious.
	9	*Extremely fat*—Cow is obviously extremely wasty and looks blocky. Tail-head and hips are buried in fatty tissue and "rounds" or "pones" of fat are protruding. Bone structure is no longer visible.

Source: Adapted from Richards et al., 1986. *J. Anim. Sci.* 62:300.

motility and morphology. **Libido** (sex drive) and mating capacity are additional important factors affecting the bull's influence on pregnancy rate. These traits are not easily measured in individual bulls before breeding or even after breeding in multiple-sire herds. The typical cow-to-bull ratio utilized in most cattle natural breeding systems is 30:1. However, some bulls can settle more than 50 cows in a 60-day breeding season. In some large pastures with rough terrain, the cow-to-bull ratio might have to be lower to assure a high calf crop percentage.

8. Crossbreeding affects calf crop percentage in several ways. Crossbred heifers usually cycle earlier and have higher conception rates than their straight bred counterparts. Crossbred calves are more vigorous and have a higher survival rate. An effective crossbreeding program can increase the calf crop by 8–12%.

9. The primary nutritional factor influencing calf crop percentage is adequate intake of energy, which can be expressed in pounds of total digestible nutrients (TDN). The quantity of TDN intake is important in helping to initiate puberty, maintaining proper body condition at calving, and keeping the postpartum interval relatively short. Other nutrients of major importance are protein, calcium, and phosphorus. Additional vitamins and minerals are important only in areas where the soil or feed is deficient.

10. Calf losses during gestation are usually low (2–3%) unless certain diseases are present in the herd. Serious reproductive diseases such as brucellosis, leptospirosis, vibriosis, trichomoniasis, and infectious bovine rhinotracheitis (IBR) can cause abortions, which may markedly reduce the calf crop percentage. These

A

B

Figure 25.3
(A) Example of a body
condition score 4.
(B) Example of a body
condition score 6.
Source: 25.3a & b: Tom Field.

diseases can be managed by blood-testing animals entering the herd or by vaccinating. Herd health programs vary for different enterprises, depending on the incidence of the diseases in the region. Details of these programs should be worked out with the local veterinarian.

11. Cows and heifers should be palpated at approximately 45 days after the end of the breeding season to determine if pregnancy has occurred. The producer should consider all marketing alternatives to maximize profits when selling open cows. Failing to check cows for pregnancy contributes to higher annual cow costs, lower calf crop percentages, and higher break-even costs of the calves produced.

12. Calf losses after 1–2 days following birth are usually small (2 – 3%) in most enterprises. Severe weather problems, such as spring blizzards, can cause high calf losses where protection from the weather is limited. In certain regions and in certain years, disease may also cause high mortality rates. Infectious calf scours and respiratory ailments can create situations where 10 to 30% of the calf crop may be lost.

Figure 25.4
Weaning weights on this ranch were increased over a 6-year period by 12% to an average of 568 pounds through the effective use of sire selection, crossbreeding, and grazing management. Source: Tom Field.

MANAGEMENT FOR OPTIMUM WEANING WEIGHTS

The primary management factors affecting calf weaning weights are as follows:

1. Calves born early in the calving season are heavier at weaning primarily because they are older. Calves are typically born over a several-week period but are weaned together on one specified day. Every time the cow cycles during the breeding season and fails to become pregnant, the weaning weight of her calf is reduced by 30–40 lb. Most commercial producers limit the breeding season to 90 days or less so the calves are heavier at weaning and can be managed in uniform groups (Fig. 25.4).

2. The amount of forage available to the cow and the calf has a marked influence on weaning weights. The cow needs feed to produce milk for the calf. The calf, after about 3 months of age, will consume forage directly in addition to the milk it receives.

3. **Growth stimulants**, approved to be given to nursing calves, will increase the weaning weight by 5–15%. Ralgro (Zeranol), Synovex C, and Compudose, common growth stimulants are implanted as pellets under the skin of the ear. Studies have shown that use of certain implants and repeated use of implants may lower percent of cattle grading USDA choice and may decrease palatability of the final product. Producers should communicate with buyers of their cattle to determine the appropriate implant protocol as some buyers prefer cattle that have not been implanted and thus may offer a premium price.

4. Providing supplemental feed to the calves where it is inaccessible to the cows will increase the weaning weight of the calves. This practice of **creep feeding** should be used with caution because it is not always profitable. It helps calves make the transition of the weaning process and is a feasible practice under drought or marginal feed-supply conditions. Creep feeding of breeding heifer calves can impair the development of their mammary system and subsequently reduce milk production. This impairment is caused by fat accumulating in the udder.

5. Diseases that affect the milk supply of the cow or growth of the calf will cause a reduction in the weaning weight of the calf (see Chapter 21).

Figure 25.5
The use of electronic identification (EID) tags is favored by enterprises where data capture and analysis is highly valued.
Source: Tom Field.

6. Genetic selection for milk production and calf growth rate will typically increase calf weaning weight. Selection based on EPDs (weaning weight and milk) is the most effective way to change weaning weights in a herd genetically. Effective bull selection will account for 80–90% of the genetic improvement in weaning weight, although weaning-weight information can also be used in culling cows and in selecting replacement heifers. It has been well demonstrated that effective selection can result in an increase in weaning weight of 4–6 lb per year on a per-calf basis. However, milk production is energetically expensive and large, high milking cows may have poor reproductive performance in times of limited feed availability. Matching the appropriate level of mature size and milk production potential of the cow herd to the feed resources of a specific ranch or region is critical to attain cost effective levels of calf production.

7. Crossbreeding for the average cow–calf producer can result in a 10–30% (average 20%) increase in pounds of calf weaned per cow exposed in the breeding herd. Most of the increase occurs due to improved reproductive performance; however, one-fourth to one-third of the 20% increase is due to the effect of heterosis on growth rate of the calf and increased milk production of the crossbred cow. The lifetime performance of crossbred females compared to straightbreds is considerably higher as a result of weaning a greater number of calves that tend to also be heavier.

8. Maintaining accurate animal identification is important to every phase of cow–calf management including inventory management, selection and culling, reproductive and nutritional management, and herd health. The use of electronic identification tags is an option that provides high value when maintaining source and age information, full supply chain performance, and ownership of calves beyond weaning (Fig. 25.5).

MANAGEMENT OF ANNUAL COW COSTS

In times of rising input costs, it may be difficult for a producer to lower annual cow costs. However, the most profitable cow–calf enterprises are those that are able to optimize the cost of production. Numerous studies show dramatic variances in cost of production across both regions and enterprises within regions. Adequate income

Table 25.5
COW–CALF PRODUCER PROFILES BY PROFIT GROUP

	Profit Group		
	Low	Medium	High
Days in breeding season	79	89	90
Percent calving	79%	89%	90%
Weaning weight	493 lb	525 lb	507 lb
Calf weight per cow exposed	413 lb	455 lb	455 lb
Cost per cow	$563	$300	$185
Calf break-even price/cwt	$136.43	$66.05	$40.63

Source: Adapted from Dunn, 2000. Note that while these are data from 2000, the relationships between profitability groups hold true across most conceivable sets of economic conditions.

Table 25.6
COMPARING ADVANTAGES OF HIGH AND MEDIUM PROFIT GROUPS

Item	Low Profit	Medium Profit ($)	High Profit ($)
Expenses per cwt	par	63.15 less	84.60 less
Dollars invested per cow	par	605 more	171 less
Calf revenue per cwt	par	6.92 less	9.78 more
Non-calf revenue per cwt	par	9.09 more	13.74 more
Total revenue percent	par	2.52 more	23.53 more

Source: Adapted from Dunn, 2000.

and expense records must be maintained so that cost areas can be carefully analyzed (Fig. 25.6). Enterprise budgets are needed to assess cow costs and make management decisions to lower cow costs.

Table 25.5 shows cow productivity, annual cow costs, and calf break-even prices for average, low-cost, and high-cost producers.

There is a difference of approximately $100 per cwt in net income when comparing low-cost producers ($41/cwt break-even) with high-cost producers ($136/cwt break-even). Table 25.6 illustrates that most of this difference is due to financial cost per cow. While pounds weaned per cow are important, its contribution to net income is far overshadowed by cost per cow. This difference illuminates where management decisions should first be focused.

The greatest cost management consideration should be given to **feed costs**, as they constitute the largest part of annual cow costs, usually 50–70%. The period from weaning a calf to the last one-third of gestation in the next pregnancy is the time when cows can be maintained on comparatively small amounts of relatively cheap, low-quality feeds. Cow–calf operations having available crop aftermath feeds (e.g., cornstalks, grain stubble, and straw) usually have the greatest opportunity to keep feed costs lower than other operations (Fig. 25.7). The most economical feed resource must be matched to the right biological type of cows—for example, usually those that display early puberty, calving ease, moderate milk, moderate mature weight, and fleshing ability (BCS) for early rebreeding. In many operations, the most

Figure 25.6
Cattle producers must keep accurate financial, forage, and herd productivity records to assess annual cow costs and the profitability of an enterprise.
Source: Courtesy of Coleman Locke.

Figure 25.7
Cornstalk fields are available to cattle after the gain has been harvested. There are millions of acres of stalk fields and other crop-aftermath feeds that can be grazed by cattle and other livestock.

economical feed resource comes from maximizing grazed forage with cows receiving minimal amounts of harvested and purchased feed.

Significant regional variation in feed costs exist for the cow–calf business with the southeastern, southern plains, and the west experiencing a nearly $200 a cow advantage due to lower feed costs as compared to the northern regions of the United States (Fig. 25.8).

Figure 25.8
Comparison of feed costs by region in the United States.
Source: USDA:ERS.

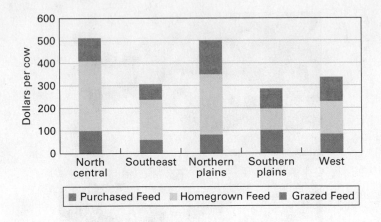

Figure 25.9
Comparison of operating costs, operating plus capitalization costs, and total economic costs by size of cow–calf enterprise.
Source: USDA:ERS.

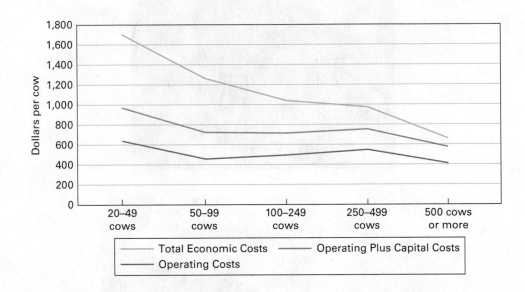

Labor costs usually compose 15–20% of the annual cow costs. Labor costs are usually lower on a per-cow basis as herd size increases and in areas where moderate weather conditions prevail. Typically 15–20 hours of labor per cow unit per year are required. Operations that use labor inefficiently require twice as much labor per cow. In larger ranching operations, optimal labor is considered to be achieved at one employee per 800 mature cows. In fact, economies of scale are substantial in cow–calf production where larger enterprises enjoy a competitive advantage in cash costs, operating costs plus capitalization, and total economic costs of operation (Fig. 25.9).

Interest charges on operating capital account for another 10–15% of the annual cow cost. Producers can reduce interest charges by carefully analyzing the costs of different credit sources.

Miscellaneous costs should not be overlooked. These smaller expense items (e.g., fuel, machinery, repair, supplies, utilities, and taxes) account for less than 20% of an average budget. Yet, they make up over 40% of the difference in cow cost between high-cost and low-cost producers. These smaller expense items should be evaluated on a regular basis.

Dramatic variation exists between cow–calf enterprises based on their marketing strategies and their region of location. Cow–calf enterprises can be categorized into those

Table 25.7

DEMOGRAPHIC AND PRODUCTION PRACTICE IMPLEMENTATION DIFFERENCES BY COW–CALF ENTERPRISES MARKETING CALVES FOLLOWING WEANING, STOCKER BACKGROUNDING, OR FINISHING

	Cow–Calf— sell at weaning	Cow–Calf—sell following stocker phase	Cow–Calf—sell following finishing	Cow–Calf - all
Demographics				
Cows (*N*)	64	93	86	79
Weaning wt. (lb)	502	499	523	502
Weaning (%)	82.6	83.0	84.5	82.9
Calves sold at weaning (%)	100	21	28	59
Calves sold following backgrounding/stocker (%)	0	79	15	36
Calves sold following finishing (%)	0	0	57	5
Farm income from cattle (%)	36	43	34	39
Production Practices				
Defined calving season	54	66	79	61
Use A.I.	4	11	19	8
Calfhood growth implant	9	17	25	14
Rotational grazing	59	62	56	60
Individual cow records	40	50	56	46

Source: USDA:ERS, 2011.

that sell their calves at weaning, choose to background or manage calves as stockers, and those that retain ownership on calves through the feedlot phase (Table 25.7).

These data illustrate that as producers own at least a portion of their calves further into the supply chain, the more likely they are to incorporate defined calving seasons, A.I., growth promotants, and individual cow records. Furthermore, those producers retaining ownership through the finishing phase are more likely to have higher weaning weights and reproductive rates.

Cow–calf enterprises are also divergent in performance and adoption of some management protocols across regions (Table 25.8). Those located in the north central states, northern plains, and west tend to have larger herd sizes, weaning weights, and weaning percentage performance as well as are more likely to retain ownership of a portion of their calves through the feedlot phase. Enterprises in these regions are also more likely to have a defined calving season, to utilize A.I., implant calves, and maintain individual cow records. The poorer levels of productivity experienced in cow–calf enterprises in the southeast and southern plains are largely offset by the feed cost advantage of these regions shown earlier in the chapter.

STOCKER-YEARLING PRODUCTION

Several alternate stocker-yearling production programs are available identified in Chapter 2, Figure 2.5. The best alternative will depend on forage and market conditions.

The primary factors affecting the costs and returns of stocker-yearling operations are cattle purchase and sale prices, cattle growth rates, the amount and quality

Table 25.8

DEMOGRAPHIC AND PRODUCTION PRACTICE IMPLEMENTATION DIFFERENCES IN COW–CALF ENTERPRISES ACROSS GEOGRAPHICAL REGIONS OF THE UNITED STATES

	North Central[a]	Southeast[b]	N. Plains[c]	S. Plains[d]	West[e]
Demographics					
Cows (N)	56	59	105	75	155
Weaning Wt. (lb)	501	480	543	493	538
Weaning %	83.6	80.9	87.3	82.9	82.8
Calves sold at weaning (%)	44	70	41	69	53
Calves sold following Backgrounding stocker (%)	45	28	49	29	39
Calves sold following finishing (%)	11	2	10	2	8
Farm income from cattle (%)	23	25	38	67	66
Production Practices					
Defined calving season	82	45	92	42	85
Use A.I.	11	4	17	6	14
Calfhood growth implants	28	7	26	8	13
Rotational grazing	54	60	58	62	71
Individual cow records	52	35	59	45	52

[a]IA and MO.
[b]VA, TN, KY, AR, MS, GA, AL, FL.
[c]KS, NE, SD, ND.
[d]OK, TX.
[e]MT, WY, CO, NM, CA, OR.
Source: USDA:ERS, 2011.

Table 25.9

PERCENTAGE BREAKDOWN OF A TYPICAL STOCKER BUDGET

Factor	Percent of Total
Calf price	81
Pasture	11
Interest	2
Purchased feed	2
Labor and overhead	2
Death loss	1
Other	1

of available feed in the form of forage and roughage, and health status of the cattle. Table 25.9 shows that management decisions affecting purchase price have the greatest impact on profitability.

Stocker-yearling producers need to be aware of current market prices for the cattle they purchase or sell. They also need to understand the loss of weight of the cattle from the time of purchase to the time the cattle are delivered to their farm or ranch. This loss in weight, called **shrink**, can sometimes reflect the difference in the profit or loss of the stocker-yearling operation. It is common for calves and yearlings to shrink 3–12% from on-farm weight to delivered weight. For example, yearlings purchased at 700 lb that shrink 4% will have a delivered weight of 672 lb. It typically takes 2–3 weeks to recover the weight loss.

The gaining ability of most stocker-yearling cattle is estimated visually. Cattle that are lightweight for their age, thin but healthy, with a relatively large skeletal frame size, usually have a high gain potential. Cattle that are light for their age are typically most profitable for the stocker-yearling operator, whereas heavier cattle are most profitable for the cow–calf producer.

Stocker-yearling cattle that are purchased and sold several times often encounter stress situations such as fatigue, hunger, thirst, and exposure to multiple disease organisms. The most common diseases include shipping fever complex and other respiratory diseases. These stress conditions make it necessary for stocker-yearling producers to have effective health programs for newly purchased cattle. Producers who have poor herd health programs typically experience higher costs of gain and higher death losses.

The primary objective of the stocker-yearling operation is to obtain the most pounds of cattle gain within economic reason while ensuring that high-quality forage yields can be obtained consistently each year. Forage management to obtain efficient production and consumption of nutritious feed is another essential ingredient of a successful stocker-yearling operation. Time of grazing and intensity of grazing (number of animals per acre) are important considerations if maximum forage production and utilization are to be maintained.

TYPES OF CATTLE FEEDING OPERATIONS

There are two basic types of cattle feeding operations: the commercial feeder and the farmer-feeder. The two types are generally distinguished by type of ownership and size of feedlot. **Commercial feedlots** are usually defined as having more than 1,000-head capacity and **farmer-feeder** feedlots as having less than 1,000-head one-time capacity.

The farmer-feeder operation is usually owned and operated by an individual or family. The commercial feedlot may be owned by an individual, a partnership, or a corporation—the last type is most common, especially as feedlot size increases. Commercial feedlots may own the cattle, feed the cattle owned by someone else (called *custom cattle feeding* or *custom feedlots*), or engage in some combination of the two. Custom-fed cattle are owned by other cattle feeders, investors, cattle producers, or packers and are fed on a contractual basis.

Even though the number of cattle fed in feedlots having more than 1,000-head capacity is increasing while feedlots less than 1,000 head are decreasing, there are some advantages to a farmer-feeder operation. Consider the following advantages and disadvantages:

1. The farmer-feeder can utilize cattle as a market for homegrown feeds.
2. The farmer-feeder can effectively utilize high roughage feeds in a backgrounding or warm-up operation.
3. The farmer-feeder can distribute available labor over several different enterprises with cattle feeding being only one of those enterprises.
4. The farmer-feeder may have advantages in flexibility. When feeding cattle is unprofitable, the farmer-feeder can totally close out the cattle feeding and divert time and dollars into other phases of the farming operation. The large commercial feedlot has much higher overhead costs, which are fixed regardless of whether the pens are filled with cattle or are empty.
5. Commercial feedlots usually obtain and analyze more records and information and make more effective management decisions. Commercial feedlots typically utilize more professional expertise (consultants) in managing nutrition, health, and marketing. Exceptional risk management is critical to the success of the

Table 25.10
THE MAJOR COMPONENT PARTS OF A FEEDLOT BUSINESS ANALYSIS

Major Component, with Primary Factors Influencing Them				
Investment in Facilities	Cost of Feeder Cattle	Feed Cost per Pound of Gain	Nonfeed Cost per Pound of Gain	Total Dollars Received
Land	Grade	Ration	Death loss	Market alternative
Pens	Weight	Rate of gain	Labor	Transportation
Equipment	Shrink	Feed efficiency	Taxes	Shrink
Feed mill	Transportation	Length of time on feed	Insurance	Dressing percentage
Office	Gain potential		Utilities	Quality grade
			Veterinary expenses	Yield grade
			Repairs	Manure value

feeding enterprise. Risk management strategies involve both price protection against rising input costs (feed, capital, feeder cattle) but also strategies to lock in fed cattle prices at profitable levels.

6. Custom cattle feeding greatly reduces the operating capital requirement for the commercial feedlot and shifts some of the risk to the customer. Commercial feed yards typically seek to have a lot capacity turnover of 2–2.5 times per annum. However, one of the greatest risks for commercial feedlots is keeping the lots full of cattle. Customers will stop putting cattle into custom lots when feeding profit margins are distinctly unfavorable in times of high input costs or when prices for calves are sufficiently high to reduce the incentive of cow–calf producers to retain ownership of their calf crop beyond the ranch. Farmer-feeders usually feed just one group of cattle per year, which requires leaving the facilities vacant for several months.

FEEDLOT CATTLE MANAGEMENT

The primary factors needed to analyze and properly manage a feedlot operation are investment in facilities, cost of feeder cattle, feed cost per pound of gain, nonfeed costs per pound of gain, and marketing. A more detailed analysis of these factors is shown in Table 25.10.

Facilities Investment

The investment in facilities varies with type and location of feedlots. In the United States, larger commercial feedlots are quite similar regardless of where they are located. The general layout is an open lot of dirt pens, with pen capacities varying from 100 to 500 heads. The pens are sometimes mounded in the center to provide a dry resting area for cattle. The fences are constructed of pole, cable, or pipe. A feed mill to process grains and other feeds is usually a part of the feedlot. Special trucks distribute feed to fence-line feed bunks. Bunker trench silos hold corn silage and other roughages. Grains might be stored in these silos; however, they are more often stored in steel bins above the ground. Investment cost per head of capacity for these types of feed yards is approximately $150.

Feedlots for farmer-feeder operations vary from unpaved, wood-fenced pens to paved lots with windbreaks, sheds, or total confinement buildings. The latter might have manure collection pits located under the cattle, which stand on slotted floors.

The feed might be stored in airtight structures. In farmer-feeder operations, however, feeds are typically stored in upright silos and grain bins, particularly where rainfall is high. Feeds are typically processed on the farm and distributed with tractor-powered equipment to feed bunks located either inside or outside the pens. Investment per head for these feed yards range from $200 to $500.

Cost of Feeder Cattle

Before buying feeder cattle, the feedlot operator first estimates anticipated feed costs and the price the fed slaughter cattle will bring. These figures are then used to project the cost of feeder cattle or what the operator can afford to pay for them. Feeder cattle are priced according to weight, sex, **fill** (content of the digestive tract), skeletal size, thickness, and body condition. Most commercial feeders prefer to buy cattle with **compensatory gain**. These cattle are thin and relatively old for their weight. They have usually been grown out on a relatively low quality of feed. When placed on feedlot rations, they gain rapidly and compensate for the previous lack of energy in their diet.

The feeder cattle buyer typically projects a high gain potential in cattle that have a large skeletal frame and little finish or body condition. However, not all cattle of this type will gain rapidly.

Heifers are usually priced a few cents a pound under steers of similar weight. The primary reason is that heifers gain more slowly, the cost per pound of gain is higher, and some feeder heifers are pregnant.

Feeder cattle of the same weight, sex, frame size, and body condition can vary several cents a pound in cost. This value difference is usually due to differences in fill. The differences in fill can amount to 10–40 lb in live weight of feeder cattle. Feed and water consumed before weighing, distance and time of shipping, temperature, and the manner in which cattle are loaded and transported are some major factors affecting the amount of shrink. Shrink results primarily from loss of fill, but weight losses can occur in other parts of the body as well.

Feed Costs

Feed cost accounts for one-third of a feedlot cost allocation (Fig. 25.10) with feeder calf purchases determining just over 60% of the total. Thus, these two factors alone are responsible for 95% of a feed yard's budget.

Figure 25.10

Purchase, processing, and delivery of feed comprises a significant cost center for a feed yard enterprise. Rations are typically delivered to pens of cattle via feed trucks that enable the efficient delivery of a consistently mixed formulated diet. Source: Tom Field.

Feed costs per pound of gain are influenced by several factors, and the knowledge of these factors is important for proper management decision making. The choice of feed ingredients and how they are processed and fed are key decisions that affect feed costs. The emergence of a highly subsidized corn-based ethanol industry has significantly raised feed costs.

Cattle that gain more rapidly and efficiently on the same feeding program have lower feed costs. Some of these differences are genetic and can sometimes be identified with the specific producer of the cattle. Most feeder cattle receive feed additives (e.g., Rumensin) and growth implants that improve gain and efficiency and eventually the cost of gain.

Feed cost per pound of gain gets progressively higher as days on feed increase. Therefore, cattle feeders should avoid feeding cattle beyond their optimum combination of slaughter weight, quality grade, and yield grade.

Nonfeed Costs

Nonfeed cost per pound of gain is sometimes referred to as **yardage** cost. Yardage cost includes costs of gain other than feed. These costs can be expressed as either cost per pound of gain or cost per head per day. Obviously, cattle that gain faster move in and out of feedlots sooner and accumulate fewer total dollar yardage costs.

Death loss and veterinary expenses caused by feeder cattle health problems can increase the nonfeed costs significantly. Most cattle feeders prefer to feed yearlings rather than calves because the death loss and health problems in yearlings are significantly lower than in calves.

Marketing

The total dollar amount received for slaughter cattle emphasizes the need for the cattle feeder to be aware of marketing alternatives and the kinds of carcasses the cattle will produce. Nearly 80% of the fed cattle are sold directly to a packer. This marketing alternative requires the cattle feeder to be aware of current market prices for the weight and grade of cattle being sold.

Many slaughter cattle at large commercial feedlots are sold on a standard shrink (pencil shrink) of 4%, with the cattle being weighed at the feedlot without being fed the morning of weigh day. Feeders who ship their cattle some distance before a sale weight is taken should manage their cattle to minimize the shrink loss.

Approximately two-thirds of fed cattle are sold on a carcass basis, an increase from the approximately 20% in 1971. Grid pricing or formula pricing systems differentiate value based on carcass characteristics such as weight, marbling score, cutability (backfat and ribeye area), and freedom from blemishes such as dark cutting, yellow fat, and bruises. As a result, feed yard managers have incorporated technologies such as ultrasound to better predict cattle composition to assure more precise management to hit specific market targets and to capture the highest value of each individual animal (Fig. 25.11). Some slaughter steers and heifers are sold on a live weight basis, with the buyer estimating the carcass weight, quality grade, and yield grade. Slaughter cattle of yield grades 4 and 5 usually have large price discounts. The price spread between the USDA Select and Choice quality grades can vary considerably over time. Some marketing alternatives will not show a price differential between Select and Choice if the cattle have been well fed for a minimum number of days (100–120 days for yearling feeder cattle).

Figure 25.12 illustrates three market steers of comparable carcass weight but different levels of fat thickness, muscularity, USDA yield grade, marbling score, and USDA quality grade. Under most market conditions, the black steer would have the highest

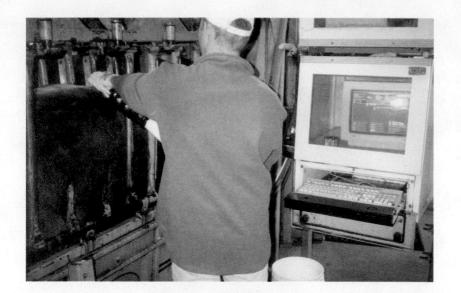

Figure 25.11
Ultrasound technology has been incorporated into feed yard management to more accurately assess composition (backfat and ribeye area) and to a lesser extent level of marbling.
Source: Tom Field.

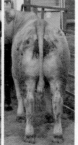

Dressing %	64.0
Carcass weight	842
Fat thickness	0.30
Ribeye area	16.6
% KPH fat	2.5
Yield grade	1.7
Maturity	A 40
Marbling score	SL 10
Quality grade	SE−

Figure 25.12
Example of three fed steers of various combinations of USDA Quality Grade and Yield Grade. Feed yard profitability is dependent on finishing cattle to optimal combinations of composition and marbling.
Source: J.D. Tatum.

Dressing %	65.4
Carcass weight	860
Fat thickness	0.60
Ribeye area	13.6
% KPH fat	2.0
Yield grade	3.4
Maturity	A 60
Marbling score	SM 50
Quality grade	CH−

Dressing %	64.2
Carcass weight	854
Fat thickness	0.80
Ribeye area	13.3
% KPH fat	3.5
Yield grade	4.2
Maturity	A 90
Marbling score	MT 10
Quality grade	CH

Figure 25.13
Average monthly returns to cattle feedlots in the Southern Plains.
Source: USDA: AMS and NASS compiled by Livestock Marketing Information Center.

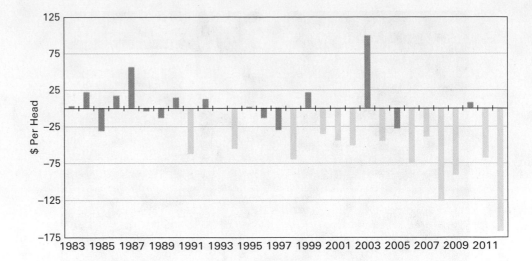

value by virtue of having sufficient marbling to quality as USDA Choice while avoiding the discounts associated with the excess fatness of USDA yield grades 4 and 5. The value differences between the crème-colored steer and the red steer would depend on the level of discounts for USDA Select and USDA Yield grade 4 carcasses. Profitable cattle feeding depends on the ability to optimize both compositional and quality traits while assuring high rates of gain, feed efficiency, and freedom from disease.

Cattle feeders who manage their cattle consistently for profitable returns know how to purchase high-performing cattle at reasonable prices. These cattle feeders formulate rations that will optimize cattle performance with feed costs. Also, their cattle have minimum health problems with a low death loss. The feeder feeds cattle for the minimum number of days required to assure carcass acceptability and palatability, and develops a marketing plan that will yield the maximum financial returns.

COSTS AND RETURNS

Analysis of the costs and returns for average fed cattle production between 1980 and 2007 shows an average fed steer price of $71.50 and an average break-even of $71. These results point to the reality that in a commodity business an average producer breaks even over time. In the long run, commodity businesses offer profits only to those who have lower than average cost and superior performance. The enterprise budget is calculated on a dollar-per-head basis. The two major cost categories were feeder cattle purchase (74.3%) and feed (19.8%). Because the market was below the break-even price, a loss of almost $56 per head was realized. Figure 25.13 shows average monthly returns for cattle feeders in the southern Great Plains. Similar returns would be expected throughout the United States. Observe the contrasts in highly profitable versus highly unprofitable years.

ENVIRONMENTAL MANAGEMENT

The four primary environmental management issues for feed yards are dust, odor, flies, and water quality. Dust management is best accomplished via regular pen maintenance with timely manure removal in late spring. The regular use of box scrapers reduces manure accumulation without cratering the pen hardpan. The use of overhead sprinklers or water truck sprayers can be utilized to maintain the recommended

Table 25.11

NUTRIENTS IN MANURE FROM FOUR SOURCES (LB/HD/YR)

Source	N	P_2O_5	K_2O
Solid manure from open lots	65	41	65
Solid manure from deep-bedding building	132	66	132
Liquid runoff from open lots	5	2	11
Liquid manure from deep pit	89	55	79

Source: Adapted from Iowa State University (2001).

moisture rate of 25–35% in 1 in. of manure requires approximately 14 gal of water per head of pen capacity. Achieving the same moisture increase in 2 in. of manure would require roughly 28 gal per head of pen capacity.

Control of odor requires regular pen maintenance, the use of correctly constructed and maintained runoff holding ponds, and proper nutritional management. The use of a more precisely balanced ration to avoid overfeeding of phosphorus is a key strategy.

Research is being conducted to use plant-oil extracts or fat extracts as a treatment for pen surfaces to control both odor and dust. Plant-oil extracts may be useful in inhibition of manure fermentation as a means to preserve nutrient value and suppress odor. These extracts may also have antimicrobial properties that help to inhibit the growth of organisms that contribute to foodborne illness.

Another strategy is to compost manure. The benefits of **composting** are to reduce volume and weight, kill weeds, concentrate nutrients, reduce odor, and to reduce fly populations. The disadvantages are labor and equipment costs, storage space, and nutrient loss. The nutrients in manure vary by type of lot construction (Table 25.11).

Fly control can be achieved via chemical, biological, or combination strategies. Chemical controls used alone are typically costly, short term in effect, and increase risk of human or environmental chemical exposure. The use of biological controls via fly-parasites has been successfully adopted in the industry.

Fly-parasites, also known as parasitic wasps, lay their eggs inside fly pupae. The fly pupae then become a food source for the fly-parasites. The use of early releases of fly-parasites before fly season followed by scheduled weekly releases is recommended. The cost of biological control ranges from $.20–$1.00 per head of cattle.

Water quality issues are likely to have a significant impact on feedlots and other intensive animal management facilities. Waste management is a primary concern as the decomposition of manure can negatively affect water quality via pathogens, nitrate, ammonia, phosphorus, salts, and organic salts. Incorrect handling, storage, or land application of manure can result in contamination of groundwater or surface water.

In an active feed yard, a layer of soil and manure becomes sufficiently compacted to create a seal that serves as a barrier to seepage. If the integrity of this layer is ensured, then water infiltration can be kept to less than 0.05 in. per day. It is critical that this layer be left undisturbed during pen cleaning.

Applying manure or wastewater directly to agricultural lands requires thorough knowledge about the following factors associated with a particular site—soil type, slope, irrigation practices, precipitation levels, crop nutrient requirements, nutrient levels of manure, and proximity to waterways or sells. The goal of manure management is essentially to collect, store, and apply wastes to lands at appropriate

Table 25.12
SUMMARY OF BEST MANAGEMENT PRACTICES FOR MANURE HANDLING, STORAGE, AND APPLICATION

Analyze for nutrient content

Account for available N from the total system

Apply to land areas large enough to accommodate manure volume

Calculate long-term manure loading rates

Maintain records of manure and soil analysis; application volume, timing, and methodology; additional fertilizer applications; and plant yields

Base manure application rates upon site-specific nutrient plans

Incorporate manure soon after application to avoid runoff

Determine application protocol based on soil composition and risk of aquifer contamination

Apply manure uniformly utilizing correct calibration

Utilize buffer zones to prevent water contamination

Use grass strips to catch and filter nutrients and sediments from runoff

Use rotational application schemes when planting high N use crops or forages

Locate manure stockpiles away from wells

Divert runoff via ditches, terraces, etc.

Maintain integrity of manure-soil seal when cleaning feedlot pens

Source: Adapted from Colorado State University (1994).

agronomic rates with the objectives of optimizing crop growth rates, economic returns, and protecting water quality.

Runoff problems can be minimized via employment of **best management practices (BMP)** such as building up-gradient ditches, dams, grass filter strips, filter fences, and other appropriate controlled drainage systems. Lagoons or ponds may be required to contain wastewater and runoff. BMP for manure handling are provided in Table 25.12.

CHAPTER SUMMARY

- Well-managed commercial cow–calf producers keep cost-effective records to achieve low break-even prices (BEs) by implementing the following formula:

$$BE = \frac{(\% \text{ calf crop} \times \text{weaning weight})}{\text{annual cow cost}}$$

- Visual scores of body condition (BCS) in cows are used to minimize costs and obtain optimum levels of percent calf crop and weaning weights.

- The primary factors affecting costs and returns of stocker-yearling and feedlot operations are (1) marketing (purchase price and sale price), (2) gaining ability of the cattle, (3) amount and quality of forage/feed, and (4) the health of the cattle.

KEY WORDS

calf crop percentage
enterprise budget (enterprise analysis)
body condition score (BCS)
dystocia
libido

growth stimulants
creep feeding
feed costs
labor costs
shrink

commercial feedlots
farmer-feeders
fill
compensatory gain

yardage
composting
best management practices (BMP)

REVIEW QUESTIONS

1. What are the seven critical management areas for a cow–calf producer?
2. Discuss the management calendar for a cow–calf enterprise.
3. What is the formula to calculate break-even price and how does each factor influence the outcome?
4. List the three critical categories of cost in a cow–calf business.
5. Discuss strategies to optimize calf crop percentage.
6. Describe the BCS system and its usefulness to managers.
7. Describe various approaches to increasing weaning weight.
8. Describe management of annual cow costs to increase profitability.
9. Compare and contrast performance of cow–calf producer profit categories.
10. Describe differences in the performance of enterprises that vary in their marketing strategy.
11. Compare differences in cow–calf enterprises based on region.
12. What are the factors affecting stocker enterprise profitability?
13. Compare various types of feed yards.
14. Describe various marketing alternatives used by feed yards.
15. What are the four major environmental issues for feed yards?
16. Describe best management practices to deal with manure.

SELECTED REFERENCES

Albin, R. C. and G. B. Thompson. 1996. *Cattle Feeding: A Guide to Management.* Amarillo, TX: Trafton Printing.

Duckett, S. K., D. G. Wagner, F. N. Owens, H. G. Dolezal, and D. R. Gill. 1996. Effects of estrogenic and androgenic implants on performance, carcass traits and meat tenderness in feedlot steers: A review. *The Professional Animal Scientist* 12:205–214.

Dunn, B.H. 2000. Characterization and Analysis of the Beef Cow-calf Enterprise of the Northern Great Plains Using Standardized Performance Analysis. Ph.D. thesis. South Dakota State University.

Field, T. G. 2007. *Beef Production and Management Decisions.* Upper Saddle River, NJ: Prentice Hall.

Field, T.G. 2006. Priorities First: Identifying Management Priorities in the Commercial Cow-Calf Business. American Angus Association, St. Joseph, MO.

Harner, J. P. and J. P. Murphy. 1998. *Planning Cattle Feedlots.* Kansas State University Ext. Publ. MF-2316. Manhattan, KS.

Iowa State University. 2001. *Beef Feedlot Systems Manual.* ISU Ext. Publ. 1867. Ames, IA.

Jarrige, R. and C. Beranger (eds.). 1981. *Beef Cattle Production.* Amsterdam: Elsevier Science Publishers.

McBride, W.D. and J. Mathews. 2011. The Diverse Structure and Organization of U.S. Beef Cow-Calf Farms. USDA:ERS. EIB #73.

National Cattlemen's Beef Association. 2001. *Cattle and Beef Handbook.* Englewood, CO.

National Research Council. 1996. *Nutrient Requirements of Beef Cattle.* Washington, DC: National Academy Press.

Richards, M. W., J. C. Spitzer, and M. B. Warner. 1986. Effect of varying levels of nutrition and body condition at calving on subsequent reproductive performance. *J. Anim. Sci.* 62:300.

Ritchie, H. D. 1992. *Calving Difficulty in Beef Cattle.* Beef Improvement Federation, BIF-FS6a and FS6b.

USDA. 2000. *Baseline References of Feedlot Management Practices.* National Animal Health Monitoring System. APHIS, VS.

26

Dairy Cattle Breeds and Breeding

Production of milk per cow has been increased markedly in the past 50 years by improvements in breeding, feeding, sanitation, and management. The application of genetic selection, coupled with the extensive use of artificial insemination, has contributed to a successful dairy herd improvement program.

CHARACTERISTICS OF BREEDS

Six major breeds of dairy cattle—**Holstein**, **Ayrshire**, **Brown Swiss**, **Guernsey**, **Jersey**, and **Red and White**—are used for milk production in the United States. These breeds are shown in Figure 26.1 where production characteristics and other information about the breeds are also given.

It is important to recognize the variation for milk yield and percent milk components that exist between breeds and within a breed. Table 26.1 shows the variation in milk production, percent fat, and percent protein for the major breeds. While the breeds are distinctly different based on breed averages, it is possible, for example, to identify Holstein cows that have higher fat percentages than Jersey cows.

Some Holstein cows produce extremely large amounts of milk—more than 100 lb/day at peak lactation. Thus, in Holsteins and other cows with high milk production, great stress is placed on udder ligaments, which can break down and no longer support the udder. If an udder breaks down, it is more susceptible to injury and disease, often necessitating culling the cow.

Ayrshires and Brown Swiss produce milk over a greater number of years than Holsteins, but their production levels are lower. Efficiency of milk production per 100 lb body weight is similar for the major breeds of dairy cows.

Guernsey and Jersey cows produce milk having high percentages of milk fat and solids-not-fat, but the total amount of milk produced is relatively low. The efficiency of energy production of different breeds varies less among the breeds than do either total quantity of milk produced or percentage of fat in the milk.

Registration Numbers

Breed popularity can be estimated from registration numbers, as shown in Table 26.2. Most dairy cows are grade cows as they do not have pedigrees (with individual registration numbers) recorded by a breed association. However, producers with registered cattle are a source of breeding

Ayrshire

Origin: Scotland
Average weight:
    Bulls—1,850 lb
    Cows—1,200 lb
Color: Mahogany and white spotted,
    may have pigmented legs
Average milk yield: 15,601 lb
Percentage of fat: 3.9%

Figure 26.1
Major breeds of dairy cows with origin, identifying characteristics, and production listed.
Source: Agri-Graphics.

Brown Swiss

Origin: Switzerland
Average weight:
    Bulls—2,000 lb
    Cows—1,400 lb
Color: Solid blackish, hairs dark with
    light tips
Average milk yield: 18,074 lb
Percentage of fat: 4.1%

Guernsey

Origin: Guernsey Island
Average weight:
    Bulls—1,600 lb
    Cows—1,100 lb
Color: Light red and white, yellow skin
Average milk yield: 15,474 lb
Percentage of fat: 4.5%

cattle for the total dairy industry. Breeds with the largest registration numbers reflect the demand from the commercial dairy industry.

DAIRY TYPE

Some descriptive terms describing **ideal dairy type** are *stature, angularity, level rump, long and lean neck, milk veins,* and *strong feet and legs.* In the past, dairy producers have placed great emphasis on dairy type, but research studies indicate that some components of dairy type may have little or no value in improving milk production and might even be deleterious if overstressed in selection. However, a properly

Figure 26.1
(continued)

Holstein

Origin: Holland
Average weight:
    Bulls—2,200 lb
    Cows—1,500 lb
Color: Black and white
Average milk yield: 20,318 lb
Percentage of fat: 3.65%

Jersey

Origin: Jersey Island
Average weight:
    Bulls—1,500 lb
    Cows—1,000 lb
Color: Blackish hairs have white tips to give
    gray color or red tips to give fawn color;
    also can be solid black or white spotted.
Average milk yield: 14,275 lb
Percentage of fat: 4.65%

Milking Shorthorn

Origin: Great Britain
Average weight:
    Bulls—2,000 lb
    Cows—1,250 lb
Color: Red and white, roan
Average milk yield: 13,930 lb
Percentage of fat: 3.6%

Red and White

Origin: Holland
Average weight:
    Bulls—2,100 lb
    Cows—1,400 lb
Color: Red and white
Average milk yield: 19,967 lb
Percentage of fat: 3.69%

attached udder and strong feet and legs are good indicators that a cow will remain productive for longer periods of time. Table 26.3 describes the components (general appearance, dairy character, body capacity, and udder) of preferred dairy type and the associated weighting for each trait under a uniform classification system. Relative

Table 26.1
AVERAGE DHI HERD PERFORMANCE BY BREED (ALL TEST CATEGORIES)

Breed	Milk (lb)	Fat (%)	Protein (%)
Ayrshire	15,418	3.9	3.2
Brown Swiss	18,703	4.1	3.4
Guernsey	15,495	4.5	3.4
Holstein	23,791	3.7	3.1
Jersey	17,302	4.8	3.6
Milking Shorthorn	14,311	3.6	3.1
Red and White	20,716	3.7	3.0

Source: Adapted from ARS:USDA, Council on Dairy Cattle Breeding.

Table 26.2
ANNUAL REGISTRATION NUMBERS FOR MAJOR U.S. BREEDS OF DAIRY CATTLE

Breed	Annual Registrations (in thousands)						Year Association Formed
	2013	2005	1995	1990	1980	1970	
Holstein	362.7	293.5	329.9	395.9	535.9	281.6	1870
Jersey	103.3	73.0	63.4	53.6	61.0	37.1	1818
Brown Swiss	10.6	10.7	10.8	11.7	12.9	16.4	1880
Guernsey	4.8	5.0	7.4	13.9	20.9	43.8	1877
Ayrshire	4.1	6.0	6.4	7.8	10.9	15.0	1875
Red and White	3.9	3.7	4.4	6.7	4.8	—	1964
Milking Shorthorn	3.0	2.9	3.2	2.3	4.9	5.4	1912

Source: Various Dairy Cattle Breed Associations.

scoring for udder structure is 40%, 20% for dairy character, frame as well as feet and leg scores receive 15% weightings with the remaining 10% assigned to body capacity classification. Classification of dairy type in bulls is weighted as follows: frame (30%), dairy character (25%), body capacity (20%), and feet and legs (25%).

Some dairy breed associations have a classification program to evaluate the type traits. Breeders use the **linear classification system** as an evaluation tool to enhance selection for high-producing cows with the durability to stay productive. The Holstein Association offers five participation programs:

Classic—all cows in the herd are evaluated; eligible for herd recognition awards.

Standard—most, but not all, cows are evaluated; eligible for herd awards.

Limited—focused on evaluation of first-calf heifers.

Introductory—offered to breeders who want to get started in the program.

Breeders' Choice—selected cows are evaluated.

A classifier approved by the association scores each cow or bull for the several traits and gives a final score if these scores are to be official records recognized by the association.

Table 26.3
RELATIVE WEIGHTING OF THE MAJOR TRAITS TO ASSESS DAIRY TYPE

Trait	Description	% of Points
Frame	Long and wide rump, height at withers and hips proportional, straight and strong back, nearly level loin, overall style and balance.	15
Dairy character	Freedom of coarseness, wide set rib structure, long and lean neck, clean design through the dewlap and brisket, loose and pliable skin	20
Body capacity	Long, deep barrel, deep flank, wide chest floor, big spring of rib	10
Feet and legs	Deep heel; short, well rounded toes; squarely placed feet, moderate hock angle, hocks free of swelling or coarseness, short pasterns that are strong and flexible	15
Udder	Moderate udder depth, squarely placed teats, rear udder attachment wide and high, evidence of strong suspensory ligament, teats cylindrical, even, and medium in size, evenly placed mammary quarters	40

Source: Adapted from Purebred Dairy Cattle Association and Holstein Association.

The final score represents the degree of physical perfection of any given animal. It is expressed in the following numbers and words:

Excellent (EX): 90–100   Good (G): 75–79
Very good (VG): 85–89   Fair (F): 65–74
Good plus (G+): 80–84   Poor (P): 50–64

The final score is used in computing predicted transmitting ability for type (PTAT). PTAT identifies genetic differences in sires that can be considered in selection programs to improve dairy cattle type.

Type has value from a sales standpoint. Although some components of type score are negatively related to milk production, type is important as a measure of the likelihood that a cow will sustain a high level of production over several years.

IMPROVING MILK PRODUCTION

Great strides have been made over the past 50 years in improving milk production through improved management and breeding. For example, the total amount of milk produced in the United States continues to increase; yet the number of dairy cows in 2012 was less than half the number in 1950 (Fig. 26.2). Average milk production per cow in 2012 was more than three times greater than the average production in 1940. In the past 10 years alone, annual milk production in the United States has increased by more than 350 lb per cow per year.

SELECTION OF DAIRY COWS

The average productive life of a dairy cow is short (approximately 3–4 years). Many cows are culled primarily because of reproductive failure, low milk yield, udder breakdown, foot and leg weaknesses, and mastitis.

Heifers whose ancestral records indicate they will be high producers should be used to replace cows that are **culled** for low production. Such heifers should also be used to replace cows that leave the herd because of infertility, mastitis, or death, although the improvement gained thereby is generally modest.

A basis for evaluating a dairy cow is the quantity of milk (pounds) and quality (total solids) that she produces. For a dairy operator to know which cows are

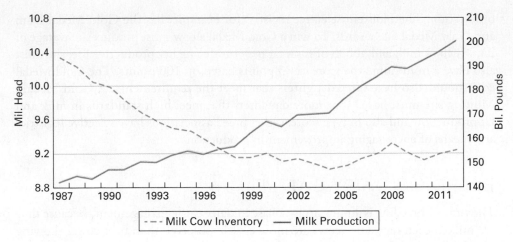

Figure 26.2
Changes in milk production in the United States, 1940–2009. Source: USDA: NASS compiled by Livestock Marketing Information Center.

good producers and which cows should be culled, a record of milk production is essential.

Nearly one-half of U.S. dairy cows are enrolled in the **National Cooperative Dairy Herd Improvement Program (NCDHIP)**, a national, industry-wide production-testing and record-keeping program. It is often referred to as the **Dairy Herd Improvement (DHI)** program. DHI is delivered via a cooperative association of 30 affiliates, six regional data processing centers, and the USDA Animal Improvement Programs Laboratory, which handles the genetic evaluation. DHI is designed to collect and process raw data into a systematic format useful in making dairy management decisions.

DHI also facilitates the development of a national database that is utilized in genetic evaluation. Of the nearly 4.5 million enrolled cows, about 55% contribute data to the genetic evaluation effort.

DHI offers numerous recording programs. Four of the most common data collection procedures are described as follows:

1. **Supervised test**—the DHIA technician weighs and samples milk for all cows from each milking during a 24-hour period.
2. **Partially supervised test**—milk weights and samples are taken alternately at A.M. and P.M. milkings by the DHIA technician and one other person.
3. **Owner conducted test**—someone other than the DHIA technician records test-day production data.
4. **Supervised electronic test**—supervised test whereby data collection is conducted electronically with the technician certifying procedures and accuracy.

Some restricted, unsupervised data collections are allowed. DHI data collection and processing is performed under strict quality standards to ensure the integrity of the information. Records obtained allow producers to compare cows within the herd and to compare herd performance to that of other herds in the region.

TriStar is the program offered by the Holstein Association for providing production records, cow and herd genetic performance reports, and recognition programs, such as the Gold Medal Dam and Dam of Merit recognition. TriStar offers four service options to meet the needs of its members.

Records are standardized to a lactation length of 305 days, 2 milkings per day (2✕), and to a mature age of cow (ME = mature equivalent). Adjustments are made to records so they can be more accurately compared on a standardized basis.

A registry association for purebred animals exists for each dairy cattle breed in the United States. In addition, some breed associations honor cows with outstanding production performance and bulls with daughters that are outstanding in production. The Holstein selective registry, for example, has the Gold Medal Dam and Gold Medal Sire awards. To win a Gold Medal a cow must produce an average of 24,881 lb of milk and 500 lb of milk fat per year during her productive life; she must also have a minimum type score of 83 points based on 100 points. The Gold Medal cow should also have three daughters that meet the requirements. A Gold Medal–winning sire must have 10 or more daughters that meet high standards in milk and fat production and also in type score. Other breed associations have selective registries as a means of encouraging improvement in production.

BREEDING DAIRY CATTLE

The time of breeding is an important phase in dairy cattle management. Because they are milked each day, dairy cows are more closely observed than beef cows, allowing visual detection of estrus. When in estrus, dairy cows may show restlessness, enlarged vulvas, and a temporary decline in milk production. Also, when cows are in standing heat they will permit other cows to mount them.

Technicians are available to artificially inseminate cattle, but well-trained dairy producers or employees are excellent inseminators. The use of semen from genetically proven sires is highly desirable, even though this semen may cost more than semen from an average bull. Considering the additional milk production that can be expected from heifers sired by a good bull, the extra investment can return high dividends. Semen costing $25–$150 per unit from genetically superior bulls may be a better investment than semen from less desirable bulls at $10–$20 per unit. Bulls should not run with the milking cows, because bulls are often dangerous. Bulls that provide semen for artificial insemination should be handled with caution as well.

The heritability of traits is indicative of the progress that can be made by selection. Susceptibilities to cystic ovaries, ketosis, mastitis, and milk fever are all low in heritability (5–10%). Percentages of fat, protein, and solids-not-fat are all high in heritability (50%). Yearly milk (ME), protein, solids-not-fat, and fat yields are medium in heritability (25–30%). Heritability estimates for traits assessed by the linear classification systems are listed in Table 26.4.

The genetic correlation between two traits is indicative of the amount of genetic change in trait A that might be expected from a certain amount of selection pressure applied to trait B. The most important genetic correlations are those that might be associated with milk yields during the first lactation. Fat, solids-not-fat, protein yield, lifetime milk yields, and length of productive life have high genetic correlations with first lactation milk yield (0.70–0.90). Overall type score, levelness of rump, udder texture, and strength of fore and rear udder are all negatively correlated with first lactation milk yield (−0.20 to −0.40). Dairy character and udder depth are positively correlated (0.35–0.40) with first lactation yield. Inbreeding tends to increase mortality rate and to reduce all production traits except fat percentage of milk and mature body weight.

There are several **inherited abnormalities** known in dairy cattle. This does not mean that dairy cattle have a higher number of inherited abnormalities than other farm animals, but that more is known about dairy cattle than about most other farm animals because dairy cattle are observed more closely. No breed of dairy cattle is free from all inherited abnormalities. Some of these abnormalities include achondroplasia (short bones), weavers, limber limbs, rectal-vaginal constriction, dumps, flexed pasterns (feet turned back), fused teats (teats on same side of udder are fused), hairlessness (almost no hair on calf), and syndactylism (only one toe on a foot). Many of these inherited abnormalities are lethal and most are recessive in their mode of inheritance.

Table 26.4
HERITABILITY OF HOLSTEIN ASSOCIATION TYPE TRAITS

Trait	h^2
Stature	0.42
Strength	0.31
Body depth	0.37
Dairy form	0.29
Rump angle	0.33
Thurl width	0.26
Rear leg score	0.21
Foot angle	0.15
Fore attachment	0.29
Rear udder height	0.28
Rear udder width	0.23
Udder cleft	0.24
Udder depth	0.28
Teat placement	0.26
Teat length	0.26
Final Score	0.29

Source: Adapted from Holstein Association Sire Summary.

Usually the occurrence of inherited abnormalities is infrequent. However, occasionally an outstanding sire might carry an abnormality or a specific dairy herd may have several genetically abnormal calves. As most inherited abnormalities result from recessive genes, both the sire and dam of genetically abnormal calves carry the undesirable gene. Breeding stock should not be kept from either parent.

One reason such rapid progress has been made in improving milk production is that genetically superior sires are identified and used widely in artificial insemination programs. Genetically superior bulls today might become the sires of 100,000 calves or more each in their productive lives by use of artificial insemination. Generally, the better sires will sire more calves than ordinary sires because dairy producers know the value of semen from outstanding bulls. Thus, selection is enhanced markedly for greater milk production by use of artificial insemination.

Traits of importance in selection of dairy cattle include milk production, milk composition, longevity, and structural correctness. These traits are usually emphasized according to their relative heritability and economic importance. Fertility is extremely important but low in heritability; so marked improvement in this trait is more likely to be accomplished environmentally through good nutrition and management rather than through selection.

Most dairy cattle in the United States are straightbred because crossing breeds has failed to lead to a significant improvement in milk production. No combination of breeds, for example, equals the straightbred Holstein in total milk production. Because many genes control milk production and the effect of each of the genes is unknown, it is impossible to manipulate genes that control milk production by the same method that would be used in the case of simple inheritance. Also, milk production is a sex-limited trait expressed only in the female. Furthermore, milk production is highly influenced by the environment. In order for milk production to be improved by genetic selection, the environment must be standardized among all animals present to ensure, insofar as possible, that differences between animals are due to inheritance rather than environment.

In addition to a high level of milk production, characteristics of longevity, regularity of breeding, ease of milking, and quiet disposition are important in dairy cows.

Bulls are evaluated for their ability to transmit the characteristic of high-level milk production both by considering the production level of their ancestors (pedigree) and by considering the production level of their daughters (progeny testing). An index is used as a predictive evaluation of the bull's ability to transmit the characteristic of high-level milk production.

Sire Genetic Evaluations, computed by the USDA, are based on comparing daughters of a given sire with their contemporary herd mates. Each sire is assigned a **predicted transmitting ability** based on the superiority or inferiority of his daughters to their herd mates. Many sires have daughters in 50–100 different herds, so the predicted transmitting ability (PTA) for milk, fat, and type is generally highly reliable and provides a sound basis for the selection of semen. The USDA publishes PTAs among sires semiannually.

SIRE SELECTION

The dairy industry (breed associations, National Association of Animal Breeders, and the USDA) uses the **best linear unbiased prediction (BLUP)** method for estimating PTA among sires. Dr. C. R. Henderson developed this method at Cornell University. The BLUP procedure accounts for genetic competition among bulls within a herd, genetic progress of the breed over generations, pedigree information available on young bulls, and differing numbers of herd mate's sires, and partially accounts for the differential culling of daughters among sires. In the BLUP method, direct comparisons are made among bulls that have daughters in the same herd. Using bulls that have daughters in two or more herds makes indirect comparisons. An example presented by the Holstein Association (1980) depicts the use of three different sires in two herds:

Herd 1	Herd 2
Daughters of sire A	Daughters of sire B
Daughters of sire C	Daughters of sire C

Direct comparisons can be made between sires B and C in herd 2. Using the common sire C as a basis for comparison can make indirect comparisons between sires A and B. PTAs are calculated for milk, protein, fat, type, and dollars returned.

The next step in the evolution of selection is the utilization of indices that combine both biological performance data on several traits as well as economic weightings for each trait as a means to calculate estimates that describe the genetic potential of an animal for a particularly important set of economically valuable traits. **Net merit index (NM$)** is an excellent example of a bioeconomic index. Net merit index is calculated by the following formula:

$$NM\$ = 0.7 (MFP\$) + \$11.30(PTA\ PL) - \$28.22(PTA\ SCS - \text{breed ave. SCS})$$

where:

$$MFP\$ \text{ (milk-fat-protein)} = \$0.031\ (PTA\ \text{milk lb}) + \$0.80\ (PTA\ \text{fat lb}) + \$2.00\ (PTA\ \text{protein lb})$$

PTA PL = Predicted transmitting ability productive live (number of months in milk until the cow is 84 months old). PTA PL is a measure of ability to avoid culling and is highly correlated with production and with type classification score, especially udders.

PTA SCS = Predicted transmitting ability somatic cell score. This estimate is *not* a replacement for excellent management and preventative care. It is useful as a tool to balance selection for increased yield without incurring excessive mastitis.

The **Type Production Index (TPI)** combines differences for milk production traits and differences for type traits into a single value. TPI is calculated via the following formula:

$$TPI = \left[3\left(\frac{PTAP}{19}\right) + \left(\frac{PTAF}{22.5}\right) + \left(\frac{PTAT}{.7}\right) + \left(\frac{UDC}{.8}\right) \right] 50 + 576$$

where:

PTAP = Predicted Transmitting Ability Protein

PTAF = Predicted Transmitting Ability Fat

PTAT = Predicted Transmitting Ability Type

UDC = Udder Composite

CHAPTER SUMMARY

- Based on registration numbers, the Holstein is the major breed of dairy cattle, followed by Jersey, Brown Swiss, Guernsey, Ayrshire, Red and White, and Milking Shorthorn.

- Selection for milk production has more than tripled the pounds of milk produced per cow from 1940 to 2013.

- The dairy industry has implemented effective production testing programs and record-keeping systems for a longer period of time compared to most other livestock industries.

KEY WORDS

Holstein
Ayrshire
Brown Swiss
Guernsey
Jersey
Red and White
ideal dairy types
linear classification system
culled
National Cooperative Dairy Herd Improvement Program (NCDHIP)
Dairy Herd Improvement (DHI)
supervised test

partially supervised test
owner conducted test
supervised electronic test
TriStar
inherited abnormalities
Sire Genetic Evaluation
predicted transmitting ability
best linear unbiased prediction (BLUP)
net merit index (NM$)
Type Production Index (TPI)

REVIEW QUESTIONS

1. Describe the strengths and weaknesses of the six major dairy breeds and their relative importance to the dairy industry.

2. Discuss the systems used to describe dairy type and the usefulness of these systems.

3. Describe trends in milk production and the role of breeds and selection in the changes.

4. What is the role and scope of DHI?

5. Define genetic correlation and its importance to dairy cattle breeding.

6. Why is the use of crossbreeding limited in the dairy industry?

7. Discuss the role of selection indexes in dairy selection.

8. Compare and contrast the net merit index and the type production index.

SELECTED REFERENCES

Bath, D. L., F. N. Dickinson, H. A. Tucker, and R. D. Appleman. 1985. *Dairy Cattle: Principles, Practices, Problems, Profits*. Philadelphia, PA: Lea & Febiger.

Holstein Association USA. 2006. *Linear Classification Program*. Brattleboro, VT.

Holstein Association USA. 2006. *Holstein Type-Production Sire Summaries*. Brattlesboro, VT.

Legates, J. E. 1990. *Breeding and Improvement of Farm Animals*. 4th ed. New York: McGraw Hill.

Schmidt, G. H., et al. 1988. *Principles of Dairy Science*. Englewood Cliffs, NJ: Prentice Hall.

Trimberger, G. W., W. E. Etgen, and D. M. Galton. 1987. *Dairy Cattle Judging Techniques*. Englewood Cliffs, NJ: Prentice Hall.

Wiggins, G. R. 1991. National genetic improvement programs for dairy cattle in the United States. *J. Anim. Sci.* 69:3853.

27

Feeding and Managing Dairy Cattle

The U.S. dairy industry has changed greatly from the days of the family milk cow. In 1850, one-third of all cattle were dairy cows. By 1940, nearly 40% of U.S. cattle were dairy animals. Currently, fewer than 10% of cattle are utilized as dairy cows. The contemporary dairy industry is highly specialized. A large investment in cows, machinery, barn, and milking parlor is necessary. Dairy operators who produce their own feed need additional money for land on which to grow the feed. They also need machinery to produce, harvest, and process crops required to meet the dietary requirements of the dairy herd.

Although the size of dairy operations varies from 30 or less to more than 5,000 milking cows, the average dairy has approximately 100 milking cows, 30 dry cows, and 100 replacement heifers. The average dairy producer farms 200–300 acres of land, raises much of the forage, and markets the milk through cooperatives of which he or she is a member. These producers sell about 3 tons of milk daily or about 2.2 million pounds annually. Their average total capital investment may exceed a half million dollars. The average producer has a partnership with a family member or another person to improve management and make labor more efficient. However, the largest dairies have increasing market share with 500+ head operators controlling almost one-half of the U.S. dairy cow inventories.

The dairy operator must provide feed and other management inputs to keep animals healthy and at a high level of efficient milk production. Feed and other inputs must be provided at a relatively low cost compared with the price of milk if the dairy operation is to be profitable.

NUTRITION OF LACTATING COWS

The average milk production per cow in the United States for a lactation period of 305 days is approximately 21,000 lb. Some herd averages exceed 25,000 lb, and some top-producing cows yield more than 40,000 lb of milk per year. Thus, some lactating cows may produce more than 150 lb of milk, more than 5 lb of milk fat, and more than 4.5 lb of protein per day. Lactation is nutritionally expensive and thus requires substantial feed intake to sustain production. For example, a cow weighing 1,400 lb that produces 40 lb of milk daily needs one and a quarter times as much energy for lactation as for maintenance. If she is producing 80 lb of milk daily, she needs 2.5 times more energy for milk production than for maintenance.

Providing adequate nutrition to lactating dairy cows is challenging and complex. Some reasons for this complexity are shown in

learning objectives

- Explain the nutrient and milk yield relationships during lactation and gestation
- List the elements of a successful dairy nutrition program
- Describe how to adjust a dairy ration for heat stress
- Compare the nutritional program for dry cows and replacement heifers
- Describe the appropriate management of dairy bulls
- Describe the various types of dairy housing
- Explain the importance of a nutrient management plan
- Describe the milking process
- List the primary reasons dairy females are culled
- Discuss mastitis prevention and preventative health practices
- Compare and contrast profitable versus unprofitable dairies
- Discuss the value of performance records benchmarks

Figure 27.1

Nutrient and milk yield relationships during lactation and gestation.

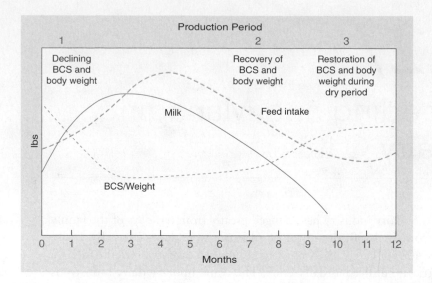

Figure 27.1. The dairy cow's nutritional needs vary widely during the production cycle. Dairy producers must formulate rations to match each period of the production cycle so as to optimize milk yield and reproduction, prevent metabolic disorders, and increase longevity. Water and energy are the two most limiting nutrients in a dairy ration.

It is difficult during the first 2–4 months after calving to provide adequate nutrition because milk yield is high and intake is limited. As the nutrient intake is less than the nutrient demand for milk, the cow uses her body fat and protein reserves to make up the difference. Thus the cow is in a negative energy balance and usually loses body weight during the months of heavy milk production (Fig. 27.1). This same period of time can also be challenging to reproduction, as conception rates are usually lower when cows are losing weight.

Body condition scores (1 = thin, 3 = average, 5 = fat) can be used to monitor nutrition, reproduction, and health programs for dairy herds. Condition scores in the low 3s and high 2s are acceptable during the first few weeks after calving, but body condition scores in the low 4s are necessary as the cows move into the dry period. Scores lower or higher than the aforementioned benchmarks suggest potential management problems. To provide adequate nutrition at effective costs, cows can be grouped by condition score and stage of lactation.

At 2–3 months into the lactation period, daily milk production peaks and then starts to decline; feed intake is adequate or higher than milk production demands. This contributes to body-weight gain (Fig. 27.1). The energy content of the ration should be monitored after approximately 5 months of lactation to prevent the cow from becoming too fat.

Different types and amounts of feeds can be used to provide the needed energy, protein, vitamins, and minerals to lactating cows. The availability, palatability, and relative costs of different feeds are the primary factors influencing ration composition. Basic nutrition principles, presented in Chapters 15, 16, and 17, provide the foundation for developing dairy cattle feeding programs.

The elements of a successful dairy nutrition program can be described as follows:

1. Assure a constant, high-quality water supply.
2. Utilize high-quality feedstuffs.
3. Maintain long fiber size in forages (20% of fibers >1.5 in. to ensure normal rumen function).

4. Maintain an optimal level of grain feeding (<60% of DM as concentrate).
5. Monitor cud-chewing to assess rumen function and sufficient fiber length (60–75% of cows chewing cud during observation at rest).
6. Assure sufficient bunk space (75% of cows need to be able to eat at once).
7. Maintain consistent, high levels of DM intake (3.8–4.0 lb of DM per 100 lb of body weight).
8. Ensure that feed is available 24 hours per day and especially when cows return from the milking parlor.
9. Do not offer free-choice ration components as this makes it impossible to control the total mixed ration.
10. Manage floors, ventilation, stalls, animal handling practices, and temperature control to assure cow comfort.
11. Monitor concentrate and forage quality to minimize variation in feedstuffs.
12. Manage to minimize occurrence of displaced abomasum and ketosis.

Monitoring cow activity is one means to avoid nutritional and reproductive problems. Ideally, cows should spend time in the following activities to assure optimal performance—lying down (50%), eating (21%), milking (13%), social (6%), lock-ups (6%), and drinking (4%). A charting spreadsheet is available from ABS Global (www.absglobal.com) to calculate cow activity.

Most dairy cattle rations are based on roughages (hay and silages). Roughages are usually the cheapest source of nutrients; often they are produced by dairy farms, but increasing herd size, greater managerial demands, higher land-tax costs, and cheaper feeds are changing this traditional role. Silages and concentrates are typically mixed together before being fed to the cows; this is known as a **total mixed ration**, and is the preferred way of feeding dairy cows (Fig. 27.2). Chopped hay and silage can be delivered to mangers on each side of an open alleyway.

Lactating dairy cows cannot obtain an adequate supply of nutrients from an all-roughage ration. Concentrates are supplied in amounts consistent with the level of milk production, body weight of the cow, amount of nutrients in the roughage, and nutrient content of the concentrates. Concentrates are usually provided to cows while they are being milked. However, high-producing cows requiring large amounts of concentrates do not have sufficient time in the milking parlor to consume all the necessary grain; therefore, additional concentrates are usually fed at another location. Another innovation that enhances management of cows in negative energy balance during the first third of lactation is the use of by-pass proteins and protected fats. These approaches protect nutrients from the action of the microbes in the rumen. For example, protein produced by the activity of the microbes in the rumen has a less desirable amino acid balance than many proteins provided in the diet, so bypassing the rumen offers the opportunity for absorption of a more desirable amino acid mix in the small intestine.

Figure 27.2

A typical forage-based total mixed ration delivered via a feed truck into a line bunk.

Photo by Justin Field.

There are economic advantages in feeding concentrates in relation to quantity of milk produced by each cow. However, this method tends to feed a cow that is declining in production too generously and to feed a cow that is increasing in production inadequately. It is a poor economic practice to allow cows in their last 2 months of lactation to have all the concentrates that they can consume. Heavy feeding at this stage of lactation does not result in increased milk production.

Young (2-year-old) cows that are genetically capable of high production should be fed large amounts of concentrates to provide the nutrition they need to grow as well as to produce milk. Without adequate nutrition, their subsequent breeding and lactation may be hindered. The dairy cow transitioning from dry to lactating, and the first-calf heifer moving into her first lactation, need to be on a rising plane of energy and protein in the 30–60 days prior to parturition, with the most rapid increase occurring in the 14 days prior to calving.

Forages vary immensely in nutrient concentrations. Legumes (alfalfa) have high calcium, protein, and potassium relative to animal nutrient requirements, whereas other forages such as grasses and corn silage are considerably lower. Concentrates (grains) usually are low in calcium and high in phosphorus. The amounts of protein supplement (soybean meal) and mineral supplement that must be added to a concentrate formulation obviously depend on the type and amount of forage being fed. Trace mineralized salt and vitamins A, D, and E are usually added to concentrates.

Table 27.1 shows nutritional management consideration of cows in various periods of production. It is recommended that higher producing cows receive rations specifically formatted to their needs. Higher producing females may require greater

---

Table 27.1
NUTRITIONAL MANAGEMENT CONSIDERATION FOR VARIOUS PRODUCTION PHASES

Production Stage	Duration	Dietary Considerations
Period 1 Early Lactation	0–70 days post-calving	• Cows will lose weight and condition • Adjust rumen to milking ration • Increase grain at start 1 lb/day to avoid acidosis • Grain should not exceed 60% of total DM • Fiber level should be a minimum of 18% ADF • 20% or more of forage should exceed 2 in. in length • CP should exceed 19% • Allow constant access to feed
Period 2 Maximum Intake	70–140 days post-calving	• Cows should maintain or increase weight • Grain intake can be maximized at 2.5% of body weight • Feed at least 2 times per day
Period 3 Mid to Late Lactation	140–305 days post-calving	• Grain feeding should be at a level to meet lactation requirements and to increase body weight • Young cows should receive a high-quality ration to meet growth requirements
Period 4 Dry Phase	60–14 days prior to subsequent lactation	• Critical to next lactation performance • Dry cows separated • DM intake of 2% of body weight • Forage should account for ½ of DM • Lower quality forage may be fed • Avoid excessive salt, calcium or phosphorus in diet
Period 5 Transition	2 weeks prior to calving	• Introduce grain to ration • Increase protein to 14–15% • Maintain long-stem hay in ration

Source: Adapted from multiple sources.

amounts of fiber, energy, and bypass protein. Feed typically accounts for about one-half of the cash costs associated with a dairy. As such, producers should allocate a comparable level of attention to the feeding aspect of the dairy.

ADJUSTING FOR HEAT STRESS

Heat stress may lead to milk production losses of 8 to 10 lb per day. Heat stress may occur at temperatures in excess of 77°F or at humidity indices of greater than 72°F. Symptoms of heat stress include a body temperature of more than 102.5°F (101.5°F is normal), excessive panting (more than 80 breaths per minute), feed intake losses, which may be as high as 15%, decreased milk yields, and declining pregnancy rates of 20% or more. Cows under heat stress may also experience higher incidences of acidosis.

Therefore, managers need to implement environmental adaptation strategies. Management protocols to minimize the effects of heat stress include provision of shade, assuring adequate air movement, providing cooling from fans or sprinklers, reformulating diets to match intake, provision of ample sources of clean water, and minimizing cow movement to avoid increasing core body temperature.

NUTRITION OF DRY COWS

How dry cows are fed and managed may influence their milk production level and health in the next lactation. Suggested nutrient densities for dry cow rations are provided in Table 27.2.

A common practice in drying off lactating cows is abruptly to stop milking the cow; with high producers, however, this may be traumatic and dry-off may need to be more gradual (intermittent milking). The buildup of pressure in the mammary gland causes the secretory tissues to stop producing milk. At the last milking, the cow should be infused with a treatment for preventing mastitis.

Dairy producers plan for a 50- to 60-day dry period. Short dry periods usually reduce future milk yield because the cow has not adequately improved body condition and the mammary tissue has not properly regenerated. Long dry periods can lower milk yield because the cow may become overly fat, and profitability may be less because feed costs are increased.

Dry cows should be separated from lactating cows so they can be fed and managed consistent with their needs. Dry cows need fewer concentrates than lactating cows. If dry cows overeat, they will likely become fat. This excessive body fat may

Table 27.2
NUTRIENT DENSITY FOR DRY COW RATIONS

Nutrient	Early Dry Period	Pre-Fresh Period
Dry matter intake (lb)	26–28	23–26
Crude protein (%)	12.5–13.0	14.0–15.0
NE² (meal/lb)	0.55–0.62	0.68–0.70
Calcium (%)	>0.60	0.45–1.50
Phosphorus (%)	0.30–0.35	0.35–0.40
Vitamin A (IU/day)	100,000	130,000
Vitamin D (IU/day)	20,000	30,000
Vitamin E (IU/day)	800	1,000–3,000

Source: Adapted from Perkins, 1998. Colorado Dairy Nutrition Conference.

lower future milk yield and cause health problems, such as fatty liver, ketosis, mastitis, retained placenta, metritis, milk fever, or even death.

NUTRITION OF REPLACEMENT HEIFERS

Young heifers (5 months of age and older) can meet most of their nutritional needs from good-quality legume or grass-legume pasture in addition to 2–3 lb of grain. If the pasture quality is poor, they will need good-quality legume hay and 3–5 lb of grain. In winter, good-quality legume hay or silage and 2–3 lb of grain are needed. If the forage is grass (hay or silage), the heifers will need 3–5 lb of grain to keep them growing well.

Heifers should be large enough to breed at about 15 months of age and calve as 2-year-olds. Weight recommendations for heifers from different breeds are shown in Table 27.3. Weight is important because it affects when the heifer reaches puberty and can be bred. Also, heifers that are heavier at first calving produce more milk. For example, the optimum weight at calving for Holstein heifers is 1,200–1,250 lb. Lighter heifers produce less milk and heavier heifers are too costly for the increased production.

MANAGEMENT OF BULLS

It is generally recommended that dairy bulls not be kept on the farm and that all breeding be done by AI. By not keeping bulls on the farm, producers (1) decrease the risk of a dangerous bull and (2) increase genetic progress by using superior AI bulls. A 1996 NAHMS survey of dairy producers found that almost half of dairy farms had no on-farm mature bulls, and another one-third had only a single sire on the farm.

Some commercial producers do keep bulls on their farms to breed cows naturally, primarily because bulls detect estrus more accurately and keep conception rates high. Clean-up bulls are typically 2 years of age or younger due to the aggressive nature of mature males. Nose rings are sometimes placed in the noses of bulls to provide a means of restraint and control. Typically these devices are nonrusting and self-piercing with application usually occurring when the bull is between 9 and 12 months of age.

Often breeding bulls are housed with lactating cows and are thus fed the same ration. Diets appropriate for the high-producing cow will exceed the nutritional requirements of young sires. The high calcium content of a diet appropriate for lactating cows may lead to calcification of muscle in the mature bull if they are fed under this regime for an extended period of time.

Table 27.3
RECOMMENDED WEIGHTS AT BREEDING FOR REPLACEMENT HEIFERS

Breed	Weight (lb)
Ayrshire	650–700
Brown Swiss	750–825
Guernsey	650–700
Holstein	750–825
Jersey	500–600

Source: Adapted from multiple sources.

CALVING OPERATIONS

Dairy cows that are close to calving should be separated from other cows, and each cow should be placed in a maternity stall that has been thoroughly cleaned and bedded with clean bedding. The cow in a maternity stall can be fed and watered there. A cow that delivers her calf without difficulty should not be disturbed; however, assistance may be necessary if the cow has not calved by 4–6 hours from the start of labor. Extreme difficulties in delivering a calf may require the services of a veterinarian, who should be called as early as possible. Approximately 20% of first-calf heifer and 10% of cows will require obstetrical assistance a birth due to mild or severe dystocia.

As soon as a calf arrives, it should be wiped dry. Any membranes covering its mouth or nostrils should be removed, and its **navel** should be dipped in a tincture of iodine solution to deter infection. Producers should be sure newborn calves have an adequate amount of colostrum because colostrum contains antibodies to help the calf resist any invading microorganisms that might cause illness. The cow should be milked to stimulate her milk production.

Many commercial dairy operators dispose of bull calves shortly after the calves are born. Some bull calves are fed for veal, while others are castrated and fed for beef. Heifer calves are grown, bred, and then milked for at least one lactation to evaluate their milk-producing ability.

Dairy calves rarely nurse their dams. Usually they are removed from their dams and fed milk or milk replacer for 4 to 8 weeks. Calves that are separated from one another usually have fewer health problems and thus a higher survival rate.

Milk that is not salable or milk substitutes are fed to the young calves. Grain and leafy hay are provided to young calves to encourage them to start eating dry feeds and to stimulate rumen development. As soon as calves are able to consume dry feeds (45–60 days of age), milk or milk replacer is removed from the diet because of its high cost.

MILKING AND HOUSING FACILITIES FOR DAIRY COWS

Dairy farming is labor-intensive, so labor-saving machines are common parts of the facilities. Dairy cows usually are managed in groups of 20–50 head; they may be housed in free-stall (cows have access to stalls but are not tied) or loose housing systems. These facilities reduce cleaning, feeding, and handling time as well as labor. In tie-stall barns, cows are tied in a stanchion and remain there much of the year; feeding and milking are done individually in the stanchion. In this case, a pipeline milking system is used. Cows are milked in place and milk is carried via a stainless-steel pipe to the bulk tank, where it is cooled and stored. In loose housing systems, such as the loaf shed or free stall (Fig. 27.3), cows are brought to a specialized milking facility, called a **parlor**, for milking. The percentages of dairy farms using various **housing systems** are listed in Table 27.4.

Many housing systems require daily or twice-daily removal of manure and waste feed; this may be done mechanically (with a tractor blade) or by flushing with water. Loose housing systems without free stalls usually have cows on a manure pack, where the pack is typically removed once or twice a year. Manure, urine, and other wastes usually are applied directly to land and/or stored in a lagoon, pit, or storage tank for breakdown before being sprayed or pumped (irrigated) onto land.

Figure 27.3
A free-stall barn with
individual stanchions.
Photo by Justin Field.

Table 27.4
HOUSING TYPE BY PRODUCTION STAGE

	Percent of Enterprise		
Type	Unweaned Dairy Heifers	Weaned Dairy Heifers	Lactating Cows
Freestall	2.5	12.1	32.6
Individual animal area	29.7	5.3	0.1
Multiple animal area housed	40.0	34.6	3.4
Tie stall or stanchion	10.5	5.9	49.2
Drylot	9.1	22.9	4.6
Pasture	7.4	10.8	9.9
Calf hutch	32.5	—	—

Source: Adapted from NAHMS, 2007.

WASTE MANAGEMENT

The increased attention of consumers and environmental protection groups toward confined animal enterprises has resulted in a multitude of local, state, and federal regulations. As a result, many dairy operators have taken a proactive stance to better manage their environmental impact. The creation of nutrient management plans coupled with the use of rations that are more precisely formulated to minimize generation of problematic levels of nitrogen and phosphorus in animal waste can effectively improve environmental impact.

A South Carolina study found that almost three-quarters of dairies evaluated were feeding excess phosphorus in the diet. This excess nutrient was not correlated with improved milk production. As such, the use of precision ration balancing had the potential to reduce the amount of P fed by about 20% on average.

Developing a **nutrient management plan** for a dairy is a critical step in assuring conformance to environmental regulations and dealing with complaints from neighbors. Nutrient management plans should be accomplished via a thorough development

Figure 27.4
Manure solids are collected from a settling tank and then spread in rows to allow for composting. The rows are turned and mixed occasionally to maintain optimal composting conditions. Dry material is then available for use as a fertilizer. Photo by Justin Field.

process and include site information (names and contact information, physical enterprise description, and emergency plans); production data (number of animals, estimated amount of waste and wastewater produced); permit information as specified by law; manure application data; test results and dead animal disposal protocols.

Records that should be maintained and retained in regards to nutrient management planning and monitoring include informational records such as soil tests, manure tests, and manure application agreements. Also, activity records such as manure storage, field application sites and quantities, and equipment calibration records.

Often times, manure is composted so that it can be made ready for application to farm ground, gardens, or other horticultural sites as a fertilizer (Fig. 27.4). When manure is applied to land, then the following pieces of information should be logged—date of application, amount applied, number of acres, how it was applied, who performed the application, the wind speed and direction, air temperature, soil conditions, and sky conditions. Finally, the cropping history should be recorded including crop yields, nutrient credits taken, use of commercial fertilizers, level of precipitation, and tillage practices.

MILKING OPERATIONS

Approximately 90% of dairies utilize a twice per day milking schedule. An example of a modern milking parlor is a concrete platform raised about 30 in. above the floor (pit) of the parlor (Fig. 27.5). It is designed to speed the milking operation, reduce labor, and make milking easier for the operator. Cows enter the parlor in groups or individually, depending on the facility. The udder and teats are washed clean, dried, and massaged for about 20–30 seconds. Milking begins about 1 minute later and continues for about 6–8 minutes. After milking is completed, the milking unit is removed manually or by computer, and the teats are dipped in a weak iodine solution. Figure 27.6 illustrates a rotary milking parlor design that allows cows to flow through the parlor in a circular pattern with the milking crew located centrally.

Modern milking machines are designed to milk cows gently, quickly, and comfortably. Some have a takeoff device to remove the milking cup once the rate of flow drops below a designated level. Removal of the milking unit as soon as milking is complete is important because prolonged exposure to the machine can cause teat and udder damage and lead to disease (mastitis) problems.

The milking machine operates on a two-vacuum system: one vacuum is located inside the rubber liner and the other outside the rubber liner of the teat cup

Figure 27.5
A herringbone designed milking parlor that allows personnel to milk 12 cows at a time (six on each side). Photo by Justin Field.

Figure 27.6
Rotary milking parlor that facilitates continuous flow of cows to and from the milking area and optimizes convenience for milking personnel. Source: Tom Field.

Figure 27.7
Milking machines use a pulsation system for automated milking. The cow's udder is cleaned before and after placement of the teat cups to assure optimal health. Photo by Justin Field.

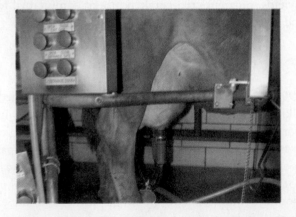

(Fig. 27.7). There is a constant vacuum on the teat to remove the milk and keep the unit on the teat. The intermittent vacuum in the pulsation chamber causes the rubber liner to collapse around the teat. This assists blood and lymph to flow out of the distal teat into the upper part of the teat and udder.

The milking parlor and its equipment must be kept sanitary. The pipes are cleaned and sanitized between milkings.

Regular milking times are established with equal intervals between milkings. Most cows are milked twice per day; however, high-producing cows produce more milk if they are milked three times as day. The latter is labor-intensive and economically feasible only in herds with high levels of productivity.

Robotic milking systems are an innovation being used by nearly 8,000 European dairies and will eventually be commercialized on a wider scale in the United States to reduce labor costs and to enhance standardization of milking protocols.

CONTROLLING DISEASES

The same diseases afflict dairy and beef cattle, but some diseases are more serious in dairy cattle than in beef cattle. Certain disease organisms can be transmitted through milk and can be a major problem for people who consume unpasteurized milk. Disease organisms are carefully monitored. Bulk tank somatic cell count (BTSCC) is the mummer of white blood cell and secretory cells per milliliter of raw milk. BTSCC is an effective measure of both udder health and milk quality. Increased levels of BTSCC would indicate rising levels of intramammary infection. The regulatory limit for BTSCC is 750,000 cells/ml although some processors have more rigorous requirements. Table 27.5 illustrates the effectiveness of milk quality management for U.S. dairies across herd size.

Several diseases, such as tuberculosis and brucellosis, can affect the health of dairy cattle and of humans consuming their milk. Requirements for the production of grade A milk include that the herd must be (1) checked regularly with the ring test for tuberculosis and (2) bled and tested for brucellosis. Further, milking and housing facilities must be clean and meet certain specifications. Grade A milk must have a low bacteria count and somatic cell count (low mastitis). Grade A milk sells at a higher price than lower grades of milk; therefore, most dairy producers strive to produce grade A milk.

Bang's disease (brucellosis) is caused by the *Brucella abortus* organism. It can markedly reduce fertility in cows and bulls and humans can contract it as a disease known as *undulant fever*. Heifers should be vaccinated in calfhood to provide immunity from disease. Since most dairy cows are bred by artificial insemination, there is little danger of a clean herd becoming contaminated from breeding. Persons drinking unpasteurized milk are taking a serious risk unless they know that the milk comes from a clean herd.

Perhaps the most troublesome disease in dairy cattle is **mastitis**—inflammation and infection of the mammary gland. The disease destroys tissue, impedes milk production, and lowers milk quality. Mastitis costs the U.S. dairy industry more than $1.5 billion each year or approximately $200 per cow annually. Reasons for culling dairy cows are listed in Table 27.6. Mastitis and reproductive failure are the leading reasons for culling.

Table 27.5
BULK TANK SOMATIC CELL COUNTS (BTSCC) BY HERD SIZE (2006)

BTSCC (cells/ml)	Herd Size		
	<100	100–499	>500
<100,000	3.7	0.3	3.2
100,000–199,000	26.1	31.4	32.3
200,000–199,000	38.4	43.5	47.6
300,000–399,000	19.8	17.0	14.1
400,000–499,000	9.6	7.8	2.3
>500,000	2.4	0.0	0.5

Source: Adapted from USDA: NAHMS, 2007.

Table 27.6
DAIRY COW CULLING PRACTICES

Reason	% Dairy Cows Culled
Udder/mastitis problems	23
Reproductive problems	26
Poor production	16
Lameness/injury	16
Disease	4
Poor temperament	1

Source : Adapted from NAHMS, 2007.

In its early stage of development, subclinical mastitis is undetectable by the human eye. This type of mastitis can be detected only with laboratory equipment (somatic cell counter at the DHI testing laboratory) or with a cow-side test called the California mastitis test (CMT). In the latter test, a reagent is added with a paddle to small wells, and then milk is squirted from each quarter into a specific well. If a reaction occurs, then the relative degree of subclinical mastitis infection can be determined. Small white clots appear in the milk of cows with advanced mastitis. At this stage it is called **clinical mastitis** and is easily observed. Milk can also be collected using a strip cup, which has a fine screen on a dark background such that white flakes, strings, or blood (from infected areas) can be seen. As mastitis progresses, the milk shows more clots and is watery, and acute infection is seen. At this stage, the signs of mastitis in the cow are obvious: the udder is swollen, red, and hot and gives pain to the cow when touched. In the final stages of mastitis, the gel formation becomes dark and is present in a watery fluid. At this stage, the udder has been so badly damaged that it no longer functions properly.

Susceptibility to mastitis may be genetically related to a certain degree, but environmental factors such as bruises, improper milking, and unsanitary conditions are more prevalent causes. High-producing animals may be more prone to stress and mastitis than low-producing animals. Also, cows with **pendulous** udders or oddly shaped teats are more likely to develop mastitis. Housing facilities and milking procedures are other important factors.

The best approach to controlling mastitis is using good management techniques to prevent its outbreak. These include using routine mastitis tests (DHI or CMT), treating infected animals, preventing infection from spreading, dipping teats after milking, using dry-cow therapy, and practicing good husbandry for cleanliness. The development of a milking parlor quality assurance program coupled with an effective employer training effort can yield positive results. Such a program should include written protocols coupled with routine audits designed to monitor conformance to the prescribed standards. Suggested protocols that should be evaluated include the following:

1. Udder prep should result in a uniformly clean surface including the teat ends.
2. Milking units should be attached to udder within one minute of the start of udder preparation.
3. Monitor teat cup attachment to minimize air leaks during milking.
4. Post-milking teat dipping should result in thorough surface coverage.

Mastitis prevention protocols and use on dairy farms are described in Table 27.7. Chronically infected cows may have to be sold due to poor performance. For example, one study found that cows with mastitis required 10 more days to first breeding, required 0.5 more services per conception, and were open 25 days longer.

Most treatments involve infusing an antibiotic into the udder through the teat canal. Depending on the antibiotic used, milk from treated cows should not be sold for human use for 3–5 days after the final treatment is given.

Vaccination and preventative health management practices for replacement heifers and cows are described in Tables 27.8 and 27.9.

Other diseases and health practices are discussed in Chapter 21.

Table 27.7
MASTITIS PREVENTATIVE MANAGEMENT ON U.S. DAIRY FARMS

Treatment	% of Operations
Forestriping—removal of small amount of milk prior to milking machine placement	
All cows	59
Some cows	34
Pre-milking disinfectant applied to teats:	
Dip cup	51
Sprayer	19
Foam	4
Post-milking disinfectant applied to teats:	
Dip cup	78
Sprayer	13
Other (foam, powder)	5

Source: Adapted from NAHMS, 2007.

Table 27.8
DAIRY HEIFER AND COW VACCINATION PROTOCOLS BY HERD SIZE

Vaccine	Herd Size		
	<100	100–499	>500
Bovine viral diarrhea	70	87	96
Infectious bovine rhinotracheitis	66	84	88
Parainfluenza (Type 3)	58	72	73
Bovine respiratory syncytial virus	60	78	79
Hemophilus somnus	31	41	41
Leptospirosis	66	81	84
Salmonella	16	38	55
Clostridia	21	43	61
E. coli mastitis	25	50	79

Source: Adapted from NAHMS, 2007.

Table 27.9
PREVENTATIVE PRACTICES ON DAIRIES

Practice	% of Enterprises	% of Cows
Deworm	63	46
Vitamin A-D-E (injection)	13	20
Vitamin A-D-E (in feed)	80	79
Selenium (injection)	15	20
Selenium (in feed)	76	73
Ionophores	27	40
Probiotics	26	34

Source: Adapted from NAHMS, 1996.

Table 27.10
COST AND RETURN FOR U.S. DAIRIES BY HERD SIZE, 2012

	Size of Herd						
	<50	50–99	100–199	200–499	500–999	>1000	All
Demographics							
Cows/farm (N)	33	68	135	312	698	2,260	183
Milk/cow (lb)	15,365	17,137	18,927	19,842	22,579	23,016	20,724
Milking > 2×/day (% farms)	2.1	2.6	8.5	30.6	59.7	55.4	9.7
Income ($ per cwt sold)							
Milk	20.25	20.04	19.57	19.38	18.82	17.55	18.64
Cattle	2.30	1.94	1.67	1.46	1.52	1.42	1.57
Other	1.23	1.17	1.03	1.02	0.94	0.88	0.98
Total	23.78	23.15	22.27	21.86	21.28	19.85	21.19
Expenses ($ per cwt sold)							
Purchased feed	7.56	6.89	7.23	8.46	8.80	8.94	8.36
Homegrown harvested feed	10.77	9.97	8.92	7.11	4.98	3.04	5.72
Grazed feed	0.52	0.25	0.15	0.12	0.02	0.02	0.09
Subtotal feed	18.85	17.11	16.30	15.69	13.80	12.00	14.17
Veterinary and medicine	0.81	0.90	0.80	0.94	0.91	0.65	0.79
Bedding	0.40	0.38	0.33	0.31	0.29	0.11	0.24
Marketing	0.24	0.21	0.22	0.23	0.28	0.22	0.23
Fuel and energy	1.36	1.23	1.02	1.00	0.75	0.59	0.82
Repairs	1.07	1.00	0.71	0.69	0.42	0.43	0.58
Subtotal	23.40	21.46	20.02	19.56	17.14	14.41	17.38
Allocated overhead ($ per cwt sold)							
Hired labor	0.59	0.91	1.32	1.97	1.97	1.55	1.54
Opportunity cost of unpaid labor	13.91	7.16	3.57	1.51	0.52	0.17	2.16
Capital recovery of machinery and equipment	8.30	6.73	4.76	3.80	2.63	2.07	3.50
Opportunity cost of land (rental)	0.12	0.06	0.04	0.02	0.01	0.00	0.02
Taxes and insurance	0.34	0.32	0.25	0.23	0.17	0.10	0.18
General farm overhead	1.07	0.97	0.77	0.73	0.45	0.44	0.61
Subtotal	24.33	16.15	10.71	8.26	5.75	4.33	8.01
Total Cost ($ per cwt sold)	47.73	37.61	30.73	27.82	22.89	18.74	25.39
Profit (total cost basis)	–23.95	–14.46	–8.46	–5.96	–1.61	1.11	–4.20
Profit (operating cost basis)	0.38	1.69	2.25	2.30	4.14	5.44	3.81

Source: USDA:ERS.

COSTS AND RETURNS

Assessment of profit opportunities requires that sufficient records be available to evaluate the farm's overall financial position, to monitor cost controls, to assure high labor efficiency, to measure production level, and to determine appropriate herd size.

Table 27.10 provides cost and return data for dairies of various sizes. Feed comprises the largest single component of cost and thus becomes the focus for most dairy managers. Productivity of a dairy is dependent on providing the appropriate ration to support not only high milk production levels, but to also assure rumen health, optimal reproductive rates, and the ability of cows to regain body condition and energy stores between lactations. Note that there are substantial scale advantages for larger enterprises particularly in regards to feed expense, opportunity cost of unpaid labor, and capital recovery of machinery and equipment. While smaller dairies have an advantage in total returns per cwt of product sold, they are less competitive on a cost basis, have lower per head productivity, and thus are less profitable than their larger counterparts.

Fortunately, the dairy industry has superior herd production and financial record-keeping systems available. Herd managers should measure their herds' productivity against regional or national **benchmarks**. For example, the competitive benchmarks for milking parlor turns per hour are 4.5, parlor labor cost per cwt. of milk at $0.50, cows milked per worker per hour at 100 with 3,000 cwts of milk per hour, and a hourly labor cost per milking stall of $0.50. A summary of typical dairy production records and the recommended goals is provided in Table 27.11.

Table 27.11
PERFORMANCE RECORDS SUMMARY

Management Area	Trait	Goal	Rationale
Lactation distribution	Days in milk (DIM)	180 days and even calving pattern	Average milk production declines with prolonged DIM
	Days dry	>80% of cows plus or minus 20% of the option for days dry	Dry periods less than 30 days or greater than 60 days result in economic loss
	Calving interval	365 days	Average calving intervals beyond 365 days yield a high percentage of cows in late lactation and below break-even production
Herd distribution	Percentage of DIM	Describes the average percentage of the month that cows spent in milk	Estimates the proportion of the month that cows contribute to cash flow
	Annual % DIM	85–87%	Each 1% increase through 91% yields a 100-lb increase in average annual lactation
Daily milk (DM) production	DM production	Depends on herd	Directly related to cash flow—trait is easily influenced by management and climate
	Average lb peaked 1st 90 days	Depends on herd	For each 1-lb increase in peak yield, the 305-day yield increases by about 220 lb
DIM at 1st breeding	Voluntary waiting period following calving prior to breeding	45–50 days	Allows for time to complete uterine involution and maintenance of a 13-month or less calving interval

Table 27.11 *(continued)*

Management Area	Trait	Goal		Rationale	
	Average age at 1st calving	22.5–25.5 mos.		>25 months yields lower relative net income per cow, profit per day of life, and lifetime profit per cow	
Heat detection status	Intervals between breeding	*Interval (days)*	*Herd Distribution (%)*		*Possible Problems*
		≤18	10		Cystic ovary, poor heat detection
		18–24	65		Normal
		25–35	10		Abnormal cycle, error in cow ID or heat detection
		36–48	10		Missed heat
		>48	5		2 missed heats or embryonic death
	Heats detected	>80%		Approximately $40 loss for each missed heat	
Breeding performance	Services per conception	<1.5		Each 0.1 service over 1.5 services per conception costs $1.50 per cow	
	First service conception	>75%		Low values indicate extended calving interval in the future	
	Service per cow	>2.0		If services per cow are greater than service per conception by ≥0.4, immediately look for cause of problem	
Somatic cell summary	Somatic cell score	*SCS*		*% Herd*	*High SCS*
		Low		>90	Indicate possible
		Med.		<7	subclinical mastitis,
		High		<3	problems with quality control or milking equipment
Culling and herd turnover	Culling guidelines or do not breed (DNB)	Depends on herd		Decisions depend on a comparison of expected income from cow and expected income from a replacement heifer	
	Selective culling	15–35%		Assure genetic rate of change by removing poor performers	
	Nonselective culling	*Annual culling rates* Reproductive culls: 5–10% Udder health culls: <5–10% Disease/injury culls: <5%		Removal of reproductive failures, diseased or injured cows, and poor udders	

Source: Adapted from Western Regional Extension Program 117.

CHAPTER SUMMARY

- Dairy farming is labor-intensive, so facilities are important in saving labor, feeding cattle, and producing wholesome, clean milk.

- Nutrition and health programs are extremely important in maintaining high milk production in cows.

- The most troublesome disease in many dairy operations is mastitis—the inflammation and infection of the mammary gland.

- Forage and concentrates must be high in nutrient content and must be provided in large amounts to express the high genetic potential for milk production.

KEY WORDS

body condition
total mixed ration
heat stress
navel
parlor
housing systems

nutrient management plan
Bang's disease (brucellosis)
mastitis
clinical mastitis
pendulous
benchmarks

REVIEW QUESTIONS

1. Describe the demographics of the U.S. dairy industry.
2. Discuss the nutritional needs of a lactating dairy cow as she moves through a lactation cycle.
3. Describe the BCS system in dairy cattle.
4. What are the 12 keys to effective nutritional management of the dairy cow?
5. Why is a total mixed ration important to a dairy?
6. Discuss how dietary considerations change as a cow moves through the production phases.
7. List several methods to mitigate heat stress.
8. Why are dry cows managed separately from the milking herd?
9. Discuss calving management issues.
10. Describe the facility needs of a commercial dairy.
11. What is the importance of a nutrient management system?
12. Describe the role of technology in milking operations.
13. Discuss the disease factors that impact milk wholesomeness.
14. Outline mastitis control alternatives.
15. Compare costs and returns for various sized dairy herds.
16. Discuss the important traits, goals, and rational for measurement that comprise an effective record keeping system.

SELECTED REFERENCES

Bath, D. L., F. N. Dickinson, H. A. Tucker, and R. D. Appleman. 1985. *Dairy Cattle: Principles, Practices, Problems, Profits*. Philadelphia, PA: Lea & Febiger.

Bertrand, J. A., J. C. Freck, and J. C. McConnell. 1999. Phosphorus intake and excretion on South Carolina dairy farms. *Prof. Ani. Sci.* 15:264–267.

Dairy Info Base. 2006. USDA:ADDS. www.adds.org.

Fiez, E., D. Grusenmeyer, E. Thomason, and B. Wiesen. 1992. Dairy Record Templates. *Western Regional Extension Publ. 117*.

Hoffmann-LaRoche, Inc. 1989–1990. *Nutritional Needs of Dairy Cows*. Nutley, NJ: Hoffmann-LaRoche.

Meyer, D. 2000. Dairying and the environment. *J. Dairy Sci.* 83:1419–1427.

National Animal Health Monitoring System. 2007. Reference of Dairy Management Practice. USDA:VS.

National Animal Health Monitoring System. 2004. Animal Disease Exclusion Practices on U.S. Dairy Operations. USDA:VS.

National Research Council. 1988. *Nutrient Requirements of Dairy Cattle*. Washington, DC: National Academy Press.

Pennsylvania State University Extension Service. 1995. Proc. Managing Dairy Farms into the 21st Century.

Perkins, B. L. 1998. Managing the 30,000 Pound Herd. Proc. Colorado Dairy Nutrition Conference, Fort Collins, CO.

28

Swine Breeds and Breeding

Breeds of swine are genetic resources available to swine producers. Most market pigs in the United States today are crossbreds, but these crossbreds derive their origin and continued perpetuation from seedstock herds that maintain pure breeds of swine. Knowledge of the breeds, their productive characteristics, and their crossing abilities is an important component of profitable production.

CHARACTERISTICS OF SWINE BREEDS

In the early stages of the formation of swine breeds, breeders distinguished one breed from another through assessment of visual characteristics. These visual distinguishing characteristics for swine were, and continue to be, hair color and erect or drooping ears. Figure 28.1 shows the major breeds of swine and their primary distinguishing characteristics. Table 28.1 provides primary traits of strength, country of origin, and descriptive characteristics for the most popular breeds of swine.

Relative importance of swine breeds is shown by registration numbers recorded by the different breed associations (Table 28.2). Although the registration numbers are for registered purebreds, the numbers do reflect the demand for breeding stock by commercial producers who produce market swine primarily by crossbreeding two or more breeds. Note how the popularity of breeds has changed with time. This change has resulted, in part, from use of more objective measurements of productivity and the breeds' ability to respond to the demand and selection for higher productivity and profitability.

Competitors of purebred breeders are large corporate seedstock suppliers who offer boars and gilts of specialized bloodlines. These lines of breeding, usually called **hybrids**, originated from crossing two or more breeds, then applying some specialized selection programs. Farmers Hybrid Company was the first swine-breeding corporation; it formed in the 1940s. Since 1960 DanBred International, Pig Improvement Company (PIC), Newsham Hybrids, Hypor, and others have joined it as genetic providers in the industry.

TRAITS AND THEIR MEASUREMENTS

Table 28.3 gives an overview of the traits preferred in market barrows and gilts. These are specific targets for measurable traits that provide a set of targets in the form of the "ideal" market hog known as Symbol. Note that as markets, genetics and industry management has changed,

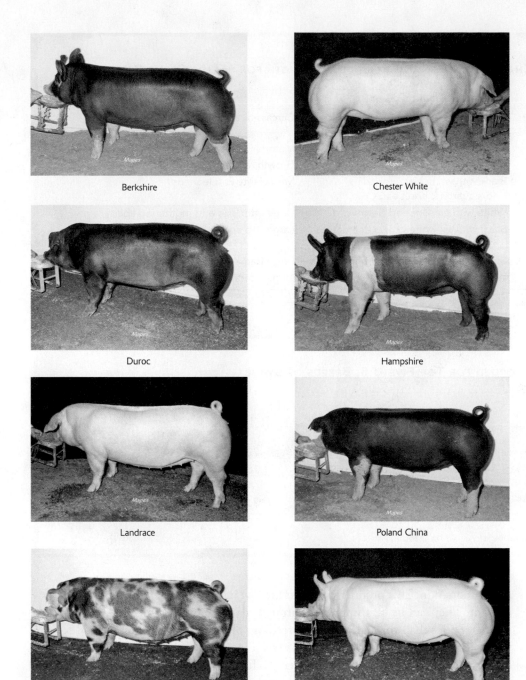

Figure 28.1
The major breeds of swine in the United States. Source: Mapes Livestock Photos (Berkshire, Chester White, Landrace, Poland China, Spotted, Duroc, and Hampshire).

Berkshire

Chester White

Duroc

Hampshire

Landrace

Poland China

Spotted

Yorkshire

Symbol's list of preferred traits have been adjusted over time due to shifts in market demands, improvements in management, and incorporation of new technologies. The ability of swine breeders to make these changes have been made possible through focused genetic selection, utilization of genetic diversity via mating systems, and superior management that allows expression of genetic potential.

Sow Productivity

Litter size, number weaned per litter, 21-day litter weight, and number of litters per sow per year measure **sow productivity**, which is of extremely high economic importance. A combination of number of pigs (born alive) and litter weight at 21 days best reflects sow productivity (Fig. 28.2). Litter weight at 21 days is used to measure

Table 28.1
ORIGIN, DESCRIPTIVE CHARACTERISTICS, AND TRAITS OF STRENGTH FOR MAJOR SWINE BREEDS

Breed	Origin	Descriptive Characteristics	Traits of Strength
Berkshire	England	Black with six white points (feet, face, tail), erect ears	Carcass quality
Chester White	USA	White with ears drooped forward	Maternal ability
Duroc	USA	Red with ears drooped forward	Growth rate and feed efficiency
Hampshire	England	Black with white belt around forequarters, erect ears	Muscularity and leanness
Landrace	Denmark	White with ears lopped downward	Large litter size and mothering ability
Poland China	USA	Black with six white points (feet, face, tail), drooped ears	Large size and muscularity
Spotted	USA	Black and white spotted, drooped ears	Large frame, growth rate and muscularity
Yorkshire	England	White with erect ears	Large litters, mothering ability and growth rate

Table 28.2
NUMBER OF LITTERS RECORDED FOR TOP FIVE U.S. BREEDS OF SWINE

Breed	2012	2008	2005	1995	1990	1985	Year Assn. Formed
Berkshire	38,696[a]	6,819	5,458	1,934	2,071	2,158	1875
Duroc	73,396[a]	76,955[a]	84,111[a]	13,709	22,179	21,852	1883
Hampshire	29,075[a]	41,311[a]	45,806[a]	16,244	18,925	14,117	1893
Landrace	59,838[a]	30,804[a]	31,650[a]	4,405	4,365	3,501	1950
Yorkshire	144,897[a]	116,280[a]	111,137[a]	19,007	23,861	24,335	1935

[a]Reported in number of head instead of number of litters.
Source: Adapted from National Swine Registry and other associations.

milk production. Sow productivity traits are low in heritability, but economically important. They can be best improved through managing the environment (e.g., feed, health), rather than through selective breeding. Eliminating sows low in productivity while keeping highly productive sows is still a logical practice for economic reasons, but not when making major genetic change. It is important for commercial producers with effective crossbreeding programs to use lines or breeds that rank high in sow and maternal productivity traits. Producers should carefully evaluate the cost of developing replacement females and understand the effect of pigs sold per litter on the number of farrowings required before a sow returns positive value. The impact of pigs marketed per litter and cost of sow development are illustrated in Table 28.4. For example, a female that is culled after only one farrowing returned no profit, while those females developed at the lowest cost and with the longest sustained production per litter returned the most profit to the enterprise.

Growth

Growth rate is economically important to most swine enterprises and has a heritability of sufficient magnitude (35%) to be included in a selection program. Seedstock producers need a scale to measure weight in relation to age (Fig. 28.3). Growth rate can be expressed in several ways and typically is adjusted to some constant basis, such as days required for swine to reach finished weight.

Table 28.3
SYMBOL: A STANDARD OF PERFORMANCE

Symbol	The graphic representation of a standard of performance and carcass composition that provides a basis for the continuing improvement of the modern hog.
Symbol	Symbolizes the commitment of the entire pork industry to making lean pork the true meat of choice of the twenty-first century.

Symbol I   A Market Barrow—adopted in 1983

Target or goal for all pork producers. These figures probably represented the top 10% of all hogs marketed in 1983.

- 250-lb market barrow
- 180-lb, 32-in.-long carcass yields 105 lb of lean pork
- Feed conversion efficiency of 2.5 from birth to a market age of 150 days
- Demonstrates a lean gain of 3/4 lb per day of age
- Last rib fat depth measurement at slaughter is 0.7 in.
- Loin muscle area is 5.8 sq in.
- Average of three backfat measurements is 1.0 in.

Symbol II   A Market Gilt—adopted in 1996

- 195-lb carcass from a live weight of 260 lb
- Intramuscular fat level greater than or equal to 2.9%
- High health production system
- Produced by an environmentally assured producer on PQA Level III
- Result of a terminal crossbreeding program
- From a maternal line capable of weaning 25 pigs per year
- Marketed at 164 days of age (gilts)
- Live weight feed efficiency of 2.40
- Fat-free lean gain efficiency of 5.90
- Fat-free lean gain of 0.78 lb per day
- Standard reference backfat of 0.6
- Fat-Free Lean Index of 52.2

Symbol III   A Market Goal—adopted in 2005

- 205-lb carcass from a live weight of 270 lb
- Intramuscular fat level greater than or equal to 3.0%
- High health production system, free of parasites and *Halothane gene*
- Produced by an environmentally assured producer on PQA program with animal care assessment
- Result of a terminal crossbreeding program
- From a maternal line capable of weaning 25 pigs per year after multiple parities
- Marketed at 156 (164) days of age
- Live weight feed efficiency of 2.4 (2.40)
- Fat-free lean gain efficiency of 5.9 (5.80)
- Fat-free lean gain of 0.95 lb per day
- Fat-Free Lean Index of 53.0 (54.7)
- Loin muscle area of 6.5 sq in. (7.1)
- Belly thickness of 1.0 in.
- 10th rib fat of 0.7 in. (0.6)
- Muscle color score of 4.0

Numbers in parentheses represent goals for gilts.
Source: Adapted from National Pork Producers Council.

Figure 28.2
The number of pigs raised per sow and the weight of the litter at 21 days of age effectively measure sow productivity. Source: National Pork Board.

Table 28.4

EFFECTS OF PIGS SOLD PER LITTER AND REPLACEMENT GILT DEVELOPMENT COST ON NUMBER OF FARROWINGS REQUIRED TO RETURN POSITIVE VALUE TO THE ENTERPRISE

Pigs Sold per Litter	Farrowing and Positive Return[a]					
	1	2	3	4	5	6
7.5	—	—	—	—	—	$17
8.0	—	—	$10	$60	$110	159
8.5	—	$7	82	156	229	300
9.0	—	56	155	252	348	442
Replacement Female Price ($/hd)						
150	—	$63	$138	$212	$284	$356
175	—	35	110	184	256	328
200	—	7	82	156	229	300
225	—	—	54	128	200	273
250	—	—	27	100	173	245

[a]Measured as Net Present Value. Assumes a $46/cwt market price.
Source: Adapted from Stadler and Lacy, 1998.

Feed Efficiency

Feed efficiency measures the pounds of feed required per pound of gain and has an associated heritability of 0.30. This trait is economically important because feed costs account for 60–70% of the total production costs for commercial producers. Obtaining feed-efficiency records requires keeping individual or group feed records. However, feed efficiency can be improved without keeping feed records, such as by testing and selecting individual pigs for gain and backfat. This improvement occurs because fast-gaining lean pigs tend to be more efficient in their feed utilization. Feed-efficiency records are obtained in some central boar-testing stations; however, costs and other factors need to be considered before similar records are taken on the farm.

Figure 28.3
Using scales to measure the growth rate of individual boars is an essential part of a sound breeding system. Digital scales improve the speed and accuracy of data collection. Source: Tom Field.

Carcass Traits

Carcass traits are used to estimate the pounds or percentage of acceptable-quality lean pork (10% fat) in the carcass. Percent lean measured on a carcass basis has a heritability of 0.48. The traits used to predict carcass composition are, in order of importance, (1) fat depth over the loin at the 10th rib, (2) loin muscle area, and (3) carcass muscling score. Percent muscle has been the focus of measuring market hog productivity. Figure 28.4 illustrates five market hogs with varying levels of percent muscle. There are substantial value differences when comparing hogs with high muscularity as opposed to those with more fat cover and lower levels of lean muscle mass.

Structural Soundness

Structural soundness is vital to the swine industry. Recent studies report unsoundness to be medium in heritability; therefore, improvement can be made through selection. Soundness can be improved through visual selection if breeders decide to cull restricted, too straight-legged, unsound boars lacking the proper flex at the hock, set of the shoulder, even size of toes, or proper curvature and cushion to the forearm and pastern.

Inherited defects and abnormalities generally occur with a low degree of frequency across the swine population. It is important to know that they exist and that certain herds may experience a relatively high incidence. Certain structural unsoundness characteristics might be categorized as inherited genetic defects. **Cryptorchidism** (retention of one or both testicles in the abdomen), umbilical and scrotal **hernias** (rupture), and **inverted nipples** (nipples or teats that do not protrude but are inverted into the mammary gland), and **pale, soft, and exudative (PSE)** carcasses are considered the most commonly occurring genetic defects and abnormalities. The PSE condition and the **porcine stress syndrome (PSS)** have occurred in recent years due to selection for extreme muscling.

The PSS gene in its homozygous mutant form *(nn)* results in either death from physical stress or PSE meat. However, the presence of the normal allele *(N)* nearly always prevents stress death or the rigidity response to testing using the anesthetic halothane.

Research studies have shown the carrier animals to be 1–4% higher than normal pigs *(NN)* in lean percentage and to be equal, if not superior, in growth rate. All of these studies, however, have suggested that the muscle quality of these pigs *(Nn)* is generally undesirable and that the frequency of PSE is in the 30–50% range.

There is evidence that the PSS gene does not account for all of the pork industry's PSE problems. However, because the gene offers only a minimal increase in lean percentage and significantly increases the PSE frequency, the gene should be eliminated from the U.S. pig population.

Figure 28.4

Comparison of five different levels of muscularity and leanness in market hogs.

Source: J.D. Tatum.

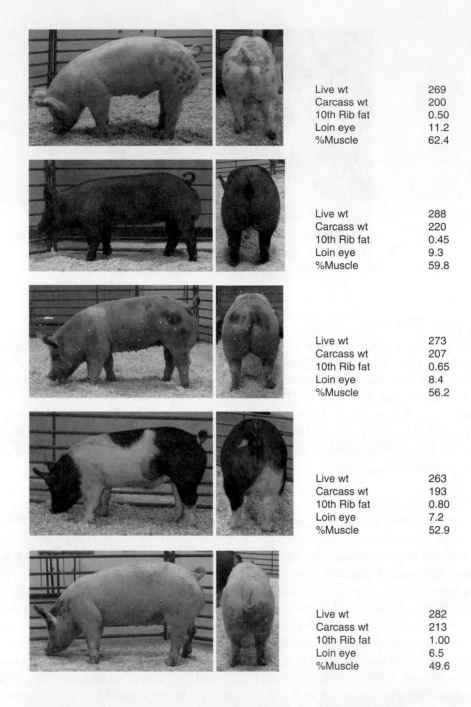

Live wt	269
Carcass wt	200
10th Rib fat	0.50
Loin eye	11.2
%Muscle	62.4

Live wt	288
Carcass wt	220
10th Rib fat	0.45
Loin eye	9.3
%Muscle	59.8

Live wt	273
Carcass wt	207
10th Rib fat	0.65
Loin eye	8.4
%Muscle	56.2

Live wt	263
Carcass wt	193
10th Rib fat	0.80
Loin eye	7.2
%Muscle	52.9

Live wt	282
Carcass wt	213
10th Rib fat	1.00
Loin eye	6.5
%Muscle	49.6

If pork is to be elevated to the "meat of choice" as noted in industry literature, then carcass leanness and muscling cannot be the entire focal point. In addition to eliminating the PSS gene, selection methods to increase pork quality appear advisable. Consumer preference studies on pork suggest that an increase in the lipid level of meat would make it more competitive with chicken and possibly other meats.

EFFECTIVE USE OF PERFORMANCE RECORDS

Commercial producers recognize that the performance levels in their herds are influenced by two factors: genetics and environment (e.g., feed, health, and housing).

One of the most important decisions in the swine enterprise is the selection of an individual, company, or cooperative to supply genetics. Determination of a seedstock supplier should be based on an assessment of individual herd performance and need, freedom from disease and genetic defects in the seedstock herd, availability of reliable data, and availability of associated herd management services. A sound genetic improvement program should include four features: (1) accurate, complete performance records including animal identification, consistent measurement of all boars and gilts (not on-again, off-again, or limited, partial performance testing), and ranking of animals within defined contemporary groups; (2) assessment of the genetic merit of economically important traits (growth rate, feed efficiency, carcass merit, and reproductive performance) based on the individual's performance relative to its contemporary group and incorporating the performance of relatives; (3) indexes weighting traits relative to their economic importance in commercial pork production (the indexes should correctly rank the individuals relative to their intended use in crossbreeding and marketing systems); and (4) selection of the highest-ranking boars and gilts based on selection indexes. Seedstock producers should utilize selection indexes as their primary selection criteria.

The rate of genetic improvement in a commercial pork producer's herd will parallel the rate of genetic progress made by the seedstock supplier the commercial producer chooses. Therefore, the commercial producer should select seedstock producers whose improvement programs have the four features previously mentioned.

STAGES (the Swine Testing and Genetic Evaluation System) sponsored by the National Swine Registry, computes genetic evaluations for several traits within a single herd. Reproductive traits include 21-day litter weight and number born alive. Postweaning traits with EPDs reported are backfat, loin eye area, pounds of fat-free lean adjusted to market weight, days to finished weight, and feed per pound of gain. A series of bioeconomic indexes are also reported and include the **sow productivity index (SPI),** the **terminal sire index (TSI)**, and the **maternal line index (MLI).** These traits are evaluated in a multibreed national sire evaluation system for the Duroc, Hampshire, Yorkshire, and Landrace breeds.

The genetic evaluation that results from STAGES is expressed as an expected progeny difference (EPD).

Several breeds each have an across-herd sire summary that is published on an annual basis. Sire summaries provide unbiased estimates of the genetic merit of boars that have a legitimate performance record or have progeny with performance records. These estimates are in the form of EPDs.

The EPD is a prediction of the progeny performance of an animal compared to the progeny of an average animal in the breed based on all information currently available. This allows for the direct comparison of all boars evaluated in the sire summary based on the predicted differences between the levels of performance of their progeny. These EPDs are expressed as plus or minus values, with the average EPD for the breed being approximately zero. Table 28.5 compares two boars assuming the EPDs have equal accuracies.

Table 28.5
COMPARISON OF EPDS FOR TWO BOARS

	Expected Progeny Difference	
Boar ID	Days to 230 lb	Backfat
A	−3.5	−0.05
B	0	+0.05

Table 28.6
COMPARISON OF EPD INDEX FOR TWO BOARS

Boar ID	Expected Progeny Difference		
	SPI	MLI	TSI
C	92	106	136
D	118	105	94

Progeny of boar A are expected to reach 250 lb 3.5 days sooner than progeny of boar B and to have 0.10 in. less backfat. This assumes the boars are bred to sows of similar genetic merit.

An accuracy value is listed with each EPD. Accuracies range from 0 to 1.0 and give a confidence estimate of the EPD. Accuracies that approach 1.0 indicate the EPD is a good estimate of the boar's true genetic potential for that trait.

The three bioeconomic indexes allow producers to focus selection pressure on specific phases of the production cycle. SPI ranks individuals for reproductive traits using a composite of traits including number born live, number weaned, and litter weight. Each unit of SPI represents $1 per litter produced by each daughter of a particular parent.

The MLI is used to evaluate replacement gilts for crossbreeding systems. Both maternal and terminal traits are included, but the maternal traits are weighted twice as much as the terminal.

The TSI is useful in selection of animals for terminal crossing programs and includes only postweaning traits (backfat, days to 250 lb, pounds of lean, and feed per pound of gain). Each TSI unit is the equivalent of $1 for every 10 pigs marketed.

Table 28.6 compares the indexes of two sires. Each litter by sire D would be expected to be worth $26 more than those of C due to heavier weaning weights and greater numbers of live pigs. Sires C and D are virtually equal in terms of maternal line index. However, every 10 pigs out of C would be expected to be worth $42 more than those sired by D due to superior growth and leanness.

The results of the National Pork Producers Council Terminal Line National Genetic Evaluation Program (NGEP) were published in 1995. This was the largest, most comprehensive genetic evaluation program ever done in the swine industry. A large representative sample of boars (795) from nine different sire lines (breeds and hybrids) were each mated to a similar genetic base of sows to produce 1,990 litters. More than 3,200 test pigs were evaluated for growth and carcass traits. Evaluations were also made for loin meat quality traits and loin eating-quality traits, with more than 30 traits measured for these three major categories. Indexes 1 and 2 were calculated, with each index combining several traits into an economic value expressed as dollars per pig. Index 1 establishes the value of a sire line to a producer selling hogs on a backfat-based carcass merit buying system. Index 2 gives the value of a sire line if retail values can be realized in addition to the values achieved in Index 1. Thus, realizing these value differences per pig is dependent on the marketing alternative utilized. Table 28.7 gives a summary of several selected traits and Indexes 1 and 2.

The differences in sire lines in the NGEP are useful in planning effective selection programs. However, individual boar selection within a breed or line using EPDs and other performance data should receive the greater emphasis.

National consumer preference studies throughout the United States were made on samples of pork from the NGEP to evaluate consumer acceptance and via comparisons of pork to chicken breast. The major findings were as follows:

Table 28.7
SELECTED TRAIT SUMMARY

| | | | | | Trait | | | | |
Sire Line	Days to Reach 250 lb	Feed Conversion (lb feed/ lb gain)	Backfat (in.)	Loin Area (sq in.)	Loin Drip Loss (%)[a]	Loin pH[b]	Tender-ness (lb)[c]	Index 1($/pig)[d]	Index 2($/pig)[e]
Berkshire	175	3.07	1.25	5.74	2.43	5.91	12.63	–$4.14	–$4.05
Danbred HD	176	2.88	0.98	6.75	3.34	5.75	12.78	2.12	1.52
Duroc	170	2.91	1.13	6.14	2.75	5.85	12.43	0.64	10.51
Hampshire	175	2.92	1.01	6.58	3.56	5.70	12.89	1.34	1.55
NGT Large White	173	2.94	1.17	5.62	2.92	5.84	13.40	–0.97	–8.87
Nebraska SPF Duroc	168	2.89	1.11	6.35	2.81	5.88	12.72	1.50	8.25
Newsham Hybrid	172	2.83	0.98	6.45	2.99	5.82	13.46	3.53	0.89
Spotted	174	3.14	1.24	5.83	2.88	5.83	13.02	–4.77	–8.20
Yorkshire	175	2.93	1.05	6.17	2.85	5.84	13.49	–0.74	–1.70

[a]Important because 1% drip loss equals a loss of 1 lb for every 100-lb meat.
[b]Higher pH associated with low drip loss, darker color, more firmness, increased tenderness—all positive attributes.
[c]Pounds of pressure to push probe into cooked loin (Instron machine).
[d]Index 1 combines days to 250 lb, feed conversion, and backfat at 10th rib.
[e]Index 2 combines intramuscular fat in loin, pH, tenderness, drip loss, and loin muscle area that were estimated using results of consumer preference study, then added to traits in Index 1.
Source: Adapted from National Genetic Swine Evaluation Program.

1. Price (58%) was the most important attribute when comparing chicken to pork. Next most important was lipid level (24%).
2. Higher lipid levels and greater tenderness were preferred by consumers when taste-testing pork.
3. Increases in lipid level have more than twice the economic value to these consumers as increased tenderness or more preferred pH levels.
4. Preferred visual characteristics must be balanced against the most preferred taste characteristics to identify the pork chop with the greatest long-term consumer appeal.

SELECTING REPLACEMENT FEMALES

Sow productivity is the foundation of commercial pork production. The sow herd also contributes half of the genetic composition to the growing-finishing pigs. These two factors show the importance of replacement gilt selection so that highly productive gilts can be retained in the swine herd. Fast-growing, sound, moderately lean replacement gilts with good body capacity should be selected from litters where 10–14 excellent pigs were weaned. Among sows that have farrowed and will rebreed, those that have structural problems, aggressive dispositions, extremely small litters (two pigs below the herd average), and poor mothering records should be culled.

As sows generally produce larger litters of heavier pigs than do gilts, replacing large numbers of sows with gilts may reduce production levels. This production differential and the low relationship among the performances of successive litters argue

Table 28.8
FACTORS CONTRIBUTING TO SOW CULLING RATE

Culling Factor	Percent Culled
Age (old)	37
Reproductive inefficiency	26
Unsoundness	15
Poor litter size/ pig survival	13
Other	9

Source: Adapted from USDA:NAHMS, 2006.

for low rates of culling based on sow performance to maintain high levels of production which has driven most commercial farms to purchase replacement females from genetic providers as opposed to selecting replacements from within the herd.

Sows are typically culled due to old age, reproductive failure, and unsoundness (Table 28.8).

BOAR SELECTION

Boar selection is extremely important in making genetic change in a swine herd. Over several generations, selected boars will contribute 80–90% of the genetic composition of the herd; however, this contribution does not diminish the importance of female selection. Boars should have dams that are highly productive sows. Productivity of replacement gilts is highly dependent on the level of sow productivity passed on by their sires. Selection standards for replacement boars will depend on the selection objectives and current performance of the herd.

CROSSBREEDING FOR COMMERCIAL SWINE PRODUCERS

The primary function of seedstock breeders is to provide breeding stock for commercial producers. Breeding programs for seedstock producers are designed to genetically improve the economically important traits. Within a specific breed, rate of improvement will be dependent on heritability of traits, selection intensity, and speed of generation turnover. The most rapid genetic changes that have occurred over the past few decades are in traits associated with growth rate and leanness.

Utilizing multiple breed inputs in a mating system yields a genetic phenomenon known as heterosis or hybrid vigor (Table 28.9). Compared to straightbreds a 40% increase in total litter market weight of crossbreds is attained. The increase in productivity is a cumulative effect of larger litters, high pig survival, and increased growth rate of individual pigs. This marked increase in productivity accounts for the fact that nearly all market hogs in commercial production are crossbreds. An effective crossbreeding program takes advantage of using hybrid vigor and selecting genetically superior breeding animals from breeds that best complement one another. A producer needs to know the relative strengths and weaknesses of the available breeds shown in Table 28.10. These breed comparisons can change over time as a result of genetic changes made in the individual breeds.

The crossbred female is an integral and important part of an effective crossbreeding program. Even though crossbreeding provides an opportunity to reap the benefits

Table 28.9
HETEROSIS ADVANTAGE FOR PRODUCTION TRAITS

Item	First-Cross Purebred Sow	Multiple-Cross Crossbred Sow	Crossbred Boar
Percentage of Advantage over Purebred			
Reproduction			
Conception rate	0.0%	8.0%	10.0%
Pigs born alive	0.5	8.0	0.0
Litter size at 21 days	9.0	23.0	0.0
Litter size, weaned	10.0	24.0	0.0
Production			
21-day litter weight	10.0	27.0	0.0
Days to 220 lb	7.5	7.0	0.0
Feed/gain	2.0	1.0	0.0
Carcass composition			
Length	0.3	0.5	0.0
Backfat thickness	−2.0	−2.0	0.0
Loin muscle area	1.0	2.0	0.0
Marbling score	0.3	1.0	0.0

Source : Adapted from *Pork Industry Handbook* (PIH-39), 1987.

Table 28.10
COMPARATIVE PERFORMANCE EVALUATIONS FOR SWINE BREEDS

Trait	Berkshire	Chester White	Duroc	Hampshire	Landrace	Poland China	Spotted	Yorkshire
Conception rate (%)	+	+	A	A	− −	A	A	−
Litter size (no. pigs raised to weaning)	−	++	A	−	++	− −	−	++
21-day litter weight (lb)	−	−	A	A	++	−	− −	+
Rate of gain (lb/day)	−	−	++	A	A	−	−	A
Feed efficiency (lb feed/lb gain)	A	A	+	+	A	−	A	A
Backfat (in.)	A	A	A	+	A	A	A	A
Loin eye area (sq in.)	A	A	A	+	A	+	A	A

Evaluation system:
++ = substantially superior to average
+ = superior to average
A = average or near average
− = below average
− − = substantially below average
Source : Adapted from multiple sources.

of many genetic sources, an unplanned crossing program will not yield success or profit for the pork producer. A well-planned crossbreeding system capitalizes on heterosis, takes advantage of breed strengths, and fits the producer's management program.

Historically the two basic types of crossbreeding systems were the **rotational cross** system and the **terminal cross** system. The rotational cross system combines

Figure 28.5
Rota-terminal crossing system.

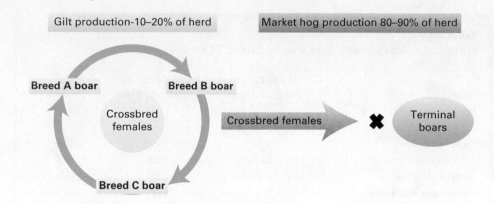

two or more breeds, with a different breed of boar being mated to the replacement crossbred females produced by the previous generation. In a simple terminal cross system, a two-breed single or rotational cross female is mated to a boar of the third breed. Figure 28.5 shows a **rota-terminal crossbreeding** system, which combines the three-breed rotational system and the terminal system of crossbreeding. Thus, all the pigs produced by terminal cross boars are sold because the gilts would lack the superiority needed in maternal traits (reproduction and milk). The female stock can be

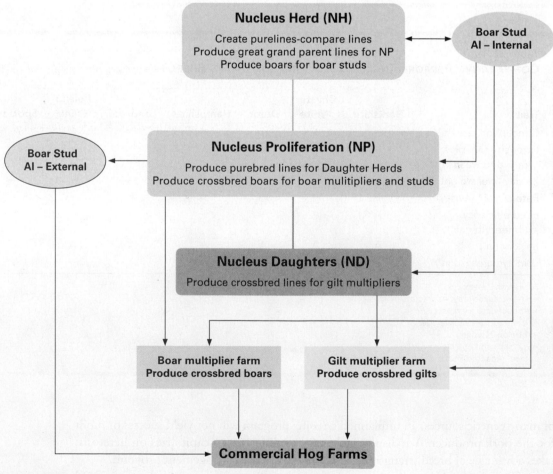

Figure 28.6
The process flow for creating improved genetics in the swine industry.

purchased from a genetic provider primarily emphasizing reproductive performance. This latter system maximizes heterosis; however, purchasing replacement females increases the risk of introducing disease into the herd.

Hybrid boars are developed from specific line crosses. These lines have been selected and developed for specific traits. When specific crosses are made, the hybrid boar is used on specific cross females to obtain the recommended breed combination and to realize a high level of heterosis in the offspring.

Creation of seedstock for the commercial swine industry is dominated by specialized genetic companies often with both domestic and international reach. The process involves multiple stages to eventually create crossbred lines of both gilts and boars ideally suited to the needs of commercial producers. This system has largely eliminated within herd gilt replacement programs at the commercial farm level. The process of creating genetics in the industry is illustrated in Figure 28.6. **Nucleus herds** are responsible for creating purelines, conducting genetic evaluations and comparison studies, and producing highly selected boars that are then sent to boar studs to produce semen for use within the company to multiply their impact. **Nucleus proliferation** produces purelines for the daughter herds and crossbred boars for the boar studs that will produce semen for use in nucleus daughter and commercial herds. **Nucleus daughter** herds produce crossbred gilts for the multiplier herds. This highly coordinated system is very effective in producing genetic lines specifically created to meet the production and market needs of the swine industry.

Commercial pork producers have many selection tools, crossing systems, and genetic breeding-stock sources for their use. Producers should capitalize on heterosis and breed strengths, and require complete performance records on all selected breeding stock. Although there is no one best system, breed, or source of breeding stock, producers must evaluate their total pork production program and integrate the most profitable combination of factors associated with a crossbreeding program. In the future, producers will have even more tools at their disposal, including the use of genetic-marker-assisted selection. Swine genetic companies have identified a number of gene markers that assist in the selection of swine that are more productive, more resistant to disease, and capable of producing more flavorful and desirable pork products. Genetic innovations will create even more opportunities in the future.

CHAPTER SUMMARY

- The most important visual characteristics in identifying breeds of swine are hair color, color pattern, and erect or drooping ears.

- Based on registration number, Yorkshire is the most numerous breed of swine, followed by Duroc and Hampshire.

- Traits of high economic importance that are included in most selection programs are (1) sow productivity (number in letter and litter weight at 21 days), (2) growth rate, (3) feed efficiency, (4) carcass traits (pounds of acceptable-quality pork), and (5) skeletal soundness.

- The use of selection indexes such as SPI, MLI, and TSI are excellent approaches to making progress in the swine herd.

- Crossbreeding in swine is widely used whereby the outstanding traits of two or more breeds are combined and heterosis increases the pounds of market pigs sold per litter.

KEY WORDS

hybrids
sow productivity
growth rate
Symbol I
Symbol II
Symbol III
feed efficiency
cryptorchidism
hernias
inverted nipples
pale, soft, and exudative (PSE)

porcine stress syndrome (PSS)
STAGES
sow productivity index (SPI)
terminal sire index (TSI)
maternal line index (MLI)
rotational cross
terminal cross
rota-terminal crossbreeding
nucleus herds
nucleus proliferation
nucleus daughter

REVIEW QUESTIONS

1. Discuss the relative strengths of the major swine breeds.
2. Outline registration trends of the major swine breeds.
3. How has the role of purebreds changed in the U.S. swine industry?
4. Describe "Symbol," the ideal market hog, and how the ideal has changed over time.
5. Outline the role of EPDs and selection indexes to progress in swine breeding.
6. Describe effective approaches to selecting replacement boars and sows.
7. Discuss the historic role of crossbreeding in the swine industry.
8. Describe the process of creating seedstock in the contemporary industry.

SELECTED REFERENCES

Boar selection guidelines for commercial pork producers. 1988. *Pork Industry Handbook,* PIH-9. Swine Extension Specialists: various state universities.

Crossbreeding systems for commercial pork production. 1988. *Pork Industry Handbook.* Swine Extension Specialists: various state universities.

Genetic evaluation programs. 1990. National Swine Improvement Federation Fact Sheet No. 12. West Lafayette, IN: Purdue University Cooperative Extension Service.

Guidelines for choosing replacement females. 1992. *Pork Industry Handbook,* PIH-27. Swine Extension Specialists: various state universities.

Harris, D. L. 2000. *Multi-Site Pig Production.* 1st ed. Ames, IA: Iowa State University Press.

McGlone, J. and W. Pond. 2003. *Pig Production.* 1st ed. Albany, NY: Delmar.. Plain, R., J. Lawrence, and G. Grimes. 2006. The structure of the U.S. pork industry. Fact Sheet 15-01-01. Des Moines, IA: Pork Information Gateway.

Pond, W. G., J. H. Maner, and D. L. Harris. 1991. *Pork Production Systems* . Westport, CT: AVI Publishing Co.

Porcine Stress Syndrome. 1992. *Pork Industry Handbook*, PIH-26. Swine Extension Specialists: various state universities.

Stadler, K., and C. Lacy. 1998. How profitable is sow longevity? *Pork '98*, February.

Stewart, T. S., et al. 1991. Genetic improvement programs in livestock: Swine testing and genetic evaluation system (STAGES). *J. Anim. Sci.* 69:3882.

29

Feeding and Managing Swine

Hog production in the United States is largely concentrated in the corn belt where the primary ingredients (corn and soybeans) in swine diets are grown in abundance. Swine production systems can be categorized into six distinct phases: breeding, gestation, farrowing, nursery, growing, and finishing. The life cycle of commercial hog production is approximately 305 days in length beginning with conception and concluding at harvest. Pregnancy lasts 114 days on average while the period from birth to harvest should be no longer than 190 days in duration. Baby pigs are typically weaned and transitioned to a nursery in less than 17 days after birth with time spent in the nursery less than 45 days. The growing–finishing phase should then be complete in less than 128 days.

Historically four primary types of swine operations comprised the commercial industry: **farrow-to-finish**, in which a breeding herd is maintained where pigs are produced and fed to harvest weight on the same farm; **farrow-to-feeder pig production**, in which feeder pigs after a short phase of postweaning growth are the primary commodity sold from the enterprise; **feeder pig finishing**, in which feeder pigs are purchased and then fed to market weight; and **farrow-to-weaner**, in which pigs are managed to weaning age and then sold or moved off-site. These historic production systems, however, were inherently inefficient requiring too many turns of ownership, too much transportation cost, lost efficiencies in feeding, and high levels of biosecurity risk.

As industry consolidation began in the 1980s, management innovations were implemented that eventually lead to **multisite rearing systems** also known as isowean or segregated production systems that incorporated multiple site production facilities where pigs were born at one facility but then weaned, grown, and finished at facilities separated from the site of sow housing. These systems were developed to reduce the risk of disease transmission, reduce the risk of whole herd depopulation due to disease outbreaks, and to maximize efficiencies of gain, growth rate, and production of lean muscle. These production plans are based on the **isowean principle**, which recognizes that preweaned pigs are free of nearly all pathogens and if they are managed as a distinct group to final marketing weights then the risk of cross-contamination and costly disease outbreaks are largely avoided.

The **three-site isowean system** where the sow herd, nursery, and finishing phases were maintained at separate sites was favored by many producers and involved a three-stage production process. Stage one involves managing a resident sow herd where baby pigs are weaned between 10 and 21 days of age. Stage two incorporates nursery production

learning objectives

- Describe the six phases of pork production and the organization of hog enterprises
- Discuss breeding female management strategies
- Outline nutritional management across the swine industry
- List the reasons for preweaned pig death losses
- Describe animal identification in the swine industry
- Compare diets as pigs grow and develop
- List the problems associated with gossypol and mycotoxins
- Explain the use of contracting in the swine industry
- Discuss the major environmental issues in the swine industry
- Outline the cyclic nature of swine profitability

in multiple barns and sites away from the sow herd where each week a nursery barn is emptied and the pigs are transitioned to finishing facilities, the nursery is then cleaned and sanitized as a new set of pigs (preferably from a single sow herd source) are started. Finally, stage 3 (growing–finishing) is conducted on yet another distinct set of sites with multiple finishing barns scheduled so that each week a contemporary group of market hogs is sold, the barn is sanitized and a new set of pigs from the nursery stage is started on a finishing diet.

In an attempt to capture production efficiencies, lower investment costs in land, buildings and manure storage, and to reduce operating costs associated with transporting pigs from site to site, a **two-site isowean system** was created. In this system sows are maintained separately just as they were in a three-site isowean system while stage 2 and 3 production was integrated at each second site with distinct nursery barns and finishing barns at each location. Merging nursery and finishing phases at a site provided the cost savings previously described. In these systems an "all in–all out" production flow is maintained by building two finishing barns for each nursery so that every 8 weeks a contemporary group of pigs leaves the nursery and transitions to a clean finishing barn. The finishing phase completes in 16 weeks or less and thus with appropriate staging allows production to proceed without having to mix groups of pigs.

The latest innovation in production systems eliminates the nursery stage altogether. The advantage of this approach is to further reduce facility and construction costs, remove another source of stress and transport expense resulting from transitioning from one phase to another, and reduce by 17 days the time from weaning to finishing. This new approach allows producers to market hogs at between 150 and 155 days from weaning to finishing.

PRODUCTION MANAGEMENT

The selection and breeding practices used to produce highly productive swine are presented in Chapter 28. The primary objective in the breeding and farrowing production phase, where a resident breeding herd is maintained, is to feed and manage sows, gilts, and boars economically to ensure a large litter of healthy, vigorous pigs is weaned from all breeding females.

Boar Management

Boars should be purchased at least 60 days before the breeding season so they can adapt to their new environment. A boar should be purchased from a herd that has an excellent herd health program (vaccinated for leptospirosis, erysipelas, porcine parvovirus, and other appropriate pathogens) and then isolated from the new owner's swine for at least 30 days. The boar should be revaccinated for erysipelas and leptospirosis. Boars not purchased from a validated brucellosis-free herd should have passed a negative brucellosis test within 30 days of the purchase date. An additional 30-day period of fence-line exposure to the sow herd is recommended to develop immunities before the boar is physically introduced.

Boars should be managed to assure excellent quality and quantity of semen without overfeeding. Boars that are fed excessive nutrients, especially energy, may become too large for useful service and often exhibit poor libido and produce poor quality semen. Young boars should receive 6–8 lb of a balanced ration up to 260 lb. Boars that are servicing females should receive a balanced diet of 3–6 lb. The amount of feed should be adjusted to keep the boar in a body condition that is neither fat nor thin. Assuring that breeding males are not heat stressed is critical, as those boars in heat stress will experience substantial declines in fertility 1 to 2 months later.

Spermatozoa produced under heat stress will be of poor quality at the conclusion of spermatogenesis and maturation 30–50 days later.

Young, untried boars should be test-mated to several gilts before the breeding season, as records show that approximately 1 of 12 young boars is infertile or sterile. The breeding aggressiveness, mounting procedure, and ability to get the gilts pregnant should be evaluated.

Successful breeding of a group of females requires an assessment of the number of boars required and the breeding ability of each boar. Generally, young boars can pen-breed 8–10 gilts during a 4-week period. A mature boar can breed 10–12 females. **Handmating** (in which producers individually mate the boar to each female) takes more labor but extends the breeding capacity of the boar. Young boars can be handmated once daily and mature boars can be used twice daily or approximately 10–12 times a week. Handmating can be used to mate heavier, mature boars with the use of a breeding crate. The breeding crate takes most of the boar's weight off the female. Breeding by handmating should correspond with time of ovulation (approximately 40 hours from the beginning of estrus) or litter size will be reduced.

Management of Breeding Females

Sow productivity is affected by milk production levels, duration of lactation, number of nonproductive days, culling rates, mortality rates, time from weaning to estrus, and conception rates. Thus, proper management of females in the breeding herd is necessary to achieve optimum reproductive efficiency. **Gilts** may start cycling as early as 5 months of age. It is recommended that breeding be avoided during their first heat cycle as the number of ova released is usually less than in subsequent heat cycles.

Breeding gilts should weigh 250–260 pounds at 8 months of age when they are exposed to mating for the first time. Prebreeding gilts should be fed individually or in groups of no more than 10 heads. Guidelines for developing breeding gilts are listed in Table 29.1.

Hot weather is detrimental to a high reproductive rate in the breeding females. High temperatures (above 85°F) will delay or prevent the occurrence of estrus, reduce ovulation rate, and increase early embryonic deaths. The animals will suffer heat stress when they are sick and have elevated body temperatures as well as when the environmental temperature is high.

Table 29.1
GUIDELINES FOR GILT DEVELOPMENT PROGRAMS

Expose gilts to first-litter or older cull sows once they reach 50–60 lb to allow adaptation to herd-level pathogens.

Segregate pigs from first-litter gilts in the nursery and finishing barns.

Feed a vitamin and mineral fortified ration.

Following first breeding, feed intake should be restricted to approximately 4 lb/day for 3–5 days.

Feed intake should then be increased to 5 lb/day to day 90 of pregnancy.

From day 90 until 2–3 days prior to parturition, feed intake should be boosted to 6.5 lb/day.

In the 48-to-72-hour time frame in advance of farrowing, intake should be limited to 5 lb/day.

During lactation, the feeding program should target maximum consumption with particular attention given to lycine, calcium, and phosphorus.

Source: Multiple sources.

Besides reducing litter size, diseases such as leptospirosis, pseudorabies, and those associated with stillbirths, mummified fetuses, and embryonic deaths may also increase the problems in getting the females pregnant. Attention to a good herd health program will assure a higher reproductive performance over the years.

The swine industry is rapidly moving to the utilization of artificial insemination with nearly two-thirds of all market hogs resulting from **artificial insemination (AI)**. As expected, as the sizes of the farms increase, use of AI becomes more common. More than 90% of hogs marketed by producers, with sow inventories greater than 500, were the result of AI. The lordosis response where a female stops and braces for mating to signal sexual receptivity is important to detect heat and can be enhanced by multisensory stimulation. A handler applying back pressure to the sow yields only about a 50% response rate, the sound of a boar grunting will increase the response to nearly three-quarters, while the opportunity for a female to smell, see, and hear the boar yields almost 100% response.

The success of an AI program is highly dependent on the timing of insemination and the ability of the technician. Mating should occur 6 to 8 hours prior to ovulation to allow for best conception rates. Appropriate training and monitoring is very important to achieving a high reproductive rate. Variation in success between individual inseminators have resulted in as much as a 1,000-pig difference in production numbers in a breeding cycle where inseminators each bred approximately 260 females.

Farrowing rate is critical to profitability. A 1,000 sow production unit with a goal of weaning 23 pigs/sow/year requires a resident herd of 1,053 sows if farrowing rate is 95%, 1,112 females if the rate falls to 90%, 1,250 head at 80%, and a 1,429 sow inventory if farrowing rate is 70%. It is important that sows and gilts obtain adequate nutrients during gestation to assure optimum reproduction. Feeding in excess is not only wasteful and costly but is also likely to increase embryonic mortality. A limited feeding system using balanced, fortified diets is recommended to ensure that each sow gets her daily requirements of nutrients without consuming excess energy. As a rule of thumb, 4–5 lb of a balanced ration provides adequate protein and energy; however, during cold weather an additional pound of feed may prove beneficial, especially for bred gilts. With limited feeding, it is important that each sow get her portion of feed and no more. The use of individual feeding stalls for each sow is best because it prevents the "boss sows" from taking feed from slower eating or timid sows. During gestation, gilts should be fed so they will gain about 100–120 lb (Fig. 29.1), and sows should gain approximately 75–100 lb. This allows gilts to farrow at 350–400 lb and sows to weigh 425–550 lb at farrowing.

Table 29.2 shows several sow rations that utilize different grains. Note how the rations are balanced for protein, specific amino acids, minerals, and vitamins. Breeding females should be closely monitored in regards to body condition score to assure optimal performance. Body condition scores of 1–5, based on visual appraisal, are recommended (1 = extremely thin, 5 = extremely fat). Females should have a body condition score of 3–4 at time of birth.

Pregnancy Detection

Determining the gestational status of the breeding female is important to good management decisions. Many producers depend on observations of sows returning to estrus as a primary test for failure to attain pregnancy. However, this practice is labor-intensive, often requires the ability to provide boar exposure, and not absolutely conclusive such as is the case with sows experiencing cystic ovaries.

Pregnancy tests that measure the progesterone concentration in a serum or urine sample provide another alternative. While accurate, the need to collect samples

Figure 29.1
Bred gilts should be fed a diet that allows them to reach to specified target weight at farrowing (usually at approximately 350 to 400 lb). Source: National Pork Board.

limits the commercial applicability of this protocol. The use of ultrasound evaluation provides a useful pregnancy test in many situations and this technology has become the preferred method of evaluation in the industry, particularly for larger herds.

Management of the Sow during Farrowing and Lactation

Farrowing is the process of the sow or gilt giving birth to her pigs. The production goal is to attain a farrowing percentage of >85%, 2.2 litters per sow per year, and >24 pigs weaned per breeding female per annum. Furthermore, production management should be in place to keep nonproductive sow days to less than 33 with breeding herd death losses lower than 4% and preweaning death loss of pigs kept below 8%. Production differences can be substantial between enterprises and even between separate barns from within a farm. Effective record keeping, personnel training, and sound management can assure high levels of productivity in the female herd.

Proper care of the sow during farrowing and lactation is necessary to ensure large litters of pigs at birth and at weaning. Sows should be dewormed about 2 weeks before being moved to the farrowing facility. The sow should be treated for external parasites at least twice, a few days prior to farrowing time. The farrowing facility should be cleaned completely of organic matter, disinfected, and left unused for 5–7 days before sows are placed in the unit.

Approximately a week prior to parturition the sow will experience reduced feed intake and her activity level will decline. Breeding records and projected farrowing dates should be used to determine the proper time of transitioning females to the farrowing facility. If farrowing is to occur in a crate or pen, the sow should be placed there no later than the 110th day of gestation. This gives some assurance that the sow will be in the facility at farrowing time, as the normal gestation is 111–115 days in length. Farrowing time can be estimated by observing an enlarged abdominal area, swollen vulva, increased respiration rate, and filled teats. The presence of milk in the teats indicates that farrowing is likely to occur within 24 hours. Eighteen hours prior to farrowing the female will cease consuming feed, will exhibit nesting behavior, and recline on her side.

Table 29.2
SELECTED SOW RATIONS (WITH CORN, GRAIN, SORGHUM, OR WHEAT AS THE GRAIN SOURCE)

Ingredient	Gestation	Pre-farrowing	Lactation 10 pigs .33 lb ADG #1	Lactation 10 pigs .33 lb ADG #2	Lactation 10 pigs .44 lb ADG #1	Lactation 10 pigs .44 lb ADG #2
Corn, yellow	1,619	1,257	1,412	—	1,323	618
Grain, sorghum	—	—	—	1189	—	—
Wheat	—	—	—	—	—	618
Soybean meal, 44%	—	—	—	535	590	—
Soybean meal (dehulled)	245	465	470	—	—	588
Dehydrated alfalfa meal, 17%	60	100	—	100	—	—
Fat (choice grease)	—	100	—	100	—	100
Calcium carbonate	16	18	22	18	28	29
Dicalcium phosphate	37	37	38	35	36	32
Salt	10	10	10	10	10	10
Vitamin-trace mineral mix	8	8	8	8	8	8
Other	5	5	5	5	5	5
Total	2,000	2,000	2,000	2,000	2,000	2,000
Protein, %	13.1	17.2	17.2	19.0	18.5	19.9
Lysine, %	0.60	0.90	0.90	0.98	1.00	1.07
Tryptophan, %	0.14	0.20	0.19	0.24	0.22	0.29
Threonine, %	0.48	0.65	0.65	0.71	0.70	0.75
Methionine + cystine, %	0.48	0.58	0.59	0.60	0.63	0.67
Calcium, %	0.80	0.90	0.91	0.89	1.02	1.00
Available phosphorus, %	0.38	0.40	0.40	0.41	0.40	0.41
Metabolizable energy, kcal/lb	1,473	1,557	1,490	1,529	1,460	1,575

Source: Adapted from *Pork Information Gateway*, Swine Diets, 2004.

Before farrowing, while the sows are in the farrowing facilities, they can be fed a ration similar to the one fed during gestation. A laxative ration can be helpful at this time to prevent constipation. The ration can be made laxative by the addition of 10 lb/ton of Epsom salts or potassium chloride, part of the protein as linseed meal, or the addition of 10% wheat bran, alfalfa meal, or beet pulp.

Sows need not be fed for 12–24 hours after farrowing, but water should be continuously available. Two or three pounds of a laxative feed may be fed at the first post farrow feeding; the amount fed should be gradually increased until the optimum feed level is reached by 10 days after farrowing. Sows that are thin at farrowing may benefit from generous feeding early after farrowing. Sows nursing fewer than 8 pigs may be fed a basic maintenance amount using a rule of thumb of 3 lb for the sow and 1 lb for each pig nursed on a daily basis. Sows and first-lactation gilts raising more than eight pigs should be fed free choice. It is unnecessary to reduce feed intake before weaning. Regardless of level of feed intake, milk secretion in the udder will cease when pressure reaches a certain threshold level.

Induced Farrowing

The practice of stipulating the time of farrowing has been utilized by many swine operations to help optimize labor rates and assure effective production scheduling. The

Figure 29.2
Sow and litter of pigs in a farrowing crate. The farrowing crate is a pen (approximately 5 × 7 ft) that restricts the sow to a fixed area so that she does not lay on her babies. A panel several inches off the floor separates the sow from the baby pigs while allowing them access for nursing. The heat lamp helps to assure optimal temperature for neonatal pigs. Source: National Pork Board.

advantages of this approach include reduced pig mortality, heavier weaning weights, more efficient use of labor resources, and more cross-fostering. Induced farrowing provides the opportunity for timely observation of the farrowing sow and can lead to reduced perinatal mortality of 0.5 pigs per litter or 1 additional pig per annum.

The standard recommendation is to induce parturition on day 112 of gestation. Typically, induction should not occur prior to 2 days before due date and as such knowing the date of breeding and the average gestation length of the sow population is required for success. Furthermore, the producer must be able to accurately compute the benefits in light of the relatively high cost of this practice. If drugs are given too early, the producer risks increased rates of abortion and stillbirth.

The preferred drugs include prostaglandin and cloprostenol (synthetic prostaglandin). Oxytocin may be administered to those sows not farrowing within 25 hours of the original injection.

Baby Pig Management from Birth to Weaning

Most sows and gilts are farrowed in farrowing stalls or pens to protect baby pigs from the female lying on them (Fig. 29.2). Producers who give attention at farrowing will decrease the number of pigs that die during birth or within the first few hours after birth. Approximately one-third of preweaning pig deaths occur within 72 hours of birth. The primary causes of preweaning **mortality** in pigs are outlined in Table 29.3. Pigs can be freed from membranes, weak pigs can be revived, and other care can be given that will reduce baby pig death loss. Manual assistance in the birth process should not be given unless obviously needed. Duration of labor ranges from 30 minutes to more than 5 hours, with pigs being born either head or feet first. The average interval between births of individual pigs is approximately 15 minutes, but this can vary from almost simultaneously to several hours in individual cases. An injection of oxytocin can be given (sows only) to speed the rate of delivery after five to six pigs have been born. This step should be preceded by palpation of the tract to assure that there is not an obstruction of the birth canal.

Manual assistance, using a well-lubricated gloved hand, can be used to aid in difficult deliveries. The hand and arm should be inserted into the reproductive tract as far as is needed to locate the pig; the pig should be grasped and gently but firmly pulled to assist delivery. Difficult births are often associated with the occurrence of lactation failure where **mastitis** (inflammation of the udder), **metritis** (inflammation of the uterus), and **agalactia** (inadequate milk supply) may occur. Antibacterial solutions such as nitrofurazone are infused into the reproductive tract after farrowing.

Table 29.3
REASONS FOR PREWEANING PIG MORTALITY

Cause	%
Crushing	54
Starvation (runt pigs/ lactation failure)	14
Various-known	10
Scours	9
Respiratory problems	5
Unknown	8

Source: NAHMS, 2006.

Table 29.4
TARGET AIR TEMPERATURES FOR SWINE

Weight/Age	Optimum (°F)	Range between UCT and LCT (°F)
Newborn	95	90–100
Preweaning	>90	No lower than 77
5–35 lb	80–90	60–95
35–80 lb	65–80	40–95
80–160 lb	60–75	25–95
160–220 lb	50–75	5–95
>220 lb	50–75	5–90

Source: Adapted from multiple sources.

Sometimes intramuscular injections of antibiotics are used to decrease or prevent some of these infections.

It is important that the newborn pig receive colostrum within 4 to 6 hours postfarrowing to give immediate and temporary protection against common bacterial infections. Within 6 hours following birth, a sow's immunoglobulin concentration in colostrum declines by 50%. Pigs from extremely large litters can be transferred to sows having smaller litters to equalize the size of all litters. Care should be given to ensure that baby pigs receive adequate colostrum before the transfer takes place.

Air temperature within the creep area for baby pigs equipped with zone heat (heat lamps, gas brooders, or floor heat) should have a temperature of 90–100°F. Other target air temperatures for swine of known weights are listed in Table 29.4. Otherwise the pigs will chill and die or become susceptible to serious health problems. Moisture should be controlled or removed from the farrowing facility without creating a draft on the pigs.

Management of environmental temperature from birth to 2 weeks of age is critical. At 2 weeks of age, the baby pig has developed the ability to regulate its own body temperature.

Soon after birth, the navel cord should be cut to 3–4 in. from the body and then treated with iodine. This disinfects an area that could easily allow bacteria to enter the body and cause joints to swell and abscesses to occur.

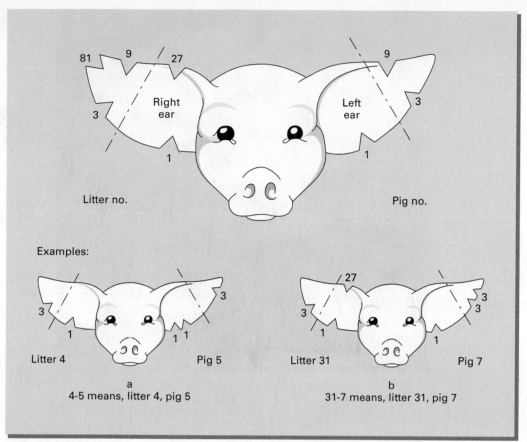

Figure 29.3

An ear-notching system for swine. Individual pig identification is necessary for meaningful production and management records. Source: Kevin Pond from Colorado State University. Used by permission of Kevin Pond.

The producer should also clip the eight sharp needle teeth of the baby pig. This prevents the sow's udder from being injured and keeps facial lacerations from occurring when the baby pigs fight one another. Approximately one-half of each tooth can be removed by using side-cutting pliers or toenail clippers.

Good records are important to the producer interested in obtaining optimum production efficiency. The basis of good production records is pig identification. Ear notching pigs at 1–3 days of age using the system shown in Figure 29.3 provides positive identification for the rest of the pig's life.

Baby pig management from 3 days to 3 weeks of age includes controlling anemia and scours, castration, and tail docking. Sow's milk is deficient in iron, so iron dextran shots are given intramuscularly at 3–4 days of age (Fig. 29.4).

Cross-fostering is the practice of transferring piglets from one sow to another within a farrowing group to reduce within litter weight variation among pigs and to assure that each sow is nursing the number of pigs best suited to her ability. This practice requires that pigs be left on their natural mother for a period of 4 to 6 hours to assure adequate colostrum intake, that the transfer occurs within 24 to 30 hours of birth, and that the foster dam farrowed within 24 hours of the donor sow. The predominant reason for this timing is that pigs establish a preferred teat to nurse within 48 hours of birth and disrupting this preference often leads to increased fighting among piglets.

Figure 29.4
An iron shot being administered to a baby pig. The preferred site of injection is the neck muscle. Injection should not be administered in the ham.
Source: National Pork Board.

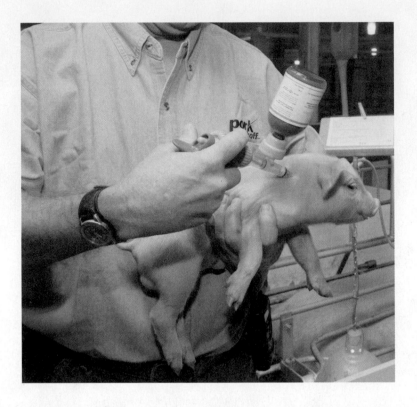

Baby pig scours are an ongoing problem for the swine producer. Colostrum and a warm, dry, draft-free, and sanitary environment are the best preventative measures against scours. Orally administered drugs determined to be effective against specific bacterial strains have been found to be the best control measures for scours. Serious diarrhea problems resulting from diseases such as **transmissible gastroenteritis (TGE)** and swine **dysentery** should be treated under the direction of a local veterinarian.

Castration is usually done before the pigs are 2 weeks old. Pigs castrated at this age are easier to handle, heal faster, and are not subjected to as much stress as those castrated later. The use of a clean, sharp instrument, low incisions to promote good drainage, and antiseptic procedures make castration a simple operation.

Tail docking has become a common management practice to prevent tail biting and the increased risk of infection that results. Tails are removed 0.5 to 0.75 in. from the body, and the stump should be disinfected along with the instrument after each pig is docked.

Pigs should be offered feed at 1–2 weeks of age in the creep area. These starter feeds can be placed on the floor or in a shallow pan. By 3–4 weeks after farrowing, the sow's milk production has likely peaked; however, the baby pigs should be eating the supplemental feed and growing rapidly (Fig. 29.5).

Rations for baby pigs (weighing 10–25, 14–20, or 25–40 lb) are shown in Table 29.5. Note the components of the rations, especially when milk products (dried whey or dried skim milk), are included in the rations of 10- to 25-lb pigs. Milk by-products are beneficial because of their desirable protein and lactose content that boosts pig performance. Table 29.6 outlines the recommended levels of feed intake for lactating gilts and sows. Females should be kept cool to assure sufficient feed intake to maintain sufficient milk production to allow pigs rapid and healthy early growth.

Figure 29.5
Weaned pigs in a modern nursery unit. The pigs were weaned from the sow and placed into these pens when they weighed 10 to 20 lb. The pigs will be moved into pens for growing and finishing when they weigh approximately 40 to 45 lb.
Source: National Pork Board.

Table 29.5
STARTER DIETS FOR EARLY WEANED PIGS

| | Ration Number and lb/Ration | | | | | |
| | Pigs 10–14 lb | | Pigs 14–20 lb | | Pigs 20–45 lb | |
Ingredient	1	2	1	2	1	2
Corn, yellow	882	682	945	766	1,156	1,229
Oat groats	—	200	—	200	—	—
Soybean oil	80	80	40	40	—	—
Fat (choice grease)	—	—	—	—	—	40
Soybean meal, 44%	265	300	480	460	590	670
Fish meal	80	80	100	100	—	—
Lactose	50	100	—	—	—	—
Dried skim milk	100	—	—	—	—	—
Dried whey	400	400	400	400	200	—
Dried blood meal	50	50	—	—	—	—
Dried plasma protein	50	60	—	—	—	—
Lysine	1	2	2	2	2	2
Methionine	2	2	1	1	—	—
Calcium carbonate	14	16	10	10	18	19
Dicalcium phosphate	11	13	7	6	19	25
Copper sulfate	2	2	2	2	2	2
Salt	5	5	5	5	7	7
Vitamin-trace mineral	8	8	8	8	8	8
Total	2,000	2,000	2,000	2,000	2,000	2,000
Protein, %	20.33	20.36	20.07	20.28	19.03	19.87
Lysine, %	1.40	1.40	1.30	1.30	1.15	1.19
Tryptophan, %	0.27	0.27	0.24	0.25	0.23	0.24
Threonine, %	0.90	0.89	0.83	0.83	0.75	0.76
Methionine + cystine, %	0.81	0.79	0.74	0.73	0.65	0.66
Lysine:calorie ratio	4.02	4.05	3.89	3.87	3.57	3.56
Calcium, %	0.86	0.86	0.75	0.75	0.71	0.73
Available phosphorus, %	0.45	0.43	0.41	0.40	0.32	0.32
Metabolizable energy, kcal/lb	1,580	1,573	1,517	1,521	1,467	1,513

Source: Adapted from *Pork Information Gateway*, Swine Diets, 2004.

Table 29.6
FEED INTAKE IN LACTATING SOWS

		Total Feed Intake/hd/day (lb)		
	Day of Lactation	1st = Litter Gilts	Sows	Feeding Frequency
	Farrowing	4	6	1×
Day	2	6	8	1×
	3	8	10	1×
	4	10	12	1×
	5	12	14	2×
	6	14	16	2×
	7	14	16	2×
	8	16	18	2×
	9	16	20	2×
	10 +	16	20	2×

Source: Adapted from D. Lewis, 2002.

Pigs can be weaned from their dams at a time that is consistent with the available facilities and the management of the producer. Most pigs are weaned between 16 and 20 days of age. Pigs weaned at young ages should receive a higher level of management. General weaning guidelines include weaning only pigs over 10 lb, weaning pigs over a 2- to 3-day period, weaning the heavier pigs in the litter first, grouping pigs according to size in pens of 30 pigs or less, providing 1 feeder hole for 2 pigs and 1 waterer for each 8–10 pigs, limiting feed for 48 hours, and using medicated water if scours develop.

Feeding and Management from Weaning to Market

A dependable and economical source of feed is the backbone of a profitable swine operation. Since approximately 60–70% of the total cost of pork production is feed related, the swine producer should be keenly aware of all aspects of swine nutrition. Given the impact of feed costs on profitability, the use of least cost ration formulation and precision feeding strategies are highly advised. Calculation of least cost rations requires that managers have access to a selection of alternative feedstuffs and ingredients, accurate feed composition data, complete accounting of feed ingredient acquisition costs including shrink and transportation expenses, as well as the upper and lower tolerances of specific limiting nutrients such as lysine or those than might result in toxicity if overfed or specific deficiencies if underfed.

Swine rations for growing–finishing market pigs as well as rations for breeding stock are formulated around cereal grains. The most common energy-rich cereal grains are corn, milo, barley, wheat, and their by-products. Because of its abundance and readily available energy, corn is used as the base cereal for comparing the nutritive value of other grains. Milo (grain sorghum) is similar in nutritional content to corn and can completely replace corn in swine rations; however, milo requires more processing (grinding, rolling, steaming, etc.). Its energy value is about 95% the value of corn. Barley contains more protein and fiber than corn, but its relative feeding value is 85–95% of corn and it is less palatable. Wheat is equal to corn in feeding value and is highly palatable. However, for best feeding results, wheat should be mixed half-and-half with some other grain. Which grain or combination of grains to use can best be determined by availability and relative cost.

Feed processing utilizes mechanical, chemical, heating, or microbial action to manipulate ration ingredients. Processing may alter particle size, enhance palatability, shift nutrient composition, improve digestibility, increase stability during storage, or to remove toxins and contaminants. Processing feeds typically involves grinding, pelleting, or the use of extrusion. Grinding corn improves feed efficiency and growth rate by 3–5%. Caution needs to be applied as feed that is too finely ground may lead to increased gastric ulcers. Pelleting the ration may increase growth rate by 3–5%, efficiency of gain by 5–10%, and reduces feed intake by up to 3%. However, the advantages of feed processing must be evaluated in light of their costs.

Cereal grains usually contain lesser quantities of proteins, minerals, and vitamins than swine require; therefore, rations must be supplemented with other feeds to increase these nutrients to recommended levels. Soybean meal has been proven to be the best single source of protein for pigs when palatability, uniformity, and economics are considered. Other feeds rich in protein (meat and bone meal or tankage) can compose as much as 25% of the supplemental protein if economics so dictate. Cottonseed meal is not recommended as a protein source unless the gossypol has been removed. **Gossypol** in relatively high levels is toxic to pigs.

Proper harvest, storage, and handling of corn are critical to minimize the occurrence of molds that may produce undesirable **mycotoxins** such as **aflatoxin**, **zearalenone**, and **vomitoxin**. These mycotoxins can result in poor growth performance, abortion, and other symptoms.

If economics dictate, a producer can supplement specific amino acids in place of some of the protein supplement. Typically, lysine may be supplemented separately because lysine is the first limiting amino acid in most swine diets.

Calcium and phosphorus are the minerals most likely to be deficient in swine diets, so care should be exercised to ensure adequate levels in the diet. The standard ingredients for supplying calcium and phosphorus in the swine diet are limestone and either dicalcium phosphate or defluorinated rock phosphate. If an imbalance of calcium and zinc exists, a skin disease called **parakeratosis** is likely to occur.

Swine producers may formulate their swine diets in several ways: (1) by using grain and a complete swine supplement; (2) by using grain and soybean meal plus a complete base mix carrying all necessary vitamins, minerals, and antibiotics; or (3) by using grain, soybean meal, vitamin premix, trace-mineral premix, salt, calcium and phosphorus sources, and an antibiotic premix. Which ration is used depends on the producer's expertise, the cost, and the availability of ingredients. The trace-mineral premix typically includes vitamins A, D, E, K, B_{12}, and other B vitamins (niacin, pantothenic acid, riboflavin, and choline).

Table 29.7 shows example diets for growing pigs. Diets for finishing pigs are noted in Table 29.8. These diets are formulated using the common grains (corn, grain sorghum, barley, or wheat).

Most swine producers use feed additives because the additives have the demonstrated ability to increase growth rate, improve feed efficiency, and reduce death and sickness from infectious organisms. Feed additives available to swine producers fall into three classifications: (1) **antibiotics** (compounds coming from bacteria and molds that kill other microorganisms), (2) **chemotherapeutics** (compounds that are similar to antibiotics but are produced chemically rather than microbiologically), and (3) **anthelmintics** (dewormers). Antibiotics and chemotherapeutics should not be substituted for poor management (e.g., unsanitary conditions and poor health care).

Feed additive selection and the level to be used will vary with the existing farm environment, management conditions, and stage of the production cycle. Disease and parasite levels will vary from farm to farm. Also, the first few weeks of a pig's life are far more critical in terms of health protection. Immunity acquired from

Table 29.7
SELECTED GROWING RATIONS FOR AVERAGE AND HIGH-LEAN BARROWS (45–75 lb) AND HIGH-LEAN GILTS (45–110 lb)

Ingredient	Ration Number and lb/Ration			
	1	2	3	4
Corn, yellow	1,407	1,494	—	—
Grain sorghum	—	—	1,470	1,275
Soybean meal, 44%	542	451	565	572
Fat (choice grease)	—	—	—	100
Calcium carbonate	18	18	19	18
Dicalcium phosphate	20	21	20	20
Salt	7	7	7	7
Trace mineral–vitamin	6	6	6	6
Lysine	—	3	3	2
Total (lb)	2,000	2,000	2,000	2,000
Protein, %	17.71	16.22	17.30	18.48
Lysine, %	0.95	0.95	0.95	1.02
Tryptophan, %	0.20	0.18	0.21	0.23
Threonine, %	0.67	0.61	0.63	0.69
Methionine + cystine, %	0.60	0.56	0.56	0.59
Calcium, %	0.65	0.65	0.65	0.65
Available phosphorus, %	0.27	0.27	0.28	0.28
Metabolizable energy, kcal/lb	1,482	1,484	1,456	1,558

Source: Adapted from *Pork Information Gateway*, Swine Diets, 2004.

Table 29.8
SELECTED FINISHING RATIONS FOR GILTS AND BARROWS (175–265)

Ingredient	Ration Number and lb/Ration			
	Gilt–1	Gilt–2	Barrow–1	Barrow–2
Corn, yellow	1,660	—	1,719	—
Grain sorghum	—	1,721	—	1,779
Soybean meal, 44%	300	235	245	181
Soybean meal, 48%	—	—	—	235
Lysine	—	3	—	3
Calcium carbonate	16	16	16	16
Dicalcium phosphate	13	14	9	10
Salt	7	7	7	7
Trace mineral–vitamin mix	4	4	4	4
Total, lb	2,000	2,000	2,000	2,000
Protein, %	13.46	13.20	12.50	12.29
Lysine, %	0.64	0.64	0.57	0.57
Tryptophan, %	0.14	0.16	0.12	0.14
Threonine, %	0.50	0.47	0.46	0.43
Methionine + cystine, %	0.50	0.45	0.47	0.42
Calcium, %	0.50	0.50	0.45	0.45
Available phosphorus, %	0.18	0.20	0.14	0.16
Metabolizable energy, kcal/lb	1,504	1,473	1,510	1,479

Source: Adapted from *Pork Information Gateway*, Swine Diets, 2004.

Figure 29.6
A group of pigs in a growing and finishing production unit. The building is enclosed and the floor is slatted. The manure from the pigs drops through the slats into a 36-in.-deep pit. The waste is removed from the pit via flushing with water. Management of waste and air quality is a priority for swine enterprises. Source: National Pork Board.

Figure 29.7
Automatic, nipple-style, waterers provide continuous water access for swine. This style of watering device helps conserve water, prevent flooring from becoming slick due to flooding, and ensures access. Source: National Pork Board.

colostrum diminishes by 3 weeks of age, and the pig does not begin producing sufficient amounts of antibodies until 5–6 weeks of age. Also during this time, the pig is subjected to stress conditions of castration, vaccinations, and weaning. Therefore, at this age the additive level of the feed is much higher than when the pig is more than 75 lb and growing rapidly toward slaughter weight. Producers should be cautious because some additives have withdrawal times, which mean the additive should no longer be included in the ration. This is necessary to prevent additives from occurring as residues in meat.

Market pigs that are fed properly prepared diets will likely have rapid and efficient liveweight gains. However, rapid and low-cost gains are also associated with excellent health programs and well-constructed facilities (Fig. 29.6). Providing adequate and consistent water supplies is also critical to the successful management of a swine enterprise (Fig. 29.7). Watering system and intake specifications are listed in Table 29.9.

The swine industry has adapted two innovative feeding approaches to more efficiently feed growing hogs—phase feeding and split-sex feeding protocols. **Phase feeding** simply means the use of specific rations for the specific needs of pigs differing in growth rate. Phase feeding typically involves the use of a sequence of four rations balanced for lysine—the most limiting amino acid for swine. **Split-sex feeding** recognizes the different nutritional requirements of barrows and gilts. The benefits of

Table 29.9
GUIDELINES FOR PROVIDING WATER TO SWINE

Stage	Head Per Water Nipple	Span between Nipples (ft)	Water Intake (gallon/day)
Nursery	10	12	
25 lbs			0.4
50 lbs			0.6
Grower	10–15	18	1.0
Finisher	15	24–36	
150 lb			1.3
200 lb			1.7
>250lb			2.1
Gestation	15	36	4.5
Lactation	1	NA	6.0

Source: Adapted from multiple sources.

split-sex management include improved growth performance, lower feed costs, and improved market timing. Barrows typically require more feed to attain similar growth rates to their female contemporaries but need less lysine.

Technological breakthroughs achieved through active research and development programs have yielded feed additives that enhance the ability of hogs to more efficiently produce lean. For example, ractopamine, marketed as Paylean™, is a nutrient that yields a repartitioning response in which feed is shifted from production of fat to production of lean. Another supplement approved for use in swine diets to enhance lean production is chromium picolinate at 200 parts per billion. It is suspected that chromium enhances carbohydrate metabolism by stimulating insulin function.

MANAGEMENT OF PURCHASED FEEDER PIGS

Feeder pigs marketed from one producer to another are subjected to more stress than pigs produced in a farrow-to-finish operation. These stresses are fatigue, hunger, thirst, temperature changes, diet changes, different surroundings, and social problems. Almost every group has a shipping–fever reaction. Care of newly arrived pigs must be directed to relieving stresses and treating shipping fever correctly and promptly.

Management priorities for newly purchased feeder pigs should include (1) a dry, draft-free, well-bedded barn or shed with no more than 50 pigs per pen; (2) a specially formulated starter ration having 12–14% protein, more fiber, and a higher level of vitamins and antibiotics than a typical starter ration; (3) an adequate intake of medicated water containing electrolytes or water-soluble antibiotics and electrolytes; and (4) prompt and correct treatment of any sick pigs.

MARKETING DECISIONS

Slaughter weights of market hogs and which market is selected can have a marked effect on income and profitability of the swine enterprise. For many years the recommended weight to sell barrows and gilts was, in most instances, 200–220 lb. The primary reasons for selling these animals at a maximum weight of 220 lb were increased cost of gain and increased fat accumulations at heavier weights. Market discounts were common on pigs weighing over 220 lb because of the increased amount of fat. Today market hogs average between 275 and 290 pounds with some contracts providing incentives for market hogs weighing in excess of 300 pounds. These heavier pigs are leaner, and they are not subject to the same market discounts as fatter pigs marketed decades ago. Most industry forecasts suggest that the trend for increased market weights will continue into the future.

The increase in the number of hogs sold on contract with final price determined on a carcass merit basis has been dramatic in the past decade. Approximately 80% of all hogs are sold on a contract basis. The most common approach is a **formula contract** where price is based on the cash market plus a predetermined premium. Other contract types include a **forward-cash contract**, a **risk-share contract** where the base price is determined from a formula that accounts for fluctuations in feed costs, and a risk-share contract where cash prices are paid within a predetermined range and adjustments are made if the cash market falls outside the range (producer and packer split gains and losses).

Several branded product initiatives have also been introduced into the marketplace in an attempt to capture more value as the result of creating product differentiation. These programs typically establish a specific set of market criteria such as lean color, lean pH, percent lean, and specific genetics such as breed or line. The American Berkshire Association has created a targeted supply chain approach that focuses on adding value to Berkshire and Berkshire cross pigs for both the domestic and export market.

COSTS AND RETURNS

Construction of a new 1,200 sow farrow-to-finish operation would require an investment of approximately $4.5 million (Fig. 29.8). Table 29.10 shows the percent breakdown of various cost factors associated with a typical farrow-to-finish enterprise with an average finished pig number of 20 per sow. Note that feed and labor constitute nearly 70% of the operational expense. Enterprise records for typical farrow-to-finish hog operations are shown in Table 29.11.

Table 29.11 outlines a sample of the performance **benchmarks** for a farrow-to-finish enterprise. Careful monitoring of these benchmarks is critical to assess progress and to identify problem areas before excessive costs are incurred.

Profitability over several years is noted for farrow-to-finish operations (Fig. 29.9). Excellent managers achieve larger positive returns and lower losses than shown in this

Figure 29.8
Farrow-to-finish operations utilize a variety of construction designs. However, most facilities utilize high quality engineering to assure appropriate airflow, animal comfort, and automated feed and water delivery.
Source: National Pork Board.

Table 29.10
COST COMPONENTS OF A FARROW-TO-FINISH OPERATION

Factor	% of Total Cost
Feed	58
Labor	9
Depreciation	6
Interest	6
Marketing/transport	5
Vet and vet supply	4
Fuel, oil, and utilities	4
Other	8

Source: Adapted from multiple sources.

Table 29.11
PERFORMANCE BENCHMARKS FOR A FARROW-TO-FINISH ENTERPRISE

Trait	Goal
Litters farrowed/sow/year	2.1 (pen mating)
	2.2 (handmating—confinement)
Farrowing rate	>85%
Pigs weaned/sow/year	>24
Mortality rate	
Breeding herd	<4% (confinement)
	<6% (outside)
Preweaning	<8%
Nursery	<2%
Growing/finishing	<1%
Days to market	<160
Average daily gain (lb/day)	
Nursery	>1.0
Finishing	>1.9
Weaning-finishing	>1.7
Feed efficiency (lb/day)	
Nursery	<1.6
Finishing	<2.6
Whole herd	<3.2

Source: Adapted from multiple sources.

figure. These top managers have low-cost production (especially low feed costs), while buying and selling pigs below and above average market prices. Even with excellent management, shifts in the market price for hogs can make it difficult to generate profits.

ENVIRONMENTAL MANAGEMENT

Management of waste and air quality factors is an important consideration for swine producers. The goals of an effective manure management plan should be to avoid or minimize (1) direct discharge, runoff, or seepage into a stream; (2) release of ammonia into the atmosphere; (3) detectable odor beyond the farm borders; (4) release of pathogens into the environment; and (5) contamination of soil and groundwater.

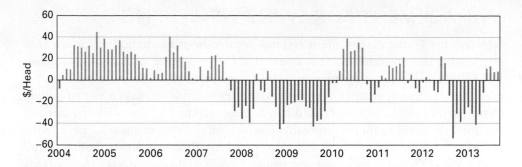

Figure 29.9
Average monthly returns for farrow-to-finish enterprises.
Source: USDA: ERS and NASS compiled by Livestock Marketing Information Center.

Table 29.12
FEED-RELATED ODOR CONTROL PRACTICES IN THE SWINE INDUSTRY

Protocol	%
Finely ground grain	27
Vegetable oil or fat to control dust	24
Low crude protein and/or synthetic amino acids	20
Pelleting	15
Phytase	11
Other feed additives	10
Add 10% fiber	8
Other	8

Source: NAHMS, 2000.

Table 29.13
NUTRITIONAL MANAGEMENT TO REDUCE EXCESS NITROGEN AND PHOSPHORUS (COMBINED EFFECTS ARE NOT ADDITIVE)

Method	Potential Decrease in: (%)	
	Nitrogen	Phosphorus
Precision ration formulation	10–15	10–15
Limit protein and amino acid supplements	20–40	—
Include highly digestible feeds	5	20–30
Include phytase in a diet with reduced phosphorus supplementation	2–5	20–30
Incorporate phase-feeding program	5–10	5–10

Source: Multiple sources.

The most frequently utilized strategies to **control odor** include dietary manipulation (50.2%), manure management (28.9%), air quality interventions (28.2%), and the addition of chemical-biological agents to lagoons or waste storage sites (16.0%) according to the U.S. Department of Agriculture (USDA). The specific feed-related odor control practices are described in Table 29.12.

Swine managers can also affect water and odor impacts by minimizing feed and water spills, improving utilization with feed processing technologies, adjusting diets for specific animal requirements, and adjusting diets for climatic variation. The specific effects of management practices on the nitrogen and phosphorus levels of waste are described in Table 29.13.

MANAGEMENT FOR SWINE WELL-BEING

Superior integrated management systems provide housing, handling, production environments, and transport that assure that the well-being of hogs is optimized. Providing excellent care for pigs, low stress environments to enhance both animal and employee performance, demonstrating responsible management to supply chain partners and consumers, and conducting business in line with the ethic of the good shepherd facilitates profitable production, delivery of superior pork to the marketplace, healthier livestock, and creation of sustainable business models. A full discussion of facilities, handling systems, and other well-being design elements are not possible in the context of this book. Students seeking career opportunities in the livestock industry should pursue additional training and information.

Hogs can be raised in both indoor and outdoor situations. Independent research demonstrates that indoor raised pigs have better reproductive rates, larger litter sizes, and weaning rates, and superior pounds weaned per sow per year. However, outdoor systems can be managed such that these performance gaps can be narrowed. A careful evaluation of both direct and indirect costs, climatic conditions, labor availability, and production site characteristics should be undertaken when determining whether to pursue housed systems with high biosecurity versus more open systems.

Housing system design considerations include animal comfort, ventilation, temperature control, waste management, flooring systems, pen layout, production flow, biosecurity, feed and water delivery, and employee comfort. For example, correct floor design should accomplish the following objectives: avoid foot erosion and injury, enhance stability of footing for both pigs and staff, enhance pig comfort, facilitate ease of cleaning to minimize pathogen survival, allow for effective manure management, and protect biosecurity.

Discussions about whether to use individual gestation stalls or group gestation pens provide an excellent case study related to animal well-being. Traditionally, gestation stalls have been used to provide individual housing for sows. This format was adopted by the industry due to the productivity advantages of individual housing systems compared to group housing, to avoid the fighting and associated animal injury that comes with establishment of social hierarchy in a group setting, and to provide more consistent access to feed and water for each individual sow. Both systems have been intensely researched and a summary of the results leads to one conclusion—either system can be well managed (or mismanaged) but each requires different strategies and adaptations to assure animal comfort and well-being. The primary issues to be dealt with include successful transitioning of the sow from the farrowing environment to a group situation, managing group interactions with special focus on aggressive or dominant females, designing feed and water delivery stations that provide access to all animals within the group, managing the inherent natural stress that accompanies sexual behavior in pigs, and providing a safe work environment for employees. Transitioning from gestation stalls to group housing is costly, requires retraining employees, and adopting new management protocols. The transition should be accomplished gradually to assure that animal care remains at optimal levels throughout.

Correct animal handling is important when moving pigs, conducting processing and health care protocols, routine observation of animals, feeding, and transport. The Pork Quality Assurance Program from the National Pork Board provides a detailed and valuable set of guidelines and training tools. As an example of specific handling guidelines, transport, and swine handling protocols are listed in Table 29.14.

Table 29.14
SWINE HANDLING AND TRANSPORT GUIDELINES

Handlers should move slowly among pigs in a pen for 1 minute per day or 5 minutes per week during the finishing period.

Feed should be withheld for 4–6 hours prior to loading or 12–18 hours prior to harvest. Ready access to water should be provided.

Eliminate shadows and assure even distribution of light.

Panels, paddles, or flags should be used to move hogs. Avoid any contact that might injure or stress the animal.

Move groups of 5–6 pigs on 3-ft-wide alleys and only 3 pigs in a 2-ft-wide alley.

Load pigs in the evening or early morning during periods when mid- or late-day heat and humidity are high. Use partitions to divide the load.

Avoid sharp turns in alleys; loading chutes should have less than 20° angles.

Minimize fighting among pigs by avoiding co-mingling of large groups of pigs in alleyways or holding areas.

The number of pigs per-running-foot of truck floor space (92 in. truck) should be 2.2 for 200-lb hogs, 1.8 for 250-lb hogs, and 1.6 for 300-lb hogs.

Avoid leaving pigs standing on a stationary, fully loaded truck.

Drivers should operate smoothly to avoid injuring pigs.

Schedule arrival at the packing plant so that hogs can be promptly off-loaded.

Source: Adapted from Pork Quality Assurance Program.

CHAPTER SUMMARY

- The incorporation of isowean principles into the design of hog production systems has created a system of multi-site, integrated production farms.

- Management of the sow during farrowing and lactation and of the baby pigs from birth to weaning are the most critical periods in a total swine management program.

- Sow rations and baby pig rations must have the proper combinations of amino acids, vitamins, and minerals (in addition to protein and energy) to assure adequate and cost-effective gains.

- Herd health programs must be well planned to prevent reproductive diseases, assure adequate lactation, and prevent scours and other diseases in baby pigs.

KEY WORDS

farrow-to-finish
farrow-to-feeder pig production
feeder pig finishing
farrow-to-weaner
multisite rearing systems
isowean principle
three-site isowean system
two-site isowean system
purebred seedstock
boar
handmating
gilt
artificial insemination (AI)
farrowing
pig mortality

mastitis
metritis
agalactia
cross-fostering
transmissible gastroenteritis (TGE)
dysentery
gossypol
mycotoxins
aflatoxin
zearalenone
vomitoxin
parakeratosis
antibiotics
chemotherapeutics
anthelmintics

phase feeding

risk-share contract (two types)

split-sex feeding

benchmarks

formula contract

odor control

forward-cash contract

REVIEW QUESTIONS

1. What are the phases of swine production and the typical amount of time spent in each?
2. What were the historic four types of swine enterprises?
3. Discuss multisite rearing systems.
4. Provide an overview of the isowean principle and its application in swine production.
5. Describe effective management of boars, breeding females, and gilt development programs.
6. Discuss the role of farrowing rate relative to profitability.
7. Compare how rations change with shifts in production phase.
8. Why might induced farrowing be utilized?
9. What are the primary causes of baby pig mortality?
10. Discuss air temperature targets for various classes of swine.
11. Describe effective baby pig management protocols.
12. What are swine feed intake goals as lactation progresses?

13. Describe least cost ration formulation.
14. Describe the feed processing approaches that can be used to improve digestibility.
15. Compare and contrast grower and finisher rations.
16. Why is phase feeding useful?
17. Outline water requirements for various classes of swine.
18. Describe the typical marketing alternatives available to hog producers.
19. What are the primary categories of cost in farrow to finish enterprises?
20. Describe the performance benchmarks for a farrow to finish enterprise.
21. What are the goals of a manure management program?
22. What strategies are available to affect odor?
23. Discuss the methods available to reduce nitrogen and phosphorus excretion.
24. Outline important considerations in managing pig well-being.

SELECTED REFERENCES

Aherne, F. 2006. Feeding the gestating sow. Fact sheet 07-01-06. Des Moines, IA: Pork Information Gateway.

Aherne, F. 2006. Feeding the lactating sow. Fact sheet 07-01-05. Des Moines, IA: Pork Information Gateway.

Baas, T. J. 1999. Key production factors to monitor the pork production business. Iowa State University (personal correspondence).

Flowers, B. 2002. Value of a good A.I. technician. *PORK*, Nov., 42.

Flowers, W. L., and H. D. Alhusen. 1991. Reproductive performance and estimates of labor requirements associated with combinations of artificial insemination and natural service in swine. *J. Anim. Sci.* 70:615.

Iowa Pork Industry Center. 1999. *Iowa Swine Business Record Summary*. Ames, IA: Iowa State University.

Lewis, D. 2002. Is your sow management costing you pigs. *PORK*, August, 20–22.

Miller, E. R., D. E. Ullrey, and A. J. Lewis. 1992. *Swine Nutrition*. Boca Raton, FL: CRC Press, Inc.

National Research Council. 1988. *Nutrient Requirements of Swine*. Washington, DC: National Academy Press.

Pond, W. G., J. H. Maner, and D. L. Harris. 1991. *Pork Production Systems*. Westport, CT: AVI Publishing Co.

Pork Industry Handbook (various years). Contains Fact Sheets (e.g., Management and Nutrition of the Newly Weaned Pig; Swine Diets; Swine Growing-Finishing Units; Pork Production Systems with Business Analyses). Cooperative Extension Services of several universities.

See, T. 2002. Water guidelines for swine. *PORK*, June 3.

30

Sheep and Goat Breeds and Breeding

Sheep and goats are considered small ruminants and play an important global role in providing meat, milk, wool, fiber and hides. Sheep were brought to the New World by Columbus in 1493 and in 1609 sheep were imported to the Virginia colonies by the British settlers. By 1662, the first woolen mills were operating in New England.

Today the economic impact of the sheep and goat industries is less than that of beef, dairy, poultry, and pork in the United States. However, profitable enterprises are attainable with both sheep and goats. The growth of certain ethnic markets, the diversity of culinary interests, and specialty niches suggest potential growth opportunities for these businesses.

MAJOR U.S. SHEEP BREEDS

Sheep have been bred for two primary purposes: production of wool and meat. Wool takes a variety of forms of variable value in the marketplace. Fine wool is utilized for making high-quality garments, production of long wool is used for heavy clothing, upholstery, and rug making. Because meat and wool traits are heritable, breeds have been developed over the years that meet the unique demands of producing protein for the human consumption as well as that of the natural fiber market. Dual-purpose sheep breeds (which produce both wool and meat) have also been developed by crossing fine-wool breeds with long-wool breeds and then selecting for both improved meat and wool production. Some breeds serve specific purposes; an example is the Karakul breed, which supplies pelts for clothing items such as caps and Persian coats.

Sheep breeds can be divided into categories or types based on wool quality (fine, medium, long, crossbred, or hair) or based on breeding purpose (ewe-type, ram-type, or dual-purpose).

Fine-wool breeds include Merino and Rambouillet, **medium-wool breeds** include the Cheviot, Dorset, Finnsheep, Hampshire, Shropshire, Southdown and Suffolk. Border Leicester, Lincoln and Romney are **long-wool breeds** while Columbia, Corridale, Polypay, and Targhee are considered **crossbred-wool breeds**. **Hair-type sheep** are of tropical origin and lack wool, instead they have a coat more similar to that of goats. Examples include the Dorper, Katahdin, and St. Croix.

Ewe breeds are usually white-faced and have fine or medium wool, long wool, or cross of these types. They are noted for reproductive efficiency, wool production, size, milking ability, and longevity. The Rambouillet, Merino, Corriedale, Targhee, Finnsheep, and Polypay are included in the ewe breed category.

Ram breeds are meat-type breeds raised primarily to produce rams for crossing with ewes of the ewe breed category. They are noted for growth rate and carcass characteristics. The Suffolk, Hampshire, Shropshire, Oxford, Southdown, and Montadale are classed as ram breeds. Recently imported breeds such as the Texel offer a high degree of muscularity and lean yield.

Dual-purpose breeds are used either as ewe breeds or as ram breeds. The Dorset and Cheviot Columbia are good examples of breeds that are often used as ewe breeds to be crossed with a ram breed. The Lincoln and Romney are often used as ram breeds to cross on other breeds of ewes to improve milking ability and fertility of the eweflock.

Characteristics

Breeds of sheep and their characteristics are listed in Table 30.1. Breeds are distinguished by wool characteristics, color of face, size, horned or polled condition, and a number of other factors.

The sheep breeds commonly used in the United States are shown in Figures 30.1 and 30.2. The Merino and Rambouillet (fine-wool) breeds and the dual-purpose breeds that possess fine-wool characteristics have a herding instinct; so one sheepherder with dogs can handle a band of 1,000 ewes and their lambs on the summer range. Winter bands of 2,500–3,000 ewes are common. Ram breeds, by contrast, tend to scatter over the grazing area. They are highly adaptable to grazing fenced pastures where feed is abundant and thus are often breeds preferred in farm flock enterprises. Rams of the ram breeds (meat-type) are often used for breeding ewes of fine-wool breeding on the range for the production of market lambs. Crossbreeding in sheep has been used for more than a half century.

The U.S. Sheep Experiment Station in Dubois, Idaho, developed the Columbia and Targhee breeds. Both breeds have proven useful on western ranges because they have the herding instinct and, when bred to a meat breed of ram, they raise better market lambs than do fine-wool ewes. The U.S. Sheep Experiment Station has made great strides in improving the Rambouillet, Columbia, and Targhee breeds.

Based on annual registration numbers, the Suffolk, Dorset, Rambouillet, and Hampshire are the most numerous breeds of sheep. The Suffolk and Hampshire are used extensively in crossbreeding programs to produce market lambs both in farm flocks and on the range. The most popular ewe breeds on the range are the Rambouillet, Columbia, and Corriedale. In a farm flock lamb-production system, the Dorset and Shropshire are important ewe breeds.

Composite Breeds

The U.S. Sheep Experiment Station in Dubois, Idaho, developed a synthetic breed of sheep, called the **Polypay**, with superior performance in lamb production and carcass quality. Four breeds (Dorset, Targhee, Rambouillet, and Finnsheep) provided the basic genetic material for the Polypay. Targhees were crossed with Dorsets, and Rambouillets were crossed with Finnsheep, after which the two crossbred groups were crossed. The offspring produced by crossing the two crossbred groups were intermated and a rigid selection program was practiced. This population was closed to outside breeding. The Rambouillet and Targhee breeds contributed hardiness, herding instinct, size, a long breeding season, and wool of high quality. The Dorset contributed good carcass, high milking quality, and a long breeding season. The Finnsheep contributed early puberty, early postpartum fertility, and high lambing rate.

Table 30.1
CHARACTERISTICS OF SHEEP BREEDS WITHIN TYPE CLASSIFICATION

Breed	Mature Size	Growth Rate[a]	Degree of Muscularity[a]	Wool Type	Fiber Diameter (micron)	Staple Length (in.)	Grease Fleece[b] Weight (lb)	Color	Origin (Imported to U.S.)
Cheviot	R 160–200 E 120–160	M	H	Medium	33–27	3–5	5–10	White with black nose	Scotland (1838)
Corriedale	R[e] 175–275 E[f] 130–180	M	M	Crossbred	31.5–24.5	3.5–6	10–17	White	New Zealand
Columbia	R 225–300 E 150–225	H	M	Medium	31–24	3.5–5	10–16	White	U.S. (1912)
Dorper	R 220–250 E 170–200	H	H	Hair	NA	NA	NA	White, black face	South Africa(1914)
Dorset[d]	R 210–250 E 140–180	MH	MH	Medium	33–27	2.5–4	5–9	White	England (1855)
Finnsheep	R 150–200 E 120–140	L	L	Medium	31–23.5	3–6	4–8	White	Finland (late 1960s)
Hampshire	R 225–325 E 175–225	H	H	Medium	33–25	2–3.5	6–10	White with wool cap and black face	England (1880)
Lincoln	R 250–350 E 200–250	M	M	Long	41–33.5	8–15	12–20	White	England(1825)
Merino[c]	R 150–225 E 110–150	LM	L	Fine to very fine	22–19	2.5–4	8–14	White	France/Spain
Oxford	R 200–300 E 150–200	MH	H	Medium	34.5–30	3–5	8–12	White with brown face and legs, wool cap	England (1846)
Rambouillet	R 250–300 E 150–200	MH	M	Fine	24.5–18.5	2–4	8–18	White	France
Romney	R 225–275 E 150–200	M	M	Long	38–31	4–6	8–12	White	England (1904)
Shropshire	R 225–250 E 150–180	MH	M	Medium	32.5–24.5	2.5–4	6–10	White with black face, wool cap	England (1855)
Southdown	R 190–230 E 130–180	M	H	Medium	29–23.5	1.5–2.5	5–8	White with grey face	England (1803)
Suffolk	R 250–350 E 180–250	H+	H	Medium	33–25.5	2–3.5	5–8	White with black face	England (1888)
Targhee	R 200–300 E 125–200	MH	M	Medium	26.5–21.5	3–4.5	10–14	White	U.S. (1926)

[a]L = low, M = medium, H = high; [b]Ewe basis; [c]Ramshorned, ewes, polled; [d]Either horned or polled; [e]Ram; [f]Ewe.

Texel

Dorset

Hampshire

Shropshire

Southdown

Suffolk

Dorper

Figure 30.1
Breeds of sheep commonly used as straightbreds or for crossbreeding to produce market lambs—Texel, Dorper, Dorset, Hampshire, Shropshire, Southdown, and Suffolk. Source: 30.01a: Ivonne Wierink/Shutterstock; 30.01b: Peter Dean/flpa.co.uk; 30.01c: Photomic/Fotolia; 30.01d: Paul Glendell / Alamy; 30.01e: Tim Scrivener / Alamy; 30.01f: Eric Isselee/Shutterstock; 30.01g; Tom Field.

Merino

Corriedale

Columbia

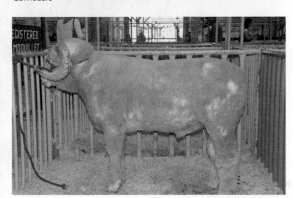
Rambouillet

Targhee

Figure 30.2

Breeds of sheep commonly used as straightbreds or for crossbreeding to improve wool production or maternal traits—Merino, Corriedale, Columbia (photo by USDA-NRCS), Rambouillet, and Targhee. Source: 30.2a: 169169/Fotolia; 30.2b: Courtesy of American Sheep Industry Association; 30.2c: USDA-NRCS; 30.2d: Courtesy of American Sheep Industry Association; 30.2e: USDA.

MAJOR GOAT BREEDS

Domesticated goats are classified into three types or uses: meat, fiber, and dairy production. The dominant breeds of each type in the United States are:

Meat goats—Spanish and Boer

Dairy goats—Alpine, LaMancha, Numbian, Oberhasli, Saanen, and Toggenburg

Fiber goats—Angora and Cashmere

The most numerous meat-type is the Spanish goat, which are able to breed out of season and are relatively small. Spanish goats make excellent range animals due to their superior hardiness and survival rates under low input management systems. Kid crops range from 75% in extensive, low-input systems to more than 150% for higher input management schemes. Spanish goats are variable in color and ear shape.

Figure 30.3
Boer goats have been imported from South Africa to improve meat production.
Source: TFoxFoto/Fotolia.

The Boer breed (Fig. 30.3) is fast becoming a popular meat goat. Developed in South Arica, the Boer is superior in growth rate and lean meat yield. Compared to other species, meat goats have lower dressing percentages, higher lean yield percentages, and less total carcass fat. Yearling male and females Boer goats average 100 and 85 lb, respectively. Adult male and female goats have average weights of 250 and 140 lb, respectively. Boers have a white body, cherry red head and neck, long, droopy ears and a convex shape to their head. A blaze face is preferred.

The six major dairy goat breeds registered in the United States are the Toggenburg, Saanan, Alpine, Nubian, LaMancha, and Oberhasli (Fig. 30.4). Breed characteristics and typical milk production benchmarks are presented in Table 30.2.

Angora goats (Fig. 30.5) produce mohair, which is a fiber that can be used to make into expensive cloth. Angoras are most prevalent in Texas where they thrive on browse typically found in rough, high brushrange sites. They prefer broadleaf plants over grass species and thus are well suited to the semiarid and arid regions of the southwest. Angora goats usually have horns that curve sideways. Mature bucks and wethers weigh 150–200 lb and mature does average 90–110 lb.

BREEDING SHEEP AND GOATS

The goal of purebred breeders (Fig. 30.6) is to make genetic progress in economically important traits. Breeders use selection as their method for genetic improvement, as crossbreeding is limited primarily to commercial producers. Purebred breeders with large operations may find it desirable to close their flocks and select ewe and ram replacements from within the closed flocks.

Normally, purebred breeders have selection programs to produce superior rams for commercial producers. At the same time, commercial producers prefer rams that contribute outstanding performance in a commercial crossbreeding program.

In general, great progress in sheep improvement can be made by selection within a breed for traits that are highly heritable and economically important. Crossbreeding can give significant genetic improvement in traits of low heritability such as fertility. Traits that are only moderately heritable can be improved by selecting genetically superior breeding animals and also by using well-designed crossbreeding systems.

Many of the traits of economic importance to sheep and goats enterprises are moderate to high in heritability. Reproductive traits tend to be lowest in heritability

Alpine

LaMancha

Nubian

Oberhasli

Saanen

Toggenburg

Figure 30.4
The six major goat breeds registered in the United States. Source: American Dairy Goat Association.

Table 30.2
CHARACTERISTICS OF DAIRY GOAT BREEDS

Breed	Origin	Size of Bucks and Does	Color	Ear and Face Structure	Milk (lb)	Fat (lb)	Protein (lb)
Alpine	France	B 175-220 D 110-200	Multiple	Erect, straight or dished	2,100	69	62
LaMancha	U.S.	B 140-180 D 110-140	Multiple	Very short, straight	1,800	66	55
Nubian	Orient	B 140-180 D 110-140	Multiple	Long and droopy, roman nose	1,750	66	52
Oberhasli	Switzerland	B 145-175 D 110-130	Bay with/ black markings	Erect, straight	1,850	64	52
Saanen	Switzerland	B 175-275 D 110-200	White	Erect, straight or dished	1,950	71	57
Toggenburg	Switzerland	B 140-180 D 110-130	Fawn to chocolate	Erect, straight or dished	1,700	57	47

Figure 30.5
A group of Angora goats showing a full fleece of mohair. Angora goats are useful for meat production, mohair production, and brush control. Source: Blackcurrent /Fotolia.

Figure 30.6
Purebred breeders use select for economically important traits to attain genetic improvement. Wool quality and production progress can be achieved because wool traits are typically highly heritable and can be measured objectively and accurately.
Source: Tom Field.

as is the case with other species. Given the influence of additive genetic effect on growth traits, wool and fiber yield and quality, as well as on carcass characteristics (Table 30.3), selection can create significant improvements in these traits.

An important growth trait in meat breeds of sheep is weight at 90 days of age, with some emphasis also given to conformation and finish of the lambs. Weight at 90 days of age can be calculated for lambs that are weaned at younger or older ages than 90 days as follows:

$$\textbf{90-day weight} = \frac{\text{weaning weight} - \text{birth weight}}{\text{age at weaning}} \times 90 + \text{birth weight}$$

Table 30.3
PERCENT HERITABILITY OF ECONOMICALLY IMPORTANT TRAITS IN THE SHEEP AND GOAT INDUSTRY

Trait	Sheep	Goats
Offspring per gestation	15	15
Weaning weight	20	25
Mature weight	40	50
Fleece/clip weight	40	30
Fiber diameter	35	15
Loineye area	35	35
Milk yield	10	30

If a lamb weighs 10 lb at birth and 100 lb at 100 days, the 90-day weight is:

$$\frac{100 \text{ lb} - 10 \text{ lb}}{100 \text{ days}} \times 90 \text{ days} + 10 \text{ lb} = 91 \text{ lb}$$

If birth weight was not recorded, an assumed birth weight of 8 lb can be used.

Approximately 85–90% of the total income from sheep of the meat breeds is derived from the sale of lambs, with only 10–15% coming from the sale of wool. By contrast, the sale of wool accounts for 30–35% of the income derived from fine-wool and long-wool breeds. Even with wool breeds, the income from lambs produced constitutes the greater portion of total income. Thus, most sheep breeders focus their attention on those traits with the greatest impact on revenue.

Production of profitable lambs requires that market lambs have a desirable degree of muscularity, level of fat thickness and yield carcasses with both a desirable level of cutability and quality. Figure 30.7 illustrates lambs of various degrees of cutability ranging from a USDA Yield Grade 1 to a USDA Yield Grade 5. The second and third lambs are the most desirable combination of muscle, ribeye, and quality. Producing lambs that meet market demand require both focused use of genetic tools as well as nutritional, health, and general management as described in Chapter 31.

The National Sheep Improvement Program was initiated in 1986 to create a unified genetic evaluation system for the industry to assist breeders in making effective. Genetic estimates are calculated for number of lambs born, litter weight at 60 days, individual weights at 30, 60, 90, 120, 180, or 360 days (three weights analyzed per flock), weight of the wool clip, wool fiber diameter, and staple length.

Sheep Reproduction

Sheep and goats (especially dairy breeds and Angoras) differ from many farm animals in having a breeding season that occurs mainly in the fall of the year. The length of the breeding season varies with breeds. Ewes of breeds with a long breeding season show heat cycles from mid- to late summer until midwinter. The breeds in this category are the Rambouillet, Merino, and Dorset. Breeds with an intermediate breeding season (Suffolk, Hampshire, Columbia, and Corriedale) start cycling in late August or early September and continue to cycle until early winter. Ewes of breeds having long or intermediate breeding seasons are more likely to fit into a program of three lamb crops in 2 years.

Ewes of breeds with short breeding seasons (Southdown, Cheviot, and Shropshire) do not start cycling until early fall and discontinue cycling at the end of the fall

Figure 30.7

Five lambs of various levels of muscularity, fatness, USDA Yield and Quality Grades. Producing profitable market lambs requires application of genetics and management.

Source: J. D. Tatum.

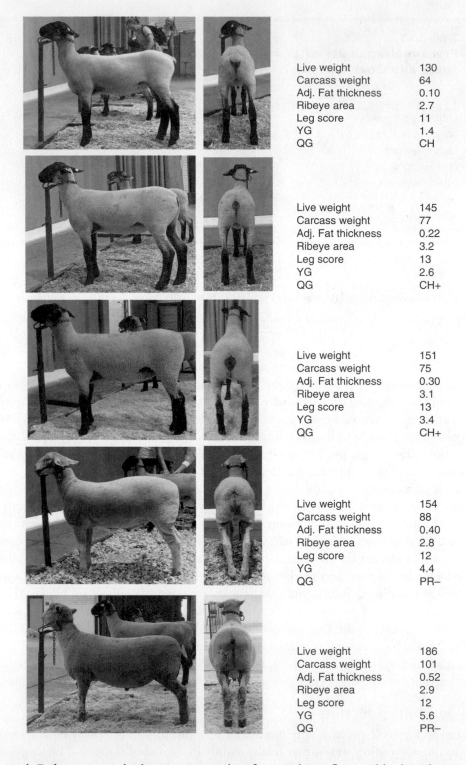

Live weight	130
Carcass weight	64
Adj. Fat thickness	0.10
Ribeye area	2.7
Leg score	11
YG	1.4
QG	CH

Live weight	145
Carcass weight	77
Adj. Fat thickness	0.22
Ribeye area	3.2
Leg score	13
YG	2.6
QG	CH+

Live weight	151
Carcass weight	75
Adj. Fat thickness	0.30
Ribeye area	3.1
Leg score	13
YG	3.4
QG	CH+

Live weight	154
Carcass weight	88
Adj. Fat thickness	0.40
Ribeye area	2.8
Leg score	12
YG	4.4
QG	PR−

Live weight	186
Carcass weight	101
Adj. Fat thickness	0.52
Ribeye area	2.9
Leg score	12
YG	5.6
QG	PR−

period. **Puberty** is reached at 5–12 months of age and is influenced by breed, nutrition, and date of birth. The average length of **estrous cycles** is slightly over 16 days, and the length of estrus (when the ewe is receptive to the ram) is about 30 hours. The length of gestation (time from breeding until the lamb is born) averages 147 days but varies; medium-wooled and meat breeds have shorter gestation periods, whereas fine-wooled breeds have longer gestation periods. Goats have comparable reproductive characteristics to sheep (Table 30.4).

Several factors affect **fertility** in sheep. Lambing rates vary both within and between breeds, with some ewes consistently producing only one lamb per year and

Table 30.4
REPRODUCTIVE CHARACTERISTICS OF SHEEP AND GOATS[a]

Characteristic	Sheep	Goats
Age at puberty	5–12 months	4–12 months
Estrous cycle duration	16 days	16 days
Length of estrus	30 hours	48 hours
Gestation length	147 days	150 days

[a]Age at puberty and gestation length are particularly affected by breed and season of year.

others producing three or four lambs each year. To improve lambing rate, producers keep replacements from ewes that consistently produce two to four lambs each year.

Examining the semen should confirm fertility of rams used in a breeding program. Semen can be collected by use of an artificial vagina and examined with a microscope. Ram fertility is evaluated by (1) checking for abnormal sperm (tailless, bent tails, no heads, and so on), (2) observing the percentage of live sperm (determined by a staining technique using fresh semen), and (3) checking the sperm motility and concentration in freshly collected semen. Two semen collections 3–4 days apart should be examined. When rams have not ejaculated for some time, there may be dead sperm in the **ejaculate**.

Goat Reproduction

Breeding season typically occurs in late September and October. Goats cycle at 18- to 21-day intervals. Does are typically bred the first time at 7 months of age or at a weight equivalent of 70 pounds. Angoras typically aren't bred for the first time until 18–24 months of age. As females come into heat they typically demonstrate behavioral signs such as bleating, tail flagging, reddening of the vulva, and vaginal discharge. Estrus can be induced by eliminating exposure to the buck for 21 days and then introducing the buck to the flock. The olfactory stimulation of this practice will induce at least a portion of the does into heat. The use of artificial lighting in January and February can be used to induce out of season breeding.

Most often, goats are mated using natural service but artificial insemination is an option. To assure high rates of pregnancy, semen should be deposited either directly into the uterus or deep into the cervix. Semen can be obtained from commercial sources in the form of 0.5 mL straws. Pregnancy detection via the use of real-time ultrasonography is highly accurate. Several hormonal pregnancy tests are available but require precise timing of sampling either serum or milk to attain high accuracy.

Kidding occurs following a gestation of approximately 150 days with 1 to 2 kids per first pregnancy. In subsequent pregnancies, dairy and meat goats typically average in excess of two kids per parturition. Angoras are more likely to have single offspring per pregnancy. Under most conditions kids are weaned at 5 months of age.

Angora goats have intermediate levels of fertility and losses of kids under range conditions can be high due to predation. A normal kid crop at weaning is approximately 80% of the number conceived.

Other Factors Affecting Reproduction of Sheep

Crossbreeding is one of the most advantageous tools available to enhance reproductive performance in sheep. Crossbred ewe lambs, when adequately fed, are usually bred to lamb at 1 year of age. Generally, crossbred lambs are more likely to conceive as lambs than are purebred lambs.

Age impacts reproductive rate as mature (3- to 7-year-old) ewes are more fertile and raise a higher percentage of lambs born than do younger or older ewes.

Light, temperature, and relative humidity also affect reproduction in sheep. Sheep respond to decreased day lengths both by showing a greater proportion of ewes in estrus and by higher conception rates. *Temperature* has a marked effect on both ewes and rams. High temperatures cause heat sterility in rams because the testicles must be at a temperature below normal body temperature for viable sperm production. Embryo survival in the ewe is also influenced by temperature. When the ambient temperature exceeds 90°F, embryo survival decreases during the first 8 days after breeding.

Environmental factors such as disease, parasites, and insufficient feed resources impact the health and well-being of sheep and when not correctly managed may reduce the number of lambs produced. Producers can increase their income and profit from a sheep operation by controlling diseases and parasites and by providing the sheep with an adequate supply of good feed. Sheep in moderate body condition are usually more productive than fat sheep. Ewes that are in moderate condition and that are gaining weight before and during the breeding season will have more and stronger lambs than ewes that are either overfat or **emaciated** at breeding.

Estrus synchronization and AI are tools that can be used to influence reproduction. In some intensively managed sheep operations, hormones can be used to synchronize estrus. Hormones can also be used to bring ewes into estrus at times other than during their normal breeding season.

Estrus can be synchronized if progesterone is given for a 12- to 14-day period in the feed, as an implant, via an intravaginal device, or by daily injections followed by injections of pregnant mare serum (PMS) the day progesterone is discontinued. Conception rates at this estrus are low; a second injection of pregnant mare serum 15–17 days later will bring the ewes into estrus and fertile matings will occur. There must be sufficient ram power to breed a large number of ewes naturally if they are synchronized, or artificial insemination can be used.

This system of synchronizing estrus can also be used to obtain pregnancies out of the normal breeding season and to obtain a normal lamb crop from breeding ewe lambs. Also, producers can use estrus synchronization for accelerated lambing if it is desired to raise two lamb crops per year.

If the ewes are to be artificially inseminated, high-quality fresh semen should be diluted with egg-yolk-citrate diluter shortly before insemination. Ram semen has been frozen, but conception rates have been low.

Estrogen content in feeds may result in a flock of sheep expressing very low fertility because of high estrogen content of the legume pasture or hay that the sheep are consuming. Producers are advised to have the legume hay or legumes in the pasture checked for estrogen if they are experiencing fertility problems in their sheep.

The Breeding Season

Sheep may be handmated or pasture mated. If they are **handmated**, some **teaser rams** are needed to locate ewes that are in heat. Either a **vasectomized** ram (each vas deferens has been severed so sperm are prevented from moving from the testicles) can be used, or an apron can be put on the ram such that a strong cloth prevents copulation.

Tagging ewes, or removing wool from the dock and vulva region before breeding, will result in a larger percentage of lambs as well as a lamb flock of more uniform age.

Ewes should be checked twice daily for **heat**. Ewes normally stay in heat for 30 hours and ovulate near the end of heat. It is desirable to breed ewes the morning after they are found in heat in the afternoon. Ewes that are found in heat in the morning should be bred in the late afternoon.

If ewes are to be handmated, the ram should breed them once. Recently bred ewes should be separated from other ewes for 1 or 2 days so that the teaser ram does not spend his energies mounting the same ewe repeatedly.

If ewes are to be **pasture mated**, they are sorted into groups according to the rams to which they are to be mated. It is best to have an empty pasture between breeding pastures so as not to entice rams to be with ewes in another breeding group.

Using numbered branding irons to apply scourable paint is a common means to identify sheep. The brisket of the breeding ram can be painted with scourable paint so that he marks the ewe when he mounts her. A light-colored paint should be used initially, and then a dark color can be used about 14 days later to detect ewes that return in heat. An orange paint can be used initially, followed, successively, with green, red, and black. The paint color should be changed each 16–17 days. Because the ewes are numbered, they can be observed daily and the breeding date of each can be recorded. Also, a ram that is not settling his ewes can be detected and replaced. Occasionally, a ram may be low in fertility even though his semen was given a satisfactory evaluation before the breeding season.

A breeding season of 40 days results in a lamb crop of uniform age and identifies ewes for culling that have an inherent tendency for late lambing.

Genetic Improvement in Commercial Sheep Production

Producers should consider the following issues when designing a breeding system for flock:

1. What are the available pasture and management resources?
2. What is the desired income percentage from the sale of lambs versus wool? (Wool and growth traits tend to be antagonistic.)
3. What are the potential benefits from heterosis?
4. How can breed differences best be utilized to meet the breeding and marketing goals of the flock?
5. What is the anticipated season of marketing?
6. How large is the flock? (Smaller flocks have fewer options than larger flocks.)

Producers must carefully balance the desire to take maximum advantage of heterosis (Table 30.5) with the need to optimize the breeds utilized to assure that the management system requirements of the flock are met. Despite its advantages, crossbreeding may not be desirable for small flocks or in flocks where the intensity of management is not sufficient to assure correct implementation of the crossbreeding plan.

Table 30.5
EFFECTS OF HETEROSIS IN CROSSBRED LAMBS

Trait	Level of Heterosis (%)
Birth weight	3
Weaning weight	5
Postweaning gain	7
Survival of weaning	10
Pounds of lamb weaned per exposed ewe	18

Source : Adapted from Schoenian, 2007.

Terminal Crossing

For example, in some extensive management conditions of the semiarid West, the use of the Rambouillet is quite common. The Rambouillet's fine-wool production, durability, and mature size are so desirable that the use of a ram from a complementary ewe breed is difficult to identify. However, the use of a black-faced breed such as the Hampshire or Suffolk is an effective terminal cross for these conditions. Under the more intensive management conditions of many farm flocks, the use of crossbreeding systems to produce ewe replacements is more common.

Most market lambs are crossbreds (produced either from crossing breeds or from mating crossbred ewes with purebred rams), but the key to the success of **crossbreeding** is the improvement in production traits made by the purebred breeders in their selection programs. Crossbreeding can be used to combine meat conformation of the sire breeds with lambing ability and wool characteristics of the ewe breeds, and also to obtain the advantage of the fast growth rate of lambs that result from heterosis. The additive as well as the heterotic effects is evident in crossbred lambs. Maximum **heterosis** is obtained in three- or four-breed crosses, either rotational or terminal.

A clear understanding of the market target and time of marketing is also useful. For example, if the spring or Easter lamb market is the target, the use of a sire breed with moderate mature size and excellent muscularity (such as the Southdown) bred to a highly fertile maternal breed (such as the Dorset) may be a desirable mating system. This system is effective because the resulting lambs finish at relatively low weights (70–80 lb) and this meets the requirements of the spring market. While the spring market has historically been a target, its importance as a niche market is declining.

Three-Breed Terminal Crossbreeding

The use of terminal crossing systems is particularly valuable in those production systems where fine-wool production is important. For example, in Australia, the Merino is the predominant wool-producing breed. While the Merino produces an outstanding quality of wool clip, Merino wethers are lightly muscled and poor in growth rate. A common crossbreeding system to produce better-quality market lambs has evolved (Fig. 30.8). In such a system, poorer-performing Merino ewes or older ewes in the flock are allocated to the process of producing F_1 females for mating to a terminal sire.

Utilizing breed differences in crossing systems has led producers to experiment with a variety of breed types. For example, the Finnsheep is highly prolific, often

Figure 30.8
Australian breeding system to produce prime lambs.

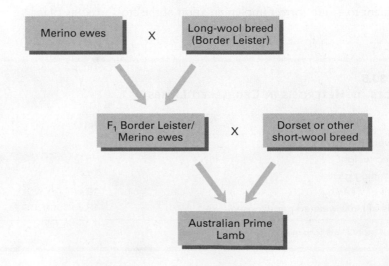

producing three to six lambs per lambing. Crossbred ewes of 25–50% Finnsheep genetics produce significantly more lambs than straightbred or crossbred ewes of the meat breeds. The use of Finnsheep in crossing programs was initiated by a desire to attain more rapid reproductive efficiency. Unfortunately, high-percentage Finnsheep market lambs lack the growth rate and muscularity required to be profitable. These producers utilizing Finnsheep have attempted to gain the efficiency improvements without the loss of performance by utilizing ¼ blood Finnsheep rams.

Some research shows that two lamb crops can be raised per year or, if feed and other conditions do not support such intensive production, three lamb crops every 2 years can be produced. To achieve intense production of this type, producers need ewes that will breed throughout the year. At present, the Polypay seem to fit into such a program fairly well. Also, the Rambouillet and Dorset have a tendency to breed out of season.

To raise three lamb crops in 2 years, ewes should be bred in late summer, probably for a 30-day period (August), for lambs to arrive in January (lamb crop 1). The ewes and lambs need to be well fed so the lambs are sufficiently finished to go to market at 90 days of age (in April). The ewes are then bred in April (30-day breeding period) to lamb in August (lamb crop 2), and the ewes and lambs need good feed so the lambs are well finished and sufficiently large to be marketed at 90 days of age (November). The ewes are bred in November to lamb in April (third lamb crop) and the lambs are marketed at 90 days of age in July.

CHAPTER SUMMARY

- Breeds of sheep are classified primarily by wool production and ewe traits (e.g., Rambouillet), meat production (e.g., Suffolk), or a combination of the two (dual-purpose—e.g., Columbia).
- Based on registration numbers, the most numerous sheep breed is the Suffolk, followed by the Dorset, Hampshire, Southdown, and Rambouillet.
- Goat breeds are classified as meat-type, dairy-type, or fiber-type.
- Under most conditions, multiple births (i.e., twinning) are highly desirable, with other economically important traits being growth rate, wool production, and carcass merit (combination of quality grades and yield grades).
- The breeding program for many large range sheep flocks involves using wool breeds or dual-purpose breeds of ewes to mate to ram breeds excelling in growth rate and carcass characteristics.

KEY WORDS

fine-wool breeds
medium-wool breeds
long-wool breeds
crossbred wool breeds
hair-type sheep
ewe breeds
ram breeds
dual-purpose breeds
Polypay
meat goat breed
dairy goat breeds
fiber goat breed
Angora goats
tight lock

flat lock
open fleece (fluffy)
Boer goats
90-day weight
puberty
estrous cycles
fertility
ejaculate
emaciated
handmated
teaser ram
vasectomized
tagging
heat

pasture mated
skin folds
crossbreeding

heterosis
three-breed terminal cross

REVIEW QUESTIONS

1. Compare strengths of sheep breeds based on wool quality and breeding purpose.
2. Compare goat breeds based on meat, dairy, and fiber type.
3. Evaluate heritability of various traits and discuss the impact on making genetic progress.
4. Describe the 90-day weight formula.
5. Describe the ideal market lamb.
6. Compare reproductive characteristics of sheep and goats.
7. Discuss impact of age, day length, and environmental factors on reproduction.
8. Discuss the questions that should be addressed before beginning a commercial sheep breeding program.
9. Describe an effective terminal crossing system.

SELECTED REFERENCES

Dickerson, G. E. 1978. *Crossbreeding Evaluation of Finnsheep and Some U.S. Breeds for Market Lamb Production.* North Central Regional Publication #246. ARS, USDA, and University of Nebraska.

Ensminger, M. E. 2002. *Sheep and Goat Science,* 6th edition. Danville, IL: Interstate Publishers.

Lasley, J. F. 1987. *Genetics of Livestock Improvement.* Englewood Cliffs, NJ: Prentice Hall.

Merck Veterinary Manual. 2012. Merck Sharp and Dohme Corp., Whitehouse Station, NJ.

Ross, C. V. 1989. *Sheep Production and Management.* Englewood Cliffs, NJ: Prentice Hall.

Schoenian, S. 2007. Breeding systems. www.sheep101. info. *The Sheepman's Production Handbook.* 1988. Denver, CO: Sheep Industry Development Program.

Wilson, D. E., and D. G. Morrical. 1991. The national sheep improvement program: A review. *J. Anim. Sci.* 69:3872.

31

Feeding and Managing Sheep and Goats

Sheep feeding and management are essential for the success of an operation. These areas must be integrated with knowledge of how breeding and environmental factors affect sheep productivity and profitability. Several areas of feeding and management that affect efficient production are discussed in this chapter. Sheep enterprises can generally be classified into one of three categories—farm flocks, range flocks, or lamb feeders. Farm flocks are typical of the sheep enterprises found in the Great Plains, the Great Lakes states, New England, and the southeastern states. Range flocks are typically found in the eleven western states while lamb feeding is concentrated in California, Colorado, and the Pacific Northwest.

PRODUCTION REQUIREMENTS FOR FARM FLOCKS

Pastures

Good pastures are essential to both range and farm flock operations (Fig. 31.1). Grass-legume mixtures, such as rye grass and clover or orchard grass and alfalfa, are ideal for sheep. Sheep can also graze on temporary pastures of such plants as Sudan grass or rape, which are often used to provide forage in the dry part of summer when permanent pastures may show no new growth. Crops of grain and grass seed may also provide pasture for sheep during autumn and early spring. Sheep do not trample wet soil as severely as do cattle; pasturing sheep on grass-seed and small-grain crops in winter and early spring when the soil is wet does not cause serious damage to the plants or soil.

A woven-wire or electric fence is necessary to contain sheep in a pasture. Forage can be best utilized by rotating (moving) sheep from one pasture to another, a practice that also assists in the control of internal parasites. Some sheep operators use temporary fencing (such as electric fencing) to divide pastures and for predator control. It is necessary to use two electrified wires, one located low enough to prevent sheep from going underneath the fence, and the other located high enough to prevent them from jumping over it.

Corrals and Chutes

It is occasionally necessary to put sheep in a small enclosure to sort them into different groups or to treat sick animals. Proper equipment is helpful in this task because sheep are often difficult to drive, particularly when ewes are being separated from their lambs. A **cutting chute**

Figure 31.1
Effective management of forages and pastures is critical to the success of the sheep enterprise.
Source: American Sheep Industry Association.

Figure 31.2
Moving sheep into a catch pen to be worked through a chute to be vaccinated.
Source: American Sheep Industry Association.

can be constructed to direct sheep into various small lots. A well-designed chute is sufficiently narrow to keep sheep from turning around and can be blocked so that sheep can be gathered together for such purposes as administering vaccines, applying parasite treatment, or reading ear tags (Fig. 31.2). Chutes should be properly spaced so that the correct width (14–16 in.) is provided. Pens used to enclose small groups of sheep can be constructed of woven-wire fencing, lumber, steel panels, or other suitable materials. A loading chute is used to place sheep onto a truck for hauling. A portable loading chute is ideal because it can be moved to different locations where loading and unloading sheep are necessary.

Shelters

Sheep do not normally suffer from cold because they have a heavy wool covering; therefore, open sheds are excellent for housing and feeding wintering ewe lambs, pregnant ewes, and rams. Although the ewes can be lambed in these sheds, the newborns need an enclosed and heated room when the weather is cold. However, sheep are vulnerable to changes in the weather following shearing. Allowances should be made to minimize stress following shearing especially in wet conditions.

Lambing Equipment

When weather permits or when natural protection is available, pasture lambing is often possible. However, small pens, about 4 × 4 ft, can be constructed for holding ewes and their newborn lambs until they are strong enough to be put with other ewes and lambs. The lambing pens (**lambing jugs**) should be equipped with heat lamps to facilitate keeping newborn lambs warm. A heat lamp above each lambing jug can be located at a height that provides a temperature of 90°F at the lamb's level.

It is advisable to identify each lamb and to record which lambs belong to which ewes. Often, a ewe and her lambs are branded with scourable paint so as to not negatively affect wool quality as a means of identification. Numbered ear tags can be applied to the lamb at birth. If newborn lambs are to be weighed, a dairy scale and a large bucket are needed. The lamb is placed in the bucket, which hangs from the scale, for weighing. It is advisable to immerse the umbilical cords of newborn lambs in a tincture of iodine to avoid infection.

Feeding Equipment

Usually sheep are given hay during the winter feeding period. Feeding mangers should be provided because hay is wasted if it is fed on the ground. Sheep may require limited amounts of concentrates. Concentrates can be fed in the same bunk as hay if the hay bunk is properly constructed, or separate feed troughs may be preferable.

Additional concentrates can be provided for lambs by placing the concentrates in a **creep**, which is constructed with openings large enough to allow the lambs to enter but small enough to keep out the ewes. In addition to concentrates, it is advisable to provide good-quality legume hay and a mineral mix containing calcium, phosphorus, and salt. Creep diets are balanced for needed nutrients, but need not be complex in nature. Wheat or barley could replace up to half of the corn or sorghum (milo) in these diets. Lambs normally eat 1.5 lb of creep diet daily from 10–120 days of age (0.10 lb at 3 weeks and 3.0 lb at 120 days).

Water is essential for sheep and providing easy access to clean, freshwater should be a priority. Water can be provided by tubs or automatic waterers. Tubs should be cleaned once each week.

A separate pen equipped with a milk feeder may be needed for orphan lambs. Milk or a milk replacer can be provided free-choice if it is kept cold. Heat lamps kept some distance from the milk feeder can be provided to help young lambs maintain critical body temperature.

TYPES OF FARM FLOCK PRODUCERS

Some producers raise purebred sheep and sell rams for breeding. Commercial producers whose pastures are productive can produce slaughter lambs on pasture, whereas those whose pastures are less desirable produce feeder lambs. Lambs that are neither sufficiently fat nor sufficiently large for harvest at weaning time usually go to feedlot operators where they are fed to market weight and condition. However, most feedlot lambs are obtained from producers of range sheep.

Purebred Breeder

Purebred sheep are usually given more feed than commercial sheep because the **purebred breeder** is interested in growing sheep that express their growth potential. Records are essential to indicate which ram is bred to each of the ewes and which ewe is the mother of each lamb. All ram lambs are usually kept together to compare and identify those most desirable for sale. The purebred breeder usually retains the ram

Table 31.1
EXAMPLE DIETS FOR FINISHING EARLY-WEANED LAMBS IN DRYLOT

Ingredient	Ration Number and % of Diet	
	1	2
Ground corn (shelled)	37.3	—
Cracked corn	37.4	—
Ground corn (ear)	—	76.9
Cottonseed hulls	10.0	—
Soybean meal, 48%	12.8	15.6
Molasses	—	5.0
Ground limestone	1.0	1.0
Trace mineral + selenium	1.0	1.0
Ammonium chloride	0.5	0.5
Vitamin premix[a]	+	+

[a]According to manufacturer's instructions.
Source: Adapted from *Sheep Production Handbook*.

lambs until they are a year of age, at which time they are offered for sale. Buyers usually seek to obtain rams from purebred breeders in early summer.

Purebred breeders have major responsibilities to the sheep industry. Because purebred breeders determine the genetic productivity of commercial sheep, they should rigidly select animals that are kept for breeding and should offer for sale only those rams that will contribute improved productivity for the commercial producer.

Commercial Market Lamb Producers

The lamb producer whose feed and pasture conditions are favorable aims for lambs to be born in late winter and early spring and to be weaned at 60–70 days of age. Weaned lambs from this system typically weigh 60 lb. These lambs are often penned and fed to gain 0.7 lb per day, with a market weight target of 120 lb. Access to a creep-fed diet in the pre-weaning phase is advisable.

Lambs can be finished on pasture or in a combined pasture-supplement feeding scenario. However, gain will be less than it would be for lot-fed lambs. Example diets for feeding early-weaned lambs are provided in Table 31.1.

Commercial producers can gain some assurance of raising heavy, well-finished market lambs at weaning by using good ram selection and crossbreeding programs. Rams that are heavy at 90 days of age are more likely to sire lambs that are heavy at this age than are rams that are light in weight at 90 days of age. The breeds used in a three-breed rotation should include one that is noted for milk-producing ability, one that is noted for rapid growth, and one that is noted for ruggedness and adaptability. The Dorset could be considered for milk production, either the Hampshire or Suffolk for rapid growth, and the Cheviot for ruggedness. All of these breeds make desirable carcasses.

Commercial producers often castrate all ram lambs unless the management plan is designed to market lambs at lighter weights and younger ages. Young ewes that are selected to replace old ewes should be structurally sound, healthy, originate from prolific dams, and have sufficient growth rate and frame size. Mature ewes that have started to decline in production and poorly productive ewes should be culled.

Commercial Feeder Lamb Producers

Some commercial sheep are produced under pasture conditions (Fig. 31.3) that are insufficient for growing the quantity or quality of feed needed for producing heavy slaughter lambs at weaning. Lambs produced under such conditions may be either fed out in the summer or carried through the summer on pasture with their dams and finished in the fall. Sudan grass or rape can be seeded so that good pasture is available in the summer, and lambs can be finished on pasture by giving them some concentrate feeds. If good summer pasture cannot be made available, it is advisable to wait until autumn, at which time the lambs are put on full feed in a feedlot.

Commercial Feedlot Operator

Lambs that come to the feedlot for finishing are typically processed upon arrival at the feedlot. Lambs are treated for internal **parasites**, and vaccinated (e.g., **overeating disease**). They are provided water and hay initially, after which concentrate feeding is introduced and increased until lambs are on full feed (Fig. 31.4).

Figure 31.3
Ewes grazing crop aftermath. Profitability depends on the ability of managers to find least cost rations and feedstuffs that meet animal requirements.
Source: Tom Field.

Figure 31.4
Sheep eating corn and pellets from a feed bunk.
Source: American Sheep Industry Association.

Table 31.2
EXAMPLE RATIONS FOR GROWING AND FINISHING LAMBS

| Ingredient | Ration Number and % of Diet | | | | | |
| | Up to 70 lb | | 70–90 lb | | 90 lb to Market | |
	1	2	1	2	1	2
Corn (ground/cracked)	52.0	—	62.0	30.0	72.0	60.5
Corn (ground ear)	—	60.5	—	30.5	—	—
Corn (ground cobs)	20.0	—	10.0	—	—	—
Soybean meal, 48%	10.5	5.5	10.5	5.5	10.5	5.5
Dehydrated alfalfa	10.0	—	10.0	—	10.0	—
Alfalfa hay (ground)	—	27.5	—	27.5	—	27.5
Molasses	5.0	5.0	5.0	5.0	5.0	5.0
Dicalcium phosphate	1.0	—	1.0	—	1.0	—
Trace mineral salt + selenium	1.0	1.0	1.0	1.0	1.0	1.0
Ammonium chloride	0.5	0.5	0.5	0.5	0.5	0.5

Source: Adapted from *Sheep Production Handbook.*

The feedlot operator hopes to profit by efficiently increasing the lambs' weight. Lambs on feed should gain about 0.5–0.8 lb per head per day. Feeder buyers prefer feeder lambs weighing 80–90 lb over larger lambs because of the need to put 30–40 lb of additional weight on the lambs to finish them. Many feeders target a final weight of 130 lb. Table 31.2 shows some example diets for growing-finishing lambs.

FEEDING EWES, RAMS, AND LAMBS

Figure 31.5 shows the expected weight changes in a 160-lb ewe raising twin lambs. A ewe of similar weight, raising a single lamb, would have about two-thirds of the weight changes shown in Figure 31.5. Economical feeding programs are implemented to correspond with these expected weight changes.

Mature, pregnant ewes usually need nothing more than lower-quality roughages (e.g., hay, wheat straw, or corn stover) during the first half of pregnancy, after which some of the ration needs to be improved with good-quality legume hay and if affordable some

Figure 31.5
Weight changes normally expected during the year for a 160-lb ewe giving birth to and raising twin lambs. Source: Sheep Production Handbook, American Sheep Industrial Association, Volume 7, 2002.

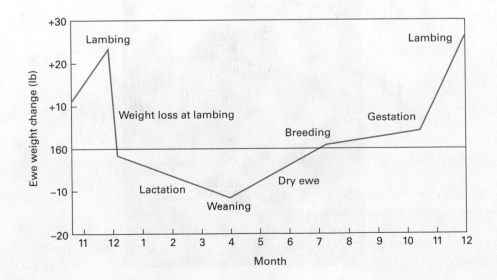

concentrates. Grains such as corn, barley, oats, milo, and wheat are satisfactory feeds. Sheep usually chew these grains sufficiently so that grinding or rolling is not essential. Rolled or cracked grains may be digested slightly more efficiently and may be more palatable, but finely ground grains are undesirable for sheep unless the grains are pelleted.

Sheep need energy, protein, salt, iodine, phosphorus, and vitamins A, D, and E. In some areas selenium is deficient and must be supplied either in the feed or by injection. Mature, ruminating sheep have little need for quality protein or B vitamins, but immature sheep have variable requirements. Rumen microorganisms can synthesize protein from nonprotein nitrogenous substances in the ration.

Sheep will perform poorly or die quickly when the water supply is inadequate, so the importance of supplying water cannot be overstated. Water that is not clean will not be well accepted by sheep. Water intake is influenced by level of feed intake, protein content of the ration, temperature, mineral intake, water temperature, odor and taste; pregnancy status; and the water content of feed consumed (including rain and dew on pastures).

Energy is perhaps the most common limiting nutrient for ewes. Dry range grasses and mature forages such as grain straws may be high in gross energy but so low in digestibility that sheep cannot obtain all their energy needs from them.

The amount of protein in the ration for sheep is of greater importance than the quality of protein because sheep can make the essential amino acids by the action of microorganisms in the rumen. The oil meals (soybean, linseed, peanut, and cottonseed) are all high in protein. Soybean meal is the most palatable and its use in a ration encourages a high feed intake. Nonprotein sources of nitrogen such as urea can be used to supply a portion, but not all, of the nitrogen needs of sheep. Not more than one-third of the nitrogen should be supplied by urea or biuret, and these materials are not recommended for young lambs with developing rumens or for range sheep on low-energy diets. If protein intake is limited by mixing with salt, adequate water must be provided. The mineral needs for sheep are calcium, phosphorus, sulfur, potassium, sodium, chlorine, magnesium, iron, zinc, copper, manganese, cobalt, iodine, molybdenum, and selenium. All of these minerals are found in varying amounts in different tissues of the body. For example, 99% of the calcium, 80–95% of the phosphorus, and 70% of the magnesium occur in the skeleton and more than 80% of the iodine is found in the thyroid gland.

Breed, age, sex, and growth rate of young animals; reproduction and lactation of the ewe; level and chemical form injected or fed; climate; and balance and adequacy of the ration influence the requirement for minerals.

Salt is important in sheep nutrition and it can become badly needed when sheep are grazing on lush pastures. Sheep may consume too much salt when they are forced to drink brackish water.

In some areas, iodine and selenium are deficient. The use of iodized salt in an area where iodine is deficient may supply the iodine needs. Selenium can be added to the concentrate mixture that is being fed or it can be given by injection.

The most critical periods of nutritional needs for the ewe are at breeding and just before, during, and shortly after lambing. A lesser but still important period for the ewe is during lactation. After the lamb is weaned, the ewe normally can perform satisfactorily on pasture or range without any additional feed. If accelerated lamb production is practiced, the ewe should be well fed after the lamb is weaned so the ewe can be bred again.

Example diets for ewes are shown in Table 31.3. Note how the amount of feed changes from maintenance to gestation to lactation. In both the gestation and lactation periods, the maintenance requirement must be met first.

Ewes are normally run on pasture or range where they obtain all their nutritional needs from grasses, browse, and forbs, except during the winter or dry periods when these plants are not growing. During these periods when there is no plant

Table 31.3
EXAMPLE DIETS (AS FED) FOR 155-lb EWES AT DIFFERENT STAGES OF PRODUCTION

Stage of Production	Ration No.	Alfalfa Hay (midbloom/lb)	Corn Silage (mature/lb)	Corn Grain (lb)	Soybean Meal (44% CP/lb)	Salt/Trace Mineral Mix (lb)[a]
Maintenance	1	3.0	—	—	—	0.05
	2	—	6.0	—	0.20	0.05
Gestation	1	3.5	—	—	—	0.05
(first 15 weeks)	2	—	6.0	—	0.25	0.05
Gestation						
(last 4 weeks)						
130–150% lambing	1	3.5	—	0.75	—	0.05
	2	—	6.0	0.75	0.40	0.05
180–225% lambing	1	3.5	—	1.25	—	0.05
	2	—	7.0	1.00	0.50	0.05
Lactation						
(first 6–8 weeks)	1	4.0	—	2.00	—	0.05
Suckling single	2	—	9.0	1.00	0.85	0.05

[a]Contains 50% trace mineral salt (for sheep) and 50% dicalcium phosphate.
Source : *Sheep Production Handbook*, 1988.

Table 31.4
BODY CONDITION SCORES FOR EWES

Score	Spinous Process[a]	Transverse Process[a]	General Description
1	Very prominent	Clearly protrude	Very thin, no fat cover
2	Prominent, but smooth	Rounded	Thin, but skeletal features do not protrude
3	Rounded, but smooth	Rounded, but smooth	Evidence of fat along rib, shoulder, back; hips protrude
4	Barely evident	Cannot be palpated	Moderate fat deposition throughout
5	Not detectable	Not detectable	Excessively fat

[a]Should be palpated down the spine, along the loin and over the rib.

growth, the sheep must be given supplemental feed. Monitoring **body condition score** is an efficient means to determine if ewes are receiving the appropriate level of feed (Table 31.4). Table 31.5 outlines target body condition scores for various phases of the production process.

Occasionally lambs are raised on milk replacer or on cow's milk if the ewe dies during lambing or if her udder becomes nonfunctional. During these two periods, frozen colostrum should be available for the lambs. If milk or milk replacer is used, it can be bottle-fed twice daily or it can be self-fed if it is kept cold. It is important to keep milk cold when it is self-fed to prevent over consumption. There should be heat lamps not far from where the lambs consume the cold milk so they can go to the heated area to become warm and to rest.

Rams should be fed to keep them healthy but not fat. A small amount of grain along with good-quality hay satisfies their nutritional needs in winter. Young bred ewes should be fed some grain along with all the legume hay they will consume because they are growing and also need some nutritional reserve for the subsequent lactation. Ewe lambs should be grown out but not fattened in the winter. Limited grain feeding along with legume hay satisfies their nutritional needs.

Table 31.5
DESIRED BODY CONDITION SCORES AT VARIOUS PRODUCTION PHASES

Production Phase	Desired Body Condition Score
Dry ewe (116–176 days)	1.5–2.0
Breeding (35–52 days)	2.5–3.0
Early gestation (first 15–17 weeks)	2.0–2.5
Late gestation[a] (last 4–6 weeks)	2.5–3.0
Early lactation[a] (first 6–8 weeks)	3.0–3.5
Late lactation, weaning (last 4–6 weeks)	2.0–2.5

[a]Add 0.5 for ewes pregnant with or nursing twins.
Source: Adapted from *Sheep Production Handbook*.

At lambing time, the grain allowance of ewes needs to be increased to assist them in producing a heavy flow of milk. Also, the lambs need to be fed a high-energy ration in the creep. Lambs obtain sufficient protein in the milk given by their mothers, but they need more energy. Rolled grains provided free choice in the creep are palatable and provide the energy needed.

Lambs on full feed are allowed some hay of good quality and all the concentrates they will consume. Some feedlot operators pellet the hay and concentrates, whereas others feed loose hay and grains. After the grasses and legumes start to grow in the spring, all sheep generally obtain their nutritional needs from the pasture.

MANAGEMENT OF FARM FLOCKS

Several systems of lamb production are available to sheep enterprises that follow the farm flock model: grass lambs, Easter lambs, and hothouse lambs. **Grass lamb** production systems require access to superior pastures capable of providing the nutrients to sustain rapid growth to market weights. The importance of the milk obtained via nursing in these systems is also critical to attaining desired market outcomes. **Easter lambs** are marketed under a relatively wide range of weights and conditions with some markets preferring lean, light lambs between 20 and 30 pounds while other niches prefer heavier lambs with a higher degree of fatness. **Hothouse lambs** are produced out of season (December to April) and are usually focused on east coast market outlets. Lambs are born in fall or early winter and are marketed within 2 to 3 months of age weighing between 30 and 60 pounds. These systems rely on ewe breeds capable of breeding out of season such as the Dorset that are mated to terminal-type rams. Hothouse production systems require ewes that are heavy milkers plus effectively feeding lambs under quality housing systems so that narrow, high quality carcass targets are met.

Handling Sheep

The correct handling of sheep is important. A sheep should never be restrained by its wool because the skin is pulled away from the flesh, causing a bruise. When a group of sheep is crowded into a small enclosure, the sheep will face away from the person who enters the enclosure. When the sheep's rear flank is grasped with one hand, the sheep starts walking backward; this allows the handler to reach out with the other hand and grasp the skin under the sheep's chin. When a sheep is being held, grasping the skin under the chin with one hand and grasping the top of the head with the other hand enables the holder to pull the sheep forward so its brisket is against the holder's knee. If a sheep is to be moved forward, the skin under the chin can be grasped with one hand and the dock (the place where the tail was removed) can be grasped with the other hand. Putting pressure on the dock makes the sheep move forward, while holding its chin with the other hand prevents it from escaping.

Lambing Operations

Before the time the ewes are due to lamb, wool should be clipped from the dock, udder, and vulva regions. This process is called **crutching** or *tagging*. If weather conditions permit, the ewes may be completely shorn. Also, all **dung tags** (small pieces of dung that stick to the wool) should be clipped from the rear and flank of pregnant ewes. Young lambs will try to locate the teat of the ewe and may try to nurse a dung tag if it is present.

Ewes should be checked periodically to locate those that have lambed. The ewe and newborn lamb should be placed in a lambing jug. A ewe that is about ready to lamb can be watched carefully but with no interference if delivery is proceeding normally. In a normal presentation, the head and front feet of the lamb emerge first. If the butt of the lamb emerges first with the rear legs tucked underneath the body (**breech presentation**), assistance may be needed if delivery is slow because the lamb can suffocate if deprived of oxygen for too long. If the front feet are presented but not the head, the lamb should be pushed back enough to bring the head forward for presentation.

As soon as the lamb is born, membranes or mucus that may interfere with its breathing should be removed. When the weather is cold, it may be necessary to dry the lamb by rubbing it with a dry cloth. If a lamb becomes chilled, it can be immersed in warm water from the neck down to restore body temperature, after which it should be wiped dry. The lamb should be encouraged to nurse as soon as possible. A lamb that has nursed and is dry should survive without difficulty if a heat lamp is provided. An ear tag, a tattoo, or both can identify the lamb.

Some ewes may not want to claim their lambs. It may be necessary to tie a ewe with a rope halter so she cannot butt or trample her lamb.

Castrating and Docking

Because birth is a period of stress for lambs, it is best to wait 3–4 days before castrating and **docking** them. To use the elastrator method of castration and docking, a tight rubber band is placed around the scrotum above the testicles (for castration) and around the tail about an inch from the buttocks (for docking). Some death losses can occur when tetanus-causing bacteria invade the tissue where the elastrator was applied. Another castration and docking practice is to remove the testicles and tail surgically. The emasculator is also useful for docking; the skin of the tail is pulled toward the lamb, the emasculator is applied about an inch from the lamb's buttocks, and the tail is cut loose next to the emasculator. A fly repellent should be applied around any wound to lessen the possibility of **fly strike** (fly eggs are deposited during warm weather).

Occasionally, ewes develop a vaginal or uterine **prolapse** (protrusion of the reproductive tract to the outside through the vulva). This condition is extremely serious and leads to death if corrective trained personnel do not take appropriate measures. The tissue should be pushed back in place, even if it is necessary to hoist the ewe up by her hind legs as a means of reducing pressure that the ewe is applying to push the tract out. After the tract is in place, the ewe can be harnessed so that external pressure is applied on both sides of the vulva. In some cases, it may be necessary to suture the tract to make certain that it stays in place. Recurrence of prolapse is likely so affected ewes should be culled.

Shearing

Sheep are usually shorn in the spring, before the hot weather months. Sheep should be kept off feed and water for 6–12 hours prior to shearing to reduce gut fill and to help reduce skin cuts. Only dry sheep should be shorn. Professional personnel typically do

Figure 31.6
Raw wool being hydraulically packed into bales. Wool of a specific grade range is packed together to help ensure the uniformity of each bale.
Source: Tom Field.

shearing, though shearing classes are available to teach the novice operator. The usual method of shearing involves clipping the fleece from the animal with power-driven shears, leaving sufficient wool covering to protect the sheep's skin. In the shearing operation, the sheep is set on its dock and cradled between the shearer's legs, which are used to maneuver the sheep into the positions needed to ease the shearing.

A fleece that has been properly clipped will remain in one large piece. It is spread with the clipped side out, rolled with the edges inside, and tied with paper twine. It is best to remove dung tags and coarse material that is clipped from the legs and put these items in a separate container. The tied fleece is put into a huge sack (Fig. 31.6). When buyers examine the wool, they can obtain core samples from the sack. The core sample is taken by inserting a hollow tube that is sharp at the end into the sack of wool. The sample obtained is examined to evaluate the wool in the sack rather than having to remove the fleeces to examine them. If undesirable material is obtained in the core sample, the price offered will be much lower than if only good wool is found.

If the wool is kept for some time before it is sold, it should be stored in a dry place and on a wooden or concrete floor to avoid damage from moisture.

Sheep that are nicked during shearing should be treated to avoid infection. Immediately following shearing, sheep should be placed into clean, dry pens and given protection from cold, wet, windy, or hot conditions.

FACILITIES FOR PRODUCTION OF RANGE SHEEP

Sheep differ from cattle in that they more readily graze weedy plants and brush as well as grasses and legumes. Because of their different grazing patterns, cattle and sheep can be effectively grazed together in some range areas. Total pounds of live weight produced can be higher than when the species are grazed separately on the same range.

Range sheep are produced in large flocks primarily in arid and semiarid regions. Sheep of fine-wool breeding tend to stay together as they graze, which makes herding possible in large range areas. Range sheep are moved about either in trucks or by trailing so they can consume available forage at various elevations. Requirements for the production of range sheep are usually different from those for farm flock operations. One type of range sheep operation is described here, but it must be noted that variations exist.

Range sheep are usually bred to lamb later than sheep in farm flocks; therefore, they can be lambed on the range. Few provisions are needed for lambing when sheep are lambed on the range, but under some conditions a tent or a lambing shed may be used to give range sheep protection from severe weather at lambing time. Temporary corrals can be constructed using snow fences and steel posts when it is necessary to contain the sheep at the lambing camp or at lambing sheds.

Sheep on range are usually wintered at relatively low elevations in areas where little precipitation occurs. Wintering sheep are provided **feed bunks** if hay is to be fed, and windbreaks to give protection from cold winds. Some producers of range sheep provide pelleted feed to supplement the winter forage. Pellets are usually placed on the ground but are sometimes dispersed in grain troughs.

A *sheep camp* or *sheepwagon* is the mobile housing used by the sheep-herder. The camp is moved by truck or by horses because the sheep need to be moved over large grazing areas. A sheepherder usually has a horse and dogs to assist in herding the sheep. The sheep are brought together to a night bedding area each evening.

MANAGING RANGE SHEEP

Range sheep are grazed in three general areas: (1) the **winter headquarters**, which is a relatively low and dry area and sometimes provides forage for winter grazing; (2) the **spring–fall range**, which is a somewhat higher elevation area and receives more precipitation; and (3) the **summer grazing** area, which is at high mountainous elevations and receives considerable precipitation, resulting in lush feeds.

The Winter Headquarters

The forage on the winter range, where there is usually less than 10 in. of precipitation annually, is composed of sagebrush and grasses. The grasses are cured on the ground from the growth of the previous summer; consequently, winter forage is of lower quality than green forage because the plants in the winter forage are mature and because they have lost nutrients. The soil in these areas is often alkaline and the water is sometimes alkaline.

Because forage in the wintering area is of poor quality, supplemental feeding that provides needed protein, carotene, and minerals (such as copper, cobalt, iodine, and selenium) is usually necessary. A pelleted mixture made by mixing sun-cured alfalfa leaf meal, grain, solvent-extracted soybean or cottonseed meal, beet pulp, molasses, bone meal or dicalcium phosphate, and trace-mineralized salt is fed at the rate of 0.25–2.0 lb per head per day, depending on the condition of the sheep. The feed may be mixed with salt to regulate intake so that feed can be placed before the sheep at all times. If the intake of feed is to be regulated through the use of salt, trace-mineralized salt should be avoided. Adequate water must also be provided at all times, because heavy salt intake is quite harmful if sheep do not have water for long periods of time.

The ewes are brought to the winter headquarters about the first of November. Rams are turned in with the ewes for breeding in November if lambing is to take place in sheds or if a spring range that is not likely to experience severe weather conditions is available for lambing. Otherwise, the rams are put with the ewes in December for breeding.

January, February, and March are critical months because the sheep are then in the process of exhausting their body stores and because severe snowstorms can occur. If sheep become snowbound, they should each be given 2 lb of alfalfa hay plus 1 lb of pellets per day containing at least 12% protein. Adequate feeding of ewes while they are being bred and afterward results in at least a 30% increase in lambs produced and about a 10% increase in wool produced. In addition, death losses are markedly reduced. Sheep that are stressed by inadequate nutrition, either as a result of insufficient feed or a ration that is improperly balanced, are highly susceptible to pneumonia and resulting heavy death losses.

The Spring–Fall Range

Pregnant ewes are shorn at the winter headquarters (usually in April). They are then moved to the spring–fall range, where they are lambed. If they are lambed on the range, a protected area is necessary. An area having scrub oak or big sagebrush on the south slopes of foothills and ample feed and water is ideal for range lambing.

Ewes that have lambed are kept in the same area for about 3 days until the lambs become strong enough to travel. The ewes that have lambs are usually fed a pelleted ration that is high in protein and fortified with trace-mineralized salt and either bone meal or dicalcium phosphate. Feeding at this time can help prevent sheep from eating poisonous plants.

Ewes with lambs are kept separate from those yet to lamb until all ewes have borne their lambs. In addition, ewes that are almost ready to lamb are separated from those that will not lamb for some time yet. Thus, after lambing gets under way, three separate groups of ewes are usually present until lambing is completed.

If the ewes are bred to lamb earlier than is usual for range lambing and a crested wheat-grass pasture is available, ewes may be lambed in open sheds. If good pasture is unavailable, the ewes may be confined in yards around the lambing sheds, starting a month before lambing. In this event, the ewes must be fed alfalfa or other legume hay and 0.50–0.75 lb of grain per head per day. The ewes should have access to a mixture of trace-mineralized salt and bone meal or dicalcium phosphate.

Although shed lambing is more expensive than range lambing, higher prices for lambs marketed earlier have made shed lambing advantageous. Fewer lambs are lost in shed lambing, and lambing can take place earlier in the year. The heavy market lambs that result produce enough income to offset the costs of shed lambing.

Summer Grazing

The Summer Range   Sheep are moved to the summer range shortly after lambing is completed if weather conditions have been such that snow has melted and lush plant growth is occurring. The sheep are put into bands of about 1,000–1,200 ewes and their lambs. In some large operations, the general practice is to put ewes with single lambs in one band and ewes with twins in another. The ewes with twin lambs are given the best range area so the lambs will have added growth from the better forage supply. Sheep are herded on the summer range to assist them in finding the best available forage.

In mid-September to October, prior to the winter storms, the lambs are weaned. Colorado producers on the range expect to exceed a 130% lamb crop. Lambs that carry sufficient finish are sent to slaughter and other lambs are sold to lamb feeders. It is the general practice among producers of Rambouillet, Columbia, and Targhee sheep to breed some of the most productive and best wooled ewes to rams of the same breed to raise replacement ewe lambs. Most of these ewe lambs are kept and grown out, and only the less desirable ones are culled. The remainder of the ewes are bred to meat-type rams, such as the Suffolk or Hampshire, and all their lambs are marketed for slaughter as feeders or as stockers (animals used in the flock for breeding).

The Fall Range   As soon as the lambs are weaned, the ewes are moved to the spring–fall range. Later they go to the winter headquarters for wintering.

The number of ewes that can be bred per ram during the breeding period of about 2 months is 15 for ram lambs, 30 for yearling rams, and 35 for mature, but not aged, rams. These numbers are general and depend greatly on the type of conditions existing on the range.

MANAGING GOATS

Goat management protocols are largely dependent on region and enterprise purpose (meat, dairy, fiber). Texas is the dominant state for goat enterprises (Table 31.6).

Simple housing for dairy goats is adequate in areas where the weather is mild because goats do best when they are outside on pasture or in an area where they can exercise freely. If the enterprise is located where the climate is characterized by frequent precipitation then an open roofed shed provides important shelter. Lactating goats can be fed in an open shed in stormy weather and taken to the milking parlor for milking. If the weather is severely cold, an enclosed barn with ample space is needed. At least 20 square feet per goat is needed if goats are to be housed in a barn, but 16 square feet per goat may suffice in open sheds.

Most meat and fiber goats thrive in pastoral or range conditions so constructed shelters might only be required in those regions where wet, cold conditions might be likely during kidding or at weaning time.

Fencing

It has been said that it is not a question of if goats will escape confinement but when. Thus good fencing is particularly important. Fences must be properly constructed with woven wire approximately 6 ft high and with 6-in. stays every 10–12 in. Particular attention should be given to those areas where uneven terrain makes it difficult to prevent gaps of sufficient size to allow goats to crawl underneath. Fencing for goats is different from fencing for cows; a woven-wire fence is best for containing goats because they can climb a rail fence. In addition, a special type of bracing at corners of the fence is necessary because goats can walk up a brace pole and climb over the fence.

With electric fences, two offset electric wires should be used at 8 in. high and 6–8 in. away from the fence. Fences should be charged to a minimum of 4,000 volts. However, an electric fence may not contain bucks of any breed; some may go through the fence despite the shock.

Another approach is to use an 8-strand barbed wire fence or a 5-strand electrified high tensile fence system. Four by four woven wire also makes an acceptable barrier. Woven-wire fences larger than these are not acceptable as they are big enough to allow goats to get their heads stuck.

Table 31.6
LEADING STATES FOR GOAT PRODUCTION IN THE UNITED STATES

All	(1,000 hd)	Meat	(1,000 hd)	Dairy	(1,000 hd)	Angora	(1,000 hd)
TX	965	TX	870	WI	46	TX	75
CO	125	TN	120	CA	38	AZ	23
OK	101	OK	95	IA	30	NM	10
MO	95	CA	87	TX	20	CA	3.4
WI	69	KY	72	NY	13	OR	2.3
U.S.	2,761	U.S.	2,275	U.S.	355	U.S.	131

Source: USDA: NASS.

Four-foot-tall fenced catch pens are useful for handling goats for routine procedures such as sorting, vaccinations, and foot trimming. Working chutes, alleys, and a head catch are useful if the number of animals to be worked is relatively large.

Dehorning

The polled condition has advantages because a polled goat is less likely to catch its head in a woven-wire fence than is a horned goat. Horned animals should be disbudded during the first week following birth by use of an electric dehorning iron. Goat breeders have been plagued by the genetic linkage between hermaphrodites and polled expression. However, polled goats with good fertility can be found.

Hoof Trimming

Goats' feet should be kept properly trimmed to prevent deformities and foot rot. In rough country, the terrain will likely keep hooves at a correct length. Under smooth conditions or under pen management, hoof trimming should be scheduled every 6–8 weeks. A good pruning shear is ideal for leveling and shaping the hoof, but final trimming can be done with a hoof knife. Foot rot should be treated with formaldehyde, copper sulfate, or iodine. Affected goats are best isolated from the others and placed on clean, dry ground after treatment.

Identification

All kids can be identified by a tattoo in the ear, except for LaMancha and pygmies, where identification is placed in the tail-web. Ear tags are also useful and are easy to read. The tattoo serves as permanent identification in case an ear tag is lost. All registered goats must be tattooed because the goat registry associations do not accept ear-tag identification.

Castration

Male kids not acceptable for breeding should be castrated at 4–6 weeks of age. This can be done by constricting the blood circulation to the testicles by use of a rubber band elastrator or by surgically removing the testicles. Some breeders nonsurgically crush the cords to the testicles with an emasculator.

Shearing

Shearing is usually done in the spring months although some producers prefer a twice a year shearing schedule. Some Angoras are not shorn for 2–3 years to allow the mohair to reach lengths of 1–2 feet. Longer fiber mohair brings higher prices and is used in making wigs and doll hair. A typical clip from a buck or wether weighs 5–8 lb, whereas a doe yields 4–6 lb. Angora goats from highly selected flocks may yield up to 15 lb of mohair per year.

At shearing time fleeces are sorted into three classes:

- Tight lock with ringlets the full length of the fiber—best value
- Flat lock which is wavy
- Open or fluffy clips—poorest quality

Milking

Some dairy goat enterprises still milk by hand but the use of milking machines in a managed parlor is desirable (Fig. 31.7). It is important that the udders of dairy goats have strong fore and rear attachments and that the two teats be well spaced. The milking stanchion should be elevated to place the goat at a convenient level for

Figure 31.7
Dairy goat milking parlor.
Source: FedeCandoniPhoto/Fotolia.

milking. Clean, sanitary conditions of the milking stanchion, the milking machine, the goat, and her udder and teats are very important prior to milking. After milking, the teats should be dipped in a weak iodine solution.

Time of Breeding

The female goat comes into heat at intervals of 18–21 days until she becomes pregnant. Young does can first be bred when they weigh 70–85 lbs at about 6–9 months of age, which means that does in good general condition may be bred to have kids at 1 year of age. Angora goats are not typically bred until they are 18–24 months of age. Most goats are seasonal breeders; the normal breeding season occurs in September, October, and November. If no effort is made to breed does at other times, most of the young will be born in February, March, or April. Normal goat lactations are 7–10 months in duration. A period of at least 2 months when no goats are lactating might occur, but with staggered breeding, the kidding and milking seasons can be extended. Housing goats in the dark for several hours each day in the spring and summer months (to simulate the onset of shorter days) causes some to come into estrus earlier than usual. Conversely, artificial additional light in the goat barn may delay estrus in the autumn.

One service is all that is needed to obtain pregnancy, but it is generally wise to delay breeding for a day after the goat first shows signs of heat. The doe stays in heat from 1 to 3 days, but the optimum period of standing heat may last only a few hours at the end of estrus. A female in heat is often noisy and restless, and her milk production may decrease sharply. She may disturb other goats in the herd, so it is advisable to keep her in a separate stall until she goes out of heat.

Male goats during the breeding season usually have a rank odor, especially when confined to a pen. Cleanliness of the male's long beard and shaggy hair helps reduce "bucky" odors.

Time of Kidding

Delivery of twins is common and delivery of triplets is not unusual, particularly among mature does. More males are conceived and born than are females; the ratio is approximately 115:100. Mature does average >2 kids, whereas younger does average 1.5 kids. Kids at birth weigh 5–8 lb, but singles may be heavier than twins or triplets, and males are usually heavier than females.

A doe almost ready to deliver young should be placed in a clean pen that is well bedded with clean straw or shavings. The doe should have plenty of clean water, some laxative feed (such as wheat bran), and fresh, soft legume hay. She should be carefully observed but not disturbed unless assistance is necessary, as indicated by excessive straining for 3 hours or more. If the kid presents the front feet and head, delivery should be easy without assistance (Fig. 31.8). If only the front feet, but not the head, are presented, the kid should be pushed back enough to bring the head forward in line with the front feet. Breech presentations are not uncommon, but these deliveries should be rapid to prevent the kid from suffocating. If assistance is needed, a qualified attendant with small, well-lubricated hands should pull when contractions occur. Gentleness is essential; harsh or ill-timed pulling can cause severe internal damage.

As soon as the kid arrives, its mouth and nostrils should be wiped clean of membranes and mucus. In cold weather, it is necessary to take the newborn to a heated room for drying. A chilled kid must be helped to regain its body temperature

A

B

C

Figure 31.8

(A) During normal presentation the head and front legs are presented first. (B) The membranes must be removed from the face of the newborn kid to avoid suffocation. (C) The licking action of the doe serves to stimulate and warm the newborn.

Source: Tom Field.

with a heat lamp or even by immersing its body up to the chin in water that is as hot as can be tolerated when the attendant's elbow is immersed in it for 2 minutes. The navel should be kept out of the water. Kids should be encouraged to nurse (Fig. 31.9) as soon as they are dry, as nursing helps the newborn kid to keep warm. In cold climates, the use of kid hutches equipped with heat lamps can be used to help neonatal goats maintain their core body temperature (Fig. 31.10).

Figure 31.9
The newborn kid should be encouraged to nurse as soon as possible following birth. Source: Tom Field.

Figure 31.10
A kid hutch can be constructed and equipped with a warming light to provide a comfortable environment for neonatal kids. Source: Tom Field.

Feeding

Goats are ruminants and thus digest roughage efficiently (Fig. 31.11). An advantage of goats is their ability to utilize low-quality forages, especially browse. Several studies have shown that grazing goats will choose up to 75% of their diet from available browse species. However, the types and proportions of feed should be related to the functions of the goats. For example, dry does and bucks that are not actively breeding perform satisfactorily on ample browse, good pasture, and good-quality grass and legume hay. If grass is short and hay is of poor quality, and goats are milking well, feeding of supplemental concentrates is necessary. Overfeeding of supplemental concentrates can cause diarrhea or obesity, which interferes with reproduction and subsequent lactation.

A summary of feed requirements for meat goats is provided in Table 31.7 and for dairy goats in Table 31.8.

Figure 31.11
Goats are efficient utilizers of forage and can be well adapted to both large and small acreage situations.
Source: Tom Field.

Table 31.7
PROTEIN AND INTAKE REQUIREMENTS FOR MEAT GOATS AT VARIOUS STAGES OF PRODUCTION

Production Phase	Protein (% CP)	Energy (% TDN)
Preweaning/creep	18	70–75
Weanlings	14–16	70
Finishing	14–16	65
Flushing (30 days prior and 30 days after breeding)	14–16	65–75
Gestation	14–16	60 (2nd and 3rd month) 65 (last 1.5 months)
Lactation (single)	14–16	60
Lactation (twins)	14–16	65
Replacement does	16	55
Bucks (mature)	11–14	60

Source: Adapted from multiple sources.

Table 31.8
PROTEIN AND INTAKE REQUIREMENTS FOR DAIRY GOATS AT VARIOUS STAGES OF PRODUCTION

Production Phase	Protein %
Preweaning/starter (2–4 mo)	18
Growing (4 mo to 6 to 8 wks before kidding)	14–16
Dry does (6 to 8 wks before kidding)	14–16
Lactating does	14–18 (early)
	12–16 (mid to late)
Bucks (mature)	12–16

Source: Adapted from multiple sources.

Heavily lactating dairy does and young does that must grow while lactating should be given good-quality hay and relatively high levels of concentrates. Good-quality pasture should supplement hay well, but even then lactating goats require some roughage in the form of hay or other source of long fiber to prevent scouring.

Pregnant does should be fed to gain weight in order to ensure adequate nutrition of the kids. A doe should be in good flesh, but not fat, when she kids, because she draws from her body reserves for milk production.

Grains such as corn, oats, barley, and milo may be fed whole because goats crack grains by chewing. If protein supplements are mixed with the grain or if the feed mix is being pelleted, the grain may be rolled, cracked, or coarsely ground. Some people prefer to mix the hay and grain and prepare the ration in a total pellet form. There is added expense in pelleting, but much less feed is wasted.

In some areas, deficiencies in minerals such as phosphorus, selenium, and iodine may exist. The use of iodized or trace-mineralized salt along with dicalcium phosphate or steamed bone meal usually provides enough minerals if legume hay or good pastures are available.

Young, growing goats and lactating goats need more protein than do bucks or dry does. A ration containing 12–15% protein is desirable for bucks and dry does, but 15–20% protein may be better for young goats and for does that are producing much milk.

Generally, kids must be allowed to nurse their dams to obtain the first milk (colostrum). After 3 days, kids may be removed from their mothers and given milk replacer by means of a lamb feeder or handheld bottle until they are large enough to eat hay and concentrates. It is advisable to encourage young kids to eat solid feed at an early age (2–3 weeks) by having leafy legume hay and palatable fresh concentrates such as rolled grain available at all times. Solid feeds are less expensive than milk replacers, and when the kids can do well on solid feeds, milk replacers should not be fed. Small kids need concentrates until their rumens are sufficiently developed to digest enough roughage to meet all their nutritional needs. At 5–6 months of age, young goats can do well on good pasture or good-quality legume hay alone.

Goats of any kind are usually run with sheep outside of the United States; this practice may be advantageous in the United States as well. Goats normally eat more browse and forbs than sheep.

Rams should never be run with doe goats, and buck goats should never be run with ewes. These animals will mate when the females come in heat (estrus); if they conceive, the pregnancy will usually terminate at about 3 or 4 months. There are reports of rare cases of sheep–goat crosses delivered alive at term.

CONTROLLING DISEASES AND PARASITES

Some common diseases of sheep and goats include the following types.

Enterotoxemia (overeating disease) is an acute, noninfectious disease that affects sheep and goats on a high nutritional plane (lush pasture, high grain feedlot rations, or when excessive milk is provided to kids/lambs). Particularly susceptible are feedlot lambs and lactating dairy goats. Symptoms include a reluctance to consume feed, diarrhea, staggering, circling, or convulsive behavior. Causative agent is a toxin produced by a specific serotype of *Clostridium perfringens, Type D.* The use of a bacterin or toxoid vaccination followed by a booster provides protection as does the use of a gradual step up to high-energy feedstuffs.

Neonatal scours accounts for nearly half of the death loss of lamb less than 2 weeks old and is the most costly disease affecting neonatal kids. Scours is a complex, multifactoral disease (infectious organisms, nutrition, environmental conditions, and animal status) characterized by excessive diarrhea and dehydration often caused by *Escherichia coli*, rotavirus, *Cryptosporidium* sp. and *Salmonella* sp. The involvement of bacterial, viral, and protozoal agents makes treatment challenging. In cases with *E. coli* involvement, **antibiotics** may be effective. Rotavirus infections can be treated with supportive therapy to maintain electrolyte balance and hydration. Excellent sanitation practices, avoiding sloppy environmental conditions, and assuring adequate intake of colostrum shortly after birth are useful in prevention. **Foot rot** is one of the most serious and common diseases affecting the sheep industry with symptoms ranging from slight to severe lameness. The disease is caused by two specific bacteria. The disease can be treated with systemic medication. It can be cured by severe trimming so that all affected parts are exposed, treating the diseased area with a solution of one part formalin solution to nine parts of water, and then turning the sheep onto a clean pasture so that reinfection does not occur. Formaldehyde must be used with caution because the fumes are damaging to the respiratory system of both the sheep and the person applying the formaldehyde. Goats tend to avoid wet areas which provide them inherent protection against the causative organisms causing foot rot.

Once all sheep in the flock are free of foot rot, making sure that it is not reintroduced into the flock can best prevent it. Rams introduced for breeding should be isolated for 30–60 days for observation. If foot rot develops, rams should continue in isolation until free of the disease. A vaccine is available for foot rot.

Sore mouth also known as contagious ecthyma usually affects young animals rather than adult sheep and goats. It is caused by a virus and can be contracted by humans. It can be controlled by vaccination, and sheep/goats should be vaccinated as a routine practice.

Sheep are subject to a nutritional disease known as **white muscle disease**. To prevent it, pregnant ewes should be given an injection of selenium during the last one-third of pregnancy and the lambs should be given an injection of selenium at birth. Selenium can be added to the feed of pregnant ewes to prevent white muscle disease in their lambs. If trace-mineralized salt is provided, it should contain sufficient selenium to supply the needs of the sheep.

Shipping fever is a highly infectious and contagious disease complex that usually affects lambs after the stress of transportation. Antibiotics and **sulfonamides** are usually effective as treatment. Care in transporting lambs helps prevent this disease.

Caseous lymphadenitis is a disease that occurs with greater frequency as sheep and goats increase in age from lambs to old animals. The pathogenic organism grows in lymph glands and causes a large development of caseous material (a thick, cheese-like accumulation) to form. It is a serious disease and one that is difficult to control. The abscesses can be opened and flushed with a solution of equal parts of 0.2%

nitrofurazone solution and 3% hydrogen peroxide. One should not open an abscess and let the thick pus go onto the floor or soil where well sheep and goats will be traveling because this may cause the disease to spread. All infected animals should be culled and should be isolated from noninfected sheep immediately on appearance of being infected.

Milk fever is due to hypocalcemia. Afflicted animals can be treated with an injection of calcium salts. A 20% solution of calcium borogluconate given **intravenously** at the rate of 100 ml per sheep should have an afflicted animal up and in good condition within 2 hours. Goats are rarely afflicted with milk fever.

Urinary calculi (kidney stones) occur when the salts in the body that are normally excreted in urine are precipitated and form stones that may lodge in the kidneys, ureters, bladder, or urethra. Affected animals will have an inability to urinate and will exhibit belly kicking and excessive stretching. Providing a constant supply of clean water helps immensely in preventing formation of calculi. The ideal ratio of calcium to phosphorus is 1.6:1.0. Ammonium chloride can be included in the ration to help prevent urinary calculi.

Pregnancy disease (ketosis) is a metabolic disease that affects ewes in late pregnancy, particularly if they are carrying twins or triplets. The problem is that ewes carrying twins or triplets must break down body fat to provide their energy needs in later pregnancy. It is possible that fat breakdown may not be sufficient for the glucose needs of these ewes, resulting in hypoglycemia. Feeding some high-energy grain such as corn, barley, or milo may prevent pregnancy disease, but one or more large lambs reducing rumen space in the mother aggravate the condition. Goats are rarely affected with ketosis.

Grass tetany (or **grass staggers**) is due to a deficiency of magnesium at a particular time, most frequently in spring when lactating ewes are put onto lush pasture where there is insufficient magnesium available. An injection of 50–100 ml of 20% calcium borogluconate or an injection of magnesium sulfate should give rapid recovery.

Annual death loss in adult sheep is approximately 5–6% of inventory while lamb death loss averages 9–10% of the lamb crop on an annual basis. About two-thirds of these losses result from nonpredator causes and losses due to predation account for the remaining one-third in adult sheep. The percent of death loss in lambs due to predators increases to approximately 43% of total losses.

The leading three causes of death loss (nonpredator) in adult sheep are old age, lambing problems, and digestive disorders. Nonpredator death losses in lambs most often result from respiratory disease, digestive problems, weather-related issues, and dystocia. These four categories account for three-quarters of nonpredator lamb mortalities.

Sheep and goats have internal and external parasites, though internal parasites are the more serious of the two. The most important internal parasites are coccidiosis, stomach worms, nodular worms, liver flukes, lungworms, roundworms, and tapeworms. Common external parasites of sheep include blowfly maggots, **keds** (sheep ticks), lice, mites, screwworms, and sheep bots. Making two applications of an effective insecticide that is not harmful to sheep can control external parasites. The two applications should be spaced so that eggs hatched after the first application will not result in egg-laying adults prior to the second application.

Bluetongue is a viral disease transmitted via a biting insect vector. Affected animals display fevers between 104 and 107°F, loss of appetite, excessive and rapid weight loss, and the oral mucous membranes redden progressively to a purplish-blue color. Lameness is typical and in extreme cases affected animals may slough hooves. While rarely fatal, sheep over four months of age should be vaccinated against the disease. Goats while often infected rarely become symptomatic.

Mastitis is an inflammation of the udder predisposed by bruising, lack of proper sanitation, and improper milking. The milk becomes curdled and stringy. It is advisable to use a milk-culture sensitivity test and engage the services of a veterinarian for treating severe cases. An udder may be treated with a suitable antibiotic by sliding a special dull plastic needle up the teat canal into the udder cistern and depositing the antibiotic, or by systemic antibiotic treatment intramuscularly or intravenously. Hot packs may help reduce the edema and, in severe cases, frequent milking is necessary and most helpful.

Tetanus or lockjaw is a common fatal bacterial disease caused by *Clostridium tetini*. The disease most frequently arises from infection through an open wound. Symptoms include muscle stiffness and spasms, uncoordinated movement, inability to eat or drink, and death is assured 72 to 96 hours after symptoms emerge. Vaccination and preventative sanitation are recommended.

DETERMINING THE AGE OF SHEEP BY THEIR TEETH

Evaluating teeth can help determine the age of a sheep. Adult sheep have eight permanent incisors on the lower jaw and 24 molars (six per side of the upper and lower jaw). Lambs have four pairs of narrow lower incisors called *milk teeth* or *baby teeth*. At approximately a year of age the middle pair of milk teeth is replaced by a pair of larger, permanent teeth. At 2 years, a second pair of milk teeth is replaced. This process continues until, at 4 years of age, the sheep has all permanent incisors and is considered **full mouthed**. Progressive loss of dental integrity then occurs and once permanent teeth begin to be lost, the animal is considered **broken-mouthed**. When all the permanent incisors are lost, the sheep has difficulty grazing and should be marketed.

COSTS AND RETURNS

The keys to profitability of enterprises involving sheep and goats is to increase the number of offspring born alive per breeding female, enhance lamb and kid survival rates, increase the prices received for market animals, milk, and fiber; and to improve the quality of products and by-products. Controlling operating costs is also important to profitability. The two areas of highest priority for cost containment are feed and labor. Total capital costs for a 50-head farm flock or meat goat startup minus the value of land and buildings would be approximately $325 to $350 per head. The components of typical budget for a farm flock enterprise are shown in Table 31.9 while those for a meat goat enterprise are provided in Table 31.10.

The sheep industry is confronted by the dual challenges of low domestic demand and not having the production capacity to meet the needs of that limited market. As Table 31.11 demonstrates, the United States is now a net importer of lamb. Reversing this trend will be critical to the economic viability of the sheep industry in the United States.

Table 31.9
PERCENTAGE OF COSTS AND RETURNS FOR A TYPICAL FARM FLOCK ENTERPRISE

Cost Factor	$ per Ewe	Percent of Total Costs/Returns
Hay, grain, salt, and mineral	$52.25	33
Supplemental feed for lambs	$37.50	23
Pasture maintenance	$8.00	5
Bedding	$6.25	4
Health and vet	$18.50	11
Shearing	$3.50	2
Ram replacement	$6.00	4
Operating interest	$4.75	3
Other	$26.00	15
Total Costs	$162.75	
Income		
Market lambs	$173.00	89
Cull breeding stock	$16.50	8
Wool	$5.25	3
Total Returns	$194.75	

Table 31.10
ESTIMATED COSTS AND RETURNS FOR A MEAT GOAT ENTERPRISE

Cost Factor	$ per doe	Percent of Total Costs/Returns
Hay	$23.25	33
Grain, salt, and mineral	$14.50	21
Pasture	$6.00	8
Health and vet	$11.00	16
Buck replacement	$3.00	4
Operating interest	$1.75	2
Other	$11.00	16
Total Costs	$70.50	
Income		
Market kids (65–80 lbs)	$105.00	90
Cull breeding stock	$11.50	10
Total Returns	$116.50	

Table 31.11
DOMESTIC VERSUS IMPORTED LAMB IN THE UNITED STATES (MILLION lb)

Year	Domestic Lamb Production	Imported Lamb	Difference
1990	363	41	322
1995	285	64	221
2000	225	131	94
2005	191	180	11
2010	164	171	−7
2013	156	173	−17

Source: USDA:ERS.

CHAPTER SUMMARY

- Sheep need good facilities for lambing and working them for sorting, branding, and treatment. Goats typically are best managed under pasture conditions.

- Farm flocks and range flocks utilize different resources and management strategies to achieve their enterprise goals.

- Sheep are usually shorn in the spring, prior to the hot weather months. Late spring storms sometimes challenge the sheep flock, especially under range conditions where there is limited protection.

- Growing demand for sheep and goat meat will be critical to the long-term sustainability of these industries.

KEY WORDS

cutting chute
lambing jugs
creep
purebred breeder
commercial producers
parasites
overeating disease
body condition score
grass lamb
Easter lambs
hothouse lambs
crutching
dung tags
breech presentation
docking
fly strike
prolapse
feed bunks
winter headquarters
spring–fall range
summer grazing

enterotoxemia
neonatal scours
antibiotics
foot rot
sore mouth
white muscle disease
shipping fever
sulfonamides
caseous lymphadenitis
milk fever
intravenously
urinary calculi
ketosis (pregnancy disease)
grass tetany (grass staggers)
keds
bluetongue
mastitis
tetanus
full mouthed
broken-mouthed

REVIEW QUESTIONS

1. Discuss the basic facility and equipment requirements for a commercial sheep enterprise.
2. Compare the types of farm flock producers.
3. Compare ration changes that occur as a lamb progresses through growing and finishing.
4. List the factors that affect water intake.
5. Describe the BCS system for sheep and discuss appropriate body condition at various production phases.
6. Compare grass, Easter, and hothouse lamb production systems.
7. Describe appropriate lambing, processing, and shearing procedures.
8. Compare shifts in management required for range sheep operations during different seasons of the year.
9. Describe regional goat production in the United States.
10. Describe fencing requirements for goat enterprises.
11. Compare goat fiber production to wool production.
12. Discuss reproductive and nutritional management of the goat.
13. Discuss disease management of sheep and goats.
14. Describe how evaluation of teeth can be used to determine age of sheep.
15. Compare costs and returns for sheep and goat enterprises.
16. Discuss the relationship of domestic lamb production and imported lamb volume.

SELECTED REFERENCES

Battaglia, R. A., and V. B. Mayrose. 1981. *Handbook of Livestock Management Techniques*. New York: Macmillan.

Botkin, M. P., R. A. Field, and C. L. Johnson. 1988. *Sheep and Wool: Science, Production, and Management*. Englewood Cliffs, NJ: Prentice Hall.

Ensminger, M. E. 2002. *Sheep and Goat Science*, 6th edition. Danville, IL: Interstate Publishers.

National Animal Health Monitoring System. 1996. *U.S. Regional Sheep and Health Management Practices*. Washington, DC: USDA: APHIS.

National Animal Health Monitoring System. 2006. *Sheep and Lamb Nonpredator Death Loss in the United States—2004*. Washington, DC: USDA: APHIS.

National Research Council. 1985. *Nutrient Requirements of Sheep*. Washington, DC: National Academy Press.

Sheep Housing and Equipment Handbook. 2005. Ames, IA: Midwest Plan Service.

Sheep Industry Council. 2004. *The Sheep Production Handbook*. Denver, CO: Sheep Industry Development Program.

32

Horse Breeds and Breeding

HORSES AND HUMANS

The horse was domesticated approximately 5,000 years ago, and thus began the relationship between humans and horses. The role of the horse through history ranges from a source of food to the focus of artisans and sculptors. The horse has played an important part in human history by changing military strategies and the division of labor in agriculture and transportation, serving as a focal point in sport, and providing satisfying recreational activities for humans, both young and old.

BREEDS OF HORSES

Horses have been used for so many different purposes that many breeds have been developed to fill specific needs. The major breeds of horses and their primary uses are listed in Table 32.1. No attempt has been made in Table 32.1 to indicate all the ways each breed is used.

The **light breeds of horses** provide pleasure for their owners through activities such as racing, riding, and exhibition in shows. Figure 32.1 show examples of these breeds. Quarter Horses are excellent for use on ranches, favored for recreation and Western equestrian competitions such as reining, Western pleasure, and stock horse events and as racing sprinters. **Ponies**, such as Shetland and Welsh, are selected for their friendliness and safety with children. The American Saddlebred and the Tennessee Walking Horse have been selected for their comfortable gaits and responsive attitudes. The American Saddlebred has regional popularity and is considered "the peacock of the show ring." The Palomino, Appaloosa, and Paint horses are color breeds developed for showing and for working livestock.

Beauty in color and markings is important in horses used for show and breeding purposes, and colored breeds have been developed accordingly. The Appaloosa, for example, has color markings of six patterns: blanket, blanket with spots, roan, roan blanket, roan blanket with spots, and solid. Appaloosas must also have striped hooves, mottled skin, and a white sclera surrounding the iris of the eye. It appears that these color patterns are under different genetic controls. In Appaloosas, Paints, and Palominos, coloration can be affected or eliminated by certain other genes, such as the gene for roan and the gene coding for gray. The gene for gray can eliminate both colors and markings, as shown by gray horses that turn white with age.

learning objectives

- List and describe the primary light and draft horse breeds
- Contrast selection in the horse industry with the other livestock enterprises
- Describe the ideal conformation of the horse
- List and describe the unsoundnesses and blemishes affecting the horse
- Describe the various gaits
- Discuss how to determine a horse's age by evaluating its teeth

Table 32.1
CHARACTERISTICS AND USES OF SELECTED BREEDS OF HORSES

Breed	Color	Origin	Height (in hands)[a]	Weight (lb)	Uses
Riding and Harness Horses					
American Quarter Horse	All colors except paint	United States	14.2–15.2	1,000–1,250	Short racing, showing, stock work
American Saddlebred	Chestnut, bay, brown, black, gray	Kentucky	15.0–16.0	1,000–1,150	Showing, pleasure riding, 3 and 5 gaited
Arabian	Bay, chestnut, white, gray, black	Arabia	14.2–15.2	850–1,000	Pleasure riding, showing
Morgan	Bay, chestnut, brown, black	New England	14.2–15.1	950–1,150	Pleasure riding, driving, showing
Standardbred	Bay most common, but all solid colors	United States	14.2–16.2	850–1,200	Harness racing
Tennessee Walking Horse	Black, bay, chestnut	Tennessee	15.0–16.0	1,000–1,200	Pleasure riding, showing
Thoroughbred	Bay, brown, chestnut, or any other solid color	England	15.2–17.0	1,000–1,300	Long racing
Ponies					
Hackney	Bay, chestnut, black, brown	England	11.2–14.2	450–850	Light harness, showing
Pony of America	Appaloosa	Iowa	11.5–13.5	700–800	Riding by children, showing
Shetland	Bay, chestnut, brown, black, gray	Shetland Islands	9.2–10.0	300–400	Riding by children, showing
Welsh	Any color except piebald and skewbald	Wales	11.0–13.0	350–850	Riding by children, showing
Draft[b]					
Belgian	Chestnut, roan usually	Belgium	15.2–17.0	1,900–2,400	Heavy pulling
Clydesdale	Bay, brown, black, roan	Great Britain	15.2–17.0	1,700–2,000	Heavy pulling
Percheron	Black, gray usually	France	15.2–17.0	1,600–2,200	Heavy pulling
Shire	Bay, brown, usually black, gray	England	16.2–17.0	1,800–2,200	Heavy pulling
Suffolk	Chestnut	England	15.2–16.2	1,500–1,900	Heavy pulling
Color Registries					
Appaloosa	Leopard, blanket, roan	Pacific Northwest	14–16	900–1,250	Pleasure riding, showing, stock work
Buckskin	Buckskin, dun, grulla	United States	14–16	900–1,250	Pleasure riding, showing, stock work
Paint	Tobiano, overo	United States	14–16	900–1,250	Pleasure riding, showing, stock work
Palomino	Palomino	United States	14–16	900–1,250	Pleasure riding, showing, stock work
Pinto	Pinto	United States (via horses brought by Spanish conquistadors)	14–16	900–1,250	Pleasure riding, showing, stock work
White and cremes	White and cream		14–16	900–1,250	Pleasure riding, showing, stock work

[a]Height is measured in inches but reported in *hands*. One hand equals 4 in.
[b]Draft horses are heavy horses used for pulling; other horses are called *light* horses and are used primarily as pleasure horses.
More detailed information can be found at www.ansi.okstate.edu/breeds/

Figure 32.1

Popular breeds of light horses. Quarter Horse, Appaloosa, Paint Horse, Arabian, Standardbred, Thoroughbred, Tennessee Walking Horse, and Clydesdales. Source: 32.1a: Lichtreflexe/Fotolia; 32.1b: Jeanma85/Fotolia; 32.1c: Zuzule/Fotolia; 32.1d: MaryPerry/Fotolia; 32.1e: Electrochris/Fotolia; 32.1f: Sportlibrary/Fotolia; 32.1g: Clarence Alford/Fotolia; 32.1h: Nicolette Wollentin/Fotolia.

Table 32.2
MAJOR U.S. DRAFT, LIGHT HORSE, AND PONY BREEDS BASED ON ANNUAL REGISTRATION NUMBERS (IN THOUSANDS), 1980–2013

Breed	2013	2008	2000	1990	1980	Year Association Formed
Light Horse						
Quarter Horse	84.0	135.9	145.9	110.6	137.1	1940
Paint	15.0	29.5	62.5	22.4	9.7	1962
Thoroughbred	21.3	36.3	36.5	43.6	39.4	1894
Tennessee Walking	3.9	10.2	14.4	8.0	6.8	1935
Standard bred	7.5	9.8	11.3	16.6	15.2	1938
Arabian	4.8	6.1	9.7	17.7	19.7	1908
Appaloosa	3.0	5.4	10.9	10.7	25.4	1938
Pinto	4.8	3.4	4.5	2.9	—	1956
Morgan	1.7	2.5	3.6	3.6	4.5	1907
American Saddlebred	1.8	2.8	2.9	3.6	3.9	1891
Draft						
Belgian	2.5	4.0	3.9	3.7	.7	1887
Percheron	1.0	2.4	2.2	1.3	.3	1876
Clydesdale	0.6	0.5	0.5	0.4	.2	1879

Sources: Adapted from various breed associations.

Another group of horses are the **draft horses**—large and powerful animals that are used for heavy work. The Percheron, Shire, Clydesdale, Belgian, and Suffolk are examples of draft horses (Table 32.1).

Popularity of Breeds

Registration numbers are a measure of breed popularity. Table 32.2 shows the registration numbers for draft and light breeds. Based on these numbers, the Quarter Horse is the most popular horse breed in the United States.

BREEDING PROGRAM

Selection

An effective breeding program is a completely objective one. There is no place in a breeding program for emotional attachment toward an animal that results in a loss of selection pressure. Every attempt should be made to see that the environment is the same for all animals in a breeding program. A particularly appealing foal that is given special care and training may develop into a desirable animal. However, many foals less appealing in early life might also develop into desirable animals if given similar attention. Environmental effects may mislead breeders resulting in the propagation of less than desirable animals whose weaknesses are masked by nutrition or other favorable management conditions. Thus, selecting a horse for breeding that has had special care and training may in fact be selecting only for special care and training. Certainly these environmentally produced differences in horses are not inherited. In fact, it is often wise to select animals that were developed under the type of environment in which they are expected to perform. If stock horses are being developed for herding cattle in rough, rugged country, selection under such conditions is more desirable than where conditions are less rigorous. Horses that possess inherited weaknesses tend to become unsound in a rugged environment and, as a result, are not used for breeding. Such animals might never show those inherited weaknesses in a less rugged environment.

An ideal environment for most horse-breeding programs has quality forage distributed over an area that requires horses to exercise as they graze. Where access to land is limited, horses may have to be kept in a small area and forced to exercise a great deal. Forced exercise tends to keep the animals from becoming too fat and gives strength to the feet and legs. Horses need regular, not sporadic, exercise. Regular exercise, even if quite strenuous, is healthy for animals that are genetically sound and may reveal the weaknesses of those that are not. Strenuous exercise can, however, be harmful to animals that have not previously exercised for a considerable period of time.

An environment should be provided that identifies horses having genetic superiority for their intended performance. For example, horses that are being bred for endurance in traveling should be made to travel long distances on a regular basis to determine if they can remain sound. Horses bred for jumping should be trained to jump as soon as they are physically mature so that those lacking the ability to jump or those that become unsound from jumping can be removed from the breeding program. Draft horses should be trained to pull heavy loads early in life (3 or 4 years of age) to determine their ability to remain sound and their willingness to pull. Performance should be measured before animals are used in a breeding program.

Any horse that is unsound should not be used for breeding regardless of the purposes for which the horse is being bred. Such abnormalities as toeing-in or toeing-out, sickle hocks, cow hocks, and contracted heels will likely lead to unsoundness and difficulties or lack of safety in traveling. Interference and forging actions are extremely objectionable because they can cause the horse to stumble or fall. Eye, mouth, and respiration defects should be selected against in all horses.

CONFORMATION OF THE HORSE

Purchasing the Horse

Buying a horse often involves a significant investment and a situation where both buyer and seller may be amateurs. However, a disciplined approach to purchasing a horse will yield favorable results (Table 32.3). Protection of the interest of both parties is an important consideration. One way to accomplish this task is through the use of the prepurchase exam. While there is variation in the way that these exams are conducted, the exam should be performed by an experienced equine veterinarian with both buyer and seller present.

The prepurchase exam usually focuses on an assessment of the horse's general health, physical condition, and structural soundness. Furthermore, it may be important to assess the horse's athleticism, temperament, and appropriateness for the horsemanship skills of the buyer. These topics are covered in more detail later in this chapter.

Table 32.3
THE PROCESS OF PURCHASING A HORSE

1. Establish your wants and needs in advance.
2. Shop around but only with reputable sellers—take your time.
3. Have a set of predetermined questions, take notes, use videotape if possible.
4. Always get a pre-purchase exam.
5. Use a thorough sales contract to protect buyer and seller.
6. Keep all records and notes relative to the sale.
7. Do not buy on impulse, and stay on budget.

Body Parts

To understand the conformation of the horse, one should be familiar with the body parts (Fig. 32.2). A more detailed drawing of the skeletal structure of the horse is shown in Chapter 18.

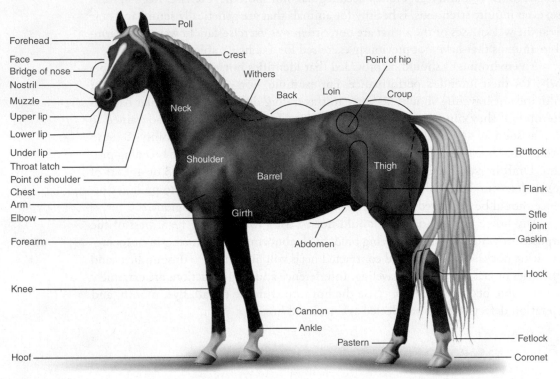

Figure 32.2

Parts of the horse. Source: IronFlame/Shutterstock.

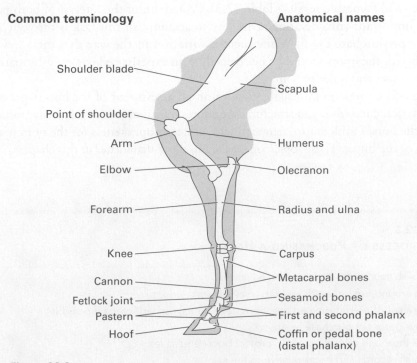

Figure 32.3

Skeletal front leg with common terminology and anatomical names.

Source: Colorado State University.

Feet and Legs   Major emphasis is placed on feet and legs in describing conformation in the horse—for identifying both correctness and conditions of unsoundness. The old saying, "No feet, no horse" is still considered valid by most horse producers.

From a front view, a vertical line from the point of the shoulder should fall in the center of the **knee**, **pastern**, **cannon**, and **foot**. Each leg is divided into two equal halves. From a side view of the front legs, a vertical line from the shoulder should fall through the center of the elbow and the center of the foot. The angle of the pastern in the ideal position is 45–50°.

Correct hind leg position from a rear view would allow a vertical line to pass from the point of the buttocks through the centers of the **hock**, cannon, pastern, and foot. The ideal position of the hind legs when viewed from the side allows a vertical line from the point of the buttocks to touch the rear edge of the cannon from the hock to the fetlock and then to meet the ground behind the heel. Deviations from these ideal positions are listed along with associated common terminology in Table 32.4.

Common terminology   **Anatomical names**

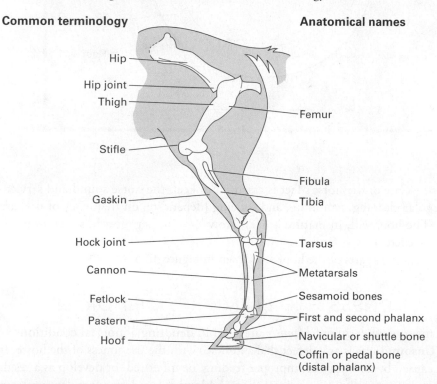

Figure 32.4
Skeletal hind leg with common terminology and anatomical names. Source: Colorado State University.

Hip
Hip joint
Thigh
Stifle
Gaskin
Hock joint
Cannon
Fetlock
Pastern
Hoof

Femur
Fibula
Tibia
Tarsus
Metatarsals
Sesamoid bones
First and second phalanx
Navicular or shuttle bone
Coffin or pedal bone (distal phalanx)

Table 32.4
DEVIATIONS FROM CORRECT CONFORMATION OF THE FRONT AND REAR LEGS OF THE HORSE—FRONT, SIDE, AND REAR VIEWS

Front View of Front Legs	Side View of Front Legs	Rear View of Rear Legs	Side View of Rear Legs
Toes outside vertical—*toed out*	Entire leg back of vertical—*camped under*	Entire leg outside of vertical—*stands wide*	Cannon and pastern in front of vertical—*sickle hocked*
Knees outside vertical—*bow legged*	Knee back of vertical—*calf kneed*	Hocks outside of vertical—*bow legged*	Hock, cannon, and pastern excessively straight—*post legged*
Entire leg inside vertical—*base narrow*	Entire leg in front of vertical—*camped out*	Entire leg inside of vertical—*stands close*	Hock, cannon and pastern back of vertical—*camped out*
Knees inside vertical—*knocked kneed*	Knee sprung forward of vertical—*buck kneed*	Hocks inside of vertical—*cow hocked*	
Toes inside vertical—*pigeon toed*			

Figure 32.5
The conformation of the hoof (viewed from the bottom side) with several parts identified. Source: Colorado State University.

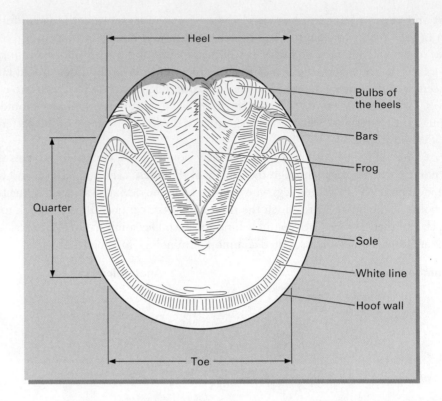

The Hoof   Care of the horse's feet is essential to keep the horse sound and serviceable. Regular cleaning, trimming, and shoeing (depending on frequency of use) are needed. The hoof will, in mature horses, grow ¼–½ in. per month, so trimming is typically needed every 6–8 weeks.

The external parts of the hoof are shown in Figure 32.5.

UNSOUNDNESS AND BLEMISHES OF HORSES

Two terms, *unsoundness* and *blemish*, are used in denoting abnormal conditions in horses. **Unsoundness** is any defect that interferes with the usefulness of the horse. It may be caused by an injury or improper feeding, be inherited, or develop as a result of inherited abnormalities in conformation. A **blemish** is a defect that detracts from the appearance of the horse but does not interfere with its usefulness. A wire cut or saddle sore, for instance, may cause a blemish without interfering with the usefulness of the horse.

Horses may have anatomical abnormalities that interfere with their usefulness. Many of these abnormalities are either inherited directly or develop because of an inherited condition. Abnormalities of the eyes, respiratory system, circulatory system, and conformation of the feet and legs are all important.

Some of the major unsoundness and blemishes include the following (several of which are identified by location on the body in Fig. 32.6).

Bog spavin is a soft swelling on the inner, anterior aspect of the hock. Although unsightly, the condition usually does not cause lameness. **Bone spavin** is a bony enlargement on the inner aspect of the hock. Both bog and bone spavin arise when stresses are applied to horses that have improperly constructed hocks. Lameness usually accompanies this condition, but most animals return to service after rest and, in some cases, surgery.

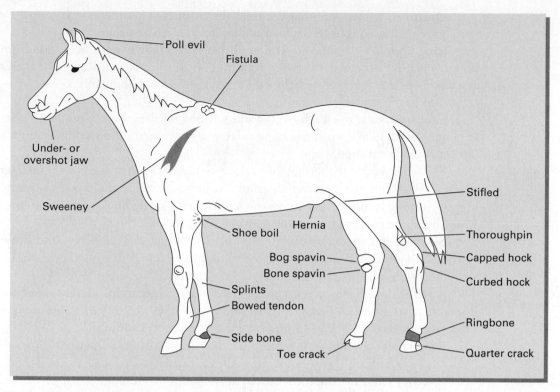

Figure 32.6
Locations of several potential unsoundnesses and blemishes. Source: Colorado State University.

Capped hock is a thickening of the skin at the point of the hock. **Curb** is a hard swelling above the cannon bone (a hand's width below the hock). The plantar ligament becomes inflamed and swollen, usually owing to poor conformation or a direct blow.

Cataracts (opacity of the lens of the eye) is inherited as a dominant trait. It can be eliminated if the afflicted horse does not produce several foals before the cataract develops.

Contracted heels are commonly an inherited conformation defect. Contracted heels may result when the frog or cushion of the foot is damaged and shrinks, allowing the heels to come together. The bottom surface of the foot becomes smaller in circumference than at the coronet band.

The term "cow hocked" indicates that the points of the hocks turn inward. Such hocks are greatly stressed when the horse is pulling, running, or jumping.

Hyperkalemic periodic paralysis (HYPP) is a muscle disease that is genetically controlled. HYPP affects some lines of Quarter Horses, Paints, and Appaloosas. Affected horses experience tremors as a result of abnormal muscle fiber activity. A mutation in the gene responsible for sodium and potassium regulation is apparently the cause. Ideally, those horses that carry the trait should not be allowed to breed.

Navicular disease is a disease complex manifested in pain in the heels of the front foot. Navicular pain may result from a variety of hoof faults including contracted heels, underslung heels, or lack of hoof symmetry. While the navicular bone was commonly considered the affected tissue, new studies suggest that this disease may be linked to damage of the tendons, ligaments, or coffin joint. Diagnosis is complicated by the multiple factors that may be involved. Early treatment, including medication, surgery, or rest is critical to assure recovery.

Quittor is a deep sore that drains at the coronet. The infection is caused by puncture wounds, corns, and the like, and results in severe lameness.

Ring bone is new bone growth on phalanges near the pastern that arises from injury, repetitive concussion over the years, or poor conformation. Ring bone shows as a hard bony enlargement encircling the areas of the pastern joint and coronet, typically affected the forefeet.

Shoe boil, or **capped elbow**, is a soft swelling on the elbow. Common causes are injury to the elbow while the horse is lying down or getting up or injury from a long heel on a front shoe.

Sickle hocked is the term used to describe the hock when it has too much set or bend. As a result, the hind feet are set too far forward. The strain of pulling, jumping, or running is much more severe on a horse with sickle hocks than on a horse whose hocks are of normal conformation.

Side bones is an abnormality that occurs when the lateral cartilages in the foot ossify. During the ossification process, lameness can occur, but some horses regain soundness with proper rest and shoeing.

A **stifled** horse is one in which the patella (the kneecap in humans) has been displaced. Older horses seldom become sound once they are stifled, while younger horses usually recover. There is a surgical operation for this condition.

String halt is an involuntary flexion of the hock during movement. It is considered a nerve disorder. Surgery can improve the condition.

Sweeney refers to atrophied muscles at any location, although many people use it to refer only to shoulder muscles. In the shoulder sweeney, the nerve crossing the shoulder blade has been injured.

A **thoroughpin** is a soft enlargement of the tendon sheath of the large tendon (tendon of Achilles) of the hock and the fleshy portion of the hind leg. Stress on the flexor tendon allows synovial fluid to collect in the depression of the hock. Lameness rarely occurs.

Toeing-in, or **pigeon-toed** refers to the turning in of the toes of the front feet. **Toeing-out** refers to the turning out of the toes of the front feet. These conditions influence the way in which the horse will move its feet when traveling. Toeing-out or moving the front feet inward is considered the more serious defect because it can lead to further interference and faults.

Tying-up, or **exertional rhabdomyolysis**, is a condition categorized as either early-form or late-form. Affecting the unconditioned or overexerted horse, tying up is characterized by a shuffling gait, heavy sweating, and overly contracted rump and thigh muscles. Severe forms may result in muscular trauma. Afflicted horses usually recover without long-term consequences. Prevention involves feeding according to a horse's workload and assuring that horses are prepared for strenuous work.

Windgalls, sometimes referred to as **wind puffs** or *puffs*, occur when the joint capsules on tendon sheaths around the pastern or fetlock joints are enlarged. The disease is common, but not serious, in hardworking horses.

GAITS OF HORSES

The major gaits of horses, along with their modifications, are as follows:

1. **Walk** is a four-beat gait in which each of the four-feet strikes the ground independently.
2. **Trot** is a diagonal, two-beat gait in which the right front and left rear feet hit the ground in unison, and the left front and right rear feet hit the ground in unison. The horse travels straight without swaying sideways when trotting.

3. **Pace** is a lateral two-beat gait in which the right front and rear feet hit the ground in unison and the left front and rear feet hit the ground in unison. There is a swaying from right to left when the horse paces.
4. **Gallop** is the fastest gait with four beats.
5. **Canter** is a three-beat gait. Depending on the lead, two diagonal legs hit the ground at the same time, with the other hind leg and lead leg hitting at different times.
6. **Rack** is a snappy four-beat gait in which the joints of the legs are highly flexed. The forelegs are lifted upward to produce a flashy effect. This is an artificial gait, whereas the walk, trot, pace, gallop, and canter are natural gaits. The rack is popular in the show ring for speed and animation.
7. **Running walk** is the fast ground-covering walk unique to the Tennessee Walking Horse. It is faster than the normal walk. The horse moves with a gliding motion as the hind leg oversteps the forefoot print by 12–18 inches or more.

EASE OF RIDING AND WAY OF GOING

When a horse's foot strikes the ground, a large shock is created that would be objectionable to the rider if no shock absorption occurred. There are several shock-absorbing mechanisms existing in horses' feet and legs. The horse has lateral cartilages on all four feet that expand outward when the foot strikes the ground. This absorbs some of the shock. The pastern on each leg absorbs some of the shock when the foot strikes the ground by bending somewhat. A pastern that is too straight will not absorb much of the shock and one that is too long and weak will let the leg go to the ground. These kinds of pasterns will soon result in unsound horses. Thus, it is very important that a pastern have the proper slope so it can absorb the optimal amount of shock without the leg going to the ground or causing too much concussion on the joints and, ultimately, the rider.

The front legs each have two joints that allow movement and absorb shock: the joint between the ulna and the humerous, and the joint between the humerous and the scapula. Also, the hind legs each have two joints that bend and thus absorb some shock. These joints are located between the metatarsus and tibia and between the tibia and femur.

If the horse's feet and legs have proper conformation, a pleasant ride can be enjoyed. If there are abnormalities due to inheritance, injury, improper nutrition, or disease, the horse will give the rider a less pleasurable ride.

Abnormalities in Way of Going

A horse that toes out with its front feet tends to dish or swing its feet inward (wing in) when its legs are in action. Swinging the feet inward can cause the striding foot to strike the supporting leg so **interference** to forward movement results. A horse that toes-in (pigeon-toed) tends to swing its front feet outward, giving a **paddling** action.

Some horses overreach with the hind leg and catch the heel of the front foot with the toe of the hind foot. This action, called **overreaching**, can cause the horse to stumble or fall. **Forging** occurs when the hind foot hits the shoe on the front foot.

Figure 32.7 shows the way of going well as the horse moves straight and true, each foot moving in a straight line. The other illustrations show the path of flight of each foot when the structure of the foot and leg deviates from the desired norm. Figure 32.8 shows how the length and slope of the hoof affects way of going.

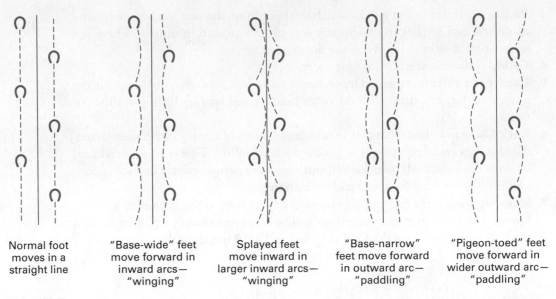

Normal foot moves in a straight line

"Base-wide" feet move forward in inward arcs— "winging"

Splayed feet move inward in larger inward arcs— "winging"

"Base-narrow" feet move forward in outward arc— "paddling"

"Pigeon-toed" feet move forward in wider outward arc— "paddling"

Figure 32.7
Path of the feet (way of going), which relates to foot and leg structure, as seen from above. Source: Colorado State University.

Figure 32.8
Illustration of how length and slope of the hoof affects way of going. Source: Colorado State University.

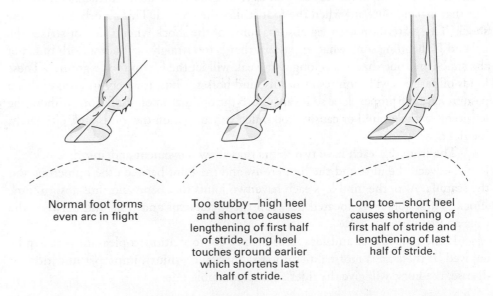

Normal foot forms even arc in flight

Too stubby—high heel and short toe causes lengthening of first half of stride, long heel touches ground earlier which shortens last half of stride.

Long toe—short heel causes shortening of first half of stride and lengthening of last half of stride.

Since few horses move perfectly true, it is important to know which movements may be unsafe. A horse that wings in (interferes) is potentially more unsafe than a horse that wings out (paddles), as the former horse may trip itself.

DETERMINING THE AGE OF A HORSE BY ITS TEETH

The age of a horse can be estimated by its teeth (Figs. 32.9 and 32.10). A foal at 6–10 months of age has 24 baby or milk teeth (12 incisors and 12 molars). The incisors include three pairs of upper and three pairs of lower incisors.

Chewing causes the incisors to become worn. The wearing starts with the middle pair and continues laterally. At 1 year of age, the center incisors show wear; at 1.5 years, the intermediates show wear; and at 2 years, the outer, or lateral, incisors show

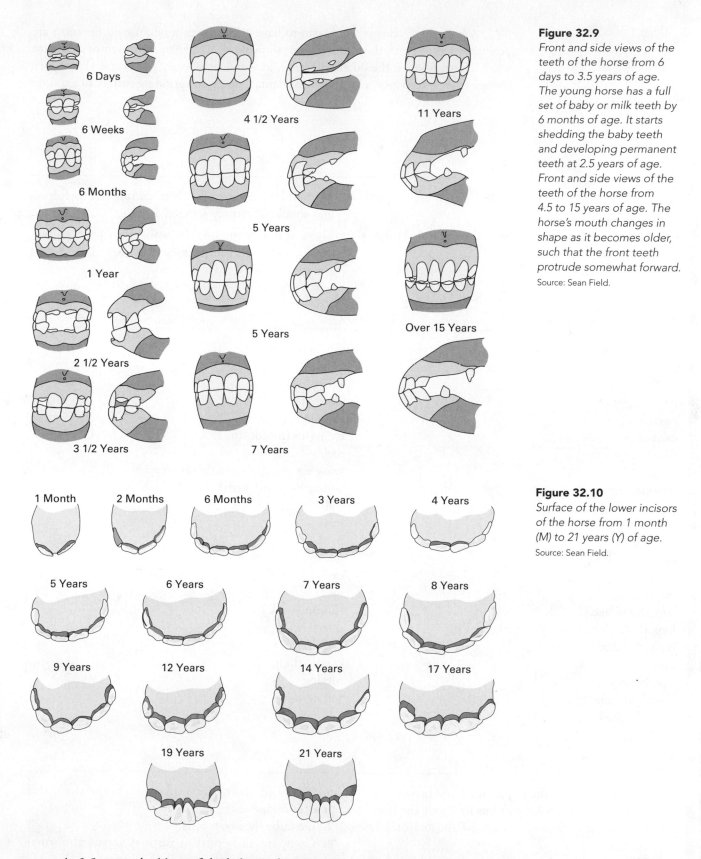

Figure 32.9
Front and side views of the teeth of the horse from 6 days to 3.5 years of age. The young horse has a full set of baby or milk teeth by 6 months of age. It starts shedding the baby teeth and developing permanent teeth at 2.5 years of age. Front and side views of the teeth of the horse from 4.5 to 15 years of age. The horse's mouth changes in shape as it becomes older, such that the front teeth protrude somewhat forward.
Source: Sean Field.

6 Days
6 Weeks
6 Months
1 Year
2 1/2 Years
3 1/2 Years

4 1/2 Years
5 Years
5 Years
7 Years

11 Years
Over 15 Years

Figure 32.10
Surface of the lower incisors of the horse from 1 month (M) to 21 years (Y) of age.
Source: Sean Field.

1 Month
2 Months
6 Months
3 Years
4 Years

5 Years
6 Years
7 Years
8 Years

9 Years
12 Years
14 Years
17 Years

19 Years
21 Years

wear. At 2.5 years, shedding of the baby teeth starts. The center incisors are shed first. Thus, at 2.5 years, the center incisors become permanent teeth; at 4 years, the intermediates are shed; at 5 years, the outer, or lateral, incisors are shed and replaced by permanent teeth.

A horse at 5 years of age is said to have a **full mouth**, because all the teeth are permanent. At 6 years, the center incisors show wear; at 7 years, the intermediates show wear; and at 8 years, the outer, or lateral, incisors show wear. Wearing is shown by a change from a deep groove to a rounded dental cup on the grinding surface of a tooth.

CHAPTER SUMMARY

- Horses are typically classified as light, draft, or pony breeds.
- Based on registration numbers, the Quarter Horse is the most popular horse breed.
- Unsoundness in horses leads to diminished usefulness and should be strictly selected against.
- Gaits of horses include the walk, trot, pace, gallop, canter, and rack. The Tennessee Walker also exhibits the running walk.

KEY WORDS

light horse breeds
ponies
draft horses
knee
pastern
cannon
foot
hock
unsoundness
blemish
bog spavin
bone spavin
capped hock
curb
cataract
contracted heels
hyperkalemic periodic paralysis (HYPP)
navicular disease
quittor
ring bone
shoe boil (capped elbow)
sickle hocked

side bones
stifled
string halt
sweeney
thoroughpin
toeing-in (pigeon-toed)
toeing-out
tying-up (exertional rhabdomyolysis)
windgalls (wind puffs)
walk
trot
pace
gallop
canter
rack
running walk
interference
paddling
overreaching
forging
full mouth

REVIEW QUESTIONS

1. List and describe the important horse breeds.
2. Describe the six color patterns in Appaloosa horses.
3. Discuss registration trends of the major horse breeds.
4. Discuss an effective process to select and purchase a horse.
5. Describe the ideal conformation of the equine.
6. List and describe common blemishes and unsoundnesses.
7. Describe the seven gaits.
8. Compare and contrast normal versus abnormal movement/way of going in the horse.
9. Describe how evaluation of teeth can be used to determine age in the equine.

SELECTED REFERENCES

Bowling, A. 1996. *Horse Genetics*. Oxon, UK: CAB International.

Butler, D. 1985. *Principles of Horseshoeing II*. Grand Prairie, TX: Equine Research.

Ensminger, M. E. 1990. *Horses and Horsemanship*. Danville, IL: Interstate Publishers.

Evans, J. W. 1989. *Horses: A Guide to Selection, Care and Enjoyment*. 4th ed. San Francisco, CA: W. H. Freeman.

Evans, J. W. (ed.). 1992. *Horse Breeding and Management*. Amsterdam: Elsevier Science Publishers.

Evans, J. W., A. Borton, H. F. Hintz, and L. D. Van Vleck. 1990. *The Horse*. 2nd ed. San Francisco, CA: W. H. Freeman.

Harris, S. 1993. *Horse Gaits, Balance and Movement*. Grand Prairie, TX: Equine Research.

Horse Industry Directory. 2006. Washington, DC: American Horse Council.

Jones, W. E. 1982. *Genetics and Horse Breeding*. Philadelphia, PA: Lea & Febiger.

Lasley, J. F. 1987. *Genetics of Livestock Improvement*. Englewood Cliffs, NJ: Prentice Hall.

Rich, G. A. 1981. *Horse Judging Guide*. Colorado State University Extension Service Publication.

Stashak, T. S. (ed.). 1987. *Adam's Lameness in Horses*. 4th ed. Grand Prairie, TX: Equine Research.

33

Feeding and Managing Horses

Humans and horses share a unique and mutually beneficial bond. Although many people enjoy horses, most own only a few; however, whether a person keeps only one horse for personal recreation or manages a large breeding farm, basic horse knowledge is vital. Information on managing horses, such as feeding, facilities, disease prevention, and parasite control, is important.

FEEDS AND FEEDING

The equine digestive tract is well suited to a forage-based diet comprised of grazed grasses/legumes, as well as mechanically harvested forages. There are significant regional differences in both availability and type of **forage** (Table 33.1). Grains also provide an important ingredient in equine rations. **Concentrate** mixtures containing grains, protein supplements, and vitamin and mineral additives are prepared and sold by commercial feed companies. U.S. Department of Agriculture (USDA) estimates suggest that approximately 90% of horse operations feed at least some grain regardless of size. Of the grain purchased for horses, nearly 80% is delivered in the form of 50- to 100-pound bags purchased at a retail feed store. Wet molasses can be added to these concentrate mixtures to make sweet feed. The forage and concentrate mixtures may be mixed and pelleted to make a complete, higher-priced convenience feed for horse owners.

Some forage varieties, such as endophyte-infected fescue, can be harmful to horses—particularly broodmares. Fescue toxicity is problematic in the southeastern region of the United States and may result in increased abortion rates, reduced rebreeding rates, diminished milk production, and significantly higher rates of stillbirth. Management approaches to solving this dilemma are multifaceted and often variable in effectiveness. Certainly, pastures with fescue should be tested for the presence of endophytes, pregnant mares should be kept off infected pastures and hay in the last 3 months of pregnancy, and reproductive rates should be carefully monitored.

Although horses spend less time chewing than ruminants, the normal, healthy horse with a full set of teeth can grind grains such as oats, barley, and corn so that cracking or rolling these feeds is unnecessary. Wheat and milo, however, should be cracked to improve digestibility. The horse's stomach is relatively small, composing only 10% of the total digestive capacity. Only a small amount of digestion takes place in the stomach, and food moves rapidly to the small intestine. From 60% to 70% of the protein and soluble carbohydrates are digested in the small intestine, and

Table 33.1
FORAGES FOR VARIOUS REGIONAL EQUINE PASTURES

Northeast, Midwest, Upper South, and Pacific Northwest:
Alfalfa
Bird's foot trefoil
Brome grass
Kentucky bluegrass
Orchard grass
Red clover
Timothy
White clover
Southeast:
Alfalfa
Bahia grass
Bermuda grass
Dallis grass
Southwest:
Alfalfa
Bermuda grass
Great Plains and Intermountain West:
Alfalfa
Wheat grass

Table 33.2
DESCRIPTION OF EQUINE BODY CONDITION SCORES

Condition Score	Description
1	Poor—extremely emaciated; ribs and bone structure easily discernable
2	Very thin—emaciated; ribs prominent; bone structure somewhat noticeable
3	Thin—slight fat cover over ribs
4	Moderately thin—faint outline of ribs, neck, shoulder, and withers not obviously thin
5	Moderate—ribs not visually apparent, but easily palpated; back level over loin; withers rounded
6	Moderately fleshy—fat over ribs and tailhead feels spongy; fat deposits generally apparent
7	Fleshy—noticeable filling of space between ribs with fat; crease down back over loin
8	Fat—deposition of fat along inner buttocks; thickening of neck; fat withers, tailhead, and behind shoulders
9	Obese—bulging fat, flank filled in flush; patchy fat appearing over ribs

Source: Adapted from Henneke et al. (1983).

about 80% of the fiber is digested in the cecum and colon. The large intestine has about 60% of the total digestive capacity, with the colon being the largest component. Bacteria that live in the cecum aid digestion there. Minerals, proteins as amino acids, lipids, and readily available carbohydrates such as glucose are absorbed in the small intestine.

Owners of pleasure horses may liberally feed horses in an attempt to attain a pleasing physical appearance. Perhaps more horses are overfed than are underfed. Proper condition of the horse can be monitored via a numerical scoring system. The range of **body condition scores** is provided in Table 33.2. Also, many people want to be kind to their animals, keeping them housed in a box stall when weather conditions are undesirable. This may not be best for the physiological state of the horse. Certainly, if any deficiency exists in the feed provided, such a deficiency is much more likely to affect horses that are not running on good pasture, where they can forage for themselves.

Young, growing foals should be fed correctly to allow for proper growth, but overfeeding and obesity are discouraged. Quality of protein and amounts of protein, minerals, and energy are important for young growing horses. Soybean meal or dried milk products in the concentrate mixture provide the amino acids and minerals that might otherwise be deficient or marginal in the weanling ration.

Good-quality pasture or hay supplemented with grain can provide the nutrition needed by young horses. An appropriate salt-mineral source should be provided at all times, and clean water is essential.

During the first 8 months of gestation, pregnant mares perform well on good pastures or on good-quality hay supplemented with a small amount of grain and an appropriate source of salt-minerals. As the fetus grows during the last trimester of pregnancy, the mare requires more concentrate and less fibrous, bulky hay. Oats, corn, or barley makes excellent feed grains for pregnant mares. Pregnant mares should be in good body condition but not obese. The pregnant mare should not be allowed to drop below a body score of 5 as a means to assure high fertility rates. Animals used for riding or working, whether they are pregnant or not, need more energy than those not working (Table 33.3). Horses being exercised heavily should be fed appropriate amounts of concentrates or high-energy forages to replace the energy used in the work. Body condition of the working horse should be carefully monitored to assure optimal health and performance. A thin horse is depicted in Figure 33.1. Horses may have insufficient body condition for a variety of reasons including insufficient feed intake, parasite infestation, or as a result of an illness.

Generally, lactating mares require more grain feeding than do geldings and pregnant or nonpregnant mares. Lactation is the most stressful nutritional period for a mare. If lactating mares are exercised, they must be fed additional grain and hay to meet the nutrient demand of the physical activity.

Stallions need to be fed as working horses during the breeding season and given maintenance ration during the nonbreeding season. Feeding good-quality hay with limited amounts of grain is usually sufficient for the stallion.

Feed companies provide properly balanced rations for horse farms of all sizes. An owner who has only one or two horses may find it highly advantageous to use a prepared feed, since it is difficult and laborious to prepare balanced rations for only a few animals. Using commercially prepared feeds or custom-blended rations can prevent nutritional errors, save time, and, for larger farms, be more cost-effective. A set

Table 33.3
MEGACALORIES BURNED AS A RESULT OF VARIOUS FORMS OF EXERCISE

Exercise—1 hour	Megacalories Burned per 1,000 lb of Body Weight
Walking	0.2
Slow trot	2.3
Fast trot–slow canter	5.7
Canter–full gallop	10.5
Strenuous (racing, reining, etc.)	17.7

Source: Adapted from NRC, *Nutrient Requirements of Horses.*

Figure 33.1
Body condition of horses should be carefully monitored to ensure that appropriate feed rations are provided. This horse is thin as the result of not receiving sufficient nutrients. Source: Tom Field.

Table 33.4
RULES OF THUMB FOR BLEND OF CONCENTRATES AND FORAGES IN EQUINE RATIONS

Class	Concentrate/Grain (%)	Forage/Hay (%)
Adults		
Maintenance	0	100
Light Work	20	80
Moderate work	40	60
Prolonged, heavy work	50	50
Brood Mares		
Last trimester	30	70
Nursing foal	30–50	50–70
Juveniles		
Weaned	50	50
Yearling	40	60

of generalized rules for the blend of concentrate and forage ingredients in a ration for horses of various production stages are listed in Table 33.4.

Table 33.5 shows some examples of horse rations and describes when and how these rations should be fed. Feed delivery to horses varies from region to region for both forages and concentrates (Tables 33.6 and 33.7). According to the NAHMS study (1998), horse owners fed their horses once per day, twice per day, three times or more daily, or less than daily at rates of 25%, 48%, 19%, and 8%, respectively.

Selection of feeds for inclusion in a ration also requires that managers make a careful assessment of the **quality characteristics** of a particular ingredient. For example, when selecting hay, it is important to evaluate the plant type composition, degree of leaf retention, stage of plant maturity at time of harvest, degree of foreign matter contamination, as well as aroma, color, and texture. Ideally, feeds should be tested at a qualified laboratory to quantify the nutrient content as well as to determine the presence of contaminants.

Table 33.5
SAMPLE RATIONS FOR HORSES OF DIFFERENT AGES AND IN VARIOUS STATES OF PRODUCTION

Creep feed for nursing foals. The grain should be fed at a rate of 0.5–0.75 lb of grain/100 lb body weight.

	Percentage in Grain Mix (%)
Corn, rolled or flaked	34.0
Oats, rolled or flaked	34.0
Soybean meal	22.0
Molasses	6.0
Dicalcium phosphate	2.0
Limestone	1.5
Trace-mineral salt	0.5

Grain mix for weanlings. The grain should be fed at a rate of 0.75–1.5 lb of grain/100 lb body weight. Select the grain according to the type of roughage fed. Allow free-choice consumption of either roughage type.

	Percentage in Grain Mix	
	Alfalfa Hay (%)	Grass Hay (%)
Corn, rolled or flaked	40.0	34.0
Oats, rolled or flaked	40.0	34.0
Soybean meal	12.0	23.0
Molasses	5.0	5.0
Dicalcium phosphate	2.5	3.0
Limestone	0	0.5
Trace-mineral salt	0.5	0.5

Grain mix for yearlings and mares. The grain should be fed at a rate of 0.5–1.0 lb of grain/100 lb of body weight. Select the grain according to the type of roughage fed. Allow free-choice consumption of either roughage.
Mares during late gestation and lactation can be fed the same grain mixes as yearlings. The grains should be fed at a rate of 0–0.5 lb/100 lb of body weight. Roughage consumption can vary from 1.5 to 2.5 lb/100 lb of body weight.

	Percentage in Grain Mix	
	Alfalfa Hay (%)	Grass Hay (%)
Corn, rolled or flaked	46.5	38.0
Oats, rolled or flaked	46.5	38.0
Soybean meal	0	15.5
Molasses	5.0	5.0
Dicalcium phosphate	1.5	2.0
Limestone	0	1.0
Trace-mineral salt	0.5	0.5

Grain mix for horses at maintenance and at work, dry mares during the first 8 months of gestation, or stallions. The grain should be fed as needed (to maintain body condition). Roughage consumption can vary from 1.5 to 2.5 lb/100 lb of body weight.

	Percentage in Grain Mix	
	Alfalfa Hay (%)	Grass Hay (%)
Corn	46.5	46.5
Oats	46.5	46.5
Molasses	5.0	5.0
Dicalcium phosphate	0	1.5
Monosodium phosphate	1.5	0
Trace-mineral salt	0.5	0.5

Source: Ginger Rich, *Horse Judging Guide*, Colorado State University Extension Service.

Table 33.6
PERCENT OF OPERATIONS THAT FED VARIOUS FORAGES IN 1997 BY REGION

| Type | Region | | | | |
	Southern	Northeast	Western	Central	All
Small bales (< 200 lbs)	82.9	92.7	92.8	83.2	86.6
Grass hay	69.5	54.9	43.8	36.0	54.2
Alfalfa hay	18.9	11.7	51.3	24.7	26.9
Grass/alfalfa mixed hay	15.3	53.8	40.4	55.9	35.1
Corn stalks, oat straw, other	5.1	4.6	12.7	8.0	7.5
Large bales (> 200 lb)	37.4	21.1	17.5	36.2	30.3
Grass hay	34.6	14.8	6.2	19.9	22.1
Alfalfa hay	2.9	1.2	5.6	4.5	3.7
Grass/alfalfa mixed hay	3.6	7.6	6.0	16.1	7.5
Corn stalks, oat straw, other	1.9	0.0	4.1	1.8	2.2
Non-baled dried forage (hay cubes, etc.)	5.9	5.4	8.8	7.6	6.9
Non-dried forage	1.4	0.0	1.2	0.6	1.0

Forage delivery systems were troughs/racks (41.4%), loose on the ground (41.1%), other individual feeders (8.5%), rubber tires (2.1%), hay nets (2.4%), and other (4.5%).

Source: NAHMS, 1998.

Table 33.7
PERCENT OF OPERATIONS FEEDING VARIOUS GRAINS/CONCENTRATES BY REGION

| Source | Region | | | | |
	Southern	Northeast	Western	Central	All
Unpelleted sweet feed	60.7	63.2	51.1	53.4	57.2
Unpelleted grain	41.9	29.2	47.5	48.1	42.9
Geriatric feed	5.5	9.6	10.8	6.9	7.6
Complete feed pellets/cubes	21.1	17.0	20.5	13.3	18.7
Grain mix with pellets	24.3	27.7	15.6	20.9	21.9
Other	8.1	7.4	11.1	6.2	8.3
None	2.2	6.2	11.9	5.0	5.6

Source: NAHMS, 1998.

One of the most potentially devastating digestive disorders of the horse is the condition known as **colic**. Colic is a term utilized to describe a broad range of abdominal pain. Abrupt changes in diet, feeding schedule, exercise regime, or housing type may lead to colic. The pain can be attributed to gas distension, decreased gut motility, parasitic infestation, ulcers, bowel displacement or twisted gut, or ingestion of sand or other foreign objects. While most cases of colic are relatively mild, affected horses typically exhibit pawing, pacing, or rolling in response to the discomfort. Treatment varies, but immediate veterinary care is advised.

The digestive system of the horse is adapted to a consistent diet of grass or hay. Table 33.8 outlines the steps managers can take to avoid colic.

Laminitis (founder) is an inflammation of the laminae of the foot resulting in severe pain and potentially chronic lameness. There are several basic causes of the

Table 33.8
MANAGEMENT STEPS TO PREVENT COLIC

Assure access to high-quality, clean water on a continuous basis.

Feed appropriate to the horse's need.

- High fiber, low carbohydrate feedstuffs of 8–10% protein are preferred.
- Feed no more than ½ of the ration in concentrate form.
- Concentrate feedings should be spaced out in several small meals.

Do not make sudden or frequent dietary changes.

Keep in mind that horses are nibblers often grazing for as much as 20 hours per day. The feeding regime should mimic this natural behavior.

Keep horses active. Those equine that spend most of their time in stalled conditions are most at risk to colic.

Avoid feeding on sandy ground to avoid sand impactions of the gut.

Maintain regular internal parasite prevention practices by assuring good sanitation in pens, paddocks, or stalls, and via regular administration of dewormers.

disease including overeating of grain that causes a change in gut pH and subsequent release of endotoxins resulting from gut bacteria death. Immunization against the endotoxins is recommended as a preventative measure. Other causes of founder include overwork on hard surfaces, overheated horses drinking excessive amounts of cold water, uterine infections caused when the mare retains placental tissue following birth, and overconsumption of lush pastures by horses already in excess body condition.

MANAGING HORSES

Proper management of horses is essential at several critical times: during the breeding season, foaling, weaning of foals, castration, and strenuous work. Enterprises that own horses are highly variable in type with about half serving as recreational experiences (Fig. 33.2), one-quarter are farms and ranches, 16% are breeding establishments, and 10% are focused on competitive showing according to the USDA.

Equine enterprises tend to have relatively small numbers of horses on inventory with the smallest category (less than nine head) dominated by recreational or farm/ranch enterprises (86%) but controlling only one-third of all horses. The larger enterprises account for less than 10% of all horse operations but control almost one-third of the inventory (Table 33.9).

Reproduction

Mares of the light breeds reach sexual maturity at 12–18 months of age, whereas draft mares are 18–24 months of age when sexual maturity is reached. Mares come into estrus every 21 days during the breeding season if they do not become pregnant. Heat lasts for 5–7 days with ovulation occurring toward the end of heat. Because of the relatively long duration of heat and because ovulation occurs toward the end of heat, horse owners often delay breeding a mare for 2 days after she has first been observed in heat. Some breeders have the mare bred every other day while she is in heat.

Although about 10% of all ovulations in mares are multiple ovulations, twinning occurs in only about 0.5% of the pregnancies that carry to term. The uterus of the mare apparently cannot support twin fetuses; consequently, most twin conceptions result in the loss of both embryos. The length of gestation is about 340 days (approximately 11 months), with usually only one foal being born. Mares usually

Figure 33.2
The vast majority of equine activities in the United States are focused on recreational activities.
Source: Coleman Locke.

Table 33.9
PERCENT OF OPERATIONS, HORSES, AND TYPE OF ENTERPRISE IN THE UNITED STATES

	Small (< 9 hd)	Medium (10–19 hd)	Large (> 19 hd)
Percent of enterprises	66	26	8
Percent of inventory	36	34	30
Boarding/training	3	10	17
Breeding	9	22	34
Farm/ranch	40	42	32
Pleasure	46	22	10
Other	2	3	6

Source: USDA, 2006.

come into heat 5–12 days following foaling, and fertile matings occur at this heat if the mare has recovered from the previous delivery.

Improving reproductive efficiency has been one of the major areas of focus in equine research. Improved technologies and management have yielded both improved reproduction performance and also the ability to allow horses with inherent fertility problems to reproduce.

Breeding Season

Fertility rates can be enhanced by dealing effectively with chronic uterine infection, identification of uterine damage due to **foaling**, waiting to breed mares until after the transition period following an estrus, and utilizing a good teasing program.

Mares should be teased daily with a stallion to determine the stage of their estrous cycle. When the mare is in standing heat (estrus), she is ready to be bred. When the stallion approaches the front of the mare, the mare reacts violently against the stallion if not in heat, but squats and urinates with a winking of the vulva if in heat. Cleanliness is paramount for both natural breeding and artificial insemination (AI) programs. Mares bred naturally should have the vulva washed and dried and the tail wrapped before being served by the stallion. If the mare can be bred twice during heat without overusing the stallion, breeding 2 and 4 days after the mare is first noticed in heat is desirable. A mature stallion can serve twice daily over a short time and once per day over a period of 1 or 2 months. A young stallion should be used lightly at about three or fewer services per week. In an AI program, the stallion's semen can be collected every other day to cover as many mares as possible, depending on the stallion's sperm numbers and motility. AI programs allow better management of the stallion.

Foaling Time

Mares normally give birth to foals in early spring and are most often foaled in a stall or barn. Prior to foaling, a mare undergoes enlargement of the udder, relaxation of the tail head, croup, and perineal region, appetite loss, and waxing of the teats where drops of sticky fluid harden at the ends of the teats giving a waxy appearance. A clean box stall that is bedded with fresh straw should be made available. If the weather is pleasant, mares can foal on clean pastures. When a mare starts to foal, she should be observed carefully but not disturbed unless assistance is necessary. If the head and front feet of the foal are being presented, it should be delivered without difficulty (Fig. 33.3). If necessary, however, a qualified person can assist by pulling the foal as the mare labors. Do not pull when the mare is not laboring and do not use a tackle to pull the foal. In situations where the front feet are presented but the head is not correctly aligned with the birth canal (Fig. 33.4), it may be necessary to push the foal back enough to get the head started along with the front feet. **Breech presentations** can endanger the foal if delivery is delayed; therefore, assistance should be given to help the mare make a rapid delivery if breech presentation occurs. If it appears that difficulties are likely to occur, a veterinarian should be called as soon as possible.

As soon as the foal is delivered, its mouth and nostrils should be cleared of membranes and mucus so it can breathe. If the weather is cold, the foal should be wiped dry and assisted in nursing. As soon as the foal nurses, its metabolic rate

Figure 33.3

A normal delivery presentation of a foal is front feet and head first.

Source: Tom Field.

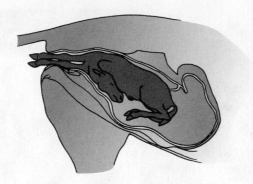

Figure 33.4
Malpresentation of foal for delivery; head and neck back. Source: *Handbook of Livestock Management Techniques* by Richard Battaglia, 4E, pp. 146, 149, 150. Copyright © 2007 by Pearson Education, Inc. Reprinted and Electronically reproduced by permission of Pearson Education Inc., Upper Saddle River, New Jersey.

increases helping it stay warm. The **umbilical cord** should be dipped in a tincture of iodine solution to prevent harmful microorganisms from invading the body.

In general, it is highly desirable to exercise pregnant mares up to the time of foaling. Mares that are exercised properly while pregnant have better muscle tone and are likely to experience less difficulty when foaling than those that get little or no exercise while pregnant. During the horsepower era, many mares used for plowing and similar work foaled in the field without difficulty. The foal was usually left with the mare for a few days, after which the foal was left in a box stall while the mare was worked.

Weaning the Foal

When weaning time arrives, it is best to remove the mare and allow the foal to remain in the surroundings to which it is accustomed. Since the foal will make every attempt to escape to find its mother, it should be left in a box stall or a secure and safe fenced lot. Fencing other than barbed wire should be used. Prior to weaning, the foal should become accustomed to eating and drinking on its own. High-quality hay or pasture and a balanced concentrate feed should be provided for the foal.

Castration

Colts that are not kept for breeding can be castrated any time after the testicles have descended; however, the stress of **castration** should not be imposed at weaning time. Some people prefer to delay castration until the colt has reached a year of age, whereas others prefer an earlier time. Genetic potential and nutrition, not time of castration, determine mature height and weight. Castration is performed because geldings can be handled more safely than stallions by owners and are less dangerous around other people. Castration should only be carried out by a licensed veterinarian.

In some cases, either one or both of the testes fail to descend into the scrotum resulting in a condition known as cryptorchid. Normal testical descent should be accomplished by 15 months of age. These individuals may require a more complicated surgery to successfully geld them.

Identification

Identification of horses varies considerably as some horses are not formally documented (29%) while others have multiple forms of identification. About one-half of equines are identified by registration paper, one-quarter via a Coggins Test document, 10% by brand, and 4% with a lip tattoo according to the USDA.

Tattooing them on the inside of the upper lip can permanently identify horses. This does not disfigure the animals and can be easily read by raising the upper lip. Other methods of identification include freeze branding, and microchip implants that are placed subcutaneously along the upper neck where the encapsulated transponder contains a unique identification sequence that can be read with a scanner.

Figure 33.5

Many horses are utilized in (A) competitive events such as jumping, (B) dressage, or as (C) working horses on farms or ranches. The working horse requires additional feed to ensure that the horse is able to perform at peak levels.

Sources: 33.5a: Ingus Evertovskis / Fotolia; 33.5b: Edu1971/Fotolia; 33.5c: John Sfondilias/Fotolia.

A

B

C

Care of Hardworking Animals

Hardworking horses that are sweating and have elevated temperature, pulse, and respiration rates should be washed (cooled down) before receiving water and feed. The animals should be given only small amounts of water and walked until they have stopped sweating and their pulse and respiration are within 10% of resting levels. At that time, water, hay, and grain can be given. This handling process is important; otherwise, horses may become colicky or founder.

Hardworking horses (Fig. 33.5) do require extra energy, usually in the form of grain. Development of an integrated nutritional, exercise, and health program to sustain high levels of performance is crucial to the performance and longevity of the horse.

HOUSING AND EQUIPMENT

Barns should be located on a higher elevation than the surrounding area to assure good drainage. They should be accessible to utilities and vehicles and, preferably, have a southeast exposure. Design should also take into account safety, air quality, labor intensity, and durability.

Although many styles of barns exist, those constructed with an aisle between two rows of stalls provide easy access and efficient use of space. Feed can be placed in the stalls on either side as the feed cart goes down the aisle. If the stalls can be opened from the outside and cleaned with mechanical equipment, labor also is saved. The hay manger in stalls should be constructed at chest height to the horse. Hayracks placed above the horse's head force the horse to inhale hay dust when reaching for the hay, and health problems can result.

To alleviate mixing hay with grain, a separate grain feeder should be placed several feet from the hayrack. The waterer or water bucket should be located along the outside wall for drainage purposes. The water source should also be placed some distance from the hay and grain so feed will not drop in the water. Electric float-controlled waterers save labor but are prone to malfunction. Although water buckets require more labor, they allow water intake to be easily monitored.

Box stalls are used for foaling and for the mare and the foal when the weather is severe. The stall should be constructed to allow complete cleaning and proper drainage. Minimum dimensions for a foaling stall are 14 × 14 ft, whereas a regular box stall should be at least 10 × 10 ft.

Feed should be stored in an area connected with or adjacent to the barn. Truck access to the feed storage area is vital. The floor should be solid (e.g., concrete) and the rodent-proof bins should be easily cleaned. Overhead (loft) storage of hay requires a great deal of labor and expensive construction. Large amounts of hay stored in the barn where the horses are housed increases dust and danger of fire.

Fences should be constructed based on assessment of cost, durability, safety, and strength. Desirable fence materials include PVC, wood, rubber/nylon, and diamond-mesh wire. If wood is used, it is important to protect material from horses that crib or chew the fence.

A dry, dust-free **tack room** is needed to house riding equipment, including saddles, blankets, bridles, and halters. In addition, it may be advisable to have a washroom where horses can be washed.

A **stock** is essential to care for horses that have been injured and need attention. If the horse is in a stock, it is unlikely that a person working around the horse can be kicked. The stock can be used for AI, teasing, palpation, or treating health problems. The stock is also useful for injections or when blood samples are being taken. Often, horses strike with a front foot when a needle is inserted for vaccinations or blood samples; a properly constructed stock will protect people as well as the horse.

CONTROLLING DISEASES AND PARASITES

Sanitation is of vital importance in controlling diseases and parasites of horses. Horses should have clean stalls and be groomed regularly. A horse that is to be introduced into the herd should be isolated for a month to prevent exposing other horses to diseases and parasites. Horses may have illnesses caused by bacteria or viruses, internal or external parasites, poisonous plants, nitrates, and/or moldy grains.

Regardless of the size of the equine enterprise, it is critical to establish a trusted working relationship with an equine veterinarian. Veterinarians can help prevent diseases as well as treat animals that are diseased. Most veterinarians prefer to assist in preventing health problems rather than treating the animals after they are ill. A variety of diseases are vaccinated against by horse owners (Table 33.10). A veterinarian should be contacted for assistance in establishing a sound preventative health plan.

Horse manure is an excellent medium for microorganisms that cause **tetanus**. Horses should be vaccinated to prevent tetanus, which can develop if an injury allows tetanus-causing microorganisms to invade through the skin. Usually two shots are given to establish **immunity**, after which a booster shot is given each year. Because

Table 33.10
PERCENT OF EQUINE OPERATIONS ADMINISTERING VACCINES

Vaccine Type	% of Operations
Flu	54
Strangles	27
Rhinopneumonitis	47
Rabies	33
West Nile virus	64
Equine encephalitis	
Western/Eastern	56
Venezuelan	18
Tetanus	61
Equine viral arteritis	12
Potomoc horse fever	11

Source: Adapted from USDA, 2006.

the same microorganisms that cause tetanus in horses also affect humans, those who work with horses should also have tetanus shots.

Strangles, also known as *distemper*, is a highly contagious bacterial disease that affects the upper respiratory tract and associated lymph glands. High fever, nasal discharge, swollen lymph glands, and a rattling sound associated with breathing are signs of strangles. The disease is spread by contamination of feed and water. Afflicted horses must be isolated and provided clean water and feed. An intranasal strangles vaccine should be given followed by an annual booster. Broodmares are subject to many infectious agents that invade the uterus and cause **abortion**; examples include *Salmonella* and *Streptococcus* bacteria and the viruses of **rhinopneumonitis** and **arteritis**. Should abortion occur, professional assistance should be obtained to determine the specific cause and to develop a preventative plan for the future.

Sleeping sickness, or **equine encephalitis**, is caused by viral infections that affect the brain of the horse. Different types, such as the eastern, western, and Venezuelan, are known. Vectors such as mosquitoes transmit encephalomyelitis. Horses rubbing noses together or sharing water and feed containers can also spread it. Vaccination against the disease consists of two intradermal injections spaced 1 week to 10 days apart. These injections should be given in April on an annual basis. Horses in the southern United States should be vaccinated twice a year.

Influenza is a common respiratory disease of horses. The virus that causes influenza is airborne, so frequent exposure may occur where horses congregate. The acute disease causes high fever and a severe cough when the horse is exercised; rest and good nursing care for 3 weeks usually gives the horse an opportunity to recover. Severe aftereffects are rare when complete rest is provided. Horse owners who plan for shows should vaccinate for influenza each spring. Two injections are required the first year, with one annual booster thereafter.

Chronic obstruction pulmonary disease (heaves) is a respiratory disease in which the horse experiences difficulty in exhaling air. The horse can exhale a certain volume of air normally, after which an effort is exerted to complete exhalation. A horse with a mild case of heaves can continue with light work, but horses with moderate to severe heaves have a limited ability to work.

Exercise-induced pulmonary hemorrage (bleeder) is a defect of circulation in which the minute blood vessels in the lungs rupture when the horse is put under the stress of exercising. Seldom fatal, the condition restricts correct lung function and inhibits breathing. The condition is most often observed in race horses.

Hyperkalemic periodic paralysis (HYPP) is a muscle disease that is genetically controlled. HYPP affects some lines of Quarter Horses, Paints, and Appaloosas. Affected horses experience tremors as a result of abnormal muscle fiber activity. A mutation in the gene responsible for sodium and potassium regulation is apparently the cause. Ideally, those horses that carry the trait should not be allowed to breed.

Moonblindness (recurrent uveitis) in which the horse is blind for a short time, regains its sight, and then again becomes blind for a time. Periods of blindness may initially be spaced as much as 6 months apart. The periods of blindness become progressively closer together until the horse is continuously blind. This condition received the name *moon blindness* because the trait is first noticed when the periods of blindness occur about a month apart, which originally led to the thought that the periods of blindness were associated with changes in the moon. More recently this condition has been associated with infections, parasites, and riboflavin deficiency.

Unfortunately, new diseases may be introduced into a region with potentially devastating consequences. A case in point is **West Nile virus**, a disease that is transmitted via mosquito or bird vectors to both equines and humans. The introduction of the virus into the United States followed outbreaks in Africa, Asia, and Europe. Affected animals exhibit depression, loss of hindquarter coordination, tremors, and paralysis. Symptoms may progress rapidly and spread widely as a result of mosquito proliferation and bird migration. A vaccine is available. Horses can become infected with internal and external parasites. External parasites such as ticks can transmit pathogens to horses. **Potomac horse fever** is an example of a tick-borne disease characterized by cessation of gut sounds followed by severe diarrhea, which may be accompanied by colic. Nearly one-third of all cases result in death with nearly one-quarter of infected equines experiencing laminitis. Vaccination and tick control are the key preventative steps.

Control of internal parasites consists of rotating horses from one pasture to another, spreading manure from stables on land that horses do not graze, and treating infested animals.

Pinworms develop in the colon and rectum from eggs that are swallowed as the horse consumes contaminated feed or water. These parasites irritate the anus, which causes the horse to rub the base of its tail against objects even to the point of wearing off hair and causing skin abrasions. Pinworms are controlled by oral administration of proper vermifuges.

Bots are the larval stage of the botfly. The female botfly lays eggs on the hairs of the throat, front legs, and belly of the horse. The irritation of the botfly causes the horse to lick itself. The eggs are then attached to the tongue and lips of the horse, where they hatch into larvae that burrow into the tissues. The larvae later migrate down the throat and attach to the lining of the stomach, where they remain for about 6 months and cause serious damage. A paste dewormer is available to control bots.

Adult **strongyles** (bloodworms) firmly attach to the walls of the large intestine. The adult female lays eggs that pass out with the feces. After the eggs hatch, the larvae climb blades of grass where they are swallowed by grazing horses. The larvae migrate to various organs and arteries where severe damage results. Blood clots, which form where arteries are damaged, can break loose and plug an artery. Strongyles are treated with ivermectin and other dewormers.

Adult **ascarids** are located in the small intestine. The adult female produces large numbers of eggs that pass out with the feces. The eggs become infective if they

Table 33.11
IMPORTANT RECORDS TO BE KEPT ON INDIVIDUAL HORSES[a]

Vaccinations

Parasite control

Medical treatments

Disease and/or injury

Dental exams/floating schedules

Shoeing and hoof trimming

Feeding schedule and ration quantity and composition

Baseline vital statistics (temperature, pulse, respiration rate)

Breeding records

Competition and training schedule

[a]Dates, products used, professionals involved should be recorded.

Table 33.12
THE INCIDENCE OF VARIOUS DISEASES AND DISORDERS IN THE HORSE

Disorder	Incidence in Foals (< 6 mo.) %	Incidence in Horses (> 6 mo.) %
Colic	4	2
Other digestive disorder	6	—
Respiratory disease	4	2
Eye conditions	1	1
Skin problems	1	1
Reproductive disease	—	1
Injury	9	5
Lameness or hoof problems	3	3
Neurological disorder	6	—

Source: USDA, 2006.

are swallowed when the horse eats them while grazing. The eggs hatch in the stomach and small intestine, and the larvae migrate into the bloodstream and are carried to the liver and lungs. The small larvae are coughed up from the lungs and swallowed. When they reach the small intestine, they mature and produce eggs. These large worms may cause an intestinal blockage in young horses. The same chemicals used for the control of strongyles are effective in the control of ascarids.

Horses should be dewormed at least twice a year and up to four times a year. The type of dewormer should be rotated every second time as a means to avoid parasites developing resistance to specific compounds.

An important component of a sound health management plan is the development of an effective record-keeping system. Table 33.11 outlines the important records that should be maintained on individual horses.

Horses are afflicted by a variety of diseases and disorders. The rate of morbidity and mortality varies by the age of the affected horse and by the particular disease in question. Table 33.12 outlines rate of disease incidence while Table 33.13 documents mortality from the most common causes of death.

Table 33.13
MORTALITY RATE ORIGINATING FROM VARIOUS DISEASES AND DISORDERS IN THE HORSE

Cause of Death	Incidence in Foals (< 6 mo.) %	Incidence in Horses (> 6 mo.) %
Old age	—	30
Injury	24	16
Colic	3	15
Lameness	8	8
Noncolic digestive disorder	8	3
Strangles	2	1
Other respiratory disease	5	2
Dystocia	—	2

Source: USDA, 2006.

CHAPTER SUMMARY

- The major digestive structures of the equine include the stomach, small intestine, cecum, and colon.

- Nutrients should be provided to the horse in accordance with its requirements as determined by age, work schedule, lactation status, and gestation status.

- Key management areas for horses include reproduction, weaning, housing, and health.

KEY WORDS

forage
concentrates
body condition scores
quality characteristics
colic
laminitis (founder)
foaling
breech presentation
umbilical cord
castration
identification
barns
box stalls
tack room
stock
tetanus
immunity

strangles
abortion
rhinopneumonitis
arteritis
sleeping sickness (equine encephalomyelitis)
influenza
Chronic obstruction pulmonary disease (heaves)
Exercise-induced pulmonary hemorrage (bleeder)
Hyperkalemic periodic paralysis (HYPP)
Moonblindness (recurrent uveitis)
West Nile virus
Potomac horse fever
pinworms
bots
strongyles
ascarids

REVIEW QUESTIONS

1. Discuss regional differences in desired pastures for horses.
2. Discuss the appropriate mix of concentrates and forages in the equine diet for various ages/stages.
3. Describe body condition scoring in horses.
4. Discuss the impact of exercise on nutritional requirements.
5. List the primary forage quality characteristics.

6. Describe management interventions that can help prevent colic.
7. Describe demographics of the U.S. horse industry.
8. Discuss reproductive management of the equine.
9. Discuss facility requirements for horse enterprises.
10. Describe the major diseases and disorders afflicting horses.
11. Describe differences in vaccination rates for various diseases.
12. List the important records to be maintained by horse owners.
13. Discuss disease incidence and mortality rates.

SELECTED REFERENCES

Ambrosiano and Harcourt. 1989. *Complete Plans for Building Horse Barns Big and Small.* Grand Prairie, TX: Equine Research, Inc.

Brown, J., S. Pilliner, and Z. Davies. 2003. *Horse and Stable Management.*

Battaglia, R.A. 2007. *Handbook of Livestock Management,* 4th ed. Upper Saddle River, NJ: Prentice Hall.

Cunha, T. J. 1991. *Horse Feeding and Nutrition*, 2nd ed. San Diego: Academic Press, Inc.

Ensminger, M. E. 1990. *Horses and Horsemanship.* Danville, IL: Interstate Publishers, Inc.

Ensminger, M. E., J. E. Oldfield, and W. W. Heinemann. 1990. *Feeds and Nutrition.* Clovis, CA: Ensminger Publishing Co.

Evans, J.W. 2001. *Horses: A Guide to Selection, Care, and Enjoyment.* W. H. Freeman, San Francisco, CA.

Henneke, O. R. 1983. An objective method for judging a horse's body condition. *Equine Veterinary Journal,* 371–372.

Loving, N. S. 1993. *Veterinary Manual for the Performance Horse.* Grand Prairie, TX: Equine Research, Inc.

National Research Council. 1989. *Nutrient Requirements of Horses.* Washington, DC: National Academy Press.

Parker, R. 2013. *Equine Science.* Clifton Park, NY: Delmar Cengage Learning.

Shideler, R. K., and J. L. Voss. 1984. *Management of the Pregnant Mare and Newborn Foal.* Colorado State University Expt. Sta. Spec. Series 35.

USDA. 2006. *Equine 2005—Baseline Reference of Equine Health and Management.* Washington, DC: National Animal Health Monitoring Systems: APHIS: VS.

34

Poultry Breeding, Feeding, and Management

The poultry industry in the United States has been transformed into a dynamic, fully coordinated system of production, processing, and marketing. The distribution of cash receipts from the production of poultry is 62%, 21%, 14%, and 3% for broilers, eggs, turkeys, and others (geese, ducks, etc.).

BREEDS AND BREEDING

The term *poultry* applies to chickens, turkeys, geese, ducks, pigeons, peafowls, and guineas. The head and neck characteristics that distinguish several poultry types are shown in Figure 34.1. Turkeys have some unusual identifying characteristics including a beard (a black lock of hair on the upper chest of the male turkey). They also have a **caruncle** —a red-pinkish fleshlike covering on the throat and neck, with the **snood** hanging over the beak.

Characteristics of Breeds

Chickens   Chickens are classified according to class, breed, and variety. A **class** is a group of birds that has been developed in the same broad geographical area. The four major classes of chickens are American, Asiatic, English, and Mediterranean. A **breed** is a subdivision of a class composed of birds of similar size and shape. Some important breeds, strains, lines, and synthetics are shown in Figure 34.2. The parts of a chicken are shown in Figure 34.3. A **variety** is a subdivision of a breed composed of birds of the same feather color and type of comb.

Factors such as egg number and size, eggshell quality, efficiency of production, fertility, and hatchability are most important to commercial egg producers. Broiler producers consider such characteristics as white plumage color and picking quality, egg production, fertility, hatchability, growth rate, carcass quality, feed efficiency, livability, and egg production. Also, breeds, strains, and lines that cross well with each other are important to broiler breeders.

The breeds of chickens listed in Table 34.1 were developed many years ago. These specific breeds are not easily identified in the commercial poultry industry because the breeds have been crossed to produce different varieties and strains (see Fig. 34.2). Fancy breeds are exhibited at shows or propagated for specialty marketing and are more a novelty than part of today's poultry industry.

learning objectives

- Describe the traditional breeds of poultry and their distinguishing characteristics
- Describe the productivity of turkeys and broilers
- Compare the use of genetic selection and crossing in the poultry industry to other livestock enterprises
- Describe incubation management
- List the primary facility and environmental management concerns in poultry enterprises
- Discuss management of laying hens
- Discuss the factors affecting profitability of poultry production
- Explain vertical integration in the poultry industry

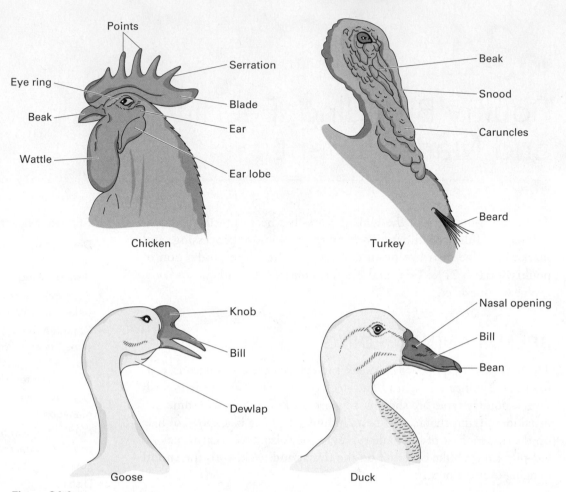

Figure 34.1
Distinguishing characteristics of several different types of poultry.

Turkeys   Turkeys are categorized into eight varieties, but most turkeys grown for meat are white. The eight varieties are (1) Bronze, (2) Narragansett, (3) White Holland, (4) Black, (5) Slate, (6) Bourbon Red, (7) Small Beltsville White, and (8) Royal Palm. Two types of turkeys are commercially important in the United States today: the Small White and the large Broad White. Since the 1950s, the emphasis has been on producing larger, white turkeys of the type shown in Figure 34.4. A corresponding decrease in the number of Small Whites has occurred.

Ducks   Ducks are classified into many domestic breeds but only a few are of economic importance. The most popular breed in the United States is the White Pekin. It is most valuable for its meat, and produces excellent carcasses at 7–8 weeks of age. Another breed, the Khaki Campbell, is used commercially in some countries for egg production.

Geese   Breeds of geese such as the Embden, Toulouse, White Chinese, and Pilgrim are all satisfactory for meat production. The characteristics preferred by most commercial goose producers include a medium-sized carcass, good livability (low mortality), rapid growth, and a heavy coat of white or nearly white feathers. The Embden

Figure 34.2

Traditional chicken breeds included the (A) Rhode Island Red and (B) Plymouth Rock that specialized as layers; and meat-type breeds such as the (C) Leghorn. Source: 34.2a: Edward Westmacott/Fotolia; 34.2b: Alkerk/Fotolia; 34.2c: Andreamangoni/Fotolia.

and White Chinese breeds meet these requirements. One variety of the Toulouse is gray; another variety is buff. The Pilgrim gander is white; the female is grayish. The White Chinese breed is pure white.

Breeding Poultry

Turkeys There is a specific breeding cycle in the production of baby poults for distribution to growers who raise them to market weight. The pedigree flocks (generation 1) or parent stocks are pure lines. The pure lines are crossed with one another to produce generation 2. Generation 2 lines are crossed with one another to produce generation 3. Eggs from these line crosses are selected into female or male lines and sent to hatcheries. Males from the male lines are selected for such meat traits as thicker thighs, meatier drumsticks, plumper breasts, faster rate of growth, and higher feed efficiency. The females of the female line are selected for greater fertility, hatchability, egg size, and meat conformation. The male-line males are crossed with female-line females; the eggs produced are hatched and the poults are sold for the production of market turkeys.

Laying hens are approximately 30 weeks of age when they reach sexual maturity and are put under controlled lighting to stimulate the start of laying. The average hen usually lays eggs for about 25 weeks and produces 88–93 eggs. The hen is considered "spent" at the end of the 25-week laying cycle, and most hens are marketed at that time for meat. Hens can be molted and stimulated to lay for another 25-week cycle.

Figure 34.3
The external parts of a chicken.

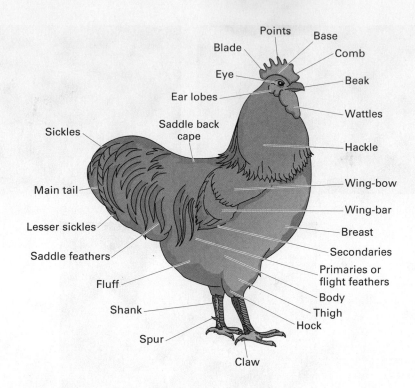

Table 34.1
CERTAIN BREEDS OF CHICKENS AND THEIR MAIN CHARACTERISTICS

Breed	Purpose	Type of Comb	Color of Egg
American Breeds			
White Plymouth Rock	Eggs and meat	Single	Brown
Wyandotte[a]	Eggs	Rose	Brown
Rhode Island Red	Eggs	Single and rose	Brown
New Hampshire	Eggs and meat	Single	Brown
Asiatic Breeds			
Brahma[a]	Meat	Pea	Brown
Cochin[a]	Meat	Single	Brown
English Breeds			
Australorp	Eggs	Single	Brown
Cornish	Meat	Pea	Brown
Orpington[a]	Meat	Single	Brown
Mediterranean Breed			
Leghorn	Eggs	Single and rose	White

[a]These breeds are of minor importance to the U.S. poultry industry.

They require 90 days of molting and produce only 75–80 eggs during this later laying cycle; therefore, producers should compare the costs of growing new layers with that of recycling the old ones before deciding if it is economical to recycle old layers. The spent hens usually go into soup products.

Turkey eggs are not produced for human consumption because it costs much more to produce food from turkey eggs than from chicken eggs. It would cost about 50 cents each to produce turkey eggs for this purpose.

Figure 34.4
Mature Tom turkey. Source: Uolir/Fotolia.

Table 34.2
AVERAGE LIVE WEIGHTS AND FEED EFFICIENCIES OF LARGE WHITE TOMS AND HENS

Year	Live Weight at 18 Weeks		Feed/Gain 0–18 Weeks		Days to 35 lb Market Weight	
	Toms	Hens	Toms	Hens	Toms	Hens
1970	16.9	9.0	3.10	3.10	235	151
1975	21.3	11.7	2.80	2.40	194	144
1980	22.4	12.2	2.70	2.30	185	135
1985	23.0	12.3	2.74	2.45	175	130
1990	27.1	14.2	2.80	2.44	156	107
1995	29.3	15.1	2.69	2.35	149	103
2000	31.9	15.5	2.62	2.36	143	100
2002	32.2	15.5	2.61	2.30	136	99

Source: Multiple sources.

All turkey hens are artificially inseminated to obtain fertile eggs. As turkeys have been selected for such broad breasts and large mature size, it is difficult for toms to mate successfully with hens. Semen is collected from the toms and used for inseminating the hens. Usually 1 tom to 10 hens is sufficient for attaining desired reproductive rates.

Compared to years past, contemporary turkeys grow more rapidly, convert feed into meat more efficiently, has more edible meat per unit of weight, and require less time to reach market. Most turkeys grown today are white-feathered, so there are no dark pin feathers to discolor the carcass. Breeders have made substantial genetic progress in the turkey industry. Over the past 30 years, the market weight of turkeys has nearly doubled while substantially decreasing the time to market and the feed required perpound of gain (Table 34.2). The average market weight of young turkeys is approximately 29 pounds per bird.

Chickens   Sophisticated selection and breeding methods have been developed in the United States for increased productivity of chickens for both eggs and meat. For example, since the late 1950s, breeders have increased 56-day body weight by over 4 lb. Improvements in feed efficiency, total yield, and breast yield have also been remarkable. The discussion here is directed primarily at improvement of chickens; however, methods described for chickens can also be applied to improve meat production in turkeys, geese, and ducks. The genome of the chicken is carried on 39 chromosome pairs with more than 1 million base pairs. The sequencing of the chicken genome in 2004 paved the way for breeders to more effectively select for traits that have been difficult or costly to measure such as disease resistance. Furthermore, breeders will be able to better match populations of birds to specific environmental and management conditions.

Early poultry breeding and selection were concentrated on qualitative traits, which, from a genetic standpoint, are more predictable than quantitative traits. **Qualitative traits**, such as color, comb type, abnormalities, and sex-linked characteristics, while important, are less important than **quantitative traits**, such as egg production, egg characteristics, growth, fertility, and hatchability.

Quantitative traits are more difficult to select for than are qualitative traits because the mode of inheritance is more complex and the role of the environment is greater. Quantitative traits differ greatly as to the amount of progress that can be attained through selection. For example, increased body weight is much easier to attain than increased egg production. Furthermore, a relationship usually exists between body size and egg size. Generally, if body size increases, a corresponding increase in egg size will occur. Most quantitative traits of chickens fall in the low-to-medium range of heritability, whereas qualitative traits fall in the high range.

Progress in selecting for egg production has been aided by the **trap nest**—a nest equipped with a door that allows a hen to enter but prevents her from leaving. The trap nest enables accurate determination of egg production of individual hens for any given period and helps to identify and eliminate undesirable egg traits and broodiness (the hen wanting to sit on eggs to hatch them). Furthermore, it allows the breeder to begin pedigree work within a flock.

According to the USDA, broiler market weight targets are broken in to four primary categories: less than 4.25 lb, 4.26–6.25 lb, 6.26–7.75 lbs, and more than 7.75 lb at time of marketing. The percent of production accounted for by each category is 32, 40, 19, and 9 percent respectively. The days on farm prior to marketing for each category equates to 39, 49, 56, and 63 days. Smaller birds are targeted to the food service sector while most heavier birds are further processed.

The U.S. broiler industry is an example of the phenomenal changes that can result from the application of genetic selection pressure and improved management (Table 34.3). The dramatic improvements in live weight, feed conversion, survivability, and time to market have been largely responsible for the remarkable growth of chicken's market share. From 1945 to 1995 the cost (adjusted for inflation) of producing live broilers declined from $2 per pound to approximately $0.25 per pound. This dramatic improvement can be attributed to the broiler multiplication process as described in Table 34.4. The global value of the grandparent, parent, and broiler inventories is estimated at $275 million, $1 billion, and $60 billion, respectively. The foundation and great grandparent stock are so valuable as to elude estimation.

The competitive nature of the genetics business has led to a high degree of market concentration. The top two genetic providers control more than 75% of the market while the top four combine to control just over 90% (Table 34.5).

Genetic providers in the future will have to refocus their emphasis on those traits that directly affect consumer demands and perception.

Consumers will continue to demand products that are affordable but also deliver superior palatability while meeting societal expectations in regards to food safety, environmental impact, and animal welfare. For example, selection emphasis will be placed on improved bird health, reduced metabolic dysfunction, improved skeletal structure, and limiting pecking behavior to meet welfare concerns.

Table 34.3
AVERAGE PERFORMANCE OF BROILERS

Year	Average Live Wt.	Feed Conversion	% Mortality	Age in Days
1925	2.2	4.7	18	112
1935	2.6	4.4	14	98
1945	3.1	4.0	10	84
1955	3.3	3.0	7	70
1965	3.5	2.4	6	63
1975	3.7	2.1	5	56
1985	4.2	2.0	5	49
1995	4.7	1.95	5	47
2000	5.0	1.95	5	46
2005	5.2	1.9	4.5	46
2009	5.6	1.85	NA	44

Source: USDA.

Table 34.4
BROILER MULTIPLICATION PROCESS

Generations	Population Size	Generation Goal
Pedigree (F)	90,000 hens	Purelines subjected to heavy culling pressure, foundation stock that is never sold.
Great grand-parents (GGP)	250,000 hens	Some selection pressure but mass production is goal. These stocks are also not sold to protect competitive advantage.
Grandparents (GP)	10,000,000 hens	Sold as baby chicks in global market.
Parents (P)	400,000,000 hens	Day-old parents are sold by a variety of companies in a broad market.
Broilers (B)	40,000,000,000/yr	Produced for consumption.

Source: Adapted from multiple sources.

Table 34.5
BIG 4 BROILER GENETICS COMPANIES

Company	% Market Share
Aviagen[a]	44
Cobb-Ventress (owned by Tyson)	33
Hubbard/ISA	10
Hybro	5
Others	8

[a]Formed by merger of Ross, Arbor Acres, and Lohmenn Indian River.
Source: Adapted from multiple sources.

The two most important types of selection applied primarily to chickens and, to a lesser extent, to other poultry today are mass selection and family selection. In **mass selection**, the older method, the best-performing males are mated to the best-performing females. Mass selection is effective in improving traits of high heritability. **Family selection** is a system whereby all offspring from a particular mating are designated as a family, and selection and culling within a population of birds are based on the performance level of the entire family. This type of selection is most adaptable for traits of low heritability. This is not to say that progress in traits of high heritability cannot be made by using family selection—quite the opposite is true. However, mass selection is effective for traits of high heritability and certainly easier.

Progress through selection is rather slow for the reproductive traits because they are low in heritability. Most breeding programs are geared toward the improvement of more than one trait at a time. For example, to improve egg-laying lines, it might be necessary to attempt simultaneously to improve egg numbers, shell thickness, interior quality of the egg, and feed efficiency.

Outcrossing   Defined as mating unrelated breeds or strains, **outcrossing** is probably more adaptable to the modern broiler industry than to the egg-production industry, though it can be used to improve egg production. Numerous experiments have established fairly accurately which breeds or strains cross well with each other. Knowledge today is sophisticated to the extent that breeders know which strain or line to use as the male line and which to use as the female line.

Crossing Inbred Lines   Development of **inbred lines** for crossing is used largely for increasing egg production. *Inbreeding* is defined as the mating of related individuals or as the mating of individuals more closely related than the average of the flock from which they originated. The mating of brother with sister is the most common system of inbreeding for poultry; however, mating parents with offspring gives the same results. Both of these types of inbreeding increase the degree of inbreeding at the same rate per generation.

Hybrid chickens are produced by developing inbred lines and then crossing them to produce chickens that exhibit hybrid vigor. The technique employed in production of hybrid chickens is essentially the same as that used in development of hybrid plants. The breeder works with many egg-laying strains or breeds while producing hybrids. The strains or breeds are inbred (mostly brother or sister) for a number of generations, preferably five or more. In the inbreeding phase, many undesirable factors are culled out—selection is most rigid at this stage.

In all phases of hybrid production, the strains are completely tested for egg production and for important egg-quality traits. After the birds are inbred for the necessary number of generations, the second step—crossing the remaining inbred lines in all possible combinations—is initiated. These crossbreds are tested for egg production and for egg-quality traits. Crossbreds that show improvement in these traits are bred in the third phase, in which the best test crosses are crossed into three-way and four-way crosses in all possible combinations. Most breeders prefer the four-way cross. The offspring that result from three-way and four-way crosses are known as **hybrid chickens**.

Strain Crossing   **Strain crossing** is more easily done than inbreeding and crossing inbred lines because it entails only the crossing of two strains that possess similar egg-production traits. It is definitely a type of crossbreeding if the two strains are not related to each other.

Selection is oftentimes based on the computation of a net merit index utilizing the heritability of each trait considered, the relative economic importance of each trait, and genetic correlations among the traits. Birds having the most desirable index are used for breeding.

In production of broiler or layer hybrid chicks, computational software is used to determine which lines or strains are most likely to yield superior chickens. In broiler hybrids, hatchability, growth rate, economy of feed use, and carcass desirability are important. In layer hybrids, hatchability, egg production, egg size, egg and shell quality, and livability are important.

Practically all line-cross layers are now free from **broodiness** because attempts have been made to eliminate genes for broodiness, and records are available on which lines produce broody chicks when crossed. Lines that produce broody chicks in a line cross are no longer used for producing commercial layers.

FEEDING AND MANAGEMENT

The success of modern poultry operations depends on many factors. Hatchery operators must care for breeding stock properly so that eggs of good quality are available to the hatchery, and they must incubate eggs under environmental conditions that ensure the hatching of healthy and vigorous birds (Fig. 34.5). Feeding, health, and financial programs are essential to sound poultry management. The broiler and egg production cycles are provided in Figures 34.6 and 34.7, respectively.

Incubation Management

Each type of poultry has an **incubation** period of definite length (Table 34.6), and incubation management practices are geared to the needs of the eggs throughout that time. The discussion here of incubation management applies specifically to chickens. Although the principles of incubation management generally apply to other types of poultry as well, specific information should be obtained from other sources. Egg production varies by species with egg-type hens producing 270 per year, broiler-type hens yielding 170 eggs annually, and turkey hens producing 105 eggs per year.

All modern commercial hatcheries have some form of forced-air incubation system. Today's commercial incubator has a forced-air (fan) system that creates a highly uniform environment inside the incubator. Forced-air incubators are available in many sizes. All sizes are equipped with sophisticated systems that control temperature and humidity, turn eggs, and bring about an adequate exchange of air between the inside and outside of the incubator. The egg-holding capacity of forced-air incubators ranges from several hundred eggs to 100,000 or more.

Temperature   Proper temperature is probably the most critical requirement for successful incubation of chicken eggs. The usual beginning incubation temperature in the forced-air incubator is 99.5–100.0°F. The temperature should usually be lowered slightly (by 0.25–0.5°F) at the end of the fourth day and remain constant until the eggs are to be transferred to the hatching compartments (approximately 3 days before

Figure 34.5
Day-old chicks. Source: Kharhan/Fotolia.

Figure 34.6
Broiler production cycle.

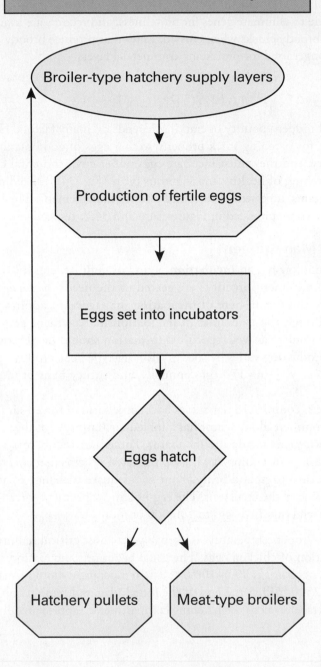

Broiler Production Cycle

Broiler-type hatchery supply layers

Production of fertile eggs

Eggs set into incubators

Eggs hatch

Hatchery pullets

Meat-type broilers

hatching). Three days before hatching, the temperature is again lowered, usually by about 1.0–1.5°F. Just before hatching, the chicks switch from embryonic respiration to normal respiration and give off considerable heat, which results in a high incubator temperature.

With reference to temperature, there are two especially critical periods during incubation—(1) the first through the fourth day and (2) the last portion of incubation. Higher-than-optimum temperatures usually speed the embryonic process and result in embryonic mortality or deformed chicks at hatching. Lower-than-optimum incubation temperatures usually slow the embryonic process and also cause embryonic mortality or deformed chicks.

Figure 34.7
Egg production cycle.

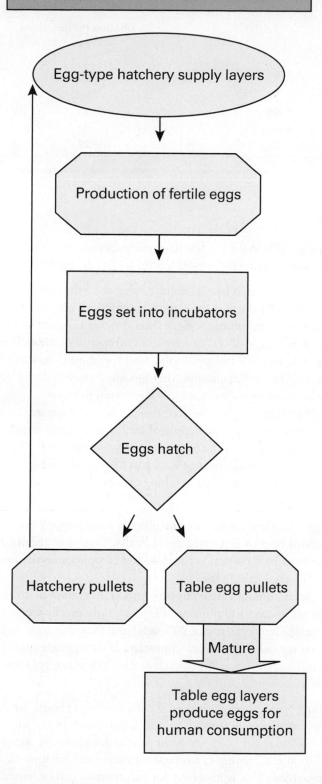

Egg Production Cycle

Egg-type hatchery supply layers

Production of fertile eggs

Eggs set into incubators

Eggs hatch

Hatchery pullets

Table egg pullets

Mature

Table egg layers produce eggs for human consumption

Humidity   A relative humidity of 60–65% is needed for optimum **hatchability**. All modern commercial forced-air incubators are equipped with sensitive humidity controls. The relative humidity should usually be raised slightly in the last few days of incubation, as it has been shown that successful hatching rates are greater when the relative

Table 34.6
INCUBATION TIMES FOR VARIOUS BIRDS

Type	Incubation Period (days)
Chicken	21
Turkey	28
Duck	28
Goose	28–32
Pheasant	24
Bobwhite quail	23
Guineas	27

humidity is close to 70% in the last few days of incubation. Providing optimum relative humidity is essential in reducing evaporation from eggs during incubation.

Position of Eggs   Eggs hatch best when incubated with the large end (the area of the space called the *air cell*) up; however, good hatchability can also be achieved when eggs are set in a horizontal position. Eggs should never be set with the small end up because a high percentage of the developing embryos will die before reaching the hatching stage. The head of the developing embryo should be situated near the air cell. In the last few days of incubation, the eggs are transferred to a different type of tray—one in which the eggs are placed in a horizontal position.

Shortly before hatching, the beak of the chick penetrates the air cell. The chick is now able to receive an adequate supply of air for its normal respiratory processes. The horny point (egg tooth) of the beak eventually weakens the eggshell until a small hole is opened, at which point the egg is said to be *pipped*. The chick is typically out of the shell within a few hours. The hatching process varies considerably among different species.

Turning of Eggs   Modern commercial incubators are equipped with time-controlled devices that permit eggs to be turned periodically. Eggs that are not turned enough during incubation have little or no chance of hatching because the embryo often becomes stuck to the shell membrane.

Commercial incubators are equipped with setting trays or compartments that allow eggs to be set in a vertical position. They also have mechanisms that rotate these trays so that the chicken eggs rotate 90° each time they are turned. The number of times the eggs are turned daily is most important. If the eggs are rotated once or twice daily, the rate of hatchability will be much lower than if the eggs are rotated five or more times daily.

Oxygen Content   The air surrounding incubating eggs should be 21% oxygen by volume. At high altitudes, however, the available oxygen in the air may be too low to sustain the physiological needs of developing chick embryos, many of which may therefore die. Good hatchability can often be attained at high altitudes if supplemental oxygen is provided. This is necessary for commercial turkey hatching, but hatchability can be improved by selecting breeding birds that hatch well in environments in which the supply of oxygen is limited.

Carbon Dioxide Content   It is vital that the incubator be properly ventilated to prevent excessive accumulation of carbon dioxide (CO_2). The CO_2 content of the air in the incubator should never be allowed to exceed 0.5% by volume. Hatchability is lowered drastically if the CO_2 content of air in the incubator reaches 2.0%. Levels of more than 2.0% almost certainly reduce hatchability to near zero.

Sanitation   The incubator must be kept as free of disease-causing microorganisms as possible. The setting and hatching compartments must be thoroughly washed or steam-cleaned and fumigated between settings. In many cases, they should be fumigated more than once for each setting. An excellent schedule is to fumigate eggs immediately after they have been set and again as soon as they have been transferred to the hatching compartment.

Candling of Eggs   In maintaining a healthy and germ-free environment, eggs must be **candled**—examined by shining a light through each egg to see if a chick embryo is developing—at least once during incubation so that infertile and dead-germ eggs (eggs containing dead embryos) can be identified. Many operators candle chicken eggs on the fourth or fifth day and turkey and waterfowl eggs on the seventh to tenth day. Some operators candle eggs a second time when transferring eggs to the hatching compartment.

Managing Young Poultry

The main objective in managing young poultry is to provide a clean and comfortable environment with sufficient feed and water (Fig. 34.8).

Housing for Broilers

Broilers raised in the South are typically housed in long (> 500 ft long × 50 ft wide) housing arrangements. Designs vary from totally enclosed walls to siding provided via curtains that can be moved in accordance with weather conditions. Flooring is usually dirt, but concrete or asphalt may be used. Most are designed to use large fans for increased ventilation. Broiler houses located in northern regions tend to be fully insulated, windowless structures with concrete floors.

Brooders, water dispensers, and feeding systems are usually suspended from the ceiling so that they can be lifted out of the way for bird penning or house cleaning. Feed and water is provided via a fully or semiautomated system on a continuous basis. Vaccinations and medications are typically provided via water or feed.

A typical broiler farm will consist of 159 acres with a small portion allotted to the broiler enterprise. The average farm has 3.2 houses with a total bird capacity of 64,000 birds housed at a density of 0.5 to 1 sq ft per bird. The broiler flock turns over 6.5 times per annum on average for a total annual production of more than 410,000 birds per farm.

House Preparation   The brooder house should be thoroughly cleaned, disinfected, and dried several days before it receives young birds. All necessary brooding

Figure 34.8
(A) Seven-week old broilers.
(B) Modern broiler house.
Source: 34.8a: Melastmohican/
Fotolia; 34.8b: Cratervalley/Fotolia.

A   B

equipment should be tested ahead of time and be in the proper place. Such advance preparation is probably as essential to success as any management practice applied to young birds. It is imperative that the disinfectant have no effect on the meat or eggs produced by birds. Food and Drug Administration regulations that govern the use of disinfectants should be strictly observed.

Litter should be placed in the house when equipment is checked. Some commonly used substances are planer shavings, sawdust, wood chips, peat moss, ground corncobs, peanut hulls, rice hulls, and sugarcane fiber. However, the use of recycled paper litter is replacing some of the more traditional types as it provides improved control of ammonia, fecal caking, litter compaction, and enhances bird welfare. The entire floor should be covered by at least 2 in. of litter, which must be perfectly dry when the birds enter.

The primary purpose of litter is to absorb moisture. Litter that gets extremely damp should be replaced. Moisture levels in litter should not exceed 35%.

Floor Space   The availability of too much or too little **floor space** can adversely affect growth and efficiency of production. If floor space is insufficient, young birds have difficulty finding adequate feed and water. This can lead to feather picking and actual cannibalism. Too much space can cause birds to become bored, which can lead to problems similar to those caused by overcrowding. However, birds should be allowed freedom of movement in the growing area.

Broilers below 4.5 lb of live weight should have a maximum stocking density of 6.5 lb per square foot. Birds weighing between 4.5 and 5.5 lb and those exceeding 5.5 lb should have a stocking density that doesn't exceed 7.5 and 8.5 lb per square foot, respectively.

Broilers, turkeys, and waterfowl should have an area of 110–145 in.^{2} for at least their first 8 weeks. In certain situations, such birds could be allowed as much as 145–220 in.^{2} before they are 8 weeks old.

Feeder Space   Chick trough feeders should be placed within the hover guard area or partially under the hover (0.8 in. per chick or 80 in. per 100 chicks of 6 weeks of age). Producers should be sure the 10-watt attraction lights under the hover are working. One linear inch of **feeder space** per bird is sufficient for most species from the age of 1 day to 3 weeks; 2 in. for 3–6 weeks; and 3 in. beyond 6 weeks. Quail require less space, whereas turkeys, geese, and ducks require slightly more. Most feeders are designed so that birds can eat on either side of them.

The operator can determine how much feeder space is needed through observation. If relatively few birds are eating at one time, the feeding area is probably too large. If too many are eating at once, the area is probably too small. Young birds are usually fed automatically after the brooder stove enclosure is removed. Automated feeding systems (Fig. 34.9) reduce labor costs and minimize bird stress due to humans moving through the house to manually fill feeders.

Water Requirements   Three fountain-type waterers should be placed within the hover guard area (one fountain-type waterer per 100 chicks). Generally, two 1-gal water fountains are adequate for 100 birds that are 1 day old. More waterers can be added as necessary. Most commercial producers switch from water fountains to an automatic watering system when birds are put on automatic feeding. If trough-type waterers are used, allow at least 1 in. of water space per bird until they reach 10–12 weeks of age. More space might be needed thereafter. Trough-type waterers designed so that birds can drink from both sides are available. A space of at least 0.5 in. per bird is needed if pan-type waterers are used.

Lighting Requirements   Manipulation of lighting is an important management tool to help control the growth rate of broilers. With the intense selection for improved

Figure 34.9
Automated poultry feeding system. Source: Branex/Fotolia.

muscularity of broilers, birds early in the growth curve partition nutrients to muscle growth at the expense of other tissues especially skeletal and internal organ systems. Lighting systems coupled with low-density feed and moderate feed restrictions are useful in overcoming this problem.

Two lighting systems are typically used in the industry—16 hours light with 8 hours of dark or increasing light protocols. The lighting system must be matched with housing design, feeding program, market weight goals, and grower contract requirements. Most programs are initiated on day 3 following receipt of chicks. Except for the first week and the final 14 days of growout, at least 4 hours of darkness should be provided in every 24-hour period during which illumination at the bird level does not exceed 50% of the light level in the remaining hours.

Other Management Factors   Temperatures in the hatchery holding room should be at least 72°F. Young birds should have access to the proper feed as soon as they are placed in the brooder house. It is important that birds such as turkey poults start eating feed early. Some young poults die for lack of feed; although feed is readily available, they have not learned to eat it.

Birds should be vaccinated and given health care according to a prescribed schedule; a veterinarian should be consulted.

Managing 10- to 20-Week-Old Poultry

Management of poultry from approximately 10–20 weeks of age is quite different from managing younger birds. Improper practices in this most critical period could adversely affect subsequent production.

Confinement rearing is used by commercial producers of replacement birds (the birds that will be kept for egg production). In the three basic systems of confinement rearing, birds are raised on solid floors, slatted floors, or wire floors or cages.

Birds raised on floors in confinement should have 1–2 ft^2 of floor space per individual. The amount of space required depends on the environmental conditions and on the condition of the house. Caged birds should be allowed 0.5 ft^2 to not more than 0.75 ft^2. Air quality is also of concern and ammonia levels should not be allowed to exceed 25 parts per million. Dust is also of concern to both the health of the flock and the people responsible for their care. The use of electrostatic space charge systems are one new approach to improving air quality in poultry houses.

Replacement chickens raised in confinement are generally fed a completely balanced growing, or developing, ration that is 15–18% protein. Most birds are kept on a full-feeding program, but in certain conditions some producers restrict the amount of feed provided. Restricted feeding can take different forms: total feed intake, protein intake, or energy intake may be limited. The main purposes of a restricted feeding program are to slow growth rate (thus delaying the onset of sexual maturity and reducing the number of small eggs produced) and to lower feeding cost. Feed intake must not be restricted whenever a restricted lighting system (which is also used to delay sexual maturity) is in effect.

Automatic feeding and watering devices are used in most confinement operations. Watering- and feeding-space requirements are practically the same as those for birds that are 6–10 weeks old. As long as birds are not crowding the feeders and waterers, there is no particular need to increase the feeding and watering space.

Water consumption is of critical importance in assuring that feed consumption and bird growth rates are optimized. Most commercial growers monitor water consumption on a house-by-house basis if not on a pen-by-pen level. Water consumption becomes of particular concern as ambient temperature rises. For reasons of water conservation and labor efficiency, the use of nipple-type watering systems is most prevalent.

Lighting conditions are of particular concern in regards to poultry. Light plays three important roles in poultry management—stimulation of physiological cycles that are light dependent, initiation of hormone release, and impact on bird vision.

Feed intake can be enhanced with a variety of specific protocols:

1. Lighting practices of 23 hr light: l hr dark or an intermittent lighting of 1 hr light: 3 hr dark with an intensity of 4- to 5-ft candles will reduce bird activity and raise consumption.
2. Maintain sufficient floor space per bird to enhance intake.
3. Assure that feed grains are free of mold and other contamination.
4. Monitor the house for unproductive birds and remove them to improve overall flock efficiency.

The lighting regime used in confinement rearing is very important. Birds raised in window-type housing receive the normal light of long-day periods unless the house is equipped with some type of light-check. During short-day periods, supplemental lighting can be used to meet the requirements of growing chickens. Replacement chickens should receive approximately 14 hours of light daily up to 12 weeks of age. To delay sexual maturity, the amount should then be reduced to about 8–9 hours daily until the birds have reached 20–22 weeks of age. The light is then either abruptly increased to 16 hours per day, or it is increased by 2–3 hours in weekly increments of 15–20 minutes added until 16 hours of light per day are reached.

Regardless of which system replacement pullets are reared under, they should be placed in the laying house when they are approximately 20 weeks of age so they can adjust to the house and its equipment before beginning to lay.

Management of Laying Hens

Prior to World War II, large numbers of small flocks of hens were the foundation of egg production. Average production per year was 100 eggs or fewer, hens suffered mortality rates approaching 40%, and the flock was subjected to disease, climatic variability, predation, and a host of other stressors. Management of the laying hen has changed significantly with the increased level of specialization in agriculture. For example, historically hens laid a clutch of eggs and then set with them until hatching. The hen and her chicks ate the same diet and by virtue of shared living space the offspring developed immunity quickly. Under modern production practices, the hen lays eggs into a clean nest, the eggs are gathered, disinfected before being set into a sterile hatchery environment, and then moved into relatively clean poultry houses as chicks.

Consolidation of hatcheries has been significant with fewer than 310 enterprises in business today as compared to approximately 5,000 in the late 1950s. However, the capacity of the U.S. hatchery complex has grown during the same time period.

The requirements of laying hens for floor space varies from 1.5–2 ft² per bird for egg-production strains and from 2.5–3.5 ft² for dual-purpose and broiler strains. Turkey breeding hens require 4–6 ft² , but less area is needed if hens are housed in cages. If turkey hens have access to an outside yard, 4 ft² of floor space is best. Game birds such as quail and pheasants require less floor space than egg-production hens. In general, 3 ft² or more of floor space is adequate for ducks, whereas geese need approximately 5 ft².

Breeder hens are housed on litter or on slatted floors, whereas birds used for commercial egg production are typically kept in cages (Fig. 34.10). Free-ranging laying hens are housed less intensively (Fig. 34.11).

Floor-type houses for breeder hens usually have 60% of the floor space covered with slats. Slatted floors are usually several feet above the base of the building and manure is often allowed to accumulate for a long time before being removed. Many slatted-floor houses are equipped with mechanical floor scrapers that remove the manure periodically. Frequent removal of manure lessens the chance that ammonia will accumulate.

Gathering of eggs in floor-type houses can be done automatically if some type of roll-away nesting equipment is used (Fig. 34.12). In houses equipped with individual nests, the eggs are gathered manually. Some operators prefer colony nests (open nests that accommodate several birds at a time).

Some type of litter or nesting material must be placed on the bottom of individual and colony nests if eggs are to be gathered manually. Birds should not be allowed to roost in the nests in darkness because dirty nests result.

Cage operations are used by most commercial egg-producing farms. Young chickens may be brooded in colony cages, whereas laying hens may be housed in 4–6 tiers of decked laying cages. A range of 67–86 square inches of space per hen is recommended. The lower end of the range is appropriate for housing smaller Leghorn strains in shallow design cages while Brown hens in deep cages require more space. Protection from predation and environmental extremes are the primary objectives of poultry housing, and management should focus on minimizing the spread of pathogens and parasites.

All cages are equipped with feed troughs that usually extend the entire length of the cage, allowing all of the hens to eat at one time. Some troughs are filled with feed manually, others automatically. It is sometimes advantageous to dub (remove the combs and wattles from) caged layers so they can obtain feed from automatic feeders by reaching their heads through the openings. Additional considerations include assuring that manure cannot drop from one cage onto others, cage floor slopes should not exceed 8 degrees, and hens should be able to stand comfortably upright. To assure favorable conditions for visual inspection of birds, light intensity at 0.5- to 1-ft candle should be provided at feeding levels of the facility.

Several types of watering devices (trough, nipple-type, or individual cup-type waterers) are available for use in individual and colony cages. The watering system should be equipped with a metering device that makes it possible to medicate the birds quickly when necessary by mixing the exact dosage of medication required with the water. Recommendations for watering space are provided in Table 34.7.

The arrangement of cages within a house varies greatly. Some producers have a step-up arrangement, such as a double row of cages at a high position, with a single row at a low position on either side. Droppings from birds caged in the double row

Figure 34.10
Stacked housing layer production system. Source: Ttstudio/Fotolia.

Figure 34.11
Laying hens in a free-ranging facility. Source: Tom Field.

fall free of the birds in the single row. An aisle approximately 3 ft wide is usually located between each group of cages. Rather than aisles, some systems have a movable ramp that can travel above the birds from one end of the house to the other. In other systems, several double cages are stacked on top of each other.

In cage operations, manure is typically collected into pits below the cages and removed by mechanical pit scrapers or belts. In stacked housing arrangements, the manure is moved via a conveyor system from each stack to a central collection point.

Figure 34.12
Egg collection system for (A) a traditional stacked housing system and (B) a free ranging facility. Source: 34.12a: Bsvowell/Fotolia; 34.12b: Tom Field.

Manure may also drop into pits that are slightly sloped from end to end and partially filled with water. The manure can be flushed out with the water into lagoons.

An alternative to cage housing is the use of free-ranging barns that allow birds more freedom of movement and the opportunity to use "brood houses" for laying. Feed and water are delivered via automatic systems. Some consumers prefer to purchase products from farms and processors employing these systems.

Housing Poultry

Factors such as temperature, moisture, ventilation, and insulation are given careful consideration in planning and managing poultry houses. An important consideration

Table 34.7
WATER SPACE REQUIREMENTS FOR LAYER HENS

Age of Bird (wks)	Linear Trough Space/ Bird (in.)	Maximum Number of Birds per Cup or Nipple[a]
0–6	0.6	20
6–18	0.8	15
>18	1.0	12

[a]Perimeter space required for round-type watering systems can be calculated by multiplying linear trough space by 0.8.

Source: Adapted from United Egg Producers' Animal Husbandry Guidelines.

is to assure that standby generators and alarm systems are in place to allow normal house operation during emergency situations.

Temperature   Most poultry houses are built to prevent sudden changes in house temperature. A bird having an average body temperature of 106.5°F usually loses heat to its environment except in extremely hot weather. Chickens perform well in temperatures between 35 and 85°F, but 55–75°F is optimal.

House temperature can be influenced by such factors as the prevailing ambient temperature, solar radiation, wind velocity, and heat production of the birds. A 4.5-lb laying hen can produce nearly 44 **British thermal units (Btu)** of heat per hour (1 Btu is the quantity of heat required to raise the temperature of 1 lb of water 1°F at or near 39°F). The amount of Btu produced varies with activity, egg-production rate, and the amount of feed consumed. Heat production by birds must be considered in designing poultry houses (about 40 of the 44 Btu produced per hour by a laying hen are available for heating). Although heat produced by birds can be of great benefit in severe cold, the house must be designed so that excess heat produced by birds in hot weather can be dissipated.

Moisture   Excess house moisture, especially in cold weather, can create an environment that is extremely uncomfortable, which can lead to a drop in production by laying birds, and that if allowed to continue for long, can cause illness.

Much of the moisture in a poultry house comes from water spilled from waterers, water in inflowing air, and water vapor from the birds themselves and their droppings. Every effort should be made, of course, to minimize spillage from watering equipment. The amount of moisture in a poultry house can be reduced by increasing the air temperature or by increasing the rate at which air is removed. Because air holds more moisture at high temperatures than at low temperatures, raising the air temperature causes moisture from the litter to enter the air and thus be dissipated. Air temperature in the house can be increased by retaining the heat produced by the birds themselves and by supplemental heat.

If incoming air is considerably colder than the air in the house, it must be warmed or it will fail to aid in moisture removal. In cold weather, most exhaust ventilation fans are run more slowly than normal, so most moisture removal is accomplished by increasing the air temperature in the house.

Ventilation   A properly designed **ventilation** system provides adequate fresh air, aids in removing excess moisture, and is essential in maintaining a proper temperature within the house. Type and amount of **insulation** and heat produced by the birds themselves must be considered in planning for ventilation.

Ventilation is accomplished by positive pressure, through which air is forced into the house to create air turbulence, or by negative pressure, through which air is removed from the house by exhaust fans. The positive pressure system is accomplished by having fans in the attic so that air is forced through holes. The negative pressure system is accomplished by locating exhaust fans near the ceiling. Some houses are ventilated so that the air is kept fairly warm and dry by the use of open walls. Also, when outside air is cold and dry, air can enter at low portions of the house and leave through vents near the ceiling as it warms. The heat created by the chickens warms the colder air, which then takes moisture from the house.

Two integral and necessary parts of the most common ventilation systems are the exhaust fan and the air-intake arrangement. The number of exhaust fans varies with size of the house, number of birds, and capacity of the fans. Most fans are rated on the basis of their cubic feet per minute (cfm) capacity, that is, how many cubic feet of air they move per minute. Fans in laying houses should operate at 4–4.5 cfm per bird when the temperature is moderate. In summer, cfm per bird could be as high as 10.

Fans also have a static pressure (water pressure for low house temperatures and a high speed for warm house temperatures). Others operate at only one speed; the amount of air removed is controlled by shutters that open, so that more air can be removed when the temperature of the house exceeds the thermostatically fixed temperature. These shutters then close at lower temperatures.

An air-intake area must be provided for the ventilation system. It is usually a slotted area in the ceiling or near the top of the walls. There should be at least 100 in.^{2} of inlet space for each 400 cfm of fan capacity.

Feeds and Feeding

Rations fed to poultry today are complex mixtures that should include, in a balanced proportion, all ingredients that have been found to be necessary for body maintenance, maximum production of eggs and meat, and optimum reproduction (fertility and hatchability).

Maximum growth potential cannot be obtained in broilers and turkeys unless maximum feed intake is coupled with correct diet formulation. Dietary intake is affected by a variety of factors including dietary energy, protein, amino acids, vitamins and minerals, antinutritional factors, and water consumption.

Antinutritional factors such as alkaloids, phytates, and other natural feed components can inhibit nutrient availability, reduce intake, and suppress growth. Meat-type poultry will consume twice as much water as they do feed on a weight basis. Thus, adequate access to water is critical to feed consumption.

Management factors may also have significant influence on the level of feed intake. The primary management protocols of note are assuring access to water and feed, environmental stress levels, and the degree of disease challenge. Managers must assure adequate feeder and waterer space so that even more submissive birds will have free access to nutrients. Feed and water must be available ad libitum. Feed availability that is disrupted for more than four hours can lead to increased susceptibility to enteric disease.

Increased levels of stress will limit feed intake. Careful attention to dealing with heat stress, poor air quality, and poor litter quality will enhance feed intake and consequent growth rate. Disease management is also critical as any factor, including vaccination, that elicits an immune response will also depress feed consumption.

Energy requirements are supplied mainly by cereal grains, grain by-products, and fats. Some important grains are yellow corn, wheat, sorghum grains (milo), barley, and oats. Most rations contain high amounts of grain (60% or higher, depending on the type of ration). A ration containing a combination of grains is generally better than a ration having only one type. Animal fats and vegetable oils are excellent sources of energy. They are usually incorporated into broiler rations or any high-energy rations.

Protein is so highly essential that most commercial poultry feeds are sold on the basis of their protein content. The types of amino acids present determine the nutritional value of protein (see Chapter 15). Excellent protein can be derived from both plants and animals. Most rations contain both plant and animal protein so that each source can supply amino acids that the other source lacks. The most common sources of plant protein are soybean meal, cottonseed meal, peanut meal, alfalfa meal, and corn gluten meal. Cereal grains contain insufficient protein to meet the needs of birds. The best sources of animal protein are fish meal, milk by-products, meat by-products, tankage, blood meal, and feather meal. The protein requirements of birds vary according to species, age, and purpose for which the birds are being raised. The protein requirements of certain birds are shown in Table 34.8. Where variable values are listed, the highest level is to be fed to the youngest individuals. Recommended nutrients for layers are provided in Table 34.9.

Whatever ration is adequate for turkeys is generally suitable for game birds. Some game bird producers feed a turkey ration at all times. Others feed a complete game bird ration. The actual nutrient requirements of game birds are not as well known as are the requirement of chickens and turkeys.

A considerable number of minerals are essential, especially calcium, phosphorus, magnesium, manganese, iron, copper, zinc, and iodine. Calcium and phosphorus, along with vitamin D, are essential for proper bone formation. A deficiency of either of these elements can lead to a bone condition known as **rickets**. Calcium is also essential for proper eggshell formation.

Table 34.8
PROTEIN REQUIREMENTS OF POULTRY

Broilers

Phase	Starter	Grower	Finisher	Pre-harvest		
Age (days)	0–15	16–27	28–39	40–44		
CP (%)	7.3	6.6	6.2	5.4		

Broiler Breeder Pullets

Phase	Starter	Grower	Developer	Pre-breeding		
Age (days)	0–4	5–12	13–19	20–22		
CP (%)	18.5	17	16	15.5		

Broiler Breeders

Phase	I	II	III			
Age (days)	22–34	34–54	54–64			
CP (%)	15	14.5	14			

Layers

Age (wks)	18–32	32–45	45–60	60–70		
CP (%)	21	19	18	17		

Turkey

Phase	Starter	Grower I	Grower II	Developer I	Developer II	Finisher
Age (wks)	0–4	5–8	9–11	12–13	14–16	17+
CP (%)	28	26	23	21	18	16

Adapted from Leeson, 2013.
Complete rations also contain specific amino acid requirements for each phase/age.

Table 34.9
NUTRIENT RECOMMENDATIONS FOR EGG-TYPE LAYERS

Age (wks)	18–32	32–45	45–60	60–70
Feed intake (g/hen/d)	95	98	105	107
CP (%)	21	19	18	17
Calcium (%)	4.4	4.5	4.5	4.6
Phosphorus (%)	0.5	0.43	0.38	0.33
Methionine (%)	0.41	0.37	0.35	0.31
Methionine + cysteine (%)	0.68	0.64	0.61	0.54
Lysine (%)	0.78	0.73	0.71	0.66
Threonine (%)	0.63	0.58	0.54	0.50
Tryptophan (%)	0.16	0.15	0.15	0.14
Arginine (%)	0.80	0.74	0.70	0.67
Valine (%)	0.70	0.65	0.61	0.57
Leucine (%)	0.48	0.44	0.39	0.36
Isoleucine (%)	0.62	0.57	0.53	0.48

All amino acids on a digestible basis.
Source: Adapted from Leesen, 2013.

The amount of calcium required varies somewhat with age, rate of egg production, and temperature. Chickens and turkeys up to 8 weeks of age require 0.8–1.2% of calcium in their diets. Calcium can be reduced from 0.8% to 0.6% in 8- to 16-week-old birds. Laying hens require at least 3.4% calcium. The amounts of phosphorus required for chickens of different ages are as follows: 0–6 weeks, 0.4%; 6–14 weeks, 0.35%; and mature, 0.32%. Turkeys require 0.3–0.6% phosphorus; game birds, approximately 0.5%. Requirements for other minerals vary greatly depending on the stage of growth and production.

The vitamins that are most important to poultry are A, D, K, and E (fat soluble), and thiamin, riboflavin, pantothenic acid, niacin, vitamin B_6, choline, biotin, folacin, and vitamin B_{12} (water-soluble). Recommended supplemental vitamin and mineral supplementation for various classes of poultry are listed in Table 34.10.

Issues Management

Affluent cultures have the luxury of choosing among a large array of food choices. Due to the wealth of resources enjoyed by consumers in developed nations, food security is not a daily concern and as such the market has created opportunities for differentiation on a number of levels such as animal handling, types of production systems, and organic/natural labels, for example. Simultaneously, activists and governmental regulators have pressured the poultry industry on topics such as bird housing, litter management, and poultry handling.

Environmental Impact

Broilers in the United States produce about 1 ton of manure per each 1,000 birds. In the context of the average broiler farm previously discussed in this chapter, each farm produces more than 825,000 lb of litter per year of which about 40% is used on farm as a fertilizer. Approximately 1,000 layers produce 40 tons of wet manure per year. Poultry manure is typically dried to less than 35% moisture to reduce volume and weight, to control odor and flies, and to maintain fertilizer value. Manure dried to 10% moisture reduces weight by almost 75% and volume by half.

Table 34.10

RECOMMENDED SUPPLEMENTAL VITAMIN AND MINERAL LEVELS FOR VARIOUS POULTRY CLASSES

	Starter Chicks (0–8 wks)	Growing Chickens (8–18 wks)	Egg-Type Layers	Starter Turkeys (0–8 wks)	Growing-Finishing Turkeys (8 wks to mkt)
Vitamin[1]					
A (miu)[2]	7.0	7.0	6.0	9.0	7.0
D_3 (miu)[2]	2.0	2.0	2.0	3.0	2.5
E (tiu)[3]	6.0	6.0	5.0	11.0	8.0
B_{12} (mg)	10.0	10.0	6.0	6.0	6.0
Riboflavin (g)	6.0	5.0	4.0	6.0	4.0
Niacin (g)	30.0	30.0	15.0	65.0	45.0
Pantothenic acid (g)	10.0	10.0	6.0	14.0	10.0
Choline (g)	450.0	450.0	250.0	600.0	550.0
Folic acid (g)	0.6	0.6	0.2	1.0	0.7
Thiamine (g)	1.0	1.0	1.0	1.0	1.0
Pyridoxine (g)	3.0	3.0	1.0	3.0	2.0
Biotin (mg)	50.0	50.0	30.0	100.0	50.0
Minerals[4]					
Manganese (mg)	25.0	25.0	50.0	50.0	50.0
Zinc (mg)	25.0	25.0	50.0	50.0	50.0
Iron (mg)	50.0	50.0	50.0	50.0	50.0
Copper (mg)	5.0	5.0	5.0	5.0	5.0
Iodine (mg)	0.2	0.2	0.2	0.2	0.2
Selenium (mg)	0.05	0.05	0.05	0.1	0.1

[1]Supplement per ton.
[2]Million International units.
[3]Thousand International units.
[4]Supplemental per pound.
Source: Adapted from Waldroup, 2002.

Poultry litter may be used as fertilizer, ruminant feed, or fuel. Unfortunately, poultry litter typically contains excessive phosphorus, copper, and zinc. After collection from bird houses, litter may be deep stacked, heated, and fermented into a silage product for use as a feed for cattle. The typical uses of litter are outlined in Table 34.11.

The management of poultry waste is critical in order to meet the requirements of regulations such as those related to concentrated animal feeding operations (CAFO) as prescribed by the Environmental Protection Agency.

About two-thirds of contract broiler growers have manure management plans in place or in the developmental stage. Nearly all company-owned broiler farms are participating in voluntary environmental management initiatives.

Animal Welfare

United Egg Producers have implemented animal husbandry guidelines that included expanding cage space to 67–86 sq in. per bird over a 12-year period. A 1999 NAHMS survey found that space allowances for laying hens were distributed as follows: less than 48 in.[2] (16.6%), 48–53.9 in.[2] (45.1%), 54–59.9 in.[2] (16.7%), and 60 in.[2] or more (21.6%). Increasing cage space per hen typically increases egg production,

Table 34.11
LITTER UTILIZATION BY APPLICATION

Use	%
Pasture fertilizer	35
Crop fertilizer	7
Sold	35
Traded	18
Other	5

Source : Adapted from Carpenter, 2001.

increases egg weight, and reduces mortality. Economic research results are more variable with disagreement as to the degree of cost: benefit as a result of the changes. Many studies show significant increases in expenses arising for lowering hen density. It is likely that consumers will have to pay more for products produced under more stringent space allocation guidelines.

Housing System Production of poultry under free-range conditions is a viable option but without superior management and the availability of a highly motivated labor force, increased mortality, and reduced productivity may be experienced. Correctly managed conventional cage systems provide humane and healthy environments. The type of housing is less critical than the quality of management.

Beak Trimming. Beak trimming is a management practice used to reduce pecking, feather pulling, cannibalism, mortality, and chronic stress of layers. **Debeaking** is not allowed in broiler production under the National Chicken Council Animal Welfare Guidelines. The egg industry places significant selection pressure on docility as a means to reduce the need for beak trimming, and the United Egg Producers recommend the practice only when necessary to prevent feather pulling and outbreaks of cannibalism. When beak trimming is utilized, the following guidelines are required:

1. Procedures should only be carried out by correctly trained personnel under a formal quality control program.
2. Beaks of chicks should be trimmed at 10 days of age or younger with an automated, precision cam-activated device equipped with a heated blade.
3. Vitamin K and sometimes Vitamin C should be provided 2 days prior and 2–3 days following the procedure to enhance clotting, minimize stress, and avoid dehydration.

Induced Molting Induced molting of egg-laying flocks has been used to extend the productive life of hens and to stimulate reproductive cycles. Induced molting allows the industry to produce eggs with 40–50% fewer new hens than if the practice were not utilized. Without molting, a flock is terminated at 75–80 weeks of age while the use of induced molting may extend productive life of a flock to 110 weeks. Molting was historically induced by fasting birds for 4–14 days. The industry invested significantly in research to establish valid alternatives to fasting. As a result of their findings the United Egg Producers established new induced molt guidelines effective in January 2006:

1. Only nonfeed withdrawal molt procedures are permitted.
2. Hens should be able to consume nutritionally adequate and palatable feed suitable for a nonproducing hen.

3. Body weight loss should not compromise hen welfare.
4. Mortality rates during the molt should not substantially exceed normal flock mortality.
5. Water must be provided at all times.
6. Reduce light period to no less than 8 hours in closed houses or to natural day length in open houses for the duration of the rest period.

Avian influenza, caused by the pathogenic strain H5N1, has generated global pandemic concerns. The poultry industry has worked diligently to develop preventative and control protocols as a means to minimizing the spread of the disease. Grower operations have been advised to limit access to farm facilities and to prevent unauthorized access. Further recommendations include disinfecting boots, using disposable gloves and overalls, disinfecting vehicles entering and leaving the premises, securing water sources, and assuring a high degree of sanitation of farm labor. Emerging pathogenic threats create the need for individual farms to develop and implement intervention and control plans.

COSTS AND RETURNS

The U.S. poultry industry has undergone significant structural and market change over the past 30 years. Over 80% of the farms producing poultry are located in the Northeast, Southeast, mid-South, Gulf Coast, and Corn Belt states. More than 75% of the total broiler production occurs in the top 10 states. In the turkey industry, 80% of the numbers are produced in the top 10 states with the top six accounting for more than two-thirds of the total. Almost two-thirds of the eggs are produced in the top 10 states. Table 34.12 illustrates the concentration of production capacity in the poultry industry.

Prior to the integration and consolidation of the poultry industry, broiler production was a secondary farm enterprise uniformly spread across the country (Fig. 34.13). However, by the early 1990s, the geographic shift had occurred (Fig. 34.14). The primary reasons for this change were the development of improved transportation and packaging systems that allowed consolidation to occur in the southern states plus California. Furthermore, the climatic and socioeconomic conditions of the South were favorable to the growing of broilers under contract for a rapidly integrating industry. The location of a majority of the broiler processing facilities in the South and Southeast hastened the consolidation of the industry. The top three firms have 48% market share in broilers (up from 35% in 1990) while the top 10 control almost 80%.

Table 34.12
TOP FIVE BROILER, EGG, AND TURKEY COMPANIES

Broiler	Broilers (mil hd)	Eggs	Layers (mil)	Turkey	Live wt. (mil lb)
Tyson Foods	1,840	Cal-Maine Foods	32.0	Butterball	1,450
Pilgrim's Corp.	1,722	Rose Acre Farms	24.6	Jennie-O-Turkey	1,342
Perdue Farms, Inc.	625	Moark, LLCFarms	16.1	Cargill Meats	1,047
Koch Foods	624	Rembrandt Enterprises	13.6	Farbest Foods	302
Sanderson Farms, Inc.	448	Daybreak Foods	13.0		

Source: Adapted from multiple sources.

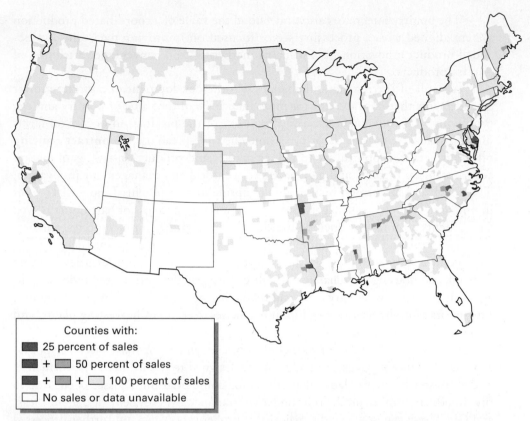

Figure 34.13
Concentration of broiler sales in the United States, 1969. Source: USDA: Economic Research Service.

Legend (Figure 34.13):
Counties with:
- 25 percent of sales
- + 50 percent of sales
- + + 100 percent of sales
- No sales or data unavailable

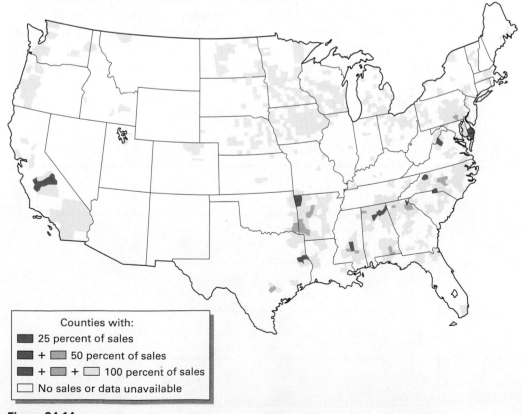

Figure 34.14
Concentration of broiler sales in the United States, 1992. Source: USDA: Economic Research Service.

Legend (Figure 34.14):
Counties with:
- 25 percent of sales
- + 50 percent of sales
- + + 100 percent of sales
- No sales or data unavailable

The poultry integrators also understood the value of a coordinated production system aligned with a processing sector focused on improving packaging, value-added product innovations, brand-name marketing, and convenience attributes of poultry products.

A result of these structural changes was the development of contract grower arrangements that now account for more than 95% of the value of poultry and egg production. Of the remainder, a vast majority of the production occurs on farms directly owned by the integrator-processor. The widespread use of **contract growing** offers farmers the opportunity to reduce market and production risk, gain market access, stabilize income, and gain access to technical and management innovation. The drawbacks for farmers include loss of upside market potential, potential loss of incentive to improve as managers and decision-makers, and loss of bargaining power. The benefits to processors include the ability to standardize input supplies, enhance quality control, and allow for more rapid response to shifts in consumer demand.

Contracts typically specify that growers provide land, labor, utilities, manure disposal, and housing in exchange for high-quality genetics, formulated rations, management expertise, veterinary care, and marketing power provided by the integrator. Integrators typically have ownership in hatcheries, feed mills, harvesting plants, and value-added processing facilities.

Contracts are structured to include a base payment, or fixed price per pound produced, an incentive-discount adjustment designed to reward better performance, and a disaster payment clause that allows for compensation if losses occur due to fire, flood, etc. Broiler production under contract is most prevalent in the Southeast and Gulf Coast states, where the majority of broiler processing and further-processed plants are located. This integrated production and marketing system has allowed the broiler industry to steadily increase both bird numbers and total production (Fig. 34.15). Emerging contract specifications include testing for pathogens such as avian influenza and salmonella. The majority require that managers use all-in, all-out flock population procedures to minimize disease transmission and to improve consistency of age and final weight at time of marketing. The majority of contracts also require operations under a specified HACCP plan, the incorporation of specified animal welfare practices, and nearly half are prohibiting the use of subtherapeutic antibiotics in feed and water.

Figure 34.15
Changes in broiler production (1955–2005).
Source: USDA.

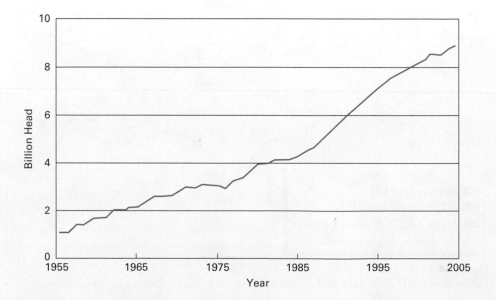

Nearly 95% of U.S. broiler production occurs on 17,440 farms in the leading 17 states. These farms have ownership of an estimated 70,000 production barns. Less than 2% of broiler production can be classified as natural/organic.

Representative characteristics of poultry farms are as follows:

Broiler Production	Turkey Production
• ⅔ of farm dedicated to other enterprises	• ¾ of farm dedicated to other enterprises
• 160-acre farm	• 225-acre farm
• 3 houses per farm	• 3 houses per farm
• 65,000 birds per farm	• 27,000 birds per farm

The investment per poultry house averages $10 per square foot. Most newly constructed barns incorporate solid panel sides as opposed to the use of side curtains, sophisticated tunnel ventilation systems, and evaporative cooling units to add more temperature control capability when the ambient temperatures become excessive (Table 34.13). Improved facility design supports better bird performance measured in improved feed efficiency, reduced mortality rates, and reduced production costs. Farms with three to six production barns on site account for two-thirds of broiler marketings.

Table 34.14 shows the production costs for eggs, broilers, and turkeys. Note that feed costs comprise 60–70% of the total costs. Management practices that monitor and lower feed costs usually have a positive effect on net returns.

Table 34.13
SIZE AND DESIGN COMPONENTS OF BROILER HOUSES OVER TIME

Construction year	Production Share (%)	Size (1,000 ft.^{2})	Side Curtains (%)	Cooling Cells (%)	Tunnel Ventilation (%)
Pre-1960	6.5	12.3	70	50	36
1966–1970	1.8	12.7	87	44	40
1976–1980	7.6	14.7	72	55	53
1986–1990	17.4	16.2	75	71	70
1996–2000	21.5	18.7	75	85	88
2001–2006	15.4	20.2	48	92	94

Source: Adapted from ERS:USDA.

Table 34.14
POULTRY MEAT AND EGG PRODUCTION COSTS AND RETURNS

Item	Production Cost (cents per doz/lb) Feed	Total	Net Return (cents per doz/lb)
Market eggs	28.8	47.0	8.9
Broilers	16.0	26.4	7.6
Turkeys	21.9	35.6	5.4

Source: USDA.

Table 34.15
INDUSTRY PERFORMANCE TARGETS

Trait	Target
Feed conversion—broilers	<2800 calories/lb of liveweight
Feed conversion—breeders	<7.5 lb per settable dozen
Average daily gain	>0.09 lb/day
Labor efficiency	>185 broilers/hour of labor[a]
Hatchability	>83%
Chicks/hen/week	>3.2

[a]Receiving through the chiller.

Table 34.15 outlines a few of the **performance targets** for profitable poultry production. It is important to recognize that the high degree of integration in the poultry industry allows for an extremely coordinated level of communication within the production and marketing chain. The use of objective goals and measurements has allowed dramatic growth in the poultry industry.

While the industry made significant strides in profitability via improved cost controls, particularly in regards to feed, an overemphasis on production costs can lead to reduced profitability. For example, if the changes made to a ration to lower costs result in lower livability or reduced gains, the net effect may be lost net income as a result of poorer productivity albeit at a lower cost. The poultry industry in the future will continue to monitor costs but will ultimately succeed because of the ability of managers to optimize costs and production tonnage.

As the animal protein market has differentiated into a variety of specialty or niche markets, some poultry producers have moved into production systems that do not utilize chemical feed additives such as antibiotics or hormones as a means to create a unique brand attribute for their products. Poultry and egg products originating from these systems often sell at premium prices, typically double that of conventionally produced poultry and eggs. However, there are significant costs associated with additive-free or natural production systems. For example, the costs associated with ration formulation, additional clean-out requirements, added payments to the grower, and loss of feed efficiency add production costs per bird of 6%, 1.5%, 8%, and 3%, respectively. While capturing market premiums are attractive to producers, the additional costs associated with these systems must be carefully evaluated in terms of net profit.

The poultry industry has also become increasingly dependent on the export of chicken meat as a primary strategy to increase profitability. While exports have shown steady growth, the ability of the United States to capture additional share of the global market has been limited by the growth of the Brazilian broiler industry spurred by significant cost of production advantages. Concerns over avian influenza add instability to the export market. Nonetheless, growth of the poultry industry will require access to global markets and the ability of the industry to remain competitive on both a price and value basis.

CHAPTER SUMMARY

- The most important poultry in the United States are chickens and turkeys.
- Breeds of chickens are relatively unimportant, as the commercial broiler and egg producers utilize hybrid chickens resulting from sophisticated breeding methods.
- The production of broilers and eggs is very intensively managed in an integrated system involving breeding, hatching, feeding, processing, and marketing name-brand products to the consumer.

KEY WORDS

caruncle
snood
class
breed
variety
qualitative traits
quantitative traits
trap nest
mass selection
family selection
outcrossing
inbred lines
hybrid chickens
strain crossing
broodiness

incubation
hatchability
candled
litter
floor space
feeder space
British thermal unit (Btu)
ventilation
insulation
rickets
debeaking
Avian influenza
contract growing
performance targets

REVIEW QUESTIONS

1. What are unusual physical characteristics of turkeys?
2. Discuss classification of chickens and turkeys.
3. Describe the turkey production cycle.
4. Describe the broiler and egg production cycle for the chicken industry.
5. Compare historic changes in turkey and broiler performance.
6. Outline breeder multiplication in the poultry industry.
7. Compare mass selection and family selection.
8. Discuss important factors in incubation management.
9. Describe housing requirements for poultry.
10. Discuss management of young chicks and layers.
11. Compare nutritional requirements at difference phases of production.
12. Discuss litter management strategies.
13. Outline the major animal well-being issues in the poultry business and strategies to address them.
14. Describe market concentration of poultry production.
15. Discuss contract grower arrangements.
16. Compare changes in poultry house design over time.
17. Discuss performance targets in the poultry business.

SELECTED REFERENCES

American Poultry History. 1996. Mount Morris, IL: Watt Publishing Co.

Carpenter, E. 2001. Nutrient management comes of age. *Poultry USA*, Dec., 60–74.

Leesen, S. 2013. Nutrition and health of poultry. *Feedstuffs*. Sept., 42-49.

Lundeen, T. 2002. Feed intake most important factor in meat-type poultry growth. *Feedstuffs*, Oct., 7,9.

National Research Council. 1994. *Nutrient Requirements of Poultry*. 9th ed. Washington, DC: National Academy Press.

North, M. O., and D. D. Bell. 1990. *Commercial Chicken Production Manual.* Westport, CT: AVI Publishing Co.

Perry, J., D. Banker, and R. Green. 1999. *Broiler Farms' Organization, Management, and Performance.* ERS: USDA. Bulletin No. 748.

Scanes, C. G., G. Brant, and M. E. Ensminger. 2004. *Poultry Science.* Upper Saddle River, NJ: Pearson Prentice Hall.

Waldroup, P. W. 2002. Dietary nutrient allowances for chickens and turkeys. *Feedstuff,* 74:28.

Watt Poultry Yearbook. 2006. Mount Morris, IL: Watt Publishing Co.

90-day weight Common measure of growth rate in sheep.

abomasum The fourth stomach compartment of ruminant animals that corresponds to the true stomach of monogastric animals.

abortion Delivery of fetus between conception and a few days before normal parturition.

abscess Localized collection of pus in a cavity formed by disintegration of tissues.

absorption The passage of liquid and digested (soluble) food across the gut wall.

accessory organs The seminal vesicles, prostate, and Cowper's glands in the male. These glands add their secretions to the sperm to form semen.

accessory sex glands Responsible for adding nutrients and fluid volume to semen.

acclimatization Series of adaptations that livestock undergo when introduced to a new set of conditions.

accuracy (ACC) of selection Numerical value, ranging from 0 to 1.0, denoting the confidence that can be placed in the EPD (expected progeny difference); for example, high (≥ 0.70), medium ($0.40–0.69$), low (≤ 0.40).

acidified Process of souring milk or crème without the addition of bacteria.

acquired resistance Form of immunity that is developed as an individual encounters antigens and disease challenge.

actin A protein that works in conjunction with myosin to contract and relax muscle fibers.

active immunity Ability of an individual to produce its own immune response.

actomyosin A protein complex of actin and myosin that comprises muscle fiber.

adipose Fat cells or fat tissue.

adjusted weaning weight A measure of weaning weight adjusted for differences in birth weight, age at weaning, and age of dam.

ad libitum Free choice; allowing animals to eat all they want.

aflatoxin Undesirable mycotoxin that may impair performance.

afterbirth The membranes attached to the fetus that are expelled afterparturition.

agalactia Inadequate milk supply.

agonistic behavior Combative behavior involving both passive and aggressive responses.

AI Abbreviation for artificial insemination.

allantois Fetal membrane responsible for oxygen transfer.

air dry Refers to feeds in equilibrium with air; they would contain approximately 10% water or 90% dry matter.

albumen The white of an egg.

alimentary tract Passageway for food and waste products through the body.

alleles Genes occupying corresponding loci in homologous chromosomes that affect the same hereditary trait but in different ways.

allelomimetic behavior Doing the same thing. Animals tend to follow the actions of other animals.

alveolus (plural *alveoli*) A hollow cluster of cells. In the mammary gland, these cells secret milk.

amino acid Any of a class of 20 molecules that are combined to form proteins in living things.

amnion A fluid-filled membrane located next to the fetus.

ampulla The dilated or enlarged upper portion of the vas deferens in bulls, bucks, and rams, where sperm are stored for sudden release at ejaculation.

anabolic A constructive, or "building up," process.

anaerobic Able to survive or function where there is no oxygen.

anatomy Science of animal body structure and the relation of the body parts.

androgen A male sex hormone, such as testosterone.

anemia Deficiency of hemoglobin, often accompanied by a reduced number of red blood cells. Usually results from an iron deficiency.

anestrous Period of time when female is not in estrus; the nonbreeding season.

Angora goats Example of a fiber goat breed.

animal agriculture The application of human creativity to the stewardship of livestock to produce food, fiber, numerous by-products and draft power.

animal liberationists Those who believe that animals should be free of any use by humans.

animal rights A philosophy that advocates for animals having equal rights under the law.

animal welfare A philosophy that advocates for the humane treatment of animals.

antemortem Before death.

anterior Situated in front of, or toward the front part of, a point of reference. Toward the head of an animal.

anterior pituitary (AP)   The part of the pituitary gland, located at the base of the brain, that produces several hormones.

anthelmintic   A drug or chemical agent used to kill or remove internal parasites.

antibiotic   A product produced by living organisms, such as yeast, which destroys or inhibits the growth of other microorganisms, especially bacteria.

antibody   A specific protein molecule that is produced in response to a foreign protein (antigen) that has been introduced into the body.

antigen   A foreign substance that, when introduced into the blood or tissues, causes the formation of antibodies. Antigens may be toxins or native proteins.

anti-inflammatory   An agent that acts to decrease inflammation and associated pain, heat, and swelling.

antiseptic   A chemical agent used on living tissue to control the growth and development of microorganisms.

antitoxin   An antibody that is capable of neutralizing poisons from animal and vegetable sources.

AP   See *anterior pituitary*.

APCS   A popular milk testing system where weight is recorded at two milkings.

aquaculture   The production of food from aquatic species.

arteriosclerosis   A disease resulting in the thickening and hardening of the artery walls.

arteritis   Virus that can cause abortion.

artery   Vessel through which blood passes from the heart to all parts of the body.

arthrogryposis   Genetic disorder characterized by abnormal fusion of the mouth and severely contracted tendons.

artificial insemination   The introduction of semen into the female reproductive tract (usually the cervix or uterus) by a technique other than natural service.

artificial vagina   A device used to collect semen from a male when he mounts in a normal manner to copulate. The male ejaculates into this device, which simulates the vagina of the female in pressure, temperature, and sensation to the penis.

ascarids   Parasite located in the small intestine, may cause an intestinal blockage in young horses.

ascaris   Any of the genus (*Ascaris*) of parasitic roundworms.

as fed   Refers to feeding feeds that contain their normal amount of moisture.

assessing risk   Process of determining the likelihood of favorable outcomes.

assimilation   The process of transforming food into living tissue.

atherosclerosis   A form of arteriosclerosis involving fatty deposits in the inner walls of the arteries.

atrophy   Shrinking or wasting away of a tissue or organ.

attenuated or chemically altered vaccines   Effective only when applied directly to the mucosal membranes.

auction   A market facility where an auctioneer sells animals to the highest bidder.

autopsy   A postmortem examination in which the body is dissected to determine the cause of death.

aves   Zoological class of poultry.

avian   Refers to birds, including poultry.

avian influenza   Caused by the pathogenic strain H5N1, has generated global pandemic concerns.

Ayrshire   See Figure 26.1.

balance sheet   A statement of assets owned and liabilities owed in dollar terms that shows the equity or net worth at a specific point in time (e.g., net worth statement).

band   (1) a relatively large group of range sheep; (2) method of identification (e.g., put a band around the leg of a chicken).

Bang's disease   See *brucellosis*.

barns   Structures designed to provide shelter.

barren   Not capable of producing offspring.

barrow   A male swine that was castrated before reaching puberty.

basal layer   Skin layer that lies between the epidermis and the dermis.

basal metabolism   The chemical changes that occur in an animal's body when the animal is in a thermoneutral environment, resting, and in a postabsorptive state. It is usually determined by measuring oxygen consumption and carbon dioxide production.

base pair   Two nitrogenous bases (adenine and thymine or guanine and cytosine) held together by weak bonds. Two strands of DNA are held together in the shape of a double helix by the bonds between base pairs.

B cells   Lymphocyte that secretes antibodies in response to specific antigens.

beef   The meat from cattle (bovine species) other than calves (the meat from calves is called *veal*).

beef industry   A coordinated set of supply chain enterprises that produce, process, and distribute beef and beef by-products to the market place.

benchmarks   Performance standards against which progress can be measured.

benchmarks   Standards of performance against which progress can be measured.

beri-beri   A disease caused by a deficiency of vitamin B_1.

best management practices (BMP)   Standards recognized by professionals as most desirable management protocols.

biologicals   Medicinal products used primarily to prevent disease, including serums, vaccines, antigens, antitoxins, etc.

biosecurity Active process to protect against the spread of infectious disease.

biotechnology The use of microorganisms, plant cells, animal cells, or parts of cells (such as enzymes) to produce industrially important products or processes.

birth weight expected progeny difference (EPD) The expected average increase or decrease in the birth weight of a beef bull's calves when compared to the other bulls in the sire summary.

bison Domesticated buffalo produced for meat, hide, and heads.

blastoderm Site of embryonic development if the egg is correctly incubated.

blemish Any defect or injury that mars the appearance of, but does not impair the usefulness of, an animal.

bloat An abnormal condition in ruminants characterized by a distention of the rumen, usually seen on an animal's upper left side, owing to an accumulation of gases.

blood spots Spots in the egg caused by a rupture of one or more blood vessels in the yolk follicle at the time of ovulation.

bluetongue Viral disease transmitted via a biting insect vector.

BLUP Best linear unbiased prediction, a method for estimating breeding values of breeding animals.

boar (1) A male swine of breeding age. (2) Denotes a male pig, which is called a *boar pig*.

boar-taint Strong, undesirable flavor often associated with pork originating from mature intact males.

body condition (1 = thin, 3 = average, 5 = fat) can be used to monitor nutrition, reproduction, and health programs for dairy herds.

body condition score A descriptive system of estimating degree of body fat.

body condition score A ranking system to distinguish differences in degree of fatness.

body condition scores A categorization system to describe differences in degree of fatness.

body maintenance The nutrients required to hold a mature animal in a steady state of well-being.

boer goats Example of a meat-type goat.

bog spavin A soft enlargement of the anterior, inner aspect of the hock.

bolus (1) Regurgitated food. (2) A large pill for dosing animals.

boluses A capsule or pill administered to an animal, typically large in size.

boneless, closely trimmed retail cuts (BCTRC) The amount of a carcass that can be fabricated into saleable retail cuts.

bone spavin A bony (hard) enlargement of the inner aspect of the hock.

bots Any of a number of related flies whose larvae are parasitic in horses and sheep.

bovine A general family grouping of cattle.

box stalls Small pens enclosed within a barn.

boxed beef Cuts of beef put in boxes for shipping from packing plant to retailers. These primal and subprimal cuts are intermediate cuts between the carcass and retail cuts.

boxed lamb See *boxed beef*. Similar process except lamb instead of beef.

boxed pork Form of merchandising case-ready or near case-ready retail pork cuts.

break joint Denotes the point on a lamb carcass where the foot and pastern are removed at the cartilaginous junction of the front leg.

bred Female has been mated to the male. Usually implies the female is pregnant.

breech The buttocks. A breech presentation at birth is where the rear portion of the fetus is presented first.

breech presentation Fetus is presented butt first with hind legs tucked under the body.

breech presentation Butt first presentation of the fetus with the hind legs tucked under the body.

breed Animals of common origin with characteristics that distinguish them from other groups within the same species.

breeding value A genetic measure for one trait of an animal, calculated by combining into one number several performance values that have been accumulated on the animal and the animal's relatives.

brisket disease A noninfectious disease of cattle characterized by congestive right heart failure. It affects animals residing at high altitudes (usually above 7,000 ft).

British breeds Breeds of beef cattle originating in England. Examples are Angus, Hereford, and Shorthorn.

British thermal unit (Btu) The quantity of heat required to raise the temperature of 1 lb of water 1°F at or near 39.2°F.

brockle-faced White-faced with other colors splotched on the face and head.

broiler A young meat-type chicken of either sex (usually up to 6–8 weeks of age) weighing 3–5 lb. Also referred to as a *fryer* or *young chicken*.

broken-mouth Some teeth are missing or broken.

brooder Fish that have reached reproductive maturity.

broodiness The desire of a female bird to sit on eggs (incubate).

Brown Swiss See Figure 26.1.

browse Woody or brushy plants. Livestock feed on tender shoots or twigs.

brucellosis   A contagious bacterial disease that results in abortions; also called *Bang's disease*.

buck   A male sheep or goat. This term usually denotes animals of breeding age.

bulbourethral (Cowper's) gland   An accessory gland of the male that secretes a fluid that constitutes a portion of the semen.

bull   A bovine male. The term usually denotes animals of breeding age.

buller-steer syndrome   A behavior problem where a steer has a sexual attraction to other steers in the pen. The steer is ridden by the other steers, resulting in poor performance and injury.

bullock   A young bull, typically less than 20 months of age.

buttermilk   The fluid remaining after butter has been made from cream. By use of bacteria, cultured buttermilk is also produced from milk.

buttons   May refer to cartilage or dorsal processes of the thoracic vertebrae. Also see *cotyledons*.

by-product   A product of considerably less value than the major product. For example in U.S. meat animals, the hide, pelt, and offal are by-products, whereas meat is the major product.

CIDR   Controlled internal drug release device designed for synchronization programs.

C-section   See *cesarean section*.

calcium   Macro mineral required for normal growth, lactation, and body maintenance.

calf   A young male or female bovine animal under a year of age.

calf crop percentage   Percentage of cows exposed to mating that weaned a calf.

calorie   The amount of heat required to raise the temperature of 1 g of water from 15°C to 16°C.

calve   Giving birth to a calf. Same as parturition.

calving interval   The amount of time (days or months) between the birth of a calf and the birth of a subsequent calf, both from the same cow.

candling   The shining of a bright light through an egg to see if it contains a live embryo.

cannon   Long bone between the knee and pastern in the forelimb and the hock and pastern in the hindlimb.

canter   A slow, easy gallop.

Campylobacter jejuni   Leading cause of bacterial diarrhea resulting from consumption of affected foods.

capon   Castrated male chicken. Castration usually occurs between 3 and 4 weeks of age.

capped hocks   Hocks that have hard growths that cover, or "cap," their points.

carbohydrates   Any foods, including starches, sugars, celluloses, and gums, that are broken down to simple sugars through digestion.

carcass merit   The value of a carcass for consumption.

caregiving behavior   Specifically related to mothering ability.

care-soliciting behavior   Behavioral response by young animals seeking sustenance, protection, or comfort.

carnivores   Depend primarily on meat based diets.

carnivorous   Subsisting or feeding on animal tissues.

carotene   The orange pigment found in carrots, leafy plants, yellow corn, and other feeds, which can be broken down to form two molecules of vitamin A.

caruncle   (1) The red and blue fleshy, unfeathered area of skin on the upper region of the turkey's neck. (2) The "buttons" on the ruminant uterus where the cotyledons on the fetal membranes attach.

casein   The major protein of milk.

caseous lymphadenitis   The pathogenic organism grows in lymph glands and causes a large development of caseous material (a thick, cheeselike accumulation) to form. It is a serious disease and one that is difficult to control.

cash-flow statement   A financial statement summarizing all cash receipts and disbursements over the period of time covered by the statement.

cash receipts   Gross sales from a specific enterprise or set of enterprises.

castrate   (1) To remove the testicles. (2) An animal that has had its testicles removed.

cataract   Opacity of the lens of the eye.

cattalo   A cross between domestic cattle and bison.

cecum (ceca)   Large, sock-shaped pouch between the horse's small and large intestines; important in cellulose digestion.

cell-mediated immunity   Form of immunity that involves the activation of phagocytes and other specialized cells.

cervix   The portion of the female reproductive tract between the vagina and the uterus. It is usually sealed by thick mucus except when the female is in estrus or delivering young.

cesarean section   Delivery of fetus through an incision in abdominal and uterine walls.

chalaza   A spiral band of thick albumen that helps hold the yolk of an egg in place.

chemotherapeutics   Chemical agents used to prevent or treat diseases.

chevon   Meat from goats.

chick   A young chicken that has recently been hatched.

cholesterol   A lipid produced by all cells of the body.

chordata   Phylum of livestock.

chorion   The outermost layer of fetal membranes.

chromosome   The self-replicating genetic structure of cells containing the cellular DNA that bears in its nucleotide sequence the linear array of genes.

chromosomes   See *chromosome*.

chronic obstruction pulmonary disease (heaves)   Respiratory disease in which the horse experiences difficulty in exhaling air.

Chronic persistent hunger (CPH)   A state of consistent malnourishment.

chyme   The thick, liquid mixture of food that passes from the stomach to the small intestine.

chymotrypsin   A milk-digesting enzyme secreted by the pancreas into the small intestine.

class   A group of animals categorized primarily by sex and age.

clinical mastitis   Stage of mastitis in which small white clumps are apparent in the milk of affected animals.

clip   One season's yield of wool.

clitoris   The ventral part of the vulva of the female reproductive tract that is homologous to the penis in the male. It is highly sensory.

cloaca   Portion of the lower end of the avian digestive tract that provides a passageway for products of the urinary, digestive, and reproductive tracts.

cloning   biotechnical process to create copies of an organism.

closed face   A condition in which sheep cannot see because wool covers their eyes.

clostridium   Genus of anaerobic bacteria. Many produce potent toxins that cause such diverse diseases as tetanus, botulism, and gas gangrene. Some of these anerobic bacteria inhabit the soil and feces.

clutch   Eggs laid by a hen on consecutive days.

coagulation (curdling)   A biochemical process used to initiate cheese making by heating and then adding a starter bacterial culture to create a semisolid state.

coccidia   A protozoan organism that causes an intestinal disease called *coccidiosis*.

coccidiosis   A morbid state caused by the presence of organisms called coccidia, which belong to a class of sporozoans.

cock   A male chicken; also called a *rooster*.

cockerel   Immature male chicken.

cod   Scrotal area of steer remaining after castration.

code of ethics   A statement of values and ethics.

codon   Triplet sequence that codes for one amino acid in mRNA.

coefficient of determination   Percentage of variation in one trait that is accounted for by variation in another trait.

coldness   Difference between effective temperature and lower critical temperature.

colic   A nonspecific pain of the digestive tract.

colon   The large intestine from the end of the ileum and beginning with the cecum to the anus.

colostrum   The first milk given by a female after delivery of her young. It is high in antibodies that protect young animals from invading microorganisms.

colt   A young male of the horse or donkey species.

comb   The fleshy outgrowth on the top of a chicken's head, usually red in color, with varying sizes and shapes.

commercial breeder   Customer of multiplier seedstock producers focused on producing market progeny.

commercial feeders   (1) a carcass grade of cattle; (2) livestock that are not registered or pedigreed by a registry (e.g., breed) association.

commercial feedlots   Professionally managed feedyards with greater than 1,000 head capacity that feed both owned and customer-owned cattle.

commercial producer   Produce lambs or kids for sale as feeder or market stock.

companion animals   Those animals kept as pets.

compensatory gain   A faster-than-normal rate of gain after a period of restricted gain.

compensatory growth   See *compensatory gain*.

complete feed   (compound) A nutritionally adequate feed for animals other than humans by specific formula compounded to be fed as the sole ration and capable to maintaining life and/or promoting production without any additional substance, except water, being consumed.

composite breed   A breed that has been formed by crossing two or more breeds.

composting   Process of accelerating natural decay.

composting   Process of accelerating natural decay of dead organisms.

concentrate   A feed used with another to improve the nutritive balance of the total and intended to be further-diluted and mixed to produce a supplement or a complete feed.

conception   Fertilization of the ovum (egg).

conceptus   Embryo and embryonic membranes taken in total.

conditioning   The treatment of animals by vaccination and other means before putting them in the feedlot.

condensed milk   A form of cow's milk where water has been removed and sugar or sweetener has been added.

conduction   Heat loss that occurs when warm molecules transfer heat to colder molecules through close association.

conformation   The physical form of an animal; its shape and arrangement of parts.

consumer   The final buyer in a supply chain.

contagious disease   Infectious disease; a disease that is transmitted from one animal to another.

contemporaries   A group of animals of the same sex and breed (or similar breeding) that have been raised under similar environmental conditions (same management group).

Continental breeds   Breeds of beef cattle originating in countries other than England. Sometimes called European or exotic breeds. Examples are Charolais, Limousin, and Simmental.

continuity   An uninterrupted succession or flow.

contract growers   Farms that provide contract services to grow and develop birds to a targeted weight for a particular processor integrator.

contracted heels   A condition in which the heels of a horse are pulled in so that expansion of the heel cannot occur when the foot strikes the ground.

convection   Process of heat transfer through air or water movement.

core samples   Sample taken from a wool or mohair sack (bale) to determine the collective grade of the fiber.

core samples   Samples (wool, feed, or meat) taken by a coring device to determine the composition of the sample.

Corn Belt   The region of the United States with the highest corn production typically considered the northern Great Plains and western Great Lakes states.

coronary band (coronet)   Boundary between the top of the hoof wall and the skin at the bottom of the pastern where hoof growth begins.

corpus luteum   A yellowish body in the mammalian ovary. The cells where follicular cells develop into the corpus luteum, which secretes progesterone. It becomes yellow in color from the yellow lipids that are in the cells.

correlation coefficient   A measure of the association of one trait with another.

cost of gain   The total cost divided by the total pound gained; usually expressed on a per-pound basis.

cotyledon   An area of the placenta that contacts the uterine lining to allow nutrients and wastes to pass from the mother to the developing young. Sometimes referred to as *button*.

cotyledonary   Type of chorionic attachment in cows, ewes, and does, characterized by specialized sites of attachment.

countercurrent blood flow action   Mechanism of maintaining lower body temperatures in the extremities to avoid excessive heat loss.

cow   A sexually mature female bovine animal—usually one that has produced a calf.

cow–calf operation   A management unit that maintains a breeding herd and produces weaned calves.

cow hocked   A condition in which the hocks are close together but the feet stand apart.

creep   An enclosure in which young can enter to obtain feed but larger animals cannot enter. This process is called *creep feeding*.

crimp   The waves, or kinks, in a wool fiber.

crop   Enlargement of the esophagus in poultry, used as temporary feed storage.

crossbred   An animal produced by crossing two or more breeds.

crossbreds   Commercial animals resulting from matings that cross two or more purebred lines.

crossbred wool breeds   Columbia and Targhee are examples.

crossbreeding   Mating animals from genetically diverse groups (i.e., breeds) within a species.

crossbreeding   The process of mating two or more breeds.

cross-fostering   The practice of transferring piglets from one sow to another within a farrowing group to reduce within litter weight variation among pigs and to assure that each sow is nursing the number of pigs best suited to her ability.

crutching   See *tagging*.

cryptorchidism   The retention of one or both testicles in the abdominal cavity in animals that typically have the testicles hanging in a scrotal sac.

cud   Bolus of feed a ruminant animal regurgitates for further chewing.

cull   To eliminate one or more animals from the beeding herd or flock.

culled   Process of removing animals from the herd usually for poor performance.

cultured   Addition of appropriate bacteria to create dairy products such as yoghurt, sour crème, and buttermilk.

curb   A hard swelling that occurs just below the point of the hock.

curd   Coagulated milk.

cutability   Fat, lean, and bone composition of meat animals. Used interchangeably with yield grade. (See also *yield grade*.)

cuticle   Causes wool fibers to cling together.

cuticle (bloom)   Protective coating of an egg shell.

cutting chute   A narrow chute where animals proceed through gates in single file such that animals can be directed into pens along the side of the chute.

cwt   An abbreviation for hundredweight (100 lb).

cycling   Infers that nonpregnant females have active estrous cycles.

dairy goat breeds   Selected for milk production.

dairy industry   A coordinated set of supply chain enterprises specifically in place to produce, process, and distribute milk products and by-products to the market place.

dam   Female parent.

dark cutter Color of the lean (muscle) in the carcass has a dark appearance, usually caused by stress (excitement, etc.) to the animal before slaughter.

dark meat See definition from Chapter 5.

dark meat Meat from aerobically functional muscles with a higher myoglobin concentration. Higher in calories and cholesterol than white meat.

debeaking To remove the tip of the beak of chickens.

dehorn To remove the horns from an animal.

deoxyribonucleic acid (DNA) A complex molecule consisting of deoxyribose (a sugar), phosphoric acid, and four nitrogen bases (a gene is a piece of DNA).

depreciation An accounting procedure by which the purchase price of an asset with a useful life of more than 1 year is prorated over time.

dermis Skin layer that lies beneath the epidermis and basal layer.

development Directive coordination of physiological and metabolic processes until maturity is achieved.

dewclaws Hard horny structures above the hoof on the rear surface of the legs of cattle, swine, and sheep.

dewlap Loose skin under the chin and neck of cattle.

DHIA Dairy Herd Improvement Association, an association which dairy producers participate in keeping dairy records. Sanctioned by the National Cooperative Dairy Herd Improvement Program.

DHIR Dairy Herd Improvement Registry, a dairy record-keeping plan sponsored by the breed associations.

diet Feed ingredients or mixture of ingredients (including water) which are consumed by animals.

diet–health relationships The association between diet and health.

diffuse Type of chorionic attachment in mares and sows, characterized by uniform villi attachment across the membrane.

digestibility The quality of being digestible. If a high percentage of a given food taken into the digestive tract is absorbed into the body, that food is said to have *high digestibility*.

digestion The reduction in particle size of feed so that the feed becomes soluble and can pass across the gut wall into the vascular or lymph system.

diploid Having the normal, paired chromosomes of somatic tissue as produced by the doubling of the primary chromosomes of the germ cells at fertilization.

disease Any deviation from a normal state of health.

disinfect To kill, or render ineffective, harmful microorganisms and parasites.

disinfectant A chemical that destroys disease-producing microorganisms or parasites.

distal Position that is distant from the point of attachment of an organ.

DM See *dry matter*.

DNA (deoxyribonucleic acid) The molecule that encodes genetic information. DNA is a double-stranded molecule held together by weak bonds between base pairs of nucleotides.

DNA fingerprint Pattern of DNA fragments unique to an individual. Often found by using restriction enzymes to cut the DNA into fragments. These fragments can be sorted and documented, forming a unique "fingerprint." This technology, the same used as a forensic tool at crime scenes, is also used to DNA parentage test animals.

DNA sequence The relative order of base pairs, whether in a fragment of DNA, a gene, a chromosome, or an entire genome.

dock (1) To cut off the tail. (2) The remaining portion of the tail of a sheep that has been docked. (3) To reduce or lower in value.

doe A female goat or rabbit.

dominance (1) A situation in which one gene of an allelic pair prevents the phenotypic expression of the other member of the allelic pair. (2) A type of social behavior in which an animal exerts influence over one or more other animals.

dominant gene A gene that overpowers and prevents the expression of its recessive allele when the two alleles are present in a heterozygous individual.

dorsal Of, on, or near the back of an animal.

double muscling A genetic trait in cattle where muscles are greatly enlarged rather than duplicate muscles.

down Soft, fluffy type of feather located under the contour feathers. Serves as insulating material.

draft horses See Figure 32.1.

draft purposes To use animals as a source of power.

drake Mature male duck.

drench To give fluid by mouth (e.g., medicated fluid is given to sheep for parasite control).

dressing percentage The percentage of the live animal weight that becomes the carcass weight at slaughter. It is determined by dividing the carcass weight by the live weight, then multiplying by 100.

drop Body parts removed at slaughter—primarily hide (pelt), head, shanks, and offal.

drop credit Value of the drop.

dry (cow, ewe, sow, mare) Refers to a nonlactating female.

dry matter Feed after water (moisture) has been removed (100% dry).

dry milk Milk that has been evaporated into a powdered form.

dual-purpose breeds   Breeds bred to have above average performance in maternal and terminal traits.

dubbing   The removal of part or all of the soft tissues (comb and wattles) of chickens.

ductus deferens   Also known as the vas deferens, responsible for sperm transport.

dung   The feces (manure) of farm animals.

dung tags   Manure contamination of wool.

duodenum   Portion of the small intestine connected to the stomach.

dwarfism   The state of being abnormally undersized. Two kinds of dwarfs are recognized; one is proportionate and the other is disproportionate.

dysentery   Severe diarrhea.

dystocia   Difficult birth.

Easter lambs   Production systems designed to produce market lambs to meet specific holiday markets.

ectoderm   The outermost layer of the three layers of the primitive embryo.

edema   Abnormal collection of fluid in body tissues that causes soft swelling.

edible   Appropriate for human consumption.

effective ambient temperature   Theoretical index of the heating or cooling power of the environment including dry bulb temperature including any environmental factor that alters heat demand.

ejaculate   The emission containing both sperm and seminal fluid.

ejaculation   Discharge of semen from the male.

electroejaculator   Rectal probe that produces mild waves of electrical stimuli that brings about ejaculation. Typically used for semen testing and collection of bulls for AI.

eliminative behavior   Behaviors associated with defecation and urination.

elite seedstock producers   Specialized farms focused on the production of genetically improved seedstock.

emaciated   In severely poor body condition.

emaciation   Thinness; loss of flesh where bony structures (hips, ribs, and vertebrae) become prominant.

embryo   Very early stage of individual development within the uterus. The embryo grows and develops into a fetus. In poultry, the embryo develops within the eggshell.

embryo transfer   The transfer of fertilized eggs from a donor female to one or more recipient females.

emu   Large flightless bird native to Australia farmed for meat, oil, feather, and egg production.

endocrine gland   A ductless gland that secretes a hormone into the bloodstream.

endoderm   The innermost layer of the three layers of the primitive embryo.

endurance   Length of time muscle can work before experiencing fatigue.

enterotoxemia   A disease of the intestinal tract caused by bacterial secretion of toxins. Its symptoms are characteristic of food poisoning.

enterprise budget   Estimated costs and returns associated with an enterprise.

entropion   Turned-in eyelids.

environment   The sum total of all external conditions that affect the well-being and performance of an animal.

environmental effects   Any effect on phenotype not of genetic origin.

enzyme   A complex protein produced by living cells that causes changes in other substances in the cells without being changed itself and without becoming a part of the product.

Eohippus   An early ancestor to the modern horse.

EPD   See *expected progeny difference*.

epididymis   The long, coiled tubule leading from the testis to the vas deferens.

epididymitis   An inflammation of the epididymis.

epiphysis   A piece of bone separated from a long bone in early life by cartilage, which later becomes part of the larger bone.

epistatic   Interaction between nonallelic genes.

epistatic interactions   The outcomes of epistatis; for example, coat color in horses.

epistasis   A situation in which a gene or gene pair masks (or controls) the expression of another nonallelic pair of genes.

equine   Refers to horses.

equine encephalomyelitis   An inflammation of the brain of horses.

eruction (or eructation)   The elimination of gas by belching.

Escherichia coli 0157:H7   A pathogenic organism of concern in food safety.

esophagus   Tubular structure leading from the mouth to the stomach.

esophageal groove   A groove in the reticulum between the esophagus and omasum. Directs milk in the nursing young ruminant directly from the esophagus to the omasum.

essential nutrient   A nutrient that cannot be synthesized by the body and must be supplied in the diet.

estrogen   Any hormone (including estradiol, estriol, and estrone) that causes the female to come physiologically into heat and to be receptive to the male. Estrogens are produced by the follicle of the ovary and by the placenta.

estrous   An adjective meaning "heat," which modifies such words as "cycle." The estrous cycle is the heat cycle, or time from one heat to the next.

estrous cycles The length of time between heats.

estrous synchronization Controlling the estrous cycle so that a high percentage of the females in the herd express estrus at approximately the same time.

estrus The period of mating activity in the female mammal. Same as heat.

ET Abbreviation for *embryo transfer*.

ethology Study of animal behavior in the animal's natural environment.

eukaryote Cell or organism with membrane-bound, structurally discrete nucleus and other well-developed subcellular compartments. Eukaryotes include all organisms except viruses, bacteria, and blue-green algae.

European breeds See *Continental breeds*.

evaporated milk Shelf stable form of milk where approximately 60% of the water has been removed.

evaporation Method of heat loss via panting or sweating.

evaporative cooling systems Systems such as misters and sprinklers designed to mitigate heat stress.

eviscerate The removal of the internal organs during the slaughtering process.

ewe A sexually mature female sheep. A ewe lamb is a female sheep before attaining sexual maturity.

ewe breeds Breeds noted for their maternal characteristics.

exercise-induced pulmonary hemorrage (bleeder) Defect of circulation in which the minute blood vessels in the lungs rupture when the horse is put under the stress of exercise.

exocrine gland Gland that secretes fluid into a duct.

exocrine gland Secrete into a series of internal ducts leading to the external environment; mammary gland is an example.

exotic breeds See *Continental breeds*.

expected progeny difference (EPD) One-half of the breeding value; the difference in performance to be expected from future progeny of a sire, compared with that expected from future progeny of an average bull in the same test.

extensive management Systems where management optimizes control over the production environment with minimal reliance on housing systems.

exterior quality factors Egg quality factors that are assessed visually with external evaluation.

fallopian tube Also known as the oviduct, site of fertilization.

family selection Selection based on performance of a family.

famine A shortage of food due to a catastrophic event that disrupts the food supply chain infrastructure.

farm-flock A sheep enterprise typically with a small inventory one part of a diversified farm.

farmer-feeder Smaller feedyards managed as part of an integrated farming operation to market high energy feedstuffs produced on the farm.

farmer-feeders Smaller feedlots that are used to market on farm produced feedstuffs and cattle.

farrow To deliver, or give birth to, pigs.

farrow-to-feeder pig production Feeder pigs after a short phase of postweaning growth are the primary commodity sold from the enterprise.

farrow-to-finish Breeding herd is maintained where pigs are produced and fed to harvest weight on the same farm.

farrow-to-weaner Pigs are managed to weaning age and then sold or moved off-site.

multisite rearing systems Segregated production systems that incorporate multiple site production facilities where pigs were born at one facility but then weaned, grown, and finished at facilities separated from the site of sow housing.

fat Adipose tissue.

FDA See *Food and Drug Administration*.

feathers Epidermal growth that provides the unique outer covering of birds.

feather picking The picking of feathers from one bird by another.

feces Bowel movements, excrement from the intestinal tract.

federal land Lands owned by the federal government of the United States.

feed additive Ingredient (such as an antibiotic or hormone-like substance) added to a diet to perform a specific role (e.g., to improve gain or feed efficiency).

feed bunk A trough or container used to feed farm animals.

feed costs Direct costs associated with grazed, harvested, and purchased feeds.

feed efficiency (1) The amount of feed required to produce a unit of weight gain or milk; for poultry, this term can also denote the amount of feed required to produce a given quantity of eggs. (2) The amount of gain made per unit of feed.

feeder Animals (e.g., cattle, lambs, pigs) that need further feeding prior to slaughter.

feeder grades Visual classifications (descriptive and/or numerical) of feeder animals. Most of these grades have been established by the USDA.

feeder pig finishing Feeder pigs are purchased and then fed to market weight.

feeder space amount of space per bird relative to access to feeder space in a housed environment.

feedlot Enterprises that utilize high energy rations to grow cattle (sheep) to ideal market weights.

felting The process of pressing wool fibers together in conjunction with heat and moisture to produce a fabric.

femininity Well-developed secondary female sex characteristics, udder development and refinement in head and neck.

feral Domesticated animals that return to nature to survive and reproduce.

fertility The capacity to initiate, sustain, and support reproduction. With reference to poultry, the term typically refers to the percentage of eggs that, when incubated, show some degree of embryonic development.

fertilization The process in which a sperm unites with an egg to produce a zygote.

fetus Later stage of individual development within the uterus. Generally, the new individual is regarded as an embryo during the first half of pregnancy, and as a fetus during the last half.

fiber goat breed Selected for production of high quality fiber.

fill The contents of the digestive tract.

filly A young female horse.

fineness A term used to describe the diameter of wool fibers.

fine-wool breeds Breeds that produce fine grade wool—Rambouillet and Merino.

fine-wool breeding Basis for determining American grades of wool.

fingerlings Usually 1–6 inches long.

finish The degree of fatness of an animal.

finishing The process of an animal attaining optimal weight and composition prior to harvest.

Finnsheep A highly prolific breed of sheep introduced into the United States in 1968.

fistula A running sore at the top of the withers of a horse, resulting from a bruise followed by invasion of microorganisms.

flank firmness and fullness Firmness of the flank muscle in lamb carcass evaluation.

flank streaking Streaks of fat in the flank muscle of lamb carcasses.

fleece The wool shorn at one time from all parts of the sheep.

fleece weight Weight of a fleece once it is removed at shearing.

flehmen A pattern of behavior expressed in some male animals (e.g., bull, ram, stallion) during sexual activity. The upper lip curls up and the animal inhales in the vicinity of the vulva or urine.

flock A group of sheep or poultry.

floor space Amount of space per bird in a poultry house.

flushing Placing females (typically sheep and swine) on a gaining level of nutrition before breeding to stimulate greater

ovulation rates; also, a behavior in fish whereby diseased fish rub against objects in tanks or ponds.

fly strike An infestation with large numbers of blowfly maggots.

foal A young male or female horse (noun) or the act of giving birth (verb).

foaling Process of the mare giving birth.

follicle A blisterlike, fluid-filled structure in the ovary that contains the egg.

follicle-stimulating hormone (FSH) A hormone produced and released by the anterior pituitary that stimulates the development of the follicle in the ovary.

Food and Drug Administration (FDA) A U.S. government agency responsible for protecting the public against impure and unsafe foods, drugs, veterinary products, and other products.

food insecure A situation of limited or uncertain access to an appropriate level of nutrients.

food safety Process to assure that food is free of pathogens, toxins, and contaminants.

food secure A situation where all family members have sufficient and predictable access to an appropriate diet.

food-size Commercially grown fish produced for food, usually ranging from 0.75 to 1.0 pounds and over 1 foot in length.

footrot A disease of the foot in sheep and cattle. In sheep it causes rotting of tissue between the horny part of the foot and the soft tissue underneath.

forage High fiber feedstuff such as grass, hay, or alfalfa.

forb Weedy or broad-leaf plants, as contrasted to grasses, that serve as pasture for animals.

forging The striking of the heel of the front foot with the toe of the hind foot by a horse in action.

formula contract Price is based on the cash market plus a predetermined premium.

forward-cash contract Base price is determined from a formula that accounts for fluctuations in feed costs.

founder Nutritional ailment resulting from overeating. Lameness in front feet with excessive hoof growth usually occurs.

frame score A numerical rating of frame size.

frame size A measure of skeletal size. It can be visual or by measurement (usually taken at the hips).

freemartin Female born twin to a bull (approximately 9 of 10 will not conceive).

freshen To give birth to young and initiate milk production. This term is usually used in reference to dairy cattle.

fry Stage from hatching until fish reach 1 inch in length.

fryer See *broiler*.

FSH See *follicle-stimulating hormone*.

full-mouth Animal has all permanent teeth fully exposed.

full sibs Animals having the same sire and dam.

gallop A three-beat gait in which each of the two front feet and both of the hind feet strike the ground at different times.

gametes Male and female reproductive cells. The sperm and the egg.

gametogenesis The process by which sperm and eggs are produced.

gander Mature male goose.

gelding A male horse that has been castrated.

gelatin A by-product created from connective tissue.

gene The fundamental physical and functional unit of heredity.

gene expression The process by which a gene's coded information is converted into the structures present and operating in the cell.

gene mapping Determination of the relative positions of genes on a DNA molecule (chromosome or plasmid) and of the distance, in linkage units or physical units, between them.

general combining ability The ability of individuals of one line or population to combine favorably or unfavorably with individuals of several other lines or populations.

generation interval Average age of the parents when off-spring are born.

generation turnover Length of time from one generation of animals to the next generation.

genetic change The rate of improving the genetic potential of a herd or flock.

genetic code The sequence of nucleotides, coded in triplets (codons) along the mRNA, that determines the sequence of amino acids in protein synthesis.

genetic defects Dysfunction resulting from heredity.

genetic engineering The technique of removing, modifying, or adding genes to a DNA molecule.

genetically adapted Process where livestock of a particular type adapt to a set of environmental and climatic conditions.

genome The sum total of a living organism's genetic material. The genome is divided into chromosomes, which contain genes, and genes are made of DNA.

genomics The study of genes and their function.

genotype The genetic constitution, or makeup, of an individual. For any pair of alleles, three genotypes (e.g., *AA, Aa,* and *aa*) are possible.

genotypic See *genotype*.

gestation The time from breeding or conception of a female until she gives birth to her young.

gilt A young female swine prior to the time that she has produced her first litter.

gizzard Site of grinding feed into smaller particle sizes in the avian.

glans penis Free end of the penis.

goat meat The primary product of goat production although milk and fiber production are options for the grower.

goiter Enlargement of the thyroid gland, usually caused by iodine-deficient diets.

gonad The testis of the male; the ovary of the female.

gonadotropic hormones Luteinizing hormone (LH) and follicle stimulating hormone (FSH).

gonadotropin-releasing hormone (GnRH) Stimulates release of LH and FSH.

gonadotrophin Hormone that stimulates the gonads.

gossypol A toxic product contained in cottonseed.

Graafian follicle Mature follicle capable of ovulating.

grade (1) a designation of live or carcass merit (e.g., choice grade); (2) livestock not registered with registry (e.g., breed) association.

grading up The continued use of purebred sires of the same breed in a grade herd or flock.

grass lambs Production system based on grazing improved, high quality pastures.

grass tetany A disease of cattle and sheep marked by staggering, convulsions, coma, and frequently death, caused by a mineral imbalance (magnesium) while grazing lush pasture.

grease wool Wool as it comes from the sheep and prior to cleaning. It contains the natural oils from the sheep.

gross energy The amount of heat, measured in calories, produced when a substance is completely oxidized. It does not reveal the amount of energy that an animal could derive from eating the substance.

growing fetus Nutrient requirements in the final trimester of development increase substantially.

growth The process of growing as the result of protein synthesis exceeding protein breakdown.

growth The increase in protein over its loss in the animal body. Growth occurs by increases in cell numbers, cell size, or both.

growth rate Economically important trait to swine industry that has a heritability of 0.35.

growth stimulants Subcutaneous implants that slowly release growth enhancing compounds.

Guernsey See Figure 26.1.

habituation The gradual adaptation to a stimulus or to the environment.

hair-type sheep Produce a fiber more like the hair of goats.

half sib   Animals having one common parent.

hand-mated   To be bred with artificial insemination or through introduction of a male only after identifying the female as being in heat.

handmated   A specific ewe/doe is chosen to be mated to a specific male.

hand mating   Same as hand breeding—bringing a female to a male for service (breeding), after which she is removed from the area where the male is located.

handmating   Situation in which producers individually mate the boar to each female.

hank   A measurement of the fineness of wool. A hank is 560 yards of yarn. More hanks of yarn are produced from fine wools than coarse wools.

haploid   One-half of the diploid number of chromosomes for a given species, as found in the germ cells.

hatchability   A term that indicates the percentage of a given number of eggs set from which viable young hatch, sometimes calculated specifically from the number of fertile eggs set.

hay   Harvested forage such as alfalfa hay.

Hazard Analysis Critical Control Point (HACCP)   Seven step process to assure quality is produced.

HDL (high-density lipoprotein)   Typically referred to as "good" cholesterol

heat   See *estrus*.

heat increment   The increase in heat production after consumption of feed when an animal is in a thermoneutral environment. It includes additional heat generated in fermentation, digestion, and nutrient metabolism.

heat stress   A state of physiological stress caused by excessive exposure to high effective ambient temperature often exacerbated by high humidity.

heat stress   Characterized by loss of production efficiency and reduced feed intake.

heat stress indices   Index of temperature and humidity to determine when management interventions should be undertaken to help keep animals cool.

heaves   A respiratory defect in horses during which the animal has difficulty completing the exhalation of inhaled air.

heifer   A young female bovine cow before the time that she has produced her first calf.

heiferette   A heifer that has calved once, after which the heifer is fed for slaughter; the calf has usually died or been weaned at an early age.

hemoglobin   The iron-containing pigment of the red blood cells. It carries oxygen from the lungs to the tissues.

hen   An adult female domestic fowl, such as a chicken or turkey.

herbivores   Depend on plant based diets.

herbivorous   Subsisting or feeding on plants.

herd   A group of animals. Used with beef, dairy, or swine.

heritability   The portion of the total variation or phenotypic differences among animals that is due to heredity.

hernia   The protrusion of some of the intestine through an opening in the body wall (also commonly called *rupture*). Two types of hernias, umbilical and scrotal, occur in farm animals.

heterosis   Performance of offspring that is greater than the average of the parents. Usually the amount of superiority of the crossbred over the average of the parental breeds. Also referred to as *hybrid vigor*.

heterozygous   A term designating an individual that possesses unlike genes for a particular trait.

hides   Skins from animals such as cattle, horses, and pigs; beef hides weigh more than 30 lb each as contrasted to calf skins which weigh less.

hinny   The offspring that results from crossing a stallion with a female donkey (jenny).

hobble   To tie two of an animal's legs together. An animal is hobbled to prevent it from kicking or moving a long distance.

hock   Joint between the gaskin and rear cannon.

Holstein   See Figure 26.1.

homeotherm   A warm-blooded animal. An animal that maintains its characteristic body temperature even though environmental temperature varies.

homogenized   Milk that has had the fat droplets broken into very small particles so that the milk fat stays in suspension in the milk fluids.

homologous   Corresponding in type of structure and derived from a common primitive origin.

homologous chromosomes   Chromosomes having the same size and shape that contain genes affecting the same characters. Homologous chromosomes occur in pairs in typical diploid cells.

homology   Similarity in DNA or protein sequences between individuals of the same species or among different species.

homozygous   A term designating an individual whose genes for a particular trait are alike.

honeycomb   Description of the lining of the reticulum.

hormone   A chemical substance secreted by a ductless gland. Usually carried by the bloodstream to other places in the body where it has its specific effect on another organ.

horse industry   A loosely coordinated set of enterprises focused on the recreational, working, and sporting use of horses.

hothouse lambs   Market lambs produced out of season typically for east coast markets.

housing systems   Options to provide animals controlled, comfortable, and healthy quarters.

humoral immunity Form of immunity that is mediated by macromolecules such as antibodies.

hybrids Originate from crossing two or more breeds and then applying specialized selection programs.

hybrid chickens Offspring that result from three-way and four-way crosses.

hybrid vigor See *heterosis*.

hydrocephalus A condition characterized by an abnormal increase in the amount of cerebral fluid, accompanied by dilation of the cerebral ventricles.

hyperkalemic periodic paralysis (HYPP) An inherited muscle disorder characterized by muscle tremors, weakness and, in severe cases, collapse and death.

hyperplasia Increase in the number of cells.

hypertension High blood pressure.

hypertrophy Increase in cell size.

hypothalamus A portion of the brain found in the floor of the third ventricle. It regulates reproduction, hunger, and body temperature and has other functions.

hypoxia A condition resulting from deficient oxygenation of the blood.

ideal dairy types Descriptive system based on stature, angularity, level rump, long and lean neck, milk veins, and strong feet and legs.

identification Methodology to document ownership.

ileum Distal portion of the small intestine.

immunity The ability of an animal to resist or overcome an infection.

impaction Obstructive lodging of food in the intestine.

implant To graft or insert material to intact tissues.

implantation The attachment of the fertilized egg to the uterine wall.

imprinting Learning associated with maturational readiness.

inbred lines Developed for crossing systems and used largely for increasing egg production.

inbreeding The mating of individuals who are more closely related than the average individuals in a population. Inbreeding increases homozygosity in the population but it does not change gene frequency.

incisor A front tooth.

incubation Period of time during which poultry embryos develop within the egg.

incubation period The time that elapses from the time an egg is placed into an incubator until the young is hatched.

independent culling level Selection method in which minimum acceptable phenotypic levels are assigned to several traits.

index (1) An overall merit rating of an animal. (2) A method of predicting the milk-producing ability that a bull will transmit to his daughters.

inedible Not appropriate for human consumption.

infection Invasion of the body tissues by microbial agents or parasites other than insects.

infectious Capable of invading and growing in living tissues. Used to describe various pathogenic microorganisms such as viruses, bacteria, protozoa, and fungi.

infectious diseases Caused by pathogenic organisms.

influenza A virus disease characterized by inflammation of the respiratory tract, high fever, and muscular pain.

infundibulum Fingerlike projections of the oviduct that "catch" the egg upon ovulation.

ingest Anything taken into the stomach.

ingestive behavior Behaviors associated with feed and water intake.

inheritance The transmission of genes from parents to offspring.

inherited abnormalities Genetic defects.

insemination Deposition of semen in the female reproductive tract.

instinct Inborn behavior.

insulation Factor that impacts ventilation design.

insulin Hormone secreted by the pancreas to control blood sugar level and utilization of sugar in the body.

integration The bringing together of all segments of a livestock or poultry production program under one centrally organized unit.

intelligence The ability to learn to adjust successfully to situations.

intensive inbreeding Mating of closely related animals whose ancestors have been inbred for several generations.

intensive management Systems where management attempts to control the environment at high levels through housing, rations, etc.

interference The striking of the supporting leg by the foot of the striding leg by a horse in action.

interior quality factors Egg quality factors inside the egg that are determined via candling.

intermuscular Between muscles.

interstitial cells The cells between the seminiferous tubules of the testicle that produce testosterone.

intravenous Within the vein. An intravenous injection is an injection into a vein.

intravenously Administered directly into the vein.

inverted nipples Nipples or teats that do not protrude but are inverted into the mammary gland.

investigating behavior Behaviors associated with exploring new or novel surroundings.

in vitro Outside the living body; in a test tube or other artificial environment.

iodine   If limited in availability it must be supplemented via iodized salt.

isowean principle   Recognizes that preweaned pigs are free of nearly all pathogens and if they are managed as a distinct group to final marketing weights then the risk of cross-contamination and costly disease outbreaks are largely avoided.

isthmus   Portion of the avian oviduct that secretes the shell membranes.

jack   A male donkey.

jackass   See *jack*.

jennet   A female donkey.

jenny   A female donkey.

Jersey   See Figure 26.1.

jowl   The area of the throat underneath the jaw.

junk science   Studies that have been incorrectly designed or interpreted and thus should be viewed as without creditability.

Karakul   A breed of fat-tailed sheep having coarse, wiry furlike hair. Used to produce Persian lambskins.

ked   An external parasite that affects sheep. Although commonly called *sheep tick*, it is actually a wingless fly.

kemp   Coarse, opaque, hairlike fibers in wool.

ketosis   A condition (also called *acetonemia*) that is characterized by a high concentration of ketone bodies in the body tissues and fluids.

kid   Young goat.

kilocalorie (kcal, Kcal)   An amount of heat equal to 1,000 calories. (See also *calorie*.)

knee   Joint between the forearm and the cannon.

known effects   Environmental effects common to a group and thus quantifiable.

kosher meat   Meat from ruminant animals with split hooves where the animals have been slaughtered according to Jewish law.

labor   Manpower requirements for an enterprise.

lactalbumin   A nutritive protein of milk.

lactation   The secretion and production of milk.

lactose   Milk sugar. When digested, it is broken down into one molecule of glucose and one of galactose.

lamb   (1) A young male or female sheep, usually less than a year of age. (2) To deliver, or give birth to, a lamb.

lamb dysentery   See *dysentery*.

lambing   Act of giving birth. Same as parturition.

lambing jug   A small pen in which a ewe is put for lambing. It is also used for containing the ewe and her lamb until the lamb is strong enough to run with other ewes and lambs.

laminitis   Inflammation of the sensitive plates of soft tissue (laminae) within the horse's foot caused by physical or physiologic injury. Severe cases of laminitis may result in founder, an internal deformity of the foot. *Acute* laminitis sets in rapidly and usually responds to appropriate, intensive treatment, while *chronic* laminitis is a persistent, long-term condition that may be unresponsive to treatment.

lard   The fat from pigs that has been produced through a rendering process.

large intestine   Primary site of water absorption.

layer   A hen that is kept for egg production.

LDL (low-density lipoprotein)   High concentrations are associated with higher risk of cardiovascular disease.

legume   Any plant of the family *leguminosae*, such as pea, bean, alfalfa, and clover.

leucocytes   White blood cells.

Leydig cells   Also known as interstitial cells, site of testosterone production.

LH   See *luteinizing hormone*.

libido   Sex drive or the desire to mate on the part of the male.

lice   Small, flat, wingless insect with sucking mouth parts that is parasitic on the skin of animals.

light horse breeds   See Figure 32.1.

linear classification system   Evaluation tool to enhance selection for high-producing cows with the durability to stay productive.

linear interaction   Interaction between genes on the same chromosome.

linebreeding   A mild form of inbreeding that maintains a high genetic relationship to an outstanding ancestor.

line crossing   The crossing of inbred lines.

linkage   The proximity of two or more markers on a chromosome. The closer together the markers are, the lower the probability that they will be separated during DNA repair or replication processes and hence the greater the probability that they will be inherited together.

linkage map   A map of the relative positions of genetic loci on a chromosome, determined on the basis of how often the loci are inherited together. Distance is measured in centimorgans.

lipid   An organic substance that is soluble in alcohol or ether but insoluble in water; used interchangeably with the term *fat*.

Listeria monocytogenes   Highly virulent foodborne bacterium.

litter   The young produced by multiparous females such as swine. The young in a litter are called *littermates*.

liver flukes   A parasitic flatworm found in the liver.

locus   The place on a chromosome where a gene is located.

long-wool breeds   Poorer quality wool breeds such as the Lincoln and Romney.

longevity   Life span of an animal. Usually refers to a long life span.

lower critical temperature (LCT) Effectiveness of vaso-constriction and behavioral responses is maximal at this limit.

luteinizing hormone (LH) A protein hormone, produced and released by the anterior pituitary, which stimulates the formation and retention of the corpus luteum. It also initiates ovulation.

lymph Transparent, nutritive yellow liquid that exudes from blood vessels into tissue spaces and is drained back into the veins through lymph vessels. Lymph plays an important role in fighting infection and maintaining the body's fluid balance.

lymphocyte Specialized white blood cells.

macroclimate The large, general climate in which an animal exists.

macromineral A mineral that is needed in the diet in relatively large amounts.

macrophages Example of cell mediated immunity, provide protection via engulfing, and destruction of pathogen.

magnum Portion of the avian oviduct that secretes the albumen or white of the egg.

maintenance A condition in which the body is maintained without an increase or decrease in body weight and with no production or work being done.

mammal Warm-blooded animals that suckle their young.

mammalia Zoological class of four legged livestock.

mammary gland Gland that secretes milk.

management The act, art, or manner of managing, handling, controlling, or directing a resource or integrating several resources.

manure management Process and protocols associated with effective management of animal waste.

manyplies The multiple tissue folds of the omasum.

marbling The distribution of fat in muscular tissue; intramuscular fat.

mare A sexually developed female horse.

marker An identifiable physical location on a chromosome whose inheritance can be monitored. Markers can be expressed regions of DNA (genes) or some segment of DNA with no known coding function but whose pattern of inheritance can be determined.

marketing Process for determining the value of a product to facilitate sale or trade.

market class Animals grouped according to the use to which they will be put, such as slaughter, feeder, or stocker.

market classes and grades Method to group commodities into more uniform categories to improve communication of value in the market often by establishing measurable sets of standards.

market grade Animals grouped within a market class according to their value.

mass selection The best-performing males are mated to the best performing females.

masticate To chew food.

mastitis Inflammation of the mammary gland.

maturity A measure of physiological age often determined by the level of ossification of the growth plates of long bones.

mean (1) Statistical term for average. (2) Term used to describe animals having bad behavior.

meat The tissues of the animal body that are used for food.

meat goat breed Noted for muscularity.

meat inspection The process of determining the safety and wholesomeness of meat.

meat spots Spots in the egg that are blood spots which have changed color or tissue sloughed off from the reproductive organs of the hen.

mechanization The application of technology and machinery to a process.

medium-wool breeds Breeds that produce wool that is of medium quality—Cheviot and Hampshire.

medium-wool breeds Breeds of sheep that produce wool quality intermediate to fine and coarse types.

medulla Inner core found in coarse and medium wool but absent from fine wool fibers.

medullated fibers These fibers are of lower value as they do not accept uniform coloration from dyes.

meiosis A special type of cell nuclear division that is undergone in the production of gametes (sperm in the male, ova in the female). As a result of meiosis, each gamete carries half the number of chromosomes of a typical body cell in that species.

melengestrol acetate (MGA) A feed additive that suppresses estrus in heifers and is widely used in the feedlot industry.

mesoderm The middle layer of the three layers of the primitive embryo.

messenger RNA (mRNA) RNA that serves as a template for protein synthesis.

metabolism (1) The sum total of chemical changes in the body, including the "building up" and "breaking down" processes. (2) The transformation by which energy is made available for body uses.

metabolizable energy Gross energy in the feed minus the sum of energy in feces, gaseous products of digestion, and energy in urine. Energy that is available for metabolism by the body.

maternal line index Bioeconomic index that measures maternal performance.

metritis Inflammation (infection) of the uterus.

MGA See *melengestrol acetate.*

microclimate A small, special climate within a macro-climate created by the use of such devices as shelters, heat lamps, and bedding.

microcomputer A small computer that has a smaller memory capacity than a larger or mainframe computer.

micromineral A mineral that is needed in the diet in relatively small amounts. The quantity needed is so small that such a mineral is often called a *trace mineral.*

milk EPD A genetic estimate of the milking ability of a beef bull's daughters when compared to the daughters of other bulls.

milk fat The fat in milk; synonymous with butterfat.

milk fever See *parturient paresis.*

milk letdown The release of milk into the teat cisterns.

Milk Only records Dairy record system similar to DHI except no milk fat samples are taken.

minimum culling level A selection method in which an animal must meet minimum standards for each trait desired in order to qualify for being retained for breeding purposes.

mites Very small arachnids that are often parasitic upon animals.

mitosis A process in which a cell divides to produce two daughter cells, each of which contains the same chromosome complement as the mother cell from which they came.

modified live vaccines Contain a modified antigen that elicits a stronger and longer lasting immune response.

modifying genes Genes that modify the expression of other genes.

mohair Fleece of the Angora goat.

monogastric Having only one stomach or only one compartment in the stomach. Examples are swine and poultry.

monoparous A term designating animals that usually produce only one offspring at each pregnancy. Horses and cattle are monoparous.

monotocous Producing a single offspring at a birth.

moon blindness Periodic blindness that occurs in horses.

moonblindness (recurrent uveitis) The horse is blind for a short time, regains its sight, and then again becomes blind for a time.

morbidity Measurement of illness; morbidity rate is the number of individuals in a group that become ill during a specified time.

mortality rate Number of individuals that die from a disease during a specified time, usually 1 year.

mouth-brooder Fish that hold eggs or newly hatched young in their mouths.

mouthed The examination of an animal's teeth.

mule The hybrid that is produced by mating a male donkey with a female horse. They are usually sterile.

mulefoot Having one instead of the expected two toes, on one or more of the feet.

multiparous Having had two or more pregnancies which resulted in viable fetuses.

mutation A change in a gene.

multiplier seedstock producers Create a higher volume of breeding stock from the genetics produced by elite breeders.

mutton The meat from a sheep that is over 1 year old.

muzzle The nose of horse, cattle, or sheep.

mycotoxins Toxic chemical products produced by fungi.

myofibrils The primary component part of muscle fibers.

myofilaments

myosin A protein that works in conjunction with actin to contract and relax muscle fibers.

nanogram One-billionth of a gram.

National Cooperative Dairy Herd Improvement Program (NCDHIP) A national, industry-wide production-testing and record-keeping program.

natural or native immunity form of immunity in place at birth.

navel The area where the umbilical cord was formerly attached to the body of the offspring.

navicular disease Disease complex manifested in pain in the heels of the front foot.

necropsy Perform a postmortem examination.

neonatal scours Diarrhea that impacts relatively young stock.

net energy Metabolizable energy minus heat increments. The energy available to the animal for maintenance and production.

net merit index (NM$) Example of a bioeconomic index that estimates differences in milk production.

nicking The way in which certain lines, strains, or breeds perform when mated together. When outstanding offspring result, the parents are said to have *nicked* well.

nipple See *teat.*

nodular worm An internal parasitic worm that causes the formation of nodules in the intestines.

nonadditive value Genetic effect due to the combination of genes.

noninfectious disease Caused by nonpathogenic factors such as injury, toxins, and nutritional deficiencies.

nonruminant Simple-stomached or monogastric animal.

NPN (nonprotein nitrogen) Nitrogen in feeds from substances such as urea and amino acids, but not from preformed proteins.

nuclear fusion Union of nuclei from two sex cells (male and female, male and male, female and female).

nucleotide The subunit of DNA composed of a five carbon sugar, a nitrogenous base, and a phosphate group.

nucleus daughter Produce crossbred lines for gilt multipliers.

nucleus herds Create and test purelines, produce great grandparent lines for nucleus proliferation herds and boars for boar studs.

nucleus proliferation Produce purebred lines for daughter herds and produce crossbred boars for multipliers and studs.

nutrient (1) A substance that nourishes the metabolic processes of the body. (2) The end product of digestion.

nutrient density Amount of essential nutrients relative to the number of calories in a given amount of food.

nutrient management plan An integrated plan to manage the nutrients and waste created in a concentrated livestock enterprise.

obesity An excessive accumulation of body fat.

odor control Integrated management protocols to reduce air quality issues.

offal All organs and tissues removed from inside the animal during the slaughtering process.

off sorts Inferior portions of a fleece that are sorted and removed just following shearing.

omasum One of the stomach components of ruminant animals that has many folds.

omnivores Consumer of a mix of plants and meat.

omnivorous Feeding on both animal and vegetable substances.

oogenesis The process by which eggs, or ova, are produced.

open Refers to nonpregnant females.

open-faced Face of sheep that is free from wool, particularly around the eyes.

opportunity costs Returns given up if debt-free resources (e.g., land, livestock, equipment) are used in their next-best level of employment.

optimum level of performance The level at which a trait or traits maximizes net profit. Resources are managed to achieve a combined balance of traits that sustains high levels of profitability.

organic Products produced under organic standards.

organs Groups of tissues that perform specific functions.

osteopetrosis Genetic disorder characterized by the marrow cavity of long bones being filled with bone tissue.

osteopetrosis Abnormal thickening, hardening, and fragility of bones, making them weaker.

osteoporosis An abnormal decrease in bone mass with an increased fragility of the bones.

ostrich Large flightless bird native to Africa farmed for meat, egg, and feather production.

outbreeding The process of continuously mating females of the herd to unrelated males of the same breed.

outcrossing The mating of an individual to another in the same breed that is not related to it. Outcrossing is a specific type of outbreeding system.

ova Plural of ovum, meaning *eggs*.

ovary The female reproductive gland in which the eggs are formed and progesterone and estrogenic hormones are produced.

overeating disease A toxic condition caused by the presence of undigested carbohydrates in the intestine, which stimulates harmful bacteria to multiply. When the bacteria die, they release toxins. Called *enterotoxemia* in some animals.

overreaching The hind leg catches the heel of the front foot with the toe of the hind foot.

overshot jaw Upper jaw is longer than lower jaw. Also called *parrot mouth*.

oviduct A duct leading from the ovary to the horn of the uterus.

ovine Refers to sheep.

ovulation The shedding, or release, of the egg from the follicle of the ovary.

ovum The egg produced by a female.

owner conducted test Someone other than the DHIA technician records test-day production data.

owner-sampler record Dairy record system similar to DHI except milk weights and samples are taken by the dairy producer instead of a DHIA supervisor.

pace A lateral two-beat gain in which the right rear and front feet hit the ground at one time and the left rear and front feet strike the ground at another time.

paddling The outward swinging of the front feet of a horse that toes in.

pale, soft, exudative (PSE) A genetically predisposed condition in swine in which the pork is very light colored, soft, and watery.

palpation Feeling by hand.

panting Form of evaporative cooling, excessive panting is a sign of heat stress.

parakeratosis Skin disease resulting from an imbalance of zinc and calcium in the diet.

parasite An organism that lives a part of its life cycle in or on, and at the expense of, another organism. Parasites of farm animals live at the expense of the farm animals.

parity Number of different times a female has had offspring.

parlor Specialized facility that facilitates efficient milking operations.

parrot mouth Upper jaw is longer than lower jaw.

partial dominance   The heterozygote produces a phenotype intermediate to either homozygote.

partially supervised test   Milk weights and samples are taken alternately at A.M. and P.M. milkings by the DHIA technician and one other person.

parturient paresis   Partial paralysis that occurs at or near time of giving birth to young and beginning lactation. The mother mobilizes large amounts of calcium to produce milk to feed newborn, and blood calcium levels drop below the point necessary for impulse transmission along the nerve tracks. Commonly called *milk fever*.

parturition   The process of giving birth.

passive immunity   Short-term form of immune protection originating from the intact of colostrum.

pastern   Structure between the fetlock and coronet.

pasteurization   The process of heating milk to 161°F and holding it at that temperature for 15 seconds to destroy pathogenic microorganisms.

pasture mated   Group mating situations where a ram or billy is exposed to a flock of females for a defined period of time.

pasture rotation   The rotation of animals from one pasture to another so that some pasture areas have no livestock on them in certain periods.

pathogen   Biologic agent (i.e., bacteria, virus, protozoa, nematode) that may produce disease or illness.

paunch   Another name for *rumen*.

pay weight   The actual weight for which payment is made. In many cases it is the shrunk weight (actual weight minus pencil shrink).

Pearson square method   An effective ration balancing tool when the number of feedstuffs is relatively small.

pedigree   The record of the ancestry of an animal.

pelt   The natural, whole skin covering, including the wool, hair, or fur (e.g., a sheep pelt has the wool left on).

pencil shrink   An arithmetic deduction (percent of liveweight) from an animal's weight to account for fill.

pendulous   Hanging loosely.

penis   The male organ of copulation. It serves both as a channel for passage of urine from the bladder as an extension of the urethra, and as a copulatory organ through which sperm are deposited into the female reproductive tract.

per capita   Per person.

per-capita calorie and protein supply   The supply of calories and protein on a per person basis specific to a region or population.

per-capita disappearance   A proxy statistic used to estimate per person consumption, it is a much better measure of per capita production of a commodity.

performance targets   Production benchmarks.

performance test   The evaluation of an animal according to it performance.

pernicious anemia   A chronic type of mycrocitic anemia caused by a deficiency of vitamin B_{12} or a failure of intestinal absorption of vitamin B_{12}.

P.G. 600   Pregnant mare serum gonadotropin.

pharmaceuticals   Medicinal products (drugs) used primarily to treat disease.

phase feeding   The use of specific rations for the specific needs of pigs differing in growth rate.

phenotype   The characteristics of an animal that can be seen and/or measured (e.g., the presence or absence of horns, the color, or the weight of an animal).

phenotypic   See *phenotype*.

pheromones   Chemical substances that attract the opposite sex.

phosphorus   Macro mineral of importance to normal function.

photoperiod   Time period when light is present.

physiology   The science that pertains to the functions of organs, organ systems, or the entire animal.

picking   The removal of feathers in dressing poultry.

pig mortality   Death loss.

pigeon-toed   See *toeing-in*.

pin bones   In cattle, the posterior ends of the pelvic bones that appear as two raised areas on either side of the tail head.

pink tooth   Congenital porphyria, teeth are pink gray and the animals tend to sunburn easily.

pinworms   A small nematode worm with unsegmented body found as a parasite in the rectum and large intestine of animals.

pituitary   Small endocrine gland located at the base of the brain.

placenta   The vascular organ that unites the fetus to the uterus.

placentomes   Specific structures of cotyledonary attachment.

pneumonia   Inflammation or infection of alveoli of the lungs caused by either bacteria or viruses.

poikilotherm   A cold-blooded animal; one whose body temperature varies with that of the environment.

polarities   competing interests.

polled   Naturally or genetically hornless.

poll evil   An abscess behind the ears of a horse.

polymer   A molecule formed by many repeating sections.

Polypay   A synthetic breed of sheep developed in the United States by combining the Dorset, Targhee, Rambouillet, and Finnsheep breeds.

polytocous   Giving birth to several offspring at one time.

ponies See Table 32.1.

porcine stress syndrome A genetic defect in swine inherited as a simple recessive. It is associated with heavily muscled animals that may suddenly die when exposed to stressful conditions. Their muscle is usually pale, soft, and exudative (PSE).

pork The meat from swine.

posterior Toward the rear end of an animal.

postgastric fermentation The fermentation of feed that occurs in the cecum, behind the area where digestion has occurred.

postmortem After death.

postnatal See *postpartum*.

postpartum After birth.

postpartum interval The length of time from parturition until the dam is pregnant again.

postweaning growth Growth that occurs following weaning till attainment of finished weight.

Potomac horse fever A tick-borne disease characterized by cessation of gut sounds followed by severe diarrhea, which may be accompanied by colic.

poult A young turkey of either sex, from hatching to approximately 10 weeks of age.

poultry This term includes chickens, turkeys, geese, pigeons, peafowls, guineas, and game birds.

Poultry by-product meal (PBPM) A high protein by-product created from rendering of wastage from poultry processing.

power Amount of work that can be performed in a specific time frame.

predicted difference Dairy bull record based on superiority or inferiority of the bull's daughters compared to their herd mates.

predicted transmitting ability (PTA) Estimate of genetic transmitting ability (i.e., one-half of the breeding value) of dairy bulls. Estimated amount by which daughters of a bull will differ from the breed average.

pregastric fermentation Fermentation that occurs in the rumen of ruminant animals. It occurs before feed passes into the portion of the digestive tract in which digestion actually occurs.

pregnancy disease A metabolic disease in late pregnancy affecting primarily ewes carrying twins or triplets. A form of ketosis. Also called *pregnancy toxemia*.

pregnancy testing Evaluation of females for pregnancy through palpation or using an ultrasound machine.

premix A uniform mixture of one or more micro-ingredients with diluent and/or carrier. Premixers are used to facilitate uniform dispersion of the micro-ingredients in a large mix.

prenatal Prior to being born; before birth.

primary follicle Wool producing follicles completely formed and present at the time of birth.

primary oocyte The functional female gamete following chromosome replication and synapsis during meiosis.

primary spermatocyte The functional male gamete following chromosome replication and synapsis during meiosis.

primiparous Bearing or having borne but one offspring.

probe A device used to measure backfat thickness in pigs and cattle.

production testing An evaluation of an animal based on its production record.

progeny testing An evaluation of an animal on the basis of performance of its offspring.

progesterone A hormone produced by the corpus luteum that stimulates progestational proliferation in the uterus of the female.

prokaryote Cell or organism lacking a membrane-bound, structurally discrete nucleus and other subcellular compartments. Bacteria are prokaryotes.

prolapsed Turned inside out.

prostaglandins Chemical mediators that control many physiological and biochemical functions in the body. One prostaglandin (PGF_2a) can be used to synchronize estrus.

prostate A gland of the male reproductive tract that is located just back of the bladder. It secretes a fluid that becomes part of semen at ejaculation.

protein A large molecule composed of one or more chains of amino acids in a specific order; the order is determined by the base sequence of nucleotides in the gene coding for the protein.

protein supplement Any dietary component containing a high concentration (at least 25%) of protein.

proventriculus Glandular stomach of avians where gastric juices and hydrochloric acid are secreted.

proximal Nearest. The position that is closest to the point of attachment for a limb or bone.

PSE See *pale, soft, and exudative*.

PSS See *porcine stress syndrome*.

PTA See *predicted transmitting ability*.

puberty The age at which the reproductive organs become functionally operative.

pullet Young female chicken from day of hatch through onset of egg production; sometimes the term is used through the first laying year.

pulmonary arterial pressure (PAP) An indicator of susceptibility to brisket or high elevation disease in cattle.

purebred An animal eligible for registry with a recognized breed association.

purebred breeders Responsible for producing improvement genetic stock for the industry.

purebred producers Specialized farms focused or producing breeding stock of a specific breed.

Punnett square Methodology used to predict the outcomes of various matings.

qualitative trait A trait expressed categorically because of a sharp distinction between phenotypes (e.g., black and red). Usually only one or a few pairs of genes are involved in the expression of a qualitative trait.

quality assurance programs A science based set of protocols designed to assure the safety and wholesomeness of food products.

quality characteristics Factors that affect the value of particular feeds.

quality grades Animals grouped according to value as Prime, Choice, etc., based on conformation and fatness of the animals.

quantitative trait A trait expressed on a continuous/numerical scale because of a gradual variation from one phenotype to another (e.g., weaning weight). Usually many gene pairs and environmental influences are involved in the expression of such traits.

quitter Deep sore that drains at the coronet.

rack (1) A rapid four-beat gait of a horse. (2) A wholesale cut of lamb located between the shoulder and loin.

radiation Process of heat transfer between two objects that are not touching.

radiation pasteurization A technology that utilizes low level gamma rays, X-rays, or electron beam to neutralize pathogens.

ram A male sheep that is sexually mature.

ram breeds Breeds notes for growth and muscularity.

range-flock A sheep enterprise typically with a large inventory that is the primary source of income for the owner.

ration The amount of total feed fed to an animal over a 24-hour period.

reach See *selection differential*.

realized heritability The portion obtained of what is reached for in selection.

realizer A feeder animal (usually cattle) that has serious health problems or injury. Economics dictate the animal be sold rather than continue the duration of the feeding program.

reasoning The ability of an animal to respond correctly to a stimulus the first time the animal encounters a new situation.

recessive gene A gene that has its phenotype masked by its dominant allele when the two genes are present together in an individual.

reciprocal recurrent selection The selection of breeding animals in two populations based on the performance of their offspring after animals from two populations are crossed.

recombinant DNA (rDNA) Isolated DNA molecules that can be inserted into the DNA of another cell. rDNA is used in the genetic engineering process.

rectal prolapse Protrusion of part of large intestine through the anus.

recurrent selection Selection for general combining ability by selecting males that sire outstanding offspring when mated to females from varying genetic backgrounds.

Red and White See Figure 26.1.

red fibers Slow twitch fibers that function under aerobic conditions and thus require higher levels of oxygenation than fast twitch fibers.

red meat Meat from cattle, sheep, swine, and goats, as contrasted to the white meat of poultry.

registered Recorded in the herdbook of a breed.

regurgitate To cast up digested food to the mouth as is done by ruminants.

reinforcement A reward for making the proper response to a stimulus or condition.

relationship The level of shared genetic relationship of potential parents.

rendering A sustainable process to recycle meat scraps, fat, bone, and offal into useful by-products.

rennet The contents of a calf's stomach, which contains the enzyme rennin and is used to thicken milk for cheese making.

replicate To duplicate, or make another exactly like, the original.

reproduction The production of live, normal offspring.

reproductive performance Trait class of highest economic importance to a cow-calf producer.

residues Contaminants.

retained placenta Placenta remains within the reproductive tract after parturition has occurred.

reticulum One of the stomach components of ruminant animals that is lined with small compartments, giving a honeycomb appearance.

rhinitis Inflammation of the mucous membranes lining the nasal passages.

rhinopneumonitis Equine herpes virus-1. It produces acute catarrh upon primary infection.

ribeye (longissimus dorsi) The surface of this muscle is used in the calculation of USDA beef yield grades.

ribonucleic acid (RNA) An essential component of living cells, composed of long chains of phosphate, ribose sugar, and several bases.

ribosome Site of protein synthesis.

ribosomal RNA Essential for ribosome structure and formation.

rickets A disease of disturbed ossification of the bones caused by a lack of vitamin D or unbalanced calcium/phosphorus ratio.

ridgling Another term for cryptorchid.

ringbone An ossification of the lateral cartilage of the foot of a horse all around the foot.

riparian An area next to water (stream, river, or lake) where more vegetation grows (compared to a greater distance from the water source) because of the added moisture from the water. Grazing animals usually inhabit this area more frequently than others, thus increasing the possibility of overgrazing.

risk-share contract Cash prices are paid within a predetermined range and adjustments are made if the cash market falls outside the range (producer and packer split gains and losses).

RNA (ribonucleic acid) A chemical found in the nucleus and cytoplasm of cells. RNA plays an important role in protein synthesis and other chemical activities of the cell.

roman nose A nose having a prominent bridge (e.g., a roman-nosed horse).

root bulb Site of wool fiber growth.

rota-terminal crossbreeding Combines the three-breed rotational system and the terminal system of crossbreeding.

rotational cross Combines two or more breeds, with a different breed of boar being mated to the replacement crossbred females produced by the previous generation.

roughage A feed that is high in fiber, low in digestible nutritents, and low in energy. Such feeds as hay, straw, silage, and pasture are examples.

rumen The large fermentation pouch of the ruminant animal in which bacteria and protozoa break down fibrous plant material that is swallowed by the animal, sometimes referred to as the *paunch*.

ruminant A mammal whose stomach has four parts (rumen, reticulum, omasum, and abomasum). Cattle, sheep, goats, deer, and elk are ruminants.

rumination The regurgitation of undigested food and chewing it a second time, after which it is again swallowed.

running walk Fast ground-covering walk unique to the Tennessee Walking Horse.

sebaceous gland Secretes sebum in sheep and causes the greasiness of raw wool.

sac-fry Fish with an external yolk sac.

sale barns Local or regional marketing facilities that provide auction services to livestock producers.

Salmonella Gram-positive, rod-shaped bacteria that cause various diseases such as food poisoning in animals.

saturated fatty acids No double bonds between the carbons in the fatty acid chain.

scale (1) Size. (2) Equipment on which an animal is weighed.

scoured wool Wool that has been cleaned of grease and other foreign material.

scours Diarrhea; a profuse watery discharge from the intestines.

screwworms Larvae of several American flies that infest wounds of animals.

scrotal circumference A measurement (usually cm or in.) of the circumference of both testicles and the scrotal sac that surrounds them.

scrotum A pouch that contains the testes. It is also a thermoregulatory organ that contracts when cold and relaxes when warm, thus tending to keep the testes at a lower temperature than that of the body.

scurs Small growths of hornlike tissue attached to the skin of polled or dehorned animals.

scurvy A deficiency disease in humans that causes spongy gums and loose teeth. It is caused by a lack of vitamin C (ascorbic acid).

secondary follicle Follicle type that emerges postnatal and is grouped with primary follicles to create fiber bundles.

seedstock Breeding animals; sometimes used interchangeably with *purebred*.

selection Differentially reproducing what one wants in a herd or flock.

selection differential The difference between the average for a trait in selected animals and the average of the group from which they come; also called *reach*.

selection index Selection method in which several traits are evaluated and expressed as one total score.

selenium May be deficient in some regions and thus requires supplementation.

semen The fluid containing the sperm that is ejaculated by the male. Secretions from the seminal vesicles, the prostate gland, the bulbourethral glands, and the urethral glands provide most of the fluid.

seminal vesicles Accessory sex glands of the male that provide a portion of the fluid of semen.

seminiferous tubules Minute tubules in the testicles in which sperm are produced. They comprise about 90% of the mass of the testes.

service To breed or mate.

settle To become pregnant.

sex-limited Existing in only one sex, such as milk production in dairy cattle.

sex-limited trait Traits expressed in only one gender (milk production for example).

shade Most cost effective strategy to abate heat stress.

shearing The process of removing the fleece (wool) from a sheep.

sheath rot Inflammation of the prepuce in male sheep.

sheep bot Any of a number of related flies whose larvae are parasitic in sheep. They usually are found in the sinuses.

shelter-seeking behavior Behaviors associated with responses to changes in weather conditions.

shipping fever A widespread respiratory disease of cattle and sheep.

shoat A young pig of either sex.

shoe boil Blemish of the horse caused by the horseshoe putting pressure on the elbow when the horse lies down.

shrink Loss of weight—commonly used in the loss in live weight when animals are marketed or loss in weight from grease wool to clean wool.

sib A brother or sister.

sickle hocks Hocks that have too much set, causing the hind feet to be too far forward and too far under the animal.

side bones Ossification of the lateral cartilages of the foot of a horse.

sigmoid flexure The S-curve in the penis of boars, rams, bucks, and bulls.

silage Forage, corn fodder, or sorghum preserved by fermentation that produces acids similar to the acids that are used to make pickled foods for people.

sire Male parent.

Sire Genetic Evaluation Computed by the USDA, are based on comparing daughters of a given sire with their contemporary herd mates.

Sire Index A dairy bull test record obtained by comparing a bull's daughters with their contemporary herd mates.

skins Skins come from smaller animals such as pigs, sheep, goats, and wild animals. A beef hide weighing less than 30 lb is called a skin.

sleeping sickness An infectious disease common in tropical Africa and transmitted by the bite of a tsetse fly.

slotted floor Floor having any kind of openings through which excreta may fall.

small intestine Primary site of nutrient absorption.

SNF See *solids-not-fat*.

snood The relatively long, fleshy extension at the base of the turkey's beak.

software Program instructions to make computer hardware function.

solids-not-fat Total milk solids minus fat. It includes protein, lactose, and minerals.

somatic cells Normal body cells carrying pairs of chromosomes.

somatotropin The growth hormone from the anterior pituitary that stimulates nitrogen retention and growth.

sore mouth A virus-caused disease affecting primarily lambs.

sow A female swine that has farrowed one litter or has reached 12 months of age.

sow productivity Economically important trait that combines litter size, number weaned per litter, 21-day litter weight, and number of litters per sow per year.

sow productivity index (SPI) Bioeconomic index that measures the combined performance of maternal and growth to weaning.

spawn Act of fish laying eggs.

spay To remove the ovaries.

species cross Crossing of two species, typically produces infertile or subfertile offspring.

specific combining ability The ability of a line or population to exhibit superiority or inferiority when combined with other lines or populations.

speculum A long tube that allows the inseminator visual access to the cervix; typically used in sheep and goats.

sperm Gamete produced by males.

spermatid The haploid germ cell prior to spermiogenesis.

spermatogenesis The process by which spermatozoa are formed.

spermiogenesis The process by which the spermatid loses most of its cytoplasm and develops a tail to become a mature sperm.

spider syndrome A recessive genetic abnormality common to black-faced sheep. The front legs are usually bent out from the knees and the hind legs typically show some deformities.

spinning count The number of hanks of yarn that can be spun from a pound of clean wool. One method of evaluating fineness of wool.

splay-footed See *toeing-out*.

split-sex feeding Rations designed to meet the different nutritional requirements of barrows and gilts.

spool joint The joint where the foot and pastern are removed from the front leg. Used to identify a mutton carcass.

spring–fall range Range sheep flocks are moved to regions with better precipitation and grazing conditions that are available at the winter headquarters.

spur A sharp projection on the back of a male bird's shank.

stock Structure designed to restrain horses for protocols, such as AI or disease treatment.

stags Castrated male sheep, cattle, goats, or swine that have reached sexual maturity prior to castration.

stallion A sexually mature male horse.

standing heat Behavior whereby an animal in estrus will stand to be mounted.

Staphylococcus aureus Pathogen with an extremely small infective dose.

staple length Length of wool fibers.

steer A castrated bovine male that was castrated early in life before puberty.

sterility Inability to produce offspring.

steroid Artificially produced drug similar to the natural hormone that controls inflammation and regulates water balance.

stewardship Careful and responsible management of resources trusted to one's care.

stifle Joint of the hind leg between the femur and tibia.

stifled Injury of the stifle joint.

stillborn Offspring born dead.

stocker-yearling (cattle) Weaned cattle that are fed high-roughage diets (including grazing) before going into the feedlot.

stocker (fish) Usually 6–12 inches in length and less than 0.75 pounds.

stockmanship Application of skill, creativity, and knowledge to sound principles of livestock management.

stomach The site of initial digestion through enzymatic and muscular activity.

stomach worms *Haemonchus contortus*, or worms of the stomach of cattle, swine, sheep, and goats.

strain crossing the crossing of two strains that possess similar egg-production traits.

strangles An infectious disease of horses, characterized by inflammation of the mucous membranes of the respiratory tract.

strength The direct result of muscle size.

streptococcus Sperical, Gram-positive bacteria that divide in only one plane and occur in chains. Some species cause serious disease.

stress An unusual or abnormal influence causing a change in an animal's function, structure, or behavior.

stringhalt A sudden and extreme flexion of the back of a horse, producing a jerking motion of the hindleg in walking.

strongyles Any of various roundworms living as parasites, especially in domestic animals.

stud Usually the same as stallion. Also a place where male animals are maintained (i.e., bull stud).

suckling gain The gain that a young animal makes from birth until it is weaned.

subcutaneous Situated beneath, or occurring beneath, the skin. A subcutaneous injection is an injection made under the skin.

sulfonamides A sulfa drug capable of killing bacteria.

summer grazing Range sheep flocks are moved to higher elevations to access high quality forages during the relatively short growing season of summer in the high country.

superovulation The hormonally induced ovulation of a greater than normal number of eggs.

supervised electronic test Supervised test whereby data collection is conducted electronically with the technician certifying procedures and accuracy.

supervised test The DHIA technician weighs and samples milk for all cows from each milking during a 24-hour period.

supplement A feed used with another to improve the nutritive balance of performance of the total and intended to be (1) fed undiluted as a supplement to other feeds, (2) offered free-choice with other parts of the ration separately available, or (3) further diluted and mixed to produce a complete feed.

sustainability Method of harvesting or using a resource so that it is not depleted or permanently damaged.

sweating Most effective in horses compared to other livestock species.

sweat gland Produces and secretes sweat.

sweeney Atrophy of muscle (typically shoulder) in horses.

sweetbread An edible by-product also known as the pancreas.

swine industry A coordinated set of supply chain enterprises that produce, process, and distribute pork and pork by-products to the market place.

switch The tuft of long hair at the end of tail (cattle and horses).

Symbol I The industry standard ideal market hog in 1983.

Symbol II The industry standard ideal market hog in 1996.

Symbol III The industry standard ideal market hog adopted in 2005.

synapsis Accurate pairing of chromosomes following duplication during meiosis.

syndactyly Union of two or more digits (e.g., in cattle, the two toes would be a solid hoof).

synthetic breeds See *composite breed*.

system A group of organs that work in concert to perform a larger, general function.

tack room Storage room for bridles, saddles, etc.

tagging Clipping wool from the dock, udder, and vulva regions of the ewe prior to breeding and lambing.

tags (1) Wool covered with manure. (2) Abbreviated form of ear tags, used for identification.

tallow The fat of cattle and sheep.

tandem selection Selection for one trait for a given period of time followed by selection for a second trait and continuing in this way until all important traits are selected.

tariffs Taxes or fees placed on imported goods—often used as political leverage.

T cells   Provides intracellular protection against disease.

TDN   Total digestible nutrients; includes the total amounts of digestible protein, nitrogen-free extract, fiber, and fat (multiplied by 2.25), all summed together.

teaser ram   A ram made incapable of impregnating a ewe by vasectomy or by use of an apron to prevent copulation, which is used to find ewes in heat.

teasing   The stallion in the presence of the mare to see if she will mate.

teat   The protuberance of the udder through which milk is drawn.

tendon   Tough, fibrous connective tissue at ends of muscle bundles that attach muscle to bones or cartilage structures.

terminal cross   A two-breed single or rotational cross female is mated to a boar of the third breed.

terminal sire   The sire used in a terminal crossbreeding program. It is intended that all offspring from a terminal sire be sold as market animals.

terminal sire index (TSI)   Bioeconomic index that measures post weaning growth and carcass performance.

testicle   The male sex gland that produces sperm and testosterone.

testosterone   The male sex hormone that stimulates the accessory sex glands, causes the male sex drive, and causes the development of masculine characteristics.

tetanus   Rigid paralytic disease caused by *Clostridium tetani*, an anaerobic bacterium that lives in soil and feces.

tetrad   A group of four similar chromotids formed by the splitting longitudinally of a pair of homologous chromosomes during meiotic prophase.

thermoneutral zone (TNZ)   Range in temperature where rate and efficiency of gain is maximized. Comfort zone.

thickness   Expression of muscle development.

thoroughpin   A hard swelling that is located between the Achilles tendon and the bone of the hock joint.

three-breed terminal cross   Breeding a male of a third-breed to F1 females created from two distinct breeds.

three-site isowean system   Sow herd, nursery, and finishing phases were maintained at separate sites.

thrush   Foot disease characterized by degeneration of the frog and a thick, foul-smelling discharge.

thyroid gland   Two-lobed endocrine gland in the neck that controls the rate at which basic body functions proceed.

thyroxine   Hormone that controls metabolic heat production during cold stress.

tissues   Specialized groups of cells that function together.

toeing-in   Toes of front feet turn in. Also called *pigeon-toed*.

toeing-out   Toes of front feet turn out. Also called *splayfooted*.

tom   A male turkey.

total quality management   Intentional, science-based systems to assure that a process yields desired quality characteristics.

total mixed ration   A complete ration that has been developed and prepared so that each bite contains the correct nutrients.

total solids   Combined percentage of protein, lipids, lactose, and minerals in milk.

TPI   Total prediction index used in dairy cattle breeding. It includes the predicted differences for milk production, fat percentage, and type into one figure in a ratio of milk production * 3:fat percentage * 1:type * 1.

transcription   The synthesis of RNA from DNA in the nucleus by matching the sequences of the bases.

transfer RNA   Identifies both an amino acid and a base triplet in mRNA.

transgenic animals   Animals that contain genes transferred from other animals, usually from a different species.

translation   Reading of the genetic code prior to protein formation.

transmissible gastroenteritis (TGE)   A serious, contagious diarrhea disease in baby pigs.

trap nest   A nest equipped with a door that allows a hen to enter but prevents her from leaving.

trial and error   Attempting different responses to a stimulus until the correct response is performed.

tripe   Edible product from walls of ruminant stomach.

triple bottom line   A framework of sustainability based on three pillars of importance—economies, ecosystems, and communities.

TriStar   Program offered by the Holstein Association for providing production records, cow and herd genetic performance reports, and recognition programs, such as the Gold Medal Dam and Dam of Merit recognition.

trot   A diagonal two-beat gait in which the right front and left rear feet strike the ground in unison, and the left front and right rear feet strike the ground in unison.

twist   Vertical measurement from top of the rump to point where hindlegs separate.

twitch   To squeeze tightly the upper lip of a horse by means of a small rope that is twisted.

two-site isowean system   Sows are maintained separately just as they were in a three-site isowean system while stages 2 and 3 production were integrated at each second site with distinct nursery barns and finishing barns at each location.

tying-up (exertional rhabdomyolysis)   Is characterized by a shuffling gait, heavy sweating, and overly contracted rump and thigh muscles. Severe forms may result in muscular trauma.

type   (1) The physical conformation of an animal. (2) All those physical attributes that contribute to the value of an animal for a specific purpose.

Type Production Index Bioeconomic index that combines differences for milk production traits and differences for type traits into a single value.

udder The encased groups travel from the fetus to and from the placenta, respectively. This cord is broken when the young are born.

ultra-high-temperature (UHT) processing A pasteurization process that yields a shelf-stable milk product that does not require refrigeration.

umbilical cord Cord connecting the foal to the womb.

undershot jaw Lower jaw is longer than upper jaw.

unknown effects Random environmental effects unique to an individual.

unsaturated At least one double bond between the carbons is present in the fatty acid chain.

unsoundness Any defect or injury that interferes with the usefulness of an animal.

upper critical temperature (UCT) Effective ambient temperature limit above which heat stress occurs and the animal employs evaporative heat loss mechanisms.

urinary calculi Disease where mineral deposits crystallize in the urinary tract. The deposits may block the tract, causing difficulty in urination.

uterus That portion of the female reproductive tract where the young develop during pregnancy.

vaccination The act of administering a vaccine or antigens.

vaccine Suspension of attenuated or killed microbes or toxins administered to induce active immunity.

vagina The copulatory portion of the female's reproductive tract. The vestibule portion of the vagina also serves for passage of urine during urination. The vagina also serves as a canal through which young pass when born.

variety Subdivision of a breed composed of birds of the same feather color and type of comb.

variety meats Edible organ by-products (e.g., liver, heart, tongue, tripe).

vas deferens Ducts that carry sperm from the epididymis to the urethra.

vasectomized Males have the vas deferens snipped to prevent passage of sperm, these males are often used in heat detection.

vasectomy The removal of a portion of the vas deferens. As a result, sperm are prevented from traveling from the testicles to become part of the semen.

veal The meat from very young cattle, under 3 months of age.

vein Vessel through which blood passes from various organs or parts back to the heart.

ventilation Systems designed to increase air flow.

ventilation Provides adequate fresh air, aids in removing excess moisture, and is essential in maintaining a proper temperature within the house.

vermifuge A chemical substance given to the animals to kill internal parasitic worms.

vertebrata Subphylum of livestock.

VFA See *volatile fatty acids*.

viability Capable of growing, practical, and able to be done.

villi Projections of the inner lining of the small intestine.

virus Ultramicroscopic bundle of genetic material capable of multiplying only in living cells. Viruses cause a wide range of disease in plants, animals, and humans, such as rabies and measles.

viscera Internal organs and glands contained in the thoracic and abdominal cavities.

vital signs Physiological indicators used to assess the health and well-being of an animal.

vitamin An organic catalyst, or component thereof, that facilitates specific and necessary functions.

vitamin A May be required as a supplement to ruminants on a dry ration.

vitamin D Important nutrient in the regulation of calcium and phosphorus in bone development.

volatile fatty acids (VFA) A group of fatty acids produced from microbial action in the rumen; examples are acetic, propionic, and butyric acids.

vomitoxin Undesirable mycotoxin that may impair performance.

vulva The external genitalia of a female mammal.

walk A four-beat gait of a horse in which each foot strikes the ground at a time different from each of the other three feet.

warble The larval stage of the heel fly that burrows out through the hide of cattle in springtime.

wattle Method of identification in cattle where strips of skin (3–6 inches) long are usually cut on the nose, jaw, throat, or brisket.

weaner An animal that has been weaned or is nearing weaning age.

weaning Separating young animals from their dams so that the offspring can no longer suckle.

weaning weight EPD A genetic estimate of the weaning weight of a beef bull's calves when compared to other bulls in the sire summary.

weanling An animal of weaning age.

West Nile virus Disease that is transmitted via mosquito or bird vectors to both equines and humans.

wet Used to describe a milking female (e.g., wet cow or wet ewe).

wether A male sheep castrated before reaching puberty.

white cells (leukocytes, white blood cells) Colorless blood cells active in the body's defense against infection or other assult. There are five types: neutrophils, lymphocytes, eosinoiphils, monocytes, and basophils.

white fibers Fast twitch fibers that function under anaerobic conditions.

white meat The more anaerobic muscles of poultry (breast and wings).

white muscle disease A muscular disease caused by a deficiency of selenium or vitamin E.

wholesale An intermediate step in the supply chain, wholesale enterprises typically sell to either retail or food service operations.

windchill The effect of wind speed on effective temperature.

windgalls (wind puffs) Occur when the joint capsules on tendon sheaths around the pastern or fetlock joints are enlarged.

winking Indication of estrus in the mare where the vulva opens and closes.

winter headquarters Range sheep operations typically move sheep in the winter to a lower elevation, drier climate to access cheaper feeds.

withdrawal time The time before slaughter that a drug should not be given to an animal.

withers Top of the shoulders.

wool The fibers that grow from the skin of sheep.

wool blindness Sheep cannot see, owing to wool covering their eyes.

wool top A continuous untwisted strand of combed wool in which the fibers lie parallel and the short fibers have been combed out.

woolens Cloth made from short wool fibers that are intermingled in the making of the cloth by carding.

worsteds Cloth made from wool that is long enough to comb and spin into yarn. The finish of worsteds is harder than woolens, and worsted clothes hold a press better.

yardage Non-feed costs associated with cattle feeding.

yearling Animals that are approximately 1 year old.

yearling weight expected progeny difference (EPD) A breeding value that measures genetic differences in yearling weight in beef cattle.

yield Used interchangeably with *dressing percentage*.

yield grades The grouping of animals according to the estimated trimmed lean meat that their carcass would provide; cutability.

yogurt A fermented dairy product sold in a gel or liquid form.

yolk (1) The yellow part of the egg. (2) The natural grease (lanolin) of wool.

yolk sac Layer of tissue encompassing the yolk of an egg.

zearalenone Undesirable mycotoxin that may impair performance.

zone of thermoneutrality The environmental temperature (about 65°F) at which heat production and heat elimination are approximately equal for most farm animals.

zygote (1) The cell formed by the union of two gametes. (2) An individual from the time of fertilization until death.